10th International Conference on Fibre Concrete (FIBRE CONCRETE 2019)

IOP Conference Series: Materials Science and Engineering
Volume 596

Prague, Czech Republic
17 – 20 September 2019

ISBN: 978-1-5108-9466-2
ISSN: 1757-8981

Printed from e-media with permission by:

Curran Associates, Inc.
57 Morehouse Lane
Red Hook, NY 12571

proceedings.com

Some format issues inherent in the e-media version may also appear in this print version.

Institute of Physics
Dirac House, Temple Back
Bristol BS1 6BE UK

Phone: 44 1 17 929 7481
Fax: 44 1 17 920 0979

techtracking@iop.org

Additional copies of this publication are available from:

Curran Associates, Inc.
57 Morehouse Lane
Red Hook, NY 12571 USA
Phone: 845-758-0400
Fax: 845-758-2633
Email: curran@proceedings.com
Web: www.proceedings.com

Table of contents

Volume 596

FIBRE CONCRETE 2019

17–20 September 2019, Prague, Czech Republic

Accepted papers received: 26 July 2019
Published online: 15 August 2019

Preface

Papers

IOP Conf. Series: Materials Science and Engineering **596** (2019) 011001 doi:10.1088/1757-899X/596/1/011001

FIBRE CONCRETE 2019

10th international conference on fibre reinforced concretes (FRC), textile reinforced concretes (TRC) and ultra-high performance cementitious composites (UHPC)

PREFACE

The Fibre Concrete Conference series is held biennially to provide a platform to share knowledge on fibre reinforced concretes, textile concretes and ultra-high performance cementitious composites regarding material properties and behaviour, technology procedures, topics of long-term behaviour, creep, durability; sustainable aspects of concrete including utilisation of waste materials in concrete production and recycling of concrete. The tradition of Fibre Concrete Conferences started in eighties of the last century. Nowadays the conference is organized by the Department of Concrete and Masonry Structures of the Czech Technical University in Prague, Faculty of Civil Engineering.

The 10th International Conference Fibre Concrete 2019 provides opportunity for presenting of 50 papers for participants from 14 countries all over the world. The keynote lectures are given by Professor Bažant and Professor Akira Hosoda.

The conference program covers wide range of topics from scientific research to practical applications. The presented contributions relate to performance and behaviour of cementitious composites, their long-term behaviour and durability, sustainable aspects, advanced analyses of structures from these composites and successful applications.

This conference was organized also to honour Professor Vladimír Křístek on the occasion of his jubilee and to appreciate his merits and discoveries in the field of concrete and cement based composites, structural mechanics and engineering.

Scientific committee:

György L. Balázs	(Hungary)
Joaquim A. O. de Barros	(Portugal)
Peter JM Bartoš	(United Kingdom)
Zdeněk P. Bažant	(USA)
Petr Bílý	(Czech Republic)
Iva Broukalová	(Czech Republic)
Frank Dehn	(Germany)
Antonio D. de Figueiredo	(Brazil)
Michaela Frantová	(Czech Republic)
Akira Hosoda	(Japan)
Alena Kohoutková	(Czech Republic)
Ludovít Naď	(Slovakia)
Martin Petřík	(Czech Republic)
Giovanni Plizzari	(Italy)
Pierre Rossi	(France)
Petr Štemberk	(Czech Republic)
Luc R. Taerwe	(Belgium)
Jan L. Vítek	(Czech Republic)

Editorial committee:

Petr Bílý	(Czech Republic)
Iva Broukalová	(Czech Republic)
Michaela Frantová	(Czech Republic)
Petr Štemberk	(Czech Republic)
Jakub Žák	(Czech Republic)

Organizing committee:

Iva Broukalová	(Czech Republic)
Josef Fládr	(Czech Republic)
Michaela Frantová	(Czech Republic)
Yuliia Khmurovska	(Czech Republic)
Martin Petřík	(Czech Republic)
Karel Šeps	(Czech Republic)
Petr Štemberk	(Czech Republic)
Helena Včelová	(Czech Republic)

IOP Conf. Series: Materials Science and Engineering **596** (2019) 011003 doi:10.1088/1757-899X/596/1/011003

CONFERENCE PARTNERS

IOP Conf. Series: Materials Science and Engineering **596** (2019) 011004 doi:10.1088/1757-899X/596/1/011004

surname	forename	Affiliation	email
Alparslan	Ugur	Kordsa Teknik Tekstil	ugur.alparslan@kordsa.com
Bazant	Zdenek	Northwestern University	z-bazant@northwestern.edu
Bibora	Petr	Research Institute for Building Materials	bibora@vustah.cz
Bily	Petr	Faculty of Civil Engineering, CTU in Prague	petr.bily@fsv.cvut.cz
Brož	Petr	University of West Bohemia, Faculty of Applied Sciences	broz.petr@tiscali.cz
De Almeida	Luiz Carlos	State University fo Campinas	almeida.unicamp@gmail.com
Del Prete	Clementina	University of Bologna	clementina.delprete2@unibo.it
ERDAL	Burak	Kordsa Teknik Tekstil A.Ş	burak.erdal@kordsa.com
Fareed	Shamsoon	NED University of Engineering & Technology	sfareed@neduet.edu.pk
Florin	Tepes-Onea	University Ovidius of Constantza , Faculty of Construction	tflorin@univ-ovidius.ro
Frantová	Michaela	Faculty of Civil Engineering, CTU in Prague	michaela.frantova@fsv.cvut.cz
Gabko	Miroslav	DAKO Brno	gabko@dakobrno.cz
Gach	Ferdinand	Považská Cementáreň, a.s., Ladce	gach.f@pcla.sk
Herka	Petr	HCE - Herka Consulting Engineering	petr.herka@gmail.com
Hoekstra	Anne	N.V. Bekaert	anne.hoekstra@bekaert.com
Holan	Jakub	Faculty of Civil Engineering, CTU in Prague	jakub.holan@fsv.cvut.cz
Horáková	Anna	Faculty of Civil Engineering, CTU in Prague	anna.horakova@fsv.cvut.cz
Horáková	Anna	Faculty of Civil Engineering, CTU in Prague	anna.horakova@fsv.cvut.cz
Horníková	Kateřina	Faculty of Civil Engineering, CTU in Prague	katerina.hornikova@fsv.cvut.cz
Chylík	Roman	Faculty of Civil Engineering, CTU in Prague	roman.chylik@fsv.cvut.cz
ilhan	izmit	Kordsa Teknik Tekstil A.Ş.	ilhan.izmit@kordsa.com
Janota	Ondřej	Faculty of Civil Engineering, CTU in Prague	ondrej.janota@fsv.cvut.cz
Jedlinský	Petr	EUROVIA CS, a.s.	petr.jedlinsky@eurovia.cz
Juhász	Péter Károly	JKP Static	office@jkp.hu
Kašpar	Tomáš	Faculty of Civil Engineering, CTU in Prague	tomas.kaspar@fsv.cvut.cz
Khmurovska	Yuliia	Czech Technical University in Prague, Department of Concrete and Masonry Structures	yuliia.khmurovska@fsv.cvut.cz
Křístek	Vladimír	Faculty of Civil Engineering CTU in Prague	kristek@fsv.cvut.cz
Kvasnička	Václav	Pontex s.r.o.	kvasnicka@pontex.cz
Labib	Wafa	Prince Sultan University, Riyadh, SA	wlabib@psu.edu.sa
Marcalíková	Zuzana	Vysoká škola báňská – Technická univerzita Ostrava	zuzana.marcalikova@vsb.cz
Mezquida-Alcaraz	Eduardo J.	ICITECH. Universitat Politècnica de València (UPV)	edmezal@alumni.upv.es
Mizerová	Cecílie	Brno University of Technology, Faculty of Civil Engineering	mizerova.c@fce.vutbr.cz
Mouta Trautwein	Leandro	Unicamp - University of Campinas	leandromt@gmai.com
Navarro-Gregori	Juan	Universitat Politècnica de València	juanagre@cst.upv.es
Navarro-Gregori	Juan	Universitat Politècnica de València	juanagre@cst.upv.es
Novák	Josef	Faculty of Civil Engineering, CTU in Prague	josef.novak.1@fsv.cvut.cz
Nývlt	Michal	Faculty of Civil Engineering, CTU in Prague	michal.nyvlt@fsv.cvut.cz
Okamoto	Takahisa	Ritsumeikan University	okatakago49@yahoo.co.jp
Orlando	Marco	Italcementi S.p.A. HeidelbergCement Group	m.orlando@italcementi.it
Pokorný	Jaroslav	Faculty of Civil Engineering, CTU in Prague	jaroslav.pokorny@fsv.cvut.cz
Příbramský	Vladimír	Faculty of Civil Engineering, CTU in Prague	vladimir.pribramsky@fsvcvut.cz
Rios	Jose D.	University of Seville	jdrios@us.es
Schreiberová	Hana	Faculty of Civil Engineering, CTU in Prague	hana.schreiberova@fsv.cvut.cz
Suárez Guerra	Fernando	Universidad de Jaén	fsuarez@ujaen.es
Sýkora	Michal	EUROVIA CS, a.s.	michal.sykora@eurovia.cz
Štefan	Radek	Faculty of Civil Engineering CTU in Prague	radek.stefan@fsv.cvut.cz
Šušteršič	Jakob	IRMA Institute for Research in Materials and Applications, Ljubljana	jakob.sustersic@irma.si
Takahisa	Okamoto	College of Science & Engineering, Ritsumeikan University	okataka1949@gmail.com
Trtík	Tomáš	Faculty of Civil Engineering, CTU in Prague	tomas.trtik@fsv.cvut.cz
Valente	Tiago	CiviTest - Pesquisa de Novos Materiais para a Engenharia Civil, Lda	tiagovalente@civitest.com
Vodička	Jan	Faculty of Civil Engineering CTU in Prague	jan.vodicka@fsv.cvut.cz
Wojnar	Richard	Bekaert Petrovice s.r.o.	richard.wojnar@bekaert.com
Záleská	Martina	Faculty of Civil Engineering, CTU in Prague	martina.zaleska@fsv.cvut.cz
Zanuy	Carlos	Universidad Politecnica de Madrid	carlos.zanuy@upm.es
Žák	Jakub	Faculty of Civil Engineering, CTU in Prague	jakub.zak@fsv.cvut.cz
Žáková	Hana	Faculty of Civil Engineering, CTU in Prague	hana.zakova@fsv.cvut.cz

IOP Conf. Series: Materials Science and Engineering **596** (2019) 011005 doi:10.1088/1757-899X/596/1/011005

Peer review statement

All papers published in this volume of *IOP Conference Series: Materials Science and Engineering* have been peer reviewed through processes administered by the proceedings Editors. Reviews were conducted by expert referees to the professional and scientific standards expected of a proceedings journal published by IOP Publishing.

IOP Conf. Series: Materials Science and Engineering **596** (2019) 012001 doi:10.1088/1757-899X/596/1/012001

Dependence of fracture size effect and projectile penetration on fiber content of FRC

Zdeněk P Bažant, Mohammad Rasoolinejad, Abdullah Dönmez and Wen Luo

Northwestern University, Evanston, IL USA

E-mail: z-bazant@northwestern.edu

Abstract. The microplane constitutive model M7f for fiber reinforced concrete (FRC), previously calibrated by extensive material test data, is used in computational simulations of the size effect in geometrically similar notched specimens, and in simulations of penetration of projectiles into FRC targets. The M7f microplane model for fiber reinforced concrete is calibrated at the material level and then used to predict structural level behavior. The results show that, for any fiber volume ratio, the Type 2 size effect must be expected.

1. Introduction

In the microplane model, the constitutive law of material including the post-peak softening in a representative volume is characterized in terms of vectors rather than tensors. The vectorial form has a powerful advantage. It allows relating the microplane deformations to microcrack opening, frictional slip, fiber pullout and possible breakage. Microplane model M7f for FRC with short discontinuous fibers was developed at Northwestern University [1, 2] and was presented at the 2017 Fibre Concrete conference in Prague. It was shown to fit well the main material tests reported for FRC. In this lecture, the M7f model is used to study two underexplored problems of FRC—the role of fibers in the energetic fracture size effect, and in the penetration of projectiles, or missiles, in FRC walls.

Steel fibers are extensively used in case of structural concrete such as ultra-high performance concrete and new construction technologies [3, 4]. The high tensile strength of steel fibers enhances the tensile behavior of FRC [3, 5, 6]. The steel fibers also enhance ductility and provide good dynamic response to the FRC [6]. The fibers also could capture early age cracking due to shrinkage [7, 8].

There are two causes of size effect, statistical (first properly characterized by Weibull in 1939), and energetic (discovered in 1984 [9, 10]). The latter dominates for reinforced concrete and is the only one studied in this lecture. It is of two types, described by equations:

$$\sigma_N = \sigma_0 \left(1 + \frac{rD_b}{D}\right)^{1/r} \quad \text{(Type 1 [11])} \tag{1}$$

$$\sigma_N = Bf_t' \left(1 + \frac{D}{D_0}\right)^{-1/2} \quad \text{(Type 2, SEL [9])} \tag{2}$$

where σ_N is the nominal strength of structure (load divided by characteristic cross section area), and $\sigma_0, D_b, r, Bf_t', D_0$ are constants. The Type II [9, 10, 12] dominates for reinforced concrete and is purely energetic, caused by the stored energy release due to large stable crack growth [13, 14]. The Type 1 occurs for plain (unreinforced concrete), in which the structure becomes unstable as soon as a continuous macro-crack starts growing from a finite damaged microcracking zone, and is caused by stress redistribution [11, 12]. For very large sizes, Type 1 transits to Weibull statistical size effect [15] (which is omitted from equation (1) and is not discussed here). The Type 2 size effect factor, equation (2), has been adopted for the new revision of the ACI code (Standard 318/2019)—for beam shear, slab punching [16], and compressions struts of strut-and-tie model.

2. Simulation of Scaled FRC Notched Beams

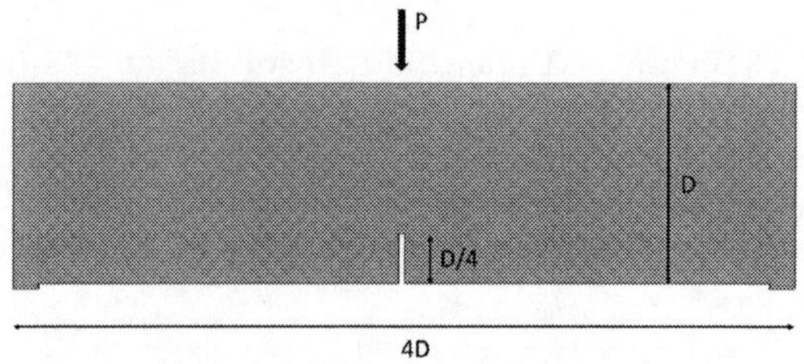

Figure 1. Specimen Geometry.

The M7 parameters have been calibrated to capture the FRC test reported by Li et al. [6]. Figure 1 shows notched specimens of FRC that have been geometrically scaled to various sizes and simulated by the FE crack band model [17], in which the element size was the same for all specimen sizes. Figure 2 shows the curves of stress versus normalized deflection. The calculated size effect plots of $\log\sigma_N$ versus $\log D$ are shown in figure 3 for fiber contents 0, 2%, 3% and 6%, and for four different specimen sizes (or depths) D = 400, 800, 1600 and 3200 mm.

3. Results for FRC Notched Test

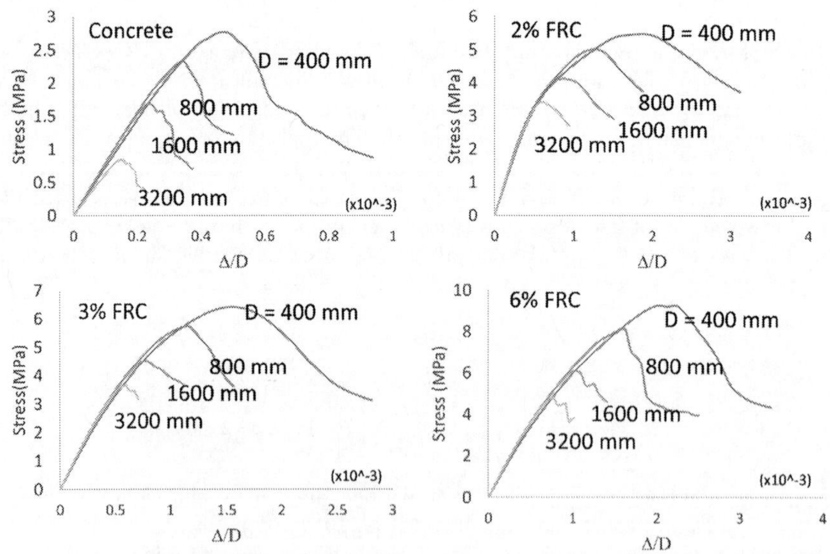

Figure 2. Three-point bending stress-strain curves.

It is seen that the Bažant SEL (shown by solid curves) fits the data points from FE simulations very well (Figure 3). It is noted that, in these notched fracture specimens, the fibers significantly elevate the maximum load, which is not the case for the compression tests of standard unnotched test cylinders. It is also interesting that, according to the simulations, the fiber content has no clear effect on the transitional size D_0 of the size effect law.

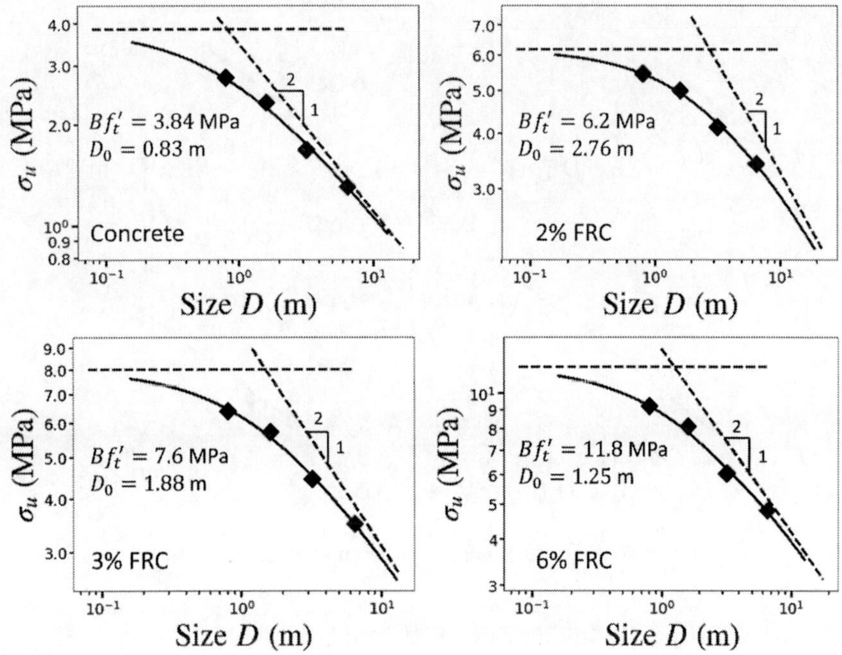

Figure 3. Size effect plots of FRC in logarithmic scale.

The measured size effect in notched fracture specimens can be exploited to determine the fracture energy, G_f, and the characteristic fracture process zone size, c_f, of FRC. To this end, the Type 2 SEL is related to equivalent linear elastic fracture mechanics (LEFM) and is converted to the linear regression plot [10, 12, 15]:

$$Y = AX + C \tag{3}$$

$$\text{where} \quad X = D, \quad Y = \frac{1}{\sigma_N^2} \tag{4}$$

$$\text{and} \quad A = \frac{g(\alpha)}{EG_f}, \quad C = c_f\,\frac{g'(\alpha)}{EG_f} \tag{5}$$

Here $g(\alpha)$ = dimensionless energy release function of LEFM and E = Young's elastic modulus. Figure 4 shows the regression plots of the computed Y versus X, for various sizes D and various fiber contents. From these regressions, one gets G_f = 279, 2424, 2491, 3994 N/m for fiber contents 0, 2%, 3% and 6%, respectively. It may be concluded that adding fibers increases the fracture energy of concrete dramatically.

The role of fibers in the Type 1 size effect of FRC has also been simulated. The plain concrete and 2% FRC showed the Type 1 size effect while 3% and 6% FRC did not. The load-displacement curves of unnotched specimens reached a plateau for 3% and 6% FRC and the maximum load became independent of the size. Calculations of projectile penetration into FRC wall as still in progress at the time of writing and will be presented at the conference.

IOP Conf. Series: Materials Science and Engineering **596** (2019) 012001 doi:10.1088/1757-899X/596/1/012001

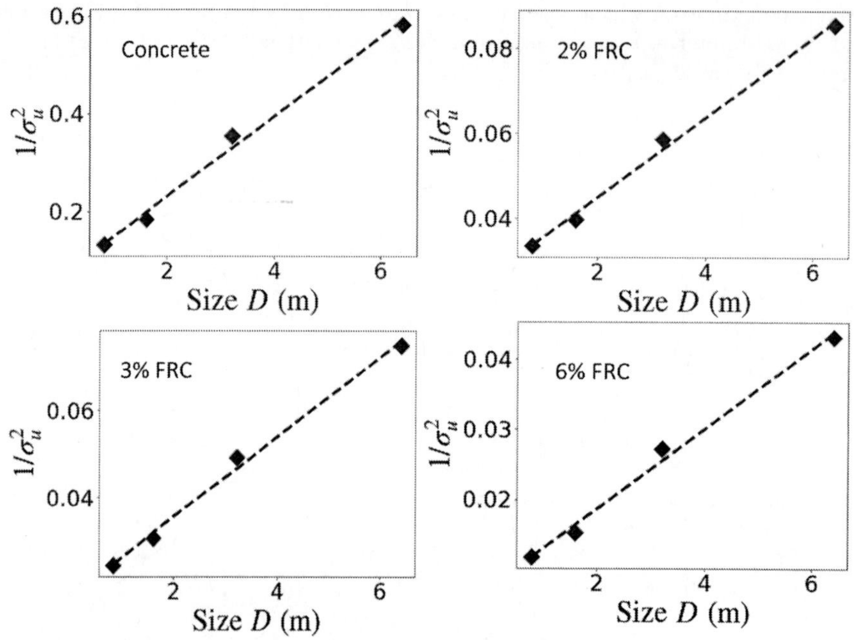

Figure 4. Linear regression size effect plots.

4. Simulation of Projectile Impact into FRC Targets

Another point of considerable interest is the effect of fibers on the resistance to impact. It was previously found at Northwestern [18, 19, 20] that comminution (or fine fragmentation) of material during penetration of projectiles through concrete has a great damping effect, decelerating the projectile, reducing penetration depth and, in the case of full penetration, reducing the exit velocity of the projectile. Most studies concluded that FRC does not significantly reduce the penetration depth or exit velocity after full penetration, but significantly reduces the size of the exit crater and generally limits the size of fragments and the extent of fragmentation [5, 21, 22, 23, 24, 25].

Fig. 5(a) shows the significant improvement of simulation predictions by considering material comminution effect. It is implemented in the microplane model M7 by scaling the stress-strain boundaries based on the local kinetic energy release from comminution. Fig. 5(b) shows the further reduction of projectile exit velocity by increasing the fiber content to 2%. The element deletion criterion has also been adjusted to a larger value for FRC based on the fact that higher fiber content reduces the concrete cracking, leading to a larger strain threshold for the homogenized material. Fig. 6 shows the damage contours of the target (in logarithmic strain) after penetration (with and without fibers). For the case of 2% fiber content, the final crater is seen to become significantly smaller than for plain concrete.

5. FRP Strenghtened Flexural Members

External strengthening of RC beams with FRP or textile composites has been shown to be quite effective for diverse structural members such as beams, columns and slabs. The bond quality between the FRP sheets and concrete is generally satisfying. So, the delamination occurs not in the interface but within the concrete immediately adjacent to the interface. The near-interface cracking, which is basically a shear failure, must be expected to lead to a size effect on structural strength. However, the size effect is not considered in the design practice, not in design code and guides for externally bonded FRP systems used for strengthening concrete structures or retrofitting them after damage. This indicates the need for a realistic analysis of the existing test data and their finite element simulations.

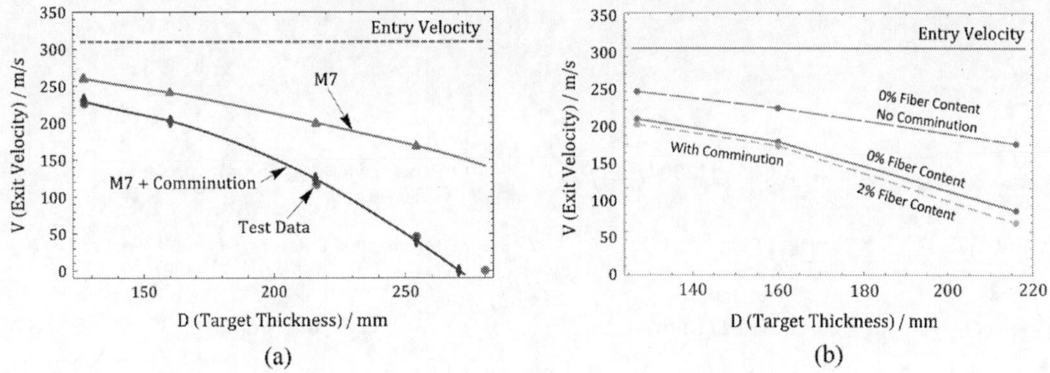

Figure 5. a) Effect of material comminution in plain concrete; b) Effect of fiber on projectile exit velocity.

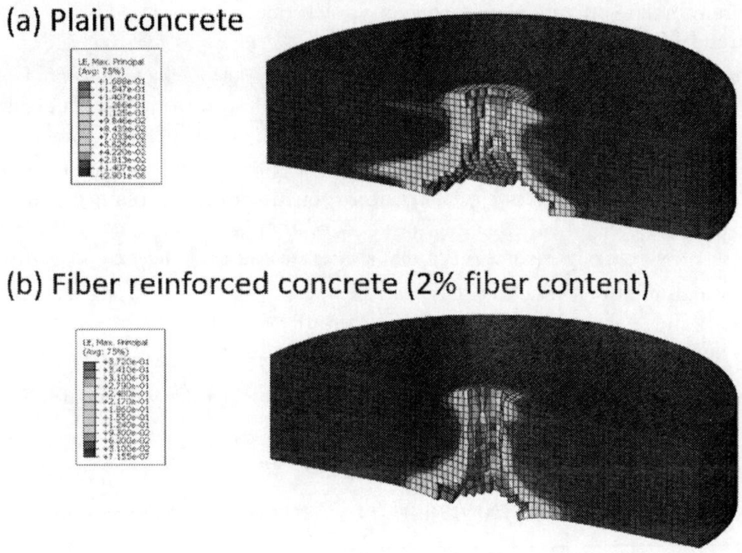

Figure 6. Crater shape after penetration.

FEM analyses, based on concrete damage model model (Microplane Model, M7), consist of the calibration by test data and the subsequent predictions (see Fig. 7a,b). The optimum size effect fits for two different FRP thicknesses are shown in Fig. 7a,b. The shear cracks within the concrete next to bonded FRP sheets propagate along the beam longitudinally, and eventually lead to delamination of the FRP sheet bonded to the RC beam. It should also be noted that the size of the reinforcement affects the cracking patterns. Stiff skin reinforcements results into a distinct size effect even for relatively smaller sizes. As can be seen in Fig. 7b, the predictions for thick FRP sheets indicate a significant size effect. And a sufficiently thin FRP strengthening might change the location of the crack initiation.

6. Conclusions

(i) Compared to concrete without fibers, short fiber reinforcement significantly increases the maximum loads in Type 2 failure of notched fracture specimens and, by implication, also the maximum loads of structures failing only after large stable crack growth (as in shear of RC

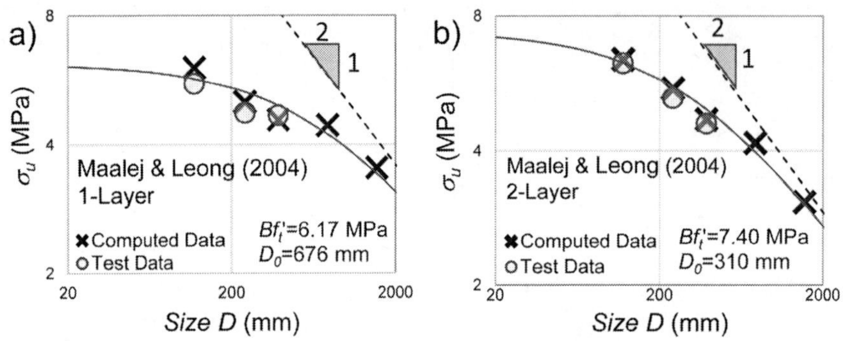

Figure 7. Size effect trends of computed data calibrated by the test data [26] in logarithmic scale.

beams).

(ii) The fiber reinforced concrete showed bigger transitional size in Type 2 size effect; but generally, no clear trend has been observed.

(iii) The values of fracture energy of FRC derived by the size effect method from the computed maximum loads of scaled notched specimens show that introducing fibers into concrete and increasing their content causes the fracture energy to increase dramatically.

(iv) The microplane model M7 with scaled boundaries is shown to be able to capture the effect of material comminution under projectile penetration not only for plain concrete but also for the FRC. The calculations indicate a negligible effect of fiber content on the exit velocity of penetrating projectile but a significant reduction of exit crater size and of the extent of fragmentation.

(v) According to simulations based on calibration by test results, a strong size effect under flexural loading exists in the fracture of concrete-FRP interfaces in RC beams strengthened by FRP. This size effect becomes more prominent for beams strengthened by thick FRP sheets.

Note: Details of the specimen and material properties and of calculations will be given in a separate refereed journal article under preparation.

Acknowledgments

Partial funding under ARO grant W911NF-19-1-0039 to Northwestern University is gratefully appreciated.

References

[1] Caner F C, Bažant Z P and Wendner R 2013 *Engineering fracture mechanics* **105** 41–57
[2] Caner F C and Bažant Z P 2012 *Journal of Engineering Mechanics* **139** 1714–1723
[3] Feng J, Sun W, Wang X and Shi X 2014 *Construction and Building Materials* **69** 403–410
[4] Alizadehsalehi S, Yitmen I, Celik T and Arditi D 2018 *International journal of occupational safety and ergonomics* 1–16
[5] Feng J, Sun W, Liu Z, Cui C and Wang X 2016 *Materials & Design* **102** 131–141
[6] Li Z, Li F, Chang T P and Mai Y W 1998 *Materials Journal* **95** 564–574
[7] Zhang J and Li V C 2001 *Journal of engineering mechanics* **127** 37–44
[8] Rasoolinejad M, Rahimi-Aghdam S and Bažant Z P 2019 *Materials and Structures* **52** 33
[9] Bažant Z P 1984 *Journal of engineering mechanics* **110** 518–535
[10] Bažant Z P and Kazemi M 1990 *International Journal of fracture* **44** 111–131
[11] Bažant Z P, Daniel I M and Li Z 1996 *Journal of engineering materials and technology* **118** 317–324
[12] Bažant Z P and Planas J 1997 *Fracture and size effect in concrete and other quasibrittle materials* vol 16 (CRC press)
[13] Rasoolinejad M and Bažant Z P 2019 *ACI Structural Journal* **116**
[14] Caner F C, Dönmez A A, Şener S and Koç V 2019 *International Journal of Solids and Structures* **162** 76–86
[15] Bažant Z P and Le J L 2017 *Probabilistic mechanics of quasibrittle structures: strength, lifetime, and size effect* (Cambridge University Press)
[16] Dönmez A and Bažant Z P 2017 *ACI Structural Journal* **114** 875–886

[17] Bažant Z P and Oh B H 1983 *Matériaux et construction* **16** 155–177
[18] Bažant Z P and Caner F C 2014 *Journal of the Mechanics and Physics of Solids* **64** 223–235
[19] Caner F C and Bažant Z P 2014 *Journal of the Mechanics and Physics of Solids* **64** 236–248
[20] Luo W, Chau V T and Bažant Z P 2019 *International Journal of Fracture* 1–11
[21] Anderson W, Watson A and Armstrong P 1984 *Structural Impact and Crashworthiness,* **2** 687–695
[22] ONeil E, Neeley B and Cargile J 1999 *Shock and Vibration* **6** 237–245
[23] Dancygier A N, Yankelevsky D Z and Jaegermann C 2007 *International Journal of Impact Engineering* **34** 1768–1779
[24] Dancygier A and Yankelevsky D 1996 *International Journal of Impact Engineering* **18** 583–599
[25] Zhang M, Shim V, Lu G and Chew C 2005 *International Journal of Impact Engineering* **31** 825–841
[26] Maalej M and Leong K 2005 *Composites Science and Technology* **65** 1148–1158

IOP Conf. Series: Materials Science and Engineering **596** (2019) 012002 doi:10.1088/1757-899X/596/1/012002

Fibre reinforced concrete with recycled concrete aggregate – inverse design approach

I Broukalová, K Šeps, J Fládr and J. Vodička

Department of Concrete and Masonry Structures, Faculty of Civil Engineering, Czech Technical University in Prague

E-mail: iva.broukalova@fsv.cvut.cz

Abstract. The paper aims at the topic of sustainable building concerning recycling of waste rubble concrete from demolition. The objective of the presented investigations was to verify the proposed breakthrough approach to utilisation of crushed waste concrete in construction of various types of structures. The traditional philosophy of the designing procedure – to find a suitable material for a given structure – is transformed into a new way of designing entitled "inverse design approach" that intends to find a suitable structure for a low-strength porous composite material that is obtained from available recycled rubble concrete. In the first stage, composition of concrete mix optimal from the point of view of mechanical properties, workability and economy is estimated based on the experience of the authors. The keynote is maximising of recycled aggregate use, minimising of cement consumption and providing competitive advantage of the composite made from concrete waste. Properties of the material are determined in laboratory tests. Possible applications of the composite in the construction industry are proposed. Finally, one of the possibilities is selected for verification. The results have shown that the introduced inverse design approach is applicable; it can lead to exploitation of significant amounts of construction waste in a useful way, to substantial savings of primary resources and to reduction of carbon footprint of the construction process.

1. Introduction

Growing demand for cutting carbon emissions, reduction of pollution emitted from building industry, saving natural resources and raw materials should target to reduction of construction waste disposal in landfills, re-use of construction materials, utilization of non-traditional materials or non-traditional use of common materials. Utilization of recycled materials is one of European Community priorities. The 7th Environmental Action Programme for Europe promotes increase of recycling and re-use of materials. In accordance with these requirements, utilization of the secondary raw materials is investigated, such as recycled concrete.

The construction and demolition waste constitutes a quarter of all types of waste production. Reuse of this waste in other areas of the economy is a step in the right direction to achieve sustainable development. Research and development institutes contribute to the topic of sustainable construction by research of recycled materials. One of the focuses is on replacing of natural aggregate by rubble concrete recycled from demolished concrete structures.

Use of RCA (recycled concrete aggregate) in concrete is restricted by code prescriptions on aggregate testing, quality of aggregate and limits of maximal amount of recycled aggregate which can be used as a substitution of natural aggregate. The restrictions are caused by the reason of worse quality and high dispersion of recycled aggregate characteristics, which are commonly known.

Background of our research was a challenge to overcome the limitations, benefit from features of the material and tribute to wider application of wastes. One of the objectives was to reduce testing of recycled aggregate in the process of manufacturing concrete composite and the attention focus on properties of the composite.

The purpose and target of our aspirations was to change the understanding in attempts to apply RCA in concrete; it should not be regarded that RCA is used to make concrete with downgraded properties compared to common concrete. Under a new perspective RCA is used to make a new composite, which is relevant in certain applications.

2. State of the art in the field of utilisation of recycled concrete

History of recycling of construction and demolition waste and reuse of materials dates back to ancient times. Modern history with comprehensive research of materials goes ahead since the fifties of the last century.

The essential characteristics of crushed concrete aggregate revealed in previous investigations are:

- downgraded material parameters compared to natural aggregate (NA),
- modified grain-size distribution, flakiness index and density,
- higher water absorption.

Many observations of particle size and shape concluded that recycled concrete aggregate (RCA) is coarser, more porous and rougher in surface but more equidimensional [1]. The density of recycled aggregate is reported to be lower than natural aggregate density [2], [3].

The biggest differences between recycled and natural aggregate are in water absorption. Common natural aggregate's water absorption value is about 0.5 – 1 %. Recycled aggregate's water absorption is higher due to cement paste fixed to the original natural aggregate. Studies showed water absorption of recycled aggregate ranging between 3 – 6.5 % [4].

Nowadays, most recycled concrete is used as aggregate in base and sub-base layers of new road pavements. Utilization of recycled concrete aggregate in new concrete mixtures and application for structural use is rather limited. The primary restrictive factor is contamination of the recycled aggregate by impurities, which causes deterioration of material parameters such as strength, elastic modulus etc. Removing the impurities increases production costs of the recycled aggregate concrete.

The problem of worse material parameters of recycled aggregate is commonly faced by blending of RCA and NA (natural aggregate) [5], [6]. The main factor adversely affecting the strength of concrete produced with RCA is supposed to be the residual mortar attached to the original aggregate in recycled waste concrete. Some investigations focused on removing and strengthening the adhered mortar in RCA. Methods for improving the properties of RCA are based on mechanical and chemical treatment [7], [8], [9].

Recent research works examined the properties of high performance concrete (HPC) with RCA anticipating that good-class recycled concrete with quality original aggregate (e.g. from prestressed structures) shall provide satisfactory properties of HPC with a blend of RCA and NA [10], [11]. These attempts are more theoretically focused and the mixes have been manufactured just in laboratories so far.

Large scale utilization solely of RCA (without blending of RCA and NA) is uncommon [12].

3. New approach to the composite design

Recycled aggregate accounts for ca 8% of aggregate use in the European production. Presented investigations focused on increasing of this ratio and consuming of a significant quantity of crushed concrete from construction and demolition waste (CDW) in production of recycled aggregate concrete (RAC) and minimising of natural aggregate depletion. Long term research of properties of recycled aggregate and concrete made from recycled aggregate resulted in proposal of changing the design philosophy. The keynotes of the approach are:

- maximising of recycled aggregate consumption
- minimising of environmental burden
- minimising of the production cost

In terms of these presumptions, a new concept of the composite mixture design and structural design procedure was introduced. Common concrete mix design begins by determining the requirements on concrete; required concrete class is set according to conditions that the concrete will be exposed to in service. In the new approach, the procedure of the structural design and concrete mix design is inverted, therefore the authors decided to call it "inverse design approach". In the initial phases a mixture is specified, subsequently material properties of hardened concrete are examined and identified. According to achieved material parameters, suitable application for the material is suggested (fig. 1).

The mix recipe is based on these principles:

1. composite is made entirely with recycled aggregate;
2. aggregate from crushed concrete is used without sorting;
3. composite is made with minimal cement content;
4. admixtures and additives are not used.

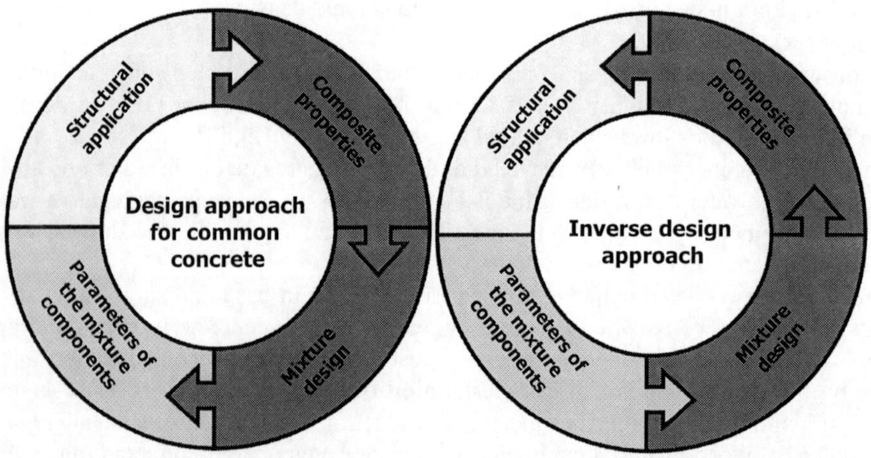

Figure 1. Flow chart of common design and inverse design approach.

The systematic investigations of the authors have proved that to meet all the above-mentioned requirements while obtaining a material with the mechanical properties sufficient for structural use, dispersed reinforcement has to be added to the mix. As a result, there are four components in the mixture: recycled concrete aggregate, cement, fibres and water

3.1. Aggregate

The coarse and fine aggregate are represented entirely by crushed recycled concrete. Using of just the recycled aggregate (not a blend of natural and recycled aggregate) tributes to higher consumption of concrete waste, generates less landfill and decreases depletion of natural resources.

Utilisation of crushed concrete aggregate with wide particle size distribution (aggregate without sorting) will provide competitive advantage of the composite.

The unsorted crushed concrete aggregate usually contains a great deal of coarse aggregate. The void content is relatively high due to the reason that the crushed material contains a great deal of large grains. Next to higher water absorption, also the porosity of RCA is higher than the natural aggregate´s ones. The scatter of porosity is high and differs according to the source of RCA and crushing method used.

IOP Conf. Series: Materials Science and Engineering **596** (2019) 012002 doi:10.1088/1757-899X/596/1/012002

3.2. Cement binder

Dosage of cement relates to the economic and ecology aspects of the research. The cement dosage is specified as minimal amount required by the codes for structural concrete.

Minimal cement dosage contributes to the environmentally friendly production. Together with leaving out of admixtures and additives, it also decreases the total costs of the resulting material.

Minimal cement dosage ensures covering of aggregate with the binder. The interlocking in the composite structure is increased by fibres.

3.3. Water

The water absorption capacity of aggregate influences the properties of the fresh concrete such as its workability.

The dosage of mixing water depends on water absorption of recycled aggregate and aggregate moisture state. The mixing water is dosed to reach the mixture consistency convenient for compaction by ramming or tamping (vibration cannot be used because of the high porosity of composite).

3.4. Fibres

Fibres in the mixture strengthen the structure of the composite and provide tensile strength and toughness. Activation of fibres after macrocracking increases the ductility of the composite.

In the choice of materials for fibres, high porosity and permeability of hardened concrete should be considered. To prevent deterioration of fibres due to penetrating moisture, synthetic fibres should be selected; steel fibres in the pervious concrete are improper because of possible corrosion. Sufficient length of fibres must provide interlocking of fibres between aggregate grains and creation of consistent composite matrix. Optimal length of fibres is usually given as proportion of fibre length to aggregate grain size ranging between 2.5 and 3. Several types of fibres were investigated in the research program, e.g. polypropylene fibres, polyolefin fibres or fibres cut from PET bottles. The concern is on choice of cheap fibres, or alternatively fibres made from waste materials.

3.5. Technology

Dry components are weighed and mixed together with synthetic fibres. After uniform dispersion of fibres, water is added. The dosage of water is adjusted to reach adequate workability for compaction by ramming or tamping. Hardened composite has porous structure with relatively large void content.

3.6. Testing material parameters, analyses

In the second phase of the inversed procedure, material properties and behaviour of the composite are identified.

Strength parameters are determined in tests commonly used for testing of concrete and fibre reinforced concrete. Compressive strength is determined on 150 mm cubes. The same type of cubes is used for testing of tensile splitting strength. Flexural behaviour and stress-strain properties are examined in four-point bending tests on 150/150/700 mm prisms.

Results of the tests are the base for finding the appropriate application of the composite. They can be used in calculations or numerical simulations to decide about the applicability of the composite for a particular structure. If a potentially suitable structure is identified, laboratory experiments on structural models made of the given composite can be performed.

In certain cases supplementary test are proposed to verify other favourable properties of the composite suitable for specific application (e.g. elastic modulus, creep properties, permeability etc.).

4. Example of the use of the inverse design approach

The text below describes one example of the design of recycled concrete mixture in accordance with the principles described in the previous chapter. Performed tests and experiments, detected parameters and properties are summarized. In the end, possible application of the material is proposed.

4.1. Mixture composition

Recycled aggregate originated from railway sleepers (fig. 2, 3) from closed precast factory siding railway located near the recycling centre. The prestressed sleepers are commonly made of concrete with strength about 45 MPa, hence the quality of the original concrete is good. The sleepers were crushed in jaw crusher, reinforcement and fasteners were removed by crushing shears before placing the sleeper on the feeding hopper with raft sorting machine. After separation of reinforcement a coarse loose material remained without great portion of metal components. The loose material was crushed by jaw crusher with electromagnetic separator to take out the remaining metal parts.

Figure 2. Concrete sleepers before crushing.

Figure 3. Recycled aggregate from the crushed sleeper.

For practical applications, conformably with the principle of cost minimizing the recycled aggregate is used without sorting, as sorting of aggregate regarding grain size distribution increases the cost of recycled aggregate.

Nevertheless, the gradation and other RCA properties were determined for research purposes. Maximum grain size was 22 mm. Results of sieve analysis are depicted in fig. 4.

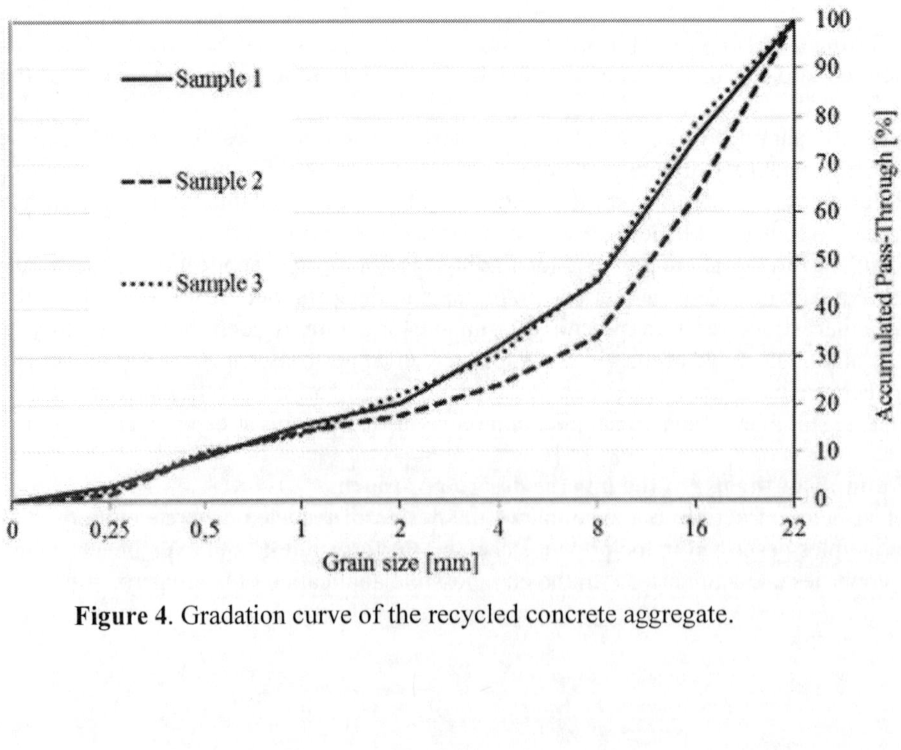

Figure 4. Gradation curve of the recycled concrete aggregate.

Each of the three samples with weight 2 kg was separated through a set of standard sieves; the weight of particles retained on each sieve was measured. Results of the sieve analysis are plotted. Grading pattern displays nearly a gap graded distribution of particles. The crushed concrete contained a great deal of coarse aggregate.
Bulk density, void content and absorbability of recycled aggregate were determined in pycnometric analysis. The void content was ca 35%. It is rather high by the reason that the crushed material contains a great deal of large grains. Absorbability was about 3.5 % which is quite a low value. Typically the absorbability of recycled aggregate is higher. The lower absorbability is given by low content of fine aggregate which provides high water absorption.

4.2. Cement binder
Type I hydraulic Portland cement was used, with 95 % of Portland cement clinker (chemical composition is in the table 1). Specific surface area of cement binder was 381 m^2/kg.

Dosage of cement relates to economic and ecology aspects of the research project and it was determined with respect to the European Standard EC 206-1 requirements as the minimal cement content for structural concrete (260 kg/m^3).

4.3. Fibres
This particular composite is strengthen with polymeric fibres. The fibres are made of special blend of polypropylene and polyethylene. The tensile strength is 610 MPa, the elastic modulus is 5 500 MPa. The length of fibres is 55 mm, aspect ratio 115 and density 910 kg/m^3.

Table 1. Mixture design.

Component	Dosage [kg/m^3]
Recycled aggregate	1 400
Cement	260
Fibres	9.1
Water	150

Dry components were weighed and mixed for about four minutes together with the fibres. After uniform dispersion of the fibres, water was added. The dosage of water was adjusted to reach adequate workability for compaction by ramming or tamping.

After hardening, the composite had a porous structure with relatively large void content (fig. 5).

Figure 5. Hardened fibre reinforced concrete with recycled aggregate.

4.4. Testing program
In laboratory tests, parameters related to the structural function of the composite – tensile splitting, flexural and compression strengths – were determined. Common tests with 150 mm cube specimens were performed at the age 28 days to determine compressive strength and tensile splitting strength.

Table 2. Results of compression tests.

Specimen	Dimensions of the specimen			Weight	Density	Load	Compressive strength
	width [mm]	length [mm]	height [mm]	[kg]	[kg/m3]	[kN]	[MPa]
A1	149	149	149	6,088	1840,4	276	12,43
A2	149	147	149	5,946	1821,9	216	9,86
A3	149	145	148	6,032	1886,5	350	16,2
A4	149	149	149	6,034	1824,1	254	11,44
A5	149	150	148	6,302	1905,2	282	12,62
A6	148	149	151	5,98	1795,9	214	9,7
A7	149	149	149	5,972	1805,3	208	9,37
A8	149	150	148	6,106	1845,9	236	10,56
Average value					1840,7		**11,52**
Conditional standard deviation					35,7		2,11

Characteristic value of compressive strength was calculated from measured data according to ISO 2394. General principles on reliability of structures were taken into account using *k*–coefficient, for eight specimens $k = 2$. The characteristic strength of recycled aggregate fibre reinforced composite was $f_k = 7.3$ MPa.

Table 3. Results of tensile splitting tests.

Specimen	1	2	3	Average value [MPa]	Conditional standard deviation
Tensile splitting strength	0.6797	0.7042	0.7231	**0.702**	0.02

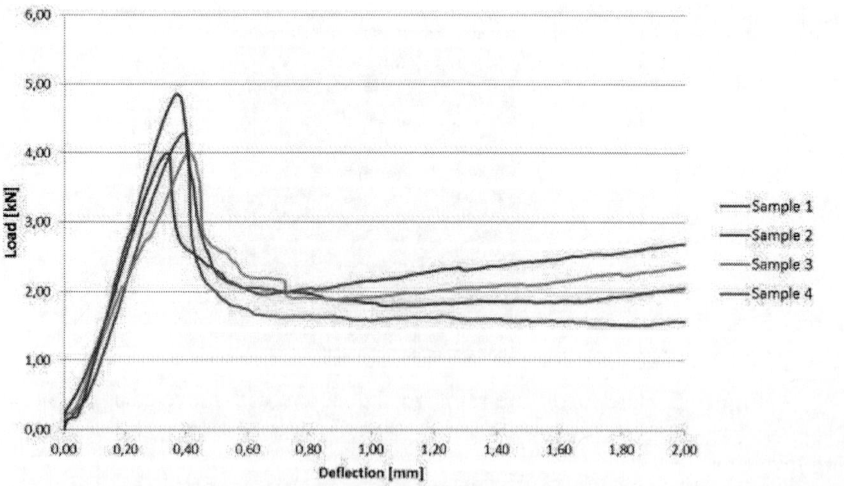

Figure 6. Load deflection diagram from four-point bending test of fibre reinforced composite with recycled aggregate on 100/100/400 mm prism.

The most efficient tests to describe the behaviour of the composites with dispersed fibre reinforcement are flexural tests. Four four-point bending test set-up was applied on 100/100/400 mm specimens. Load-deflection curves are depicted in figure 6.

Compressive strength is lower than strength of common concretes as anticipated; the tensile properties and ductility are acceptable and comparable to dense concrete.

High porosity of the composite led to the assumption of efficacious thermal properties. Volumetric thermal capacity and thermal conductivity were measured.

Table 4. Thermal parameters.

Specimen	thermal conductivity λ [W/mK]	volumetric heat capacity c [10^6 J/m^{3}K]
1	0.6741	1.4244
2	0.7505	1.5288
3	0.5892	1.3925
4	0.7689	1.5213
5	0.6005	1.4068
6	0.5926	1.4083
Mean value	0.663	1.447
Standard deviation	0.082	0.061

Thermal conductivity of the developed porous composite is comparable to conductivity of concretes with lightweight aggregate (e.g. expanded clay, typically 0,3 – 0,8 W/mK) and less than thermal conductivity of dense concrete and solid brick masonry (usually about 1,4 W/mK). The volumetric heat capacity is lower than the volumetric heat capacity of dense and lightweight concrete and solid brick masonry. The potential of thermal benefits of the composite was proved.

4.5. Composite application

To select an appropriate application of the composite, strength and thermal properties were considered. In the end, wall structures were selected as the most prospective option. To verify this possibility, laboratory tests with compressed members were executed. Two non-standard types of prism specimens with dimensions 150/150/600 mm and 150/150/1 200 were loaded with given eccentricity in the direction of their axis and their behaviour was also modelled in numerical simulation (fig. 7).

Results of these tests and simulations were satisfactory and proved sufficient resistance for combination of compression and bending. Subsequent probabilistic analysis revealed problems given by relatively high scatter of material characteristics which implies low characteristic value of strength and high material safety factor. Hence the resistance of the structure is adequate just for very low loading. For application of the material in load-bearing walls, the material characteristics would have to be homogenized and the dispersion of material parameters decreased. Results of tests and related analyses were presented in [13] and [14].

The thermal properties of the material are favourable and application in non-loadbearing walls is prospective. There are two potential forms of application in the wall structure – in-situ variant (composite compacted by ramming in the formwork) and masonry units or hollow blocks made of the composite.

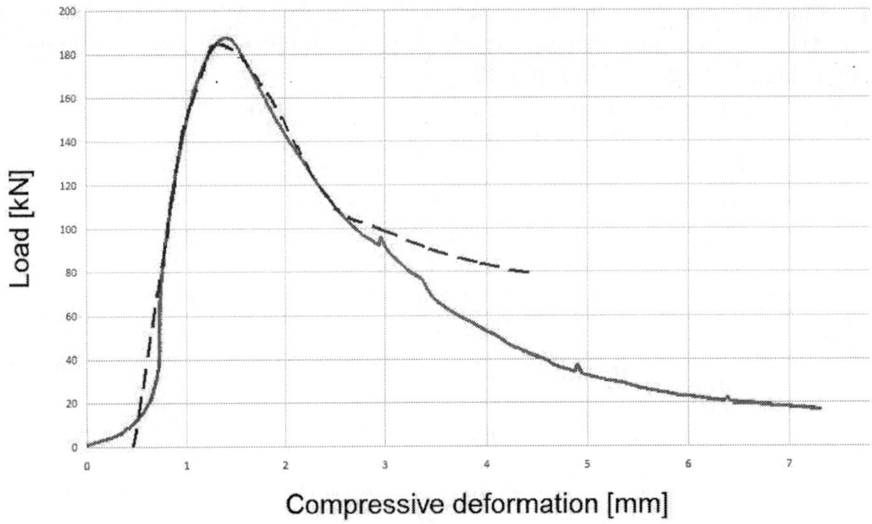

Figure 7. Comparison of load – deflection diagram of the specimen 150/150/600 mm recorded in compression laboratory test (full line in red) and load – deflection diagram calculated in numerical simulation of the test (dashed line in blue).

5. Further examples of possible applications of the composite

The target in the development of the new composite was not an effort to copy or imitate existing material but to find a new material and according to its properties and behaviour propose an appropriate application.

A number of mixes with various combinations of aggregate (of different sources and with different way of crushing), fibres (common fibres available on the market and alternative fibres, e.g. fibres cut from plastic bottles) and binders (various types of cement or blend of cement and alternative binders) were inquired into and examined. Material characteristics obtained for particular mixes – porosity, strengths characteristics, ductility, thermal and acoustic properties – were considered according to principles of inverse design approach. This section presents examples of proposed applications for several types of composite:

- sub-layer in pavements, roadways, parking areas, sport grounds;
- stabilizing layers in earth bodies;
- strengthening layers in earthen dams;
- noise barriers.

Particular applications were assessed concerning technical feasibility. Computational models were developed and numerical simulations performed, for some types of applications loading was performed on scale models.

5.1. Fibre reinforced composite from RCA in sub-layers

The composite can be efficiently used for stabilizing layers in pavements, cycle routes, roadways, parking areas and sport grounds. This application takes the advantage of permeability of the composite that will tribute to draining of sub-layers in these types of structures. Technology of the composite compacting complies with the road and pavements construction technology of roller compacting. Vast amounts of material are consumed in sub-layers, thus a big portion of waste material will be utilised and not deposited in the waste landfills.

5.2. Fibre reinforced composite from RCA in earthworks

Use of the composite as a strengthening layer in earthworks benefits from tensile strength, ductility and permeability of the composite. The pervious layer allows water from precipitation to pass through, prevents runoff and contributes to groundwater recharge.

Numerical simulations showed that stabilizing FRC layers in the earthen body enable to increase the steepness of the slope and thus to decrease farmland occupation (Hruby, 2011).

Numerical simulation of 15 m high embankment with three stabilizing layers confirmed that sloping can be steeper by 11 degrees or narrower by 10 meters compared to un-stiffened earth body. The reduction of the embankment width by 2x10 m determined in the studied case would bring significant savings. Considering the purchasing price of farmland in the Czech Republic as 1 €/m^2, saving of 20x1x1000 = 20 000 € per one kilometre of the dam can be estimated. The savings can be even much higher in case that the structure is built on a more expensive estate such as forest or building land.

5.3. Fibre reinforced composite from RCA in earthen flood dams

Fibre reinforced composite with recycled aggregate can be utilized for stabilizing layers in earthen flood dams.

This application was proposed for recycled aggregate that incorporated higher amount of fine aggregate and the porosity of resulting composite was lower.

To prove the benefits of the stiffening composite layers, two laboratory models in scale 1:30 were prepared. The first model of dam was made entirely from soil; the stiffening layer was only used at the dam crest to prevent damage of the crest in case of overflow and to enable observing of the erosion of the downstream side of the dam.

The other model contained three more stiffening layers except the one in the crest (fig. 8, 9).

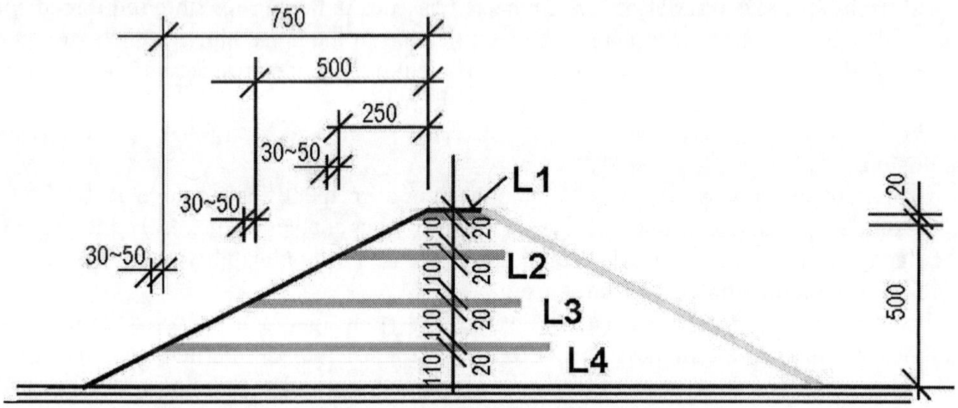

Figure 8. Model of the dam stiffened with four stiffening layers made of fibre reinforced composite with recycled aggregate (dimensioning in mm).

The experimental small-scale models proved benefits of the stiffening layers from fibre reinforced composite with recycled aggregate in case of overtopping of the dam.

Figure 9. Dam model with water running over the crest.

5.4. Noise barriers

Porous materials are generally exploited as sound absorbing materials. On the premise that sound energy will be dissipated in a series of interconnected cavities in the porous fibre reinforced composite with recycled aggregate is based next proposed application in the noise barrier. The assumption must be verified by laboratory test. Sound absorbing coefficient must be determined and reduction of sound pressure level must be verified for particular frequencies. [15]

6. Conclusions

The paper introduced a new approach to the design of cementitious composites with recycled aggregate content – so called "inverse design approach" – that enables to maximize the exploitation of secondary aggregates in concrete for structural applications. The applicability and possible benefits of this approach were demonstrated on several examples.

The inverse design approach is based on the idea that for each recycled concrete mix, an application that will gain the highest profit from the properties of the particular material should be selected. The whole method is therefore in accordance with the principles of the sustainable development which is conditioned by the effective use of the resources. The advantages of the "inverse design approach" include also simplification of the composite manufacturing; as the recycled aggregate is used without sorting and other additional treatment. The main benefit is changing the perspective and offering alternative view of designing cement composites from RCA – it should not be regarded that RCA is used to make concrete with downgraded properties compared to common concrete. In the proposed approach RCA is used to make a new composite, which is relevant in certain applications.

Certainly the limitations must be considered, too. The input material – recycled concrete aggregate possesses heterogeneous properties, what affects the scatter of resulting composite characteristics. The manufactured composite is a porous low-strength material with higher dispersion of material parameters, what must be considered in the choice of appropriate application.

The resistance, safety and the economic aspects still play the main role in the decision-making process when the construction material for the planned structure is selected. However, new aspects of the designing process such as environmental issues and depletion of natural resources are becoming more and more important. Increased utilization of recycled concrete contributes to saving of raw materials, elimination of the need for disposal of construction waste and low carbon production.

Acknowledgments
This paper has been prepared with the financial support of the Technology Agency of the Czech Republic within the project TH02030649 EESDOK – Environmentally Efficient Construction and Demolition Waste for Structures and the Ministry of Education, Youth and Sports, project: Durability of concrete structure and assessment of its life cycle, SGS19/149/OHK1/3T/11.

References

[1] Dhir R K, Limbachiya M C, and Leelawat T 1999 Suitability of recycled concrete aggregate for use in BS 5328 designated mixes. *Proceedings of the Institution of Civil Engineers - Structures and Buildings*. **134** (3): 257 -274.

[2] Limbachiya M C, Leelawat T, and Dhir R K 2000 Use of recycled concrete aggregate in high-strength concrete. *Materials and Structures*, **33**, 574–580.

[3] Sagoe-Crentsil K K, Brown T, and Taylor A H 2001 Performance of concrete made with commercially produced coarse recycled concrete aggregate. *Cement and Concrete Research*, **31**, 701–712

[4] McNeil K and Kang T H K 2013 *Int J Concr Struct Mater* **7**: 61. doi:10.1007/s40069-013-0032-5

[5] Angulo S C, Carrijo P M, Figueiredo A D, Chaves A P and John V M 2010 On the classification of mixed construction and demolition waste aggregate by porosity and its impact on the mechanical performance of concrete *Materials and Structures* (2010)

[6] Dosho Y 2007 Development of a sustainable concrete waste recycling system: Application of recycled aggregate concrete produced by aggregate replacing method. In: Rudolph N. Kraus, Tarun.R. Naik, Peter Claisse, Sadeghi-Pouya, ed. Proc. Int. *Conf: Sustainable construction materials and technologies*, 11-13 June 2007 Coventry, Special papers proceedings., Pub. UW Milwaukee CBU, p142-156

[7] Shi C, et al., 2015 Performance enhancement of recycled concrete aggregate A review, *Journal of Cleaner Production* (2015)

[8] Ryou J S and Lee Y S 2014 *Int J Concr Struct Mater* (2014) **8**: 165. doi:10.1007/s40069-014-0067-2

[9] Choi H, Choi H, Lim M et al. 2016 *Int J Concr Struct Mater* (2016) **10**: 87. doi:10.1007/s40069-015-0124-5

[10] Ajdukiewicz A and Kliszczewicz A 2002 Influence of recycled aggregates on mechanical properties of HS/HPC *Cement and Concrete Composites*, Volume **24**, Issue 2, April 2002, Pages 269-279, ISSN 0958-9465

[11] Bílý P and Kohoutková A 2017 A Numerical Analysis of the Stress-strain Behavior of Anchorage Elements and Steel Liner of a Prestressed Concrete Containment Wall. *Structures* **12**, pp. 24-39.

[12] Yehia S, Helal K, Abusharkh A et al. *International Journal of Concrete Structures and Materials* (2015) 9: 219. doi:10.1007/s40069-015-0100-0

[13] Broukalová I, Šeps K, Vodička J and Novák, J 2015 A Feasibility Study on the Structural Use of Fibre Reinforced Concrete with Recycled Aggregate. *In Proceedings of the Fifteenth International Conference on Civil, Structural and Environmental Engineering Computing. Stirling: Civil-Comp Press Ltd*, 2015, ISSN 1759-3433. ISBN 978-1-905088-63-8.

[14] Broukalová I and Šeps K 2017 Probabilistic Analysis of Structural Member from Recycled Aggregate Concrete, *In: Fibre Concrete 2017. Bristol: IOP Publishing Ltd, 2017. IOP Conference Series: Materials Science and Engineering*. vol. **246**. ISSN 1757-899X.

[15] Broukalová I and Fládr J 2018 Noise barrier panel, Czech Republic. *Utility Model CZ* 32117. 2018-10-02.

IOP Conf. Series: Materials Science and Engineering **596** (2019) 012003 doi:10.1088/1757-899X/596/1/012003

Analytical model for rigid (steel) and flexible (synthetic) fibre mixing in concrete

K P Juhász

Department of Mechanics, Materials & Structures, Budapest University of Technology, Műegyetem rkp. 3, 1111 Budapest, Hungary

E-mail: office@jkp.hu

Abstract. Quantity of the fibres intersecting the unit cross section is a relevant parameter. Firstly, the residual tension strength, which is the base of the material model of the fibre reinforced concrete, depends most on the number of fibres intersecting the cracked cross section. Secondly, by counting the fibres on the cross section of the broken test specimen the non-uniform distribution of the fibres can be examined and taken into account during the evaluation. Therefore, for most of the material model the initial parameter is the quantity of fibres intersecting the unit cross section.

1. Introduction

Romualdi and Mandel already worked on determining this quantity at the early scientific research of fibre reinforced concrete [1], where the behaviour of the fibre reinforced concrete and the quantity of fibres intersecting the cracked cross section of the specimen are analysed through experiments. In 1972 Naaman [2] determined the quantity of fibres intersecting the cross section by supposing uniform distribution and applying probability analysis in his doctoral thesis. Krenchel, who provided a calculation method based on empirical results [3], introduced the orientation factor for describing the orientation of the fibres, which became the base for fibre reinforced material models. Among others, Soroushian and Lee [4], Ng, Foster and Htut [5], Lee, Cho and Vecchio [6] and Dupont and Vandewalle [7] were all engaged in determining this orientation factor.

Orientation of the fibres changes near the formworks, and by that the orientation factor is altered as well, which affects the quantity of the fibres intersecting the unit cross section. The literature calls this phenomenon wall-effect. In literature, beams with square cross section are usually examined, in which case the wall-effect is present at the sides (one side of the mould) and at the corners (two sides of the mould). Dupont and Vandewalle [7] and Stroven [8] investigated how wall-effect influenced the orientation factor in the case of steel fibres. Zerbino et al. [9] examined the location of steel fibres in self compacting concrete elements, in such elements as slab, wall and beam, which was similar to Sarmiento et al [10], who compared the experimental results with the numerical ones in the case of steel fibres located in the beams.

The wall-effect is different in the case of synthetic fibres: while steel fibres rotate when coming in contact with the formwork, then the synthetic fibres react by bending. Oh, Kim and Choi [11] did not take this effect into consideration when analysing synthetic fibre reinforced beams, however Alberti et al. [12] suggested to consider flexibility of the fibres in modelling.

In this paper various methods will be demonstrated that help determining the quantity of the fibres intersecting the cross section. The wall-effect and the variation of the orientation factor will be examined. The rigid (steel) and the flexible (synthetic) fibres will be differentiated. Furthermore, calculation methods of the orientation factor will be presented, by which the quantity of fibres intersecting the cross section can be determined. Predictions of the analytical model will be compared to experimental results.

2. Determining the quantity of fibres intersecting the unit cross section

Quantity of the fibres intersecting the unit cross section is the basis for most of the material models that consider fibres discreetly. Orientation of the fibres is also a relevant parameter for evaluating the experiments, due to which on one hand, the quality of the mixing can be concluded [7], while on the other hand the effect of non-uniform location of the fibres on the cracked cross section can also be considered when evaluating the test results [13]. Numerous researchers examined the impact of mixing on the experimental results regarding different types of elements [9, 10].

2.1. Derivation by Romualdi and Mandel (1964)

Fibres parallel or nearly parallel to the tensile stress are effective in crack control. Thus, some correction must be made for those fibres that are ineffectively oriented. Romualdi and Mandel [1] assumed that the ratio of the average of the projected lengths in one direction to the total length is a proper correction. Based on this the average length of fibres with a given midpoint and length of lf projected to axis x with uniform distribution of α and γ is given by:

$$l_{\text{f,x,m,1}} = \frac{\int_0^{\frac{\pi}{2}}\int_0^{\frac{\pi}{2}} l_{\text{f}} \cos\alpha \cos\gamma \, d\alpha \, d\gamma}{\left(\frac{\pi}{2}\right)^2} = 0.405 l_{\text{f}} \tag{1}$$

Romualdi and Mandel derived the quantity of fibres intersecting the unit cross section from the average distance of midpoint of the fibres. This is given by the following:

$$n = \frac{l_{\text{f,x,m,1}} N}{V} = 0.405 l_{\text{f}} \frac{N}{V} \tag{2}$$

where n is the quantity of the fibres intersecting the unit cross section [pc/m^2], and N is the quantity of the fibres in the volume V.

If both α and γ are distributed uniformly, the endpoints on the sphere densify around the axis z (figure 1b). This is unrealistic. It is well-known, that uniform distribution on the surface of the sphere can be achieved by uniform distributions of the azimuthal angle and the height of the endpoint (figure 1d).
In this case average length of the fibres in the projection of direction x with uniform distribution of α and Z is given by:

$$l_{\text{f,x,m,2}} = \frac{\int_0^{\frac{\pi}{2}}\int_0^{\frac{l_{\text{f}}}{2}} l_{\text{f}} \cos\alpha \cos\left(\arcsin\frac{Z}{l_{\text{f}}}\right) dZ \, d\alpha}{\frac{\pi}{2} l_f} = 0.5 l_{\text{f}} \tag{3}$$

The formula of Romualdi and Mandel, equation (3) has to be modified for the distribution of the end point of the fibres on the sphere to be uniform, thus the quantity of the fibres intersecting the unit cross section are modified as given below:

$$n = \frac{N}{V} 0.5 l_{\text{f}} \tag{4}$$

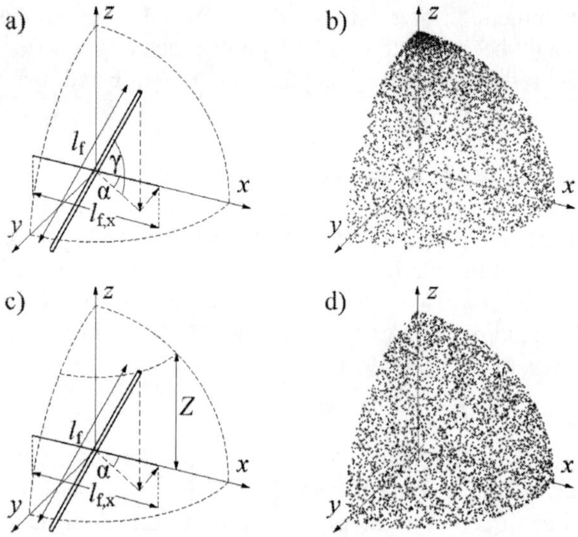

Figure 1. a) Determining spatial distribution of the fibres by polar coordinates α and γ b) Endpoints of the fibres with uniform distribution of α and γ c) Determining spatial distribution of the fibres by parameters α and Z d) Endpoint of fibres with uniform distribution of α and Z.

2.2. Derivation by Naaman (1972)

Naaman [2] determined the quantity of fibres intersecting the cross section by probability analysis. A fibre with a midpoint on axis x is to be considered here (figure 2).

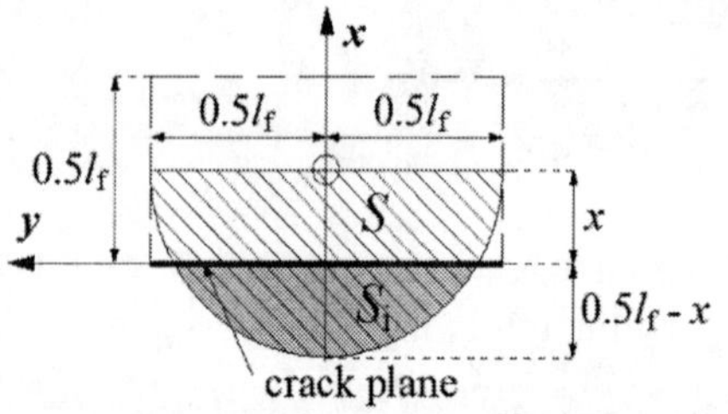

Figure 2. Probability of the fibre intersecting the crack plane.

If the midpoint of the fibre is located at a distance smaller than $0.5l_f$, then the crack plane can be intersected by the fibre. The probability of this is the quotient of surface of spherical cap S_i and half-sphere S drawn around the midpoint of fibre according to figure 2.

$$q_{\text{Naaman}} = \frac{S_i}{S} = \frac{0.5l_f - x}{0.5l_f} = 1 - \frac{2x}{l_f} \tag{5}$$

Considering unit volume at both sides of the crack plane, the quantity of fibres intersecting the unit cross section is similar to the obtained formula by uniform distribution on the sphere (equation (4)):

$$n = 2\int_0^{\frac{l_f}{2}} \left(1 - \frac{2x}{l_f}\right)\frac{N}{V}dx = \frac{N}{V}2\int_0^{\frac{l_f}{2}} \left(1 - \frac{2x}{l_f}\right)dx = \frac{N}{V}0.5l_f \tag{6}$$

2.3. Orientation factor of Krenchel (1975)

Supposing an ideal state each fibre is perpendicular to the crack plane, therefore in the current case each fibre is facing direction x. Considering the volume fraction of the fibres: V_f (weight of the fibres in concrete [kg/m^3]/bulk density of fibres [kg/m^3]), and the cross section area of a discrete fibre: A_f. In this case the quantity of the fibres intersecting cross section A is given by the following formula:

$$n_i = \frac{V_f}{A_f}A \tag{7}$$

where n_i indicates the ideal quantity [pc] in cross section A.

As orientation of the fibres is not ideal in reality Krenchel [3] introduced the orientation factor with the following formula:

$$\theta_{\mathrm{Krenchel}} = \frac{n_a}{n_i} = n_a \frac{A_f}{V_f A} \tag{8}$$

where n_a is the quantity [pc] of fibres intersecting cross section A, determined by experiment by counting at the cracked surface or sliced section.

It is therefore apparent that orientation factor of Krenchel is no other than the factor modifying fibre length in the derivation by Romualdi and Mandel and Naaman, which, supposing uniform distribution, is 0.5.

3. Wall-effect

In the case of rigid (steel) and flexible (synthetic) fibres the effect of the formwork to the orientation of the fibres is different, as fibres of different materials behave distinctly when coming in contact with the formwork. This distinct behaviour will be presented with its effects to the quantity of fibres intersecting the cross section.

3.1. Wall-effect in the case of rigid (steel) fibres

3.1.1. Model of Dupont and Vandewalle for steel fibres

The researchers divided a square cross section of a beam to three regions considering the effect of formworks: 1: undisturbed zone, 2: disturbed zone – one side of the mould, 3: disturbed zone – two sides of the mould (corner) (figure 3).

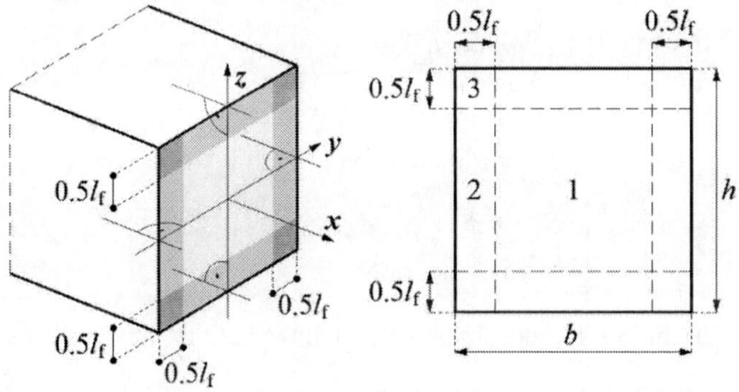

Figure 3. Cross section zones [7].

It was supposed that the midpoint of the fibres near the formwork does not change, and thus their distribution stays uniform, however those fibres that meet the formwork swift around their midpoint in such a way that they contact the formwork. This is true for those fibres having their midpoint closer to the formwork than a distance of $0.5l_f$. These factors were considered with various orientation factors and then the quantity of fibres intersecting the cross section was determined according to formula of Krenchel, equation (7).

The formula of Romualdi and Mandel equation (4) and Krenchel equation (7) start with the condition that the direction of the fibres is vertical at the cross section, then a modified fibre length is used for calculation instead of the actual one. The former is using the effective fibre length and the latter is applying the fibre length modified by the orientation factor. Both of the researchers had similar results. Based on this model only those fibres are influenced by the wall-effect which are able to come in contact with the formwork, thus the midpoint of the fibre is closer to the formwork than a distance of $0.5l_f$.

3.1.2. Suggested model

Coming back again to the derivation of Naaman, it should be examined what happens if the fibres are analysed in an infinite rectangular cylinder. Let us consider a space named T_1 that is $0.5l_f$ wide, $0.5l_f$

deep and has infinite length in direction z, in which the probability of its fibres intersecting the face A_1 is examined according to figure 4. Midpoints of the fibres are located in this space T_1, the origin is located at the edge of the face A_1. The probability of intersecting the face A_1 by fibres with midpoint coordinates of x and y is given by:

$$\theta(x,y)=\frac{S_i(x)-\left(S_{n1}(x,y)+S_{n2}(x,y)\right)}{S} \tag{9}$$

where $S_{n,1}$ and $S_{n,2}$ is a surface that belongs to those fibres which are intersecting the plane of the A_1, but not the A_1 according to figure 4a. Let us consider two adjacent spaces with a size similar to T_1, named T_{2a} and T_{2b}. Fibres with a midpoint of $(x;\ 0.5l_f+y)$ and $(x;\ -y)$ located here and the surfaces belonging to fibres with these midpoints, which extend to space T_1 will similarly be S_{n1} and S_{n2} as can be seen in figure 4b.

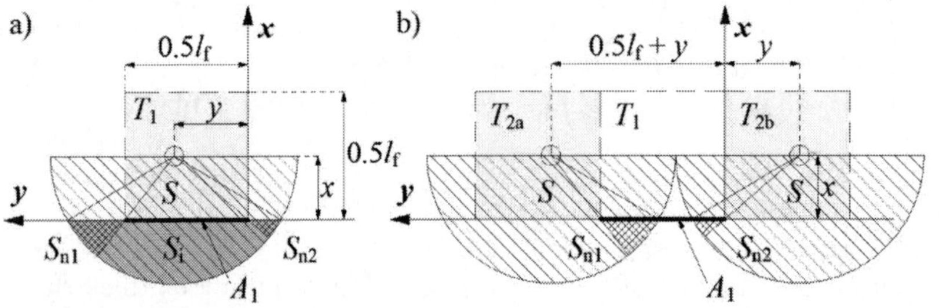

Figure 4. a) interpretation of fibres intersecting the face A_1 from space T_1 b) penetration from the adjacent spaces T_{2a} and T_{2b}.

In the case of fibres located in space T_1 with coordinate y the probability of intersecting the face of A_1 is given by:

$$\theta(y)=2\int_0^{0.5l_f}\frac{S_i(x)-\left(S_{n1}(x,y)+S_{n2}(x,y)\right)}{S}dx \tag{10}$$

and in the case of fibres located at space T_{2a} and T_{2b} with coordinate y the probability of intersecting is:

$$\theta(y)=2\int_0^{0.5l_f}\frac{S_{n,i}(x,y)}{S}dx \tag{11}$$

The probability of certain fibres with midpoint coordinates of y intersecting the face A_1 is demonstrated in figure 5.

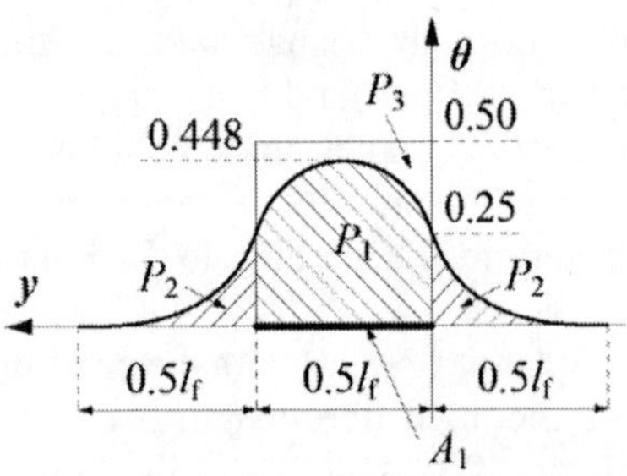

Figure 5. The probability of intersecting of the fibres.

Moving away from the centre of face A_1 until the edge in space T_1 the probability is decreasing as the sum of S_{n1}+S_{n2} is increasing, respectively, moving away from the edge of face A_1 the probability is also decreasing in space T_2. The area below the curve of face A_1 is named P_1 and the area above the curve completing up to 0.50 is named P_3. However the face A_1 can also be intersected by fibres from adjacent spaces T_{2a} and T_{2b}. The corresponding area in figure 5 will be named P_2. Carrying out the integration the result is $2P_2=P_3$, so by considering intersecting of fibres reaching over from adjacent spaces T_{2a} and T_{2b}, the outcome is given below:

IOP Conf. Series: Materials Science and Engineering **596** (2019) 012003 doi:10.1088/1757-899X/596/1/012003

$$\theta_{T_1:T_{2a}-T_{2b}} = \frac{P_1 + 2P_2}{0.5l_f} = \frac{0.196947 + 2 \cdot 0.026527}{0.5} = 0.50 \tag{12}$$

which matches the orientation factor of Naaman derived to infinite space: equation (6).

It should be examined what happens if there is a mould next to the adjacent space T_1, at the border of the similar sized space T_3, as it can be seen in figure 6a. Fibres located at this space can contact the formwork, so the formwork can impact their orientation, fibres rotate. Due to the changing of the orientation reaching over of the fibres from space T_3 to T_1 is limited, therefore fewer fibres are able to intersect face A_1. Space T_3 can be divided into three clearly separable parts, interpreted in the plane of xy. If midpoint of the fibre is in field I, then it does not intersect face A_1, if it is in field II, then it does, but due to the wall-effect there are fewer fibres than without the mould. If midpoint of the fibre is in field III they are not affected by the wall-effect at all, but can intersect the face A_1. The explanation and demonstration of these fields can be seen in figure 6c-f.

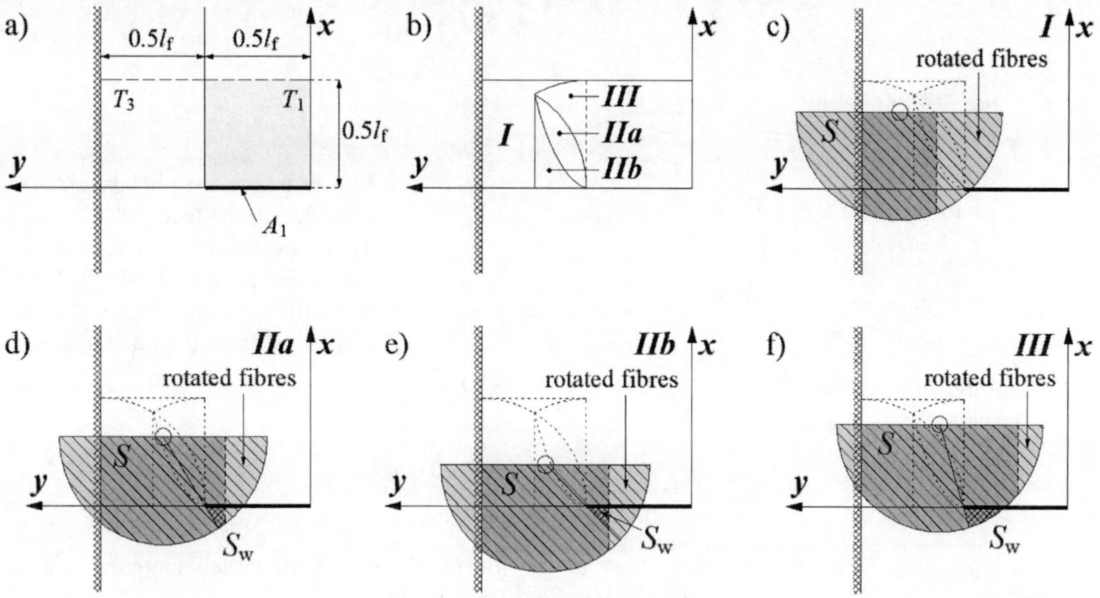

Figure 6. a) T_1 and T_3, mould at border of T_3 b) Defining fields I, IIa, IIb and III c) Explanation of fields I d) IIa e) IIb f) III.

In this case the probability of intersecting face A_1:

$$\theta(y) = 2 \int_{II-III} \frac{S_w(x,y)}{S} dx \tag{13}$$

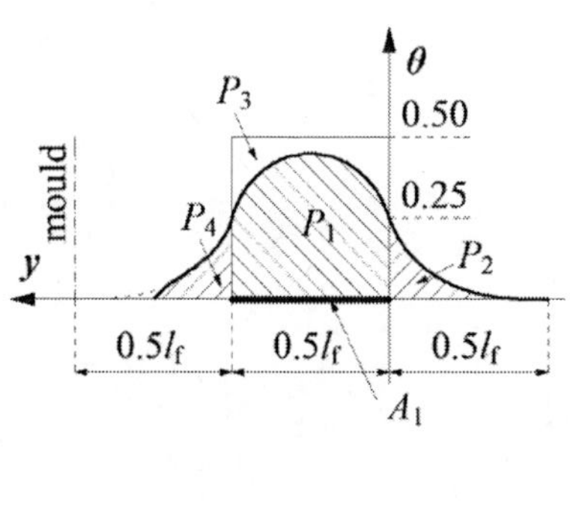

Figure 7. The probability of intersecting face A_1 if the space of A_1 is bounded by a continuous space from one side and by a space concerned by wall effect from the other side.

Probability depicted at axis y is in figure 7. The probability of fibres in space T_1 intersecting face A_1 considering the adjacent and wall-effect free space T_2 and also space T_3 that is affected by the wall-effect from the other side can be explained as the following:

$$\theta_{T_1:T_3-T_2} = \frac{P_1 + P_2 + P_4}{0.5l_f} = \frac{0.196947 + 0.026527 + 0.025418}{0.5} = 0.497 \qquad (14)$$

Fibres in space T_1 are not able to contact the mould, therefore the formwork does not have an impact on their orientation, only on the fibres in space T_3. However due to the limited over-reaching of the fibres still fewer fibres intersect face A_1 in this case, thus we are talking about a *quasi-wall-effect* here. It should be examined what happens in space T_3 where the formwork affects orientation of the fibres directly (figure 8). Fibres in this space can contact the formwork and their orientation changes. In this case probability of the intersecting is:

$$\theta(y) = 2\int_0^{0,5l_f} \frac{S_i(x)}{S(x,y)}dx \qquad (15)$$

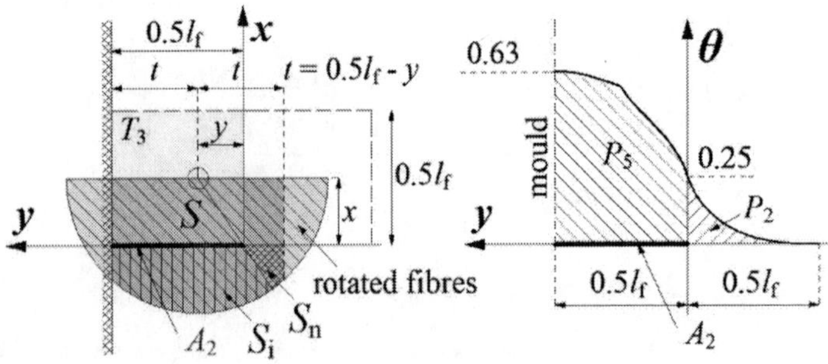

Figure 8. a) Fibres in space T_3 intersecting the face A_2
b) The probability of intersecting face A_2 if space T_3 is bounded by a continuous space from one side and a wall from the other side.

The probability of fibres in space T_3 intersecting face A_2 considering the fibres reaching over from the adjacent space T_1 is given below:

$$\theta_{T_3:zs-T_1} = \frac{P_5 + P_2}{0.5l_f} = \frac{0.262373 + 0.026527}{0.5} = 0.578 \qquad (16)$$

The orientation factor of the fibres next to the formwork is θ=0.63, which matches with the orientation factor of fibres located in the plane θ=2/π=0.63 [4, 14].

3.2. *Wall-effect in the case of flexible (synthetic) fibres*

It is supposed that synthetic fibres bend when coming in contact with the formwork. Alberti et al [12] considered this by decreasing the length of fibres, in current dissertation it is considered by the bending of the fibres. The effect of the formwork in this case is only present at those fibres which have their midpoint closer to the formwork than the distance of $0.5l_f$. In the case of synthetic fibres the reaching over is not limited due to the bending of the fibres, thus the *quasi-wall-effect* does not exist (figure 9).

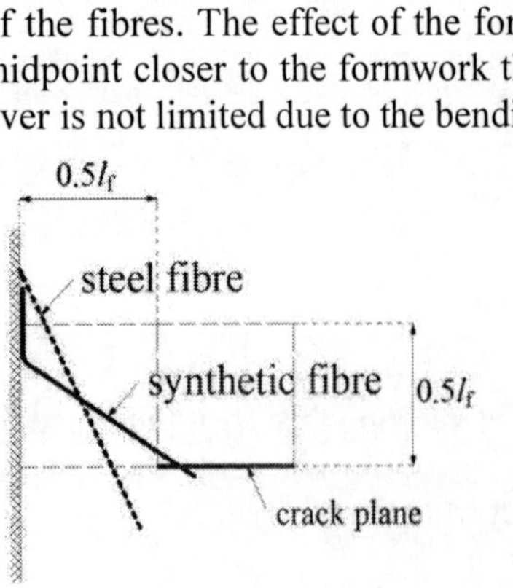

Figure 9. Over-reaching of steel and synthetic fibres to the adjacent cross section zone when the wall-effect is active.

IOP Conf. Series: Materials Science and Engineering **596** (2019) 012003 doi:10.1088/1757-899X/596/1/012003

As endpoints of the fibres will not be on the surface of sphere after the bending, the probability analysis method of Naaman cannot be used in this case. At the same time its utilization is not required, as the wall-effect does not have an impact on the adjacent space. Thus for determining the quantity of fibres intersecting the cross section in the case of synthetic fibres the modified formula of Romualdi and Mandel derived from the average fibre length equation (4) is more suitable. For this the average length of projection in direction x of a fibre needs to be determined. Fibres contacting the formwork bend, their projection length of direction x is considered by modifying equation (3):

$$l_{\mathrm{f,x,m,3}} = \frac{\int_0^{\frac{\pi}{2}}\int_0^{W} l_\mathrm{f}\cos\alpha\cos\left(\arcsin\frac{Z}{l_\mathrm{f}}\right)dZ\,d\alpha + \int_0^{\frac{\pi}{2}}\int_W^{\frac{l_\mathrm{f}}{2}}\left[l_\mathrm{f}\frac{W}{Z}\left(\cos\left(\arcsin\frac{Z}{l_\mathrm{f}}\right)-1\right)+l_\mathrm{f}\right]\cos\alpha\,dZ\,d\alpha}{\frac{\pi}{2}l_f} \tag{17}$$

where W is the distance of the fibres from the formwork.

In the case of synthetic fibres the wall-effect is not symmetrical as it is in the case of steel fibres (figure 10), so the average of the projection length in direction x will be the following:

$$l_{\mathrm{f,x,m,4}} = \frac{l_{\mathrm{f,x,m,2}} + l_{\mathrm{f,x,m,3}}}{2} \tag{18}$$

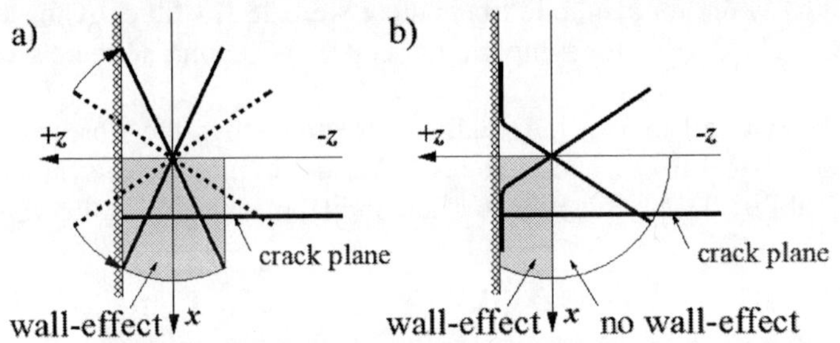

Figure 10. The impact of wall-effect on the quantity of fibres intersecting the cross section in the case of a) steel fibres and b) synthetic fibres.

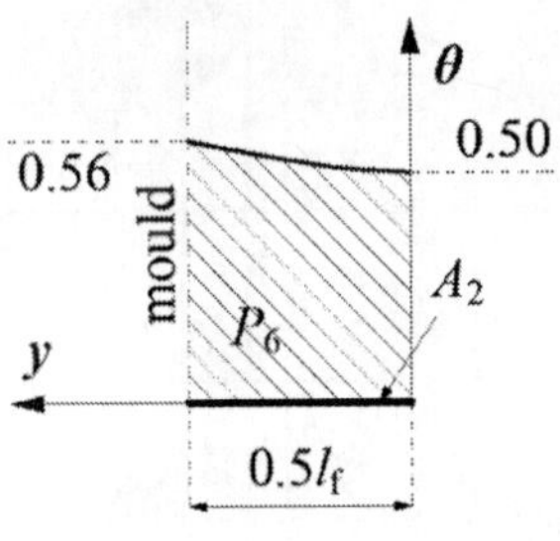

Figure 11. Probability of intersecting of face A_2 in the case of flexible (synthetic) fibre.

Probability is demonstrated in figure 11. Thus the probability of fibres in space T_3 intersecting face A_2 in the case of flexible (synthetic) fibres is:

$$\theta_{T_3:zs} = \frac{P_6}{0.5 l_\mathrm{f}} = 0.53 \tag{19}$$

4. Orientation factors of rigid (steel) and flexible (synthetic) fibres

Steel fibre zones of Dupont and Vandewalle [7] can be seen in figure 12a, modified zones based on current study can be seen in figure 12b, and synthetic fibre zones can be seen in figure 12c, accompanied by their respective orientation factors in each case.

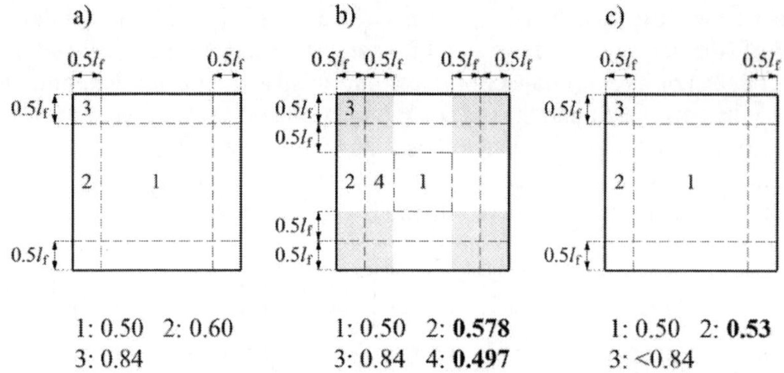

Figure 12. Cross section zones and their orientation factors.

5. Verification

Four steel and four synthetic fibre reinforced concrete beams were made. The pouring method was different: two of the beams were poured parallel to the longitudinal axis of the beam, and two of the beams were poured perpendicular. With this the effect of the different pouring method was taken into account. Cross section size of the beams was 150x150 mm, their lengths was 1 meter. The determination of the fibre dosage was carried out in order to have an equal number of fibres in each of the beams. For steel fibres there were 3181 fibres in 1 kg, while for synthetic fibres there were 35714 fibres. Thus the chosen dosage for steel fibres was 33.73 kg/m^3, while for synthetic fibres it was 3 kg/m^3, so there were 107346 fibres in 1 m^3 of concrete.

The complete beams were cut to slices of 50 mm in longitudinal direction, then the fibres were counted in zones 1, 2, 3 and 4 in the case of steel fibres, and in zones 1, 2 and 3 in the case of synthetic fibres. Histograms, standard normal probability density function, test and model mean value of the fibres intersecting the cross sections can be seen in figure 13.

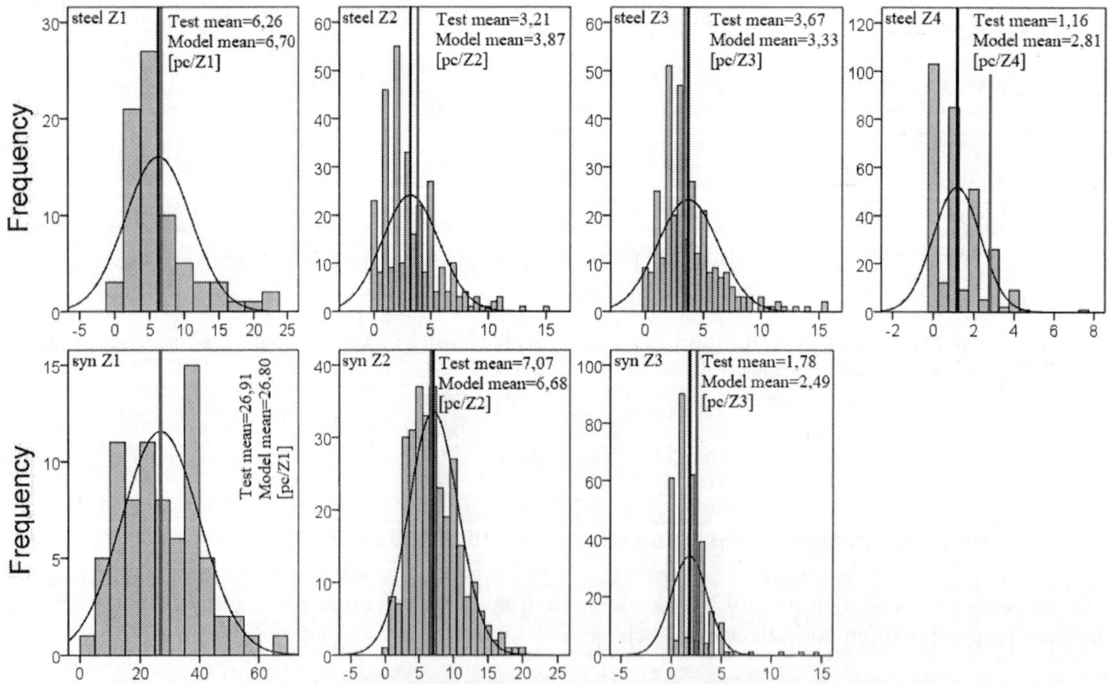

Figure 13. Histograms of steel and synthetic fibers intersecting the cross sections.

IOP Conf. Series: Materials Science and Engineering **596** (2019) 012003 doi:10.1088/1757-899X/596/1/012003

6. Summary

The major difference between the steel and synthetic fibres regarding their behaviour in concrete is the rigidity of discrete fibres: while steel fibres can be regarded as rigid, then synthetic fibres can be flexible. Based on current model fibres are described by their midpoint and orientation by supposing uniform distribution, from which the quantity of fibres intersecting the cracked surface can be derived. Orientation changes near the formwork, it becomes partly directional. This directionality is different for steel fibres as it is for synthetic ones, therefore the quantity of fibres intersecting the cross section also changes in a different way near the formwork.

In this paper a suggested model to steel fibres was presented, introducing the *quasi-wall-effect*, indicating that the impact of the wall influences wider areas of the cross section. Quantity of fibres intersecting the cross section can be determined by the model. Wall-effect works in a different way in the case of the synthetic fibres than it does for steel fibres. It was demonstrated that there is no quasi-wall-effect in the case of the synthetic fibres, and due to their flexibility when coming in contact with the formwork the orientation factors also change. Orientation factors for synthetic fibres was provided as well. The analytical model was verified both for steel and synthetic fibres.

References

[1] Romualdi J P and Mandel J A 1964 Tensile strength of concrete affected by uniformly distributed and closely spaced short lengths of wire reinforcement *ACI Journal* **61(6)** pp 657-71

[2] Naaman A E 1972 *A statistical theory of strength for fiber reinforced concrete* Doctoral thesis, Massachusetts Institute of Technology

[3] Krenchel H 1975 Fibre spacing and specific fibre surface, in *Fibre reinforced cement and concrete* ed A Neville, London: The Construction Press pp 67-5

[4] Soroushian P and Lee C D 1990 Distribution and orientation of fibers in steel fiber reinforced concrete *ACI Material Journal* **87(5)** pp 433-9

[5] Ng T S, Foster S J and Htut T N S 2012 *Fracture of Steel Fibre Reinforced Concrete – the Unified Variable Engagement Model* UNICIV Report R-460, The University of New South Wales

[6] Lee S C, Cho J Y and Vecchio F J 2011 Diverse embedment model for steel fiber-reinforced concrete in tension: model development *ACI Materials Journal* **108(5)** pp 516-25

[7] Dupont D and Vandewalle L 2004 Comparison between The Round Plate Test and The RILEM 3-Point Bending Test, in *6th International RILEM Symposium on Fibre-Reinforced Concretes* (BEFIB'2004) ed M di Prisco, R Felicetti and G A Plizzari pp 101-10

[8] Stroven P 2010 Methodology of modelling fiber reinforcement in concrete elements, in *Fracture mechanics of concrete structures - high performance, fiber reinforced concrete, special loadings and structural applications* ed B H Oh et al pp 1418-24

[9] Zerbino R, Tobes J M, Bossio M E and Giaccio G 2012 On the orientation of fibres in structural members fabricated with self-compacting fibre reinforced concrete *Cement & Concrete Composites* **34** pp 191-200

[10] Sarmiento E V, Zirgulis G, Sandbakk S, Geiker M R and Kanstad T 2012 Influence of concrete flow on fibre distribution, orientation and mechanical properties of fibre reinforced concrete in *BEFIB2012 - Fibre reinforced concrete* ed J Barros pp 418-30

[11] Oh B H, Kim J C and Choi Y C 2007 Fracture behavior of concrete members reinforced with structural synthetic fibers *Engineering Fracture Mechanics* **74** pp 243-57

[12] Alberti M G, Enfedaque A and Gálvez J C 2017 On the prediction of the orientation factor and fibre distribution of steel and macro-synthetic fibres for fibre-reinforced concrete *Cement and Concrete Composites* **77** pp 29-48

[13] Juhász K P 2015 Evaluation of fibre reinforced concrete beam test results based on the examination of the real fibre distribution, in *Proceedings of Fibre Concrete 2015* ed A Kohoutková et al

[14] Lee C and Kim H 2010 Orientation factor and number of fibers at failure plane in ring-type steel fiber reinforced concrete *Cement and Concrete Research* **40(5)** pp 810-19

Appraisal of MC2010 shear resistance approaches coupled with a residual flexural strength prediction model

T D S Valente[1], J A O Barros[2], B N Moraes-Neto[3] and Y H Moussa[4]

[1]PhD Candidate, CiviTest, Lda, V.N. Famalicão, Portugal
[2]Full Professor, ISISE, IB-S, Dep. Civil Eng. Minho University, Portugal
[3]Assistant Prof., Dep. Civil Eng. Federal University of Pará, Brazil
[4]MSC Student, Dep. Civil Eng. Federal University of Pará, Brazil

E-mail: tiagovalente@civitest.com, tel: +351252315199

Abstract. In the present work the predictive performance of the two approaches proposed by Model Code 2010 for the evaluation of the shear capacity of fiber reinforced concrete (FRC) elements flexurally reinforced with conventional steel bars is assessed considering a database (DBs) constituted by 80 FRC beams do not including conventional transverse reinforcements. The accuracy of these shear models is evaluated by statistical analysis of the prediction ratio between the experimental and estimated shear capacity of the beams of the DBs, and applying the Demerit Points Classification approach for further information about the reliability of the two approaches in design context. Due to the absence of the post-cracking experimental characterization of the FRC used in several beams considered in the DBs, an approach was developed for estimating the residual flexural strength parameters from the most relevant known variables of steel fiber reinforcement mechanisms for concrete, namely the fiber volume and aspect ratio, and the concrete compressive and tensile strength. The residual flexural strength prediction model is assessed and its influence on the performance of the shear resistance models is evaluated.

1. Introduction

Regarding the use of discrete fibers in concrete reinforcement, previous research has already pointed that it increases concrete shear resistance and deformation capacity, decreases crack width and spacing, and can change the type of fracture mode for members without transverse conventional reinforcements, from brittle to a ductile one [1–5]. In addition, the fiber reinforcement can be adopted for partial or total replacement of steel stirrups in structural elements preserving the required shear capacity to the member [2,6,7].

Based on the extensive research already carried out regarding the experimental assessment of the shear behavior of fiber reinforced concrete (FRC) members, a database (DB_s) that collects the shear tests data of beams flexurally reinforced with conventional steel bars and made by steel fiber reinforced concrete (SFRC) has been developed [8]. These beams are herein abbreviated by the acronym R/SFRC. The DB_s has already been used for the appraisal of R/SFRC shear resistance prediction models [9–11]. Currently, the DB_s is being extended, and some difficulties have been encountered in using the data of some shear tests due to the lack of some key information, namely the residual flexural strength of the SFRC, which is used in both approaches proposed by CEB *fib* Model Code 2010 (MC2010) [12] for the evaluation of the shear capacity of FRC elements flexurally reinforced with conventional steel bars.

In fact, the final version of MC2010 [12], published in 2013, covers the design guidelines and recommendations for the dimensioning process of FRC structural members, including two shear resistance prediction models. Considering the expansion of the FRC shear tests DB_s, a new opportunity has been found to appraise two of the most widely used FRC shear resistance prediction models, coupled with a residual flexural strength prediction model. In the context of a model for predicting the punching capacity of SFRC slabs, Moraes-Neto *et al.* [13] have mounted a database (DB_f) formed by results from three point notched beam bending tests (3PNBBT) with specimens of concrete reinforced with hooked ends steel fibers (SFRC), executed according to the RILEM TC 16-TDF [14] and MC2010 recommendations. Treating statistically the results from the DB_f, the authors derived simple equations for determining the residual flexural tensile strength parameters ($f_{Ri}, (i=1,2,3,4)$) from the volume content (V_f) and aspect ratio (l_f/d_f) of the hooked ends steel fibers. The authors have recognized that one of the debilities of the proposed equations is their independence of the quality of the matrix, since the fiber reinforcement mechanisms, and consequently the post-cracking tensile capacity of a FRC, depend on this property. In the present work the DB_f is extended in order to have enough information for including the concrete strength as a variable making part of a new generation of more reliable and consistent equations for determining f_{Ri}. These equations are used to estimate the contribution of the fiber reinforcement for the SFRC shear capacity according to the two approaches of the MC2010 whenever f_{Ri} are not known experimentally.

2. MC2010 FRC shear resistance prediction models

The Model Code 2010 presents two models to predict the shear resistance of FRC structural members based on: (i) the concept of the residual flexural strength of FRC (f_{Ri}); (ii) the simplified modified compression field theory (SMCFT) [15] coupled with the Variable Engagement Model (VEM) [16].

The resistance of FRC members using the model based on the concept of the residual flexural strength, denoted in this paper as MC2010_EEN, is determined by the expression [12]:

$$V_{Rd,F} = \left\{ \frac{0.18}{\gamma_c} \cdot k \cdot \left[100 \cdot \rho_{sl} \left(1 + 7.5 \cdot \frac{f_{Ftuk}}{f_{ctk}} \right) \cdot f_{ck} \right]^{1/3} + 0.15 \cdot \sigma_{cp} \right\} b_w \cdot d \;\; ; \;\; \sigma_{cp} < 0.2 \cdot f_{cd} \tag{1}$$

where γ_c is the partial safety factor for concrete without fibers; $k = 1 + \sqrt{200/d} \le 2.0$ is a factor that takes into account the beam's size effect; d is the effective depth of the cross section [mm]; $\rho_{sl} = A_{sl}/(b_w \cdot d)$ is the longitudinal reinforcement ratio; A_{sl} is the cross-sectional area of the longitudinal reinforcement [mm^2]; f_{Ftuk} is the characteristic value of the ultimate residual tensile strength of the FRC [MPa]; f_{ctk} is the characteristic tensile strength for the concrete without fibers [MPa]; f_{ck} is the concrete characteristic compressive strength [MPa]; σ_{cp} is the average axial stress acting in the cross-section [MPa], f_{cd} is the design value of the concrete compressive strength; and b_w is the smallest width of the tensile zone of the cross-section [mm].

The shear resistance, $V_{Rd,F}$, is assumed to be not smaller than the minimum value, $V_{Rd,F,\min}$, obtained from [12]:

$$V_{Rd,F,\min} = \left(v_{\min} + 0.15 \cdot \sigma_{cp} \right) \cdot b_w \cdot d \;\; ; \;\; v_{\min} = 0.035 \cdot k^{3/2} \cdot f_{ck}^{1/2} \tag{2}$$

For evaluation of the shear resistance, the ultimate residual tensile strength is computed from equation (3) considering the residual flexural strength parameters of FRC, $f_{Ri}\,(i = 1,3)$ and the ultimate crack width equal to $w_u = 1.5mm$. For the design situations of FRC members, in equation (3) should be considered the characteristic values of the residual flexural strength of the FRC.

$$f_{Ftu} = f_{Fts} - \frac{w_u}{CMOD_3(=2.5mm)} \cdot (f_{Fts} - 0.5 \cdot f_{R3} + 0.2 \cdot f_{R1}) \geq 0 \;;\; f_{Fts} = 0.45 \cdot f_{R1} \tag{3}$$

The MC2010_EEN model is based on the shear contribution of plain concrete members without transverse reinforcements, as proposed in Eurocode 2 [17]. The increased post-cracking toughness and crack-opening restriction provided by the fiber reinforcement to the concrete is empirically considered by multiplying the longitudinal reinforcement ratio by the factor $(1+7.5 f_{Ftuk}/f_{ctk})$. Thus, the fiber reinforcement contribution is considered as an additional flexural reinforcement [18].

According to the model based on the SMCFT/VEM theory, hereafter denoted as MC2010_MCFT, the general expression that predicts the shear resistance of FRC member is equal to:

$$V_{Rd} = V_{Rd,F} + V_{Rd,s} \tag{4}$$

where $V_{Rd,F}$ is the FRC shear resistance contribution, while $V_{Rd,s}$ is the added shear resistance provided by the transverse conventional reinforcement.

The FRC shear resistance, $V_{Rd,F}$, is the shear resistance contribution of the concrete matrix, $V_{Rd,c}$, plus of the contribution of the fibers bridging the shear failure crack, $V_{Rd,f}$, being determined by the following expression[12]:

$$V_{Rd,F} = V_{Rd,c} + V_{Rd,f} = \frac{1}{\gamma_F} \cdot \left[k_v \cdot \sqrt{f_{ck}} + k_{fd} \cdot f_{Tk}(w_u) \cdot \cot\theta\right] \cdot z \cdot b_w \;,\; \sqrt{f_{ck}} \leq 8MPa \tag{5}$$

where k_v is the parameter that determines the contribution of the aggregate interlock mechanism for the shear strength of the cross section ((6)); k_{fd} is a fiber dispersion reduction factor [19], assuming the value of $k_{fd} = 0.8$; $f_{Tk}(w_u)$ is the characteristic value of the post-cracking tensile capacity of FRC, determined from direct tensile tests, evaluated at the ultimate crack width of w_u; θ is the inclination of the compressive stress field, corresponding to the inclination of the critical diagonal crack; z is the internal lever arm that can be estimated as $z = 0.9 \cdot d$ [mm] [12].

The k_v is function of the parameter that considers the aggregate size influence, k_{dg} (equation (7)), of the longitudinal strain at the mid depth of the effective shear area, ε_x, and of the transverse reinforcement ratio of the cross-section, ρ_w. In equation (7) the term d_g is the maximum aggregate dimension in the concrete matrix [mm].

$$k_v = \begin{cases} \dfrac{0.4}{1+1500 \cdot \varepsilon_x} \cdot \dfrac{1300}{1000 + z \cdot k_{dg}} & \text{for } \rho_w < 0.08 \cdot \sqrt{f_{ck}/f_{yk}} \\ \dfrac{0.4}{1+1500 \cdot \varepsilon_x} & \text{for } \rho_w \geq 0.08 \cdot \sqrt{f_{ck}/f_{yk}} \end{cases} > 0.0 \tag{6}$$

$$k_{dg} = \begin{cases} \dfrac{32}{16 + d_g} \geq 0.75 & \text{for } f_{ck} \leq 70MPa \\ 2.0 & \text{for } f_{ck} > 70MPa \text{ and for ligthweight concrete} \end{cases} \tag{7}$$

The strain, ε_x, at mid-depth of the effective shear area is defined as half the tension chord strain, $\varepsilon_x = \varepsilon_s/2$, namely [12,20]:

$$\varepsilon_x = 0 \le \frac{1}{2 \cdot E_s \cdot A_{sl}} \left[\frac{M_{Ed}}{z} + \frac{V_{Ed}}{2} \cot\theta + N_{Ed} \left(\frac{1}{2} \pm \frac{\Delta e}{z} \right) \right] \le 0.003 \tag{8}$$

where Δe is the eccentricity of the beam axis with respect to its mid-depth of the effective shear area; M_{Ed} and V_{Ed} correspond to the acting bending moment and shear force at the cross-section (assumed positive values) and N_{Ed} is the axial force at the cross-section (positive for tension and negative for compression).

According to MC2010, the value of θ can be freely chosen in the interval of $\theta_{min} \le \theta \le 45°$, while the value of θ_{min} is related with the longitudinal strain level in the mid-depth of the cross-section, ε_x, [12]:

$$\theta_{min} = \left(29° + 7000 \cdot \varepsilon_x\right) \tag{9}$$

Based on the work of [21], where a $\sigma - w$ relationship of FRC was derived from inverse analysis using the results of prism bending tests, the direct post-cracking tensile capacity of FRC can be estimated from the expression [9]:

$$f_{Tk}(w) = k_G \min\left(0.4 f_{R2k} + 1.2\left(f_{R4k} - f_{R2k}\right) \xi(w),\ f_{ctk,min}\right) \ ; \ \xi(w) = \alpha w - 0.25 \tag{10}$$

The factor k_G takes into account fibre alignment due to casting bias and wall influences that occur in the prism bending test. In Table 1 is presented the value of k_G considering the different prism bending test standards. The value for the factor α also depends on the prism bending test configuration and is presented in Table 1.

Table 1. Value of k_G and α depending on prism bending test standard [9].

Prism bending test standard	k_G	α
ASTM C1609 [22]	0.70	1/3
EN 14651 [23]	0.60	5/12
RILEM TC 162-TDF [14]	0.60	5/12
UNI 11039 [24]	0.60	43/84

The ultimate crack width, w_u, orthogonal to the critical diagonal crack is determined according to equation (11) [12].

$$w_u = \left(0.2 + 1000 \cdot \varepsilon_x\right) \ge 0.125\ mm \tag{11}$$

The shear reinforcement resistance, $V_{Rd,s}$, is determined according to the following equation [12]:

$$V_{Rd,s} = A_{sw}/s_w \cdot z \cdot f_{ywd} \cdot \left(\cot\theta + \cot\alpha\right) \cdot \sin\alpha \tag{12}$$

where A_{sw} is the shear reinforcement area [mm^2], f_{ywd} is the design value of the yield stress of the shear reinforcement, and α is the inclination of the transverse reinforcement.

The design shear resistance cannot exceed the crushing capacity of concrete in the web, determined as [12]:

$$V_{Rd,max} = k_c \cdot \frac{f_{ck}}{\gamma_c} \cdot b_w \cdot z \cdot \frac{\cot\theta + \cot\alpha}{1 + \cot^2\theta} \tag{13}$$

where k_c is a strength reduction factor defined by $k_c = k_\varepsilon \cdot \eta_{fc}$; k_ε is a factor that takes into account the strain in the web of the structural element, being determined according to the level of approximation defined in MC2010 (level of approximation I: $k_\varepsilon = 0.55$; level of approximation II and III: equation (14)); η_{fc} is a factor to consider the influence of the concrete strength class on the level of the shear failure brittleness , being determined from equation (15).

$$k_\varepsilon = \left(1.2 + 55 \cdot \varepsilon_1\right)^{-1} \le 0.65 \; ; \; \varepsilon_1 = \varepsilon_x + \left(\varepsilon_x + 0.002\right)\cot^2\theta \tag{14}$$

$$\eta_{fc} = \left(30/f_{ck}\right)^{1/3} \le 1.0 \;\; \left(f_{ck} \text{ in MPa}\right) \tag{15}$$

Both models can only be adopted if conventional steel longitudinal reinforcement is present in the cross-section of the FRC member.

3. FRC beams shear tests database (DB_s)

An extended version of a database (DB_s) comprising shear tests of FRC beams [8] is considered for the assessment of the MC2010 shear prediction models.

The DB_s includes the data of 113 steel fiber reinforced concrete (R/SFRC) beams submitted to shear, collected from the following sources [1,5,21,25–38]. In terms of cross section configuration of the beams considered in the DB_s, 99 beams are of rectangular shape, while 14 are of T-shape. Within the 113 R/SFRC beams, 99 are of deep type cross section $\left(d/b_w > 1.0\right)$ and 14 of current and shallow type cross section $\left(d/b_w \le 1.0\right)$. The effective shear span ratio, a/d, ranges from 2.0 to 4.0, therefore it is assured the applied load is not directly conducted to the closest support of the beam in any of the R/SFRC of the DB_s.

All the R/SFRC of the DB_s are flexurally reinforced with passive steel bars (prestress effect is not treated in the present work). The flexural reinforcement ratio varies between 1.0 to 3.1%, and the effective depth of the cross-section, *d*, is within the interval 150 to 1440 mm. None of the R/SFRC beams is shear reinforced with conventional reinforcements (like stirrups). In the 113 R/SFRC beams of the DB_s only hooked end type of fibers were used, with a fiber aspect ratio, l_f/d_f (l_f and d_f are the length and diameter of the fiber, respectively) varying from 48 to 80. The fiber volume, V_f, ranges from 0.25% to 1.5%. The concrete maximum aggregate size, d_g, varies between 10 and 25 mm, while the average compressive strength of the SFRC ranges between 19.2 to 64.6 MPa. Regarding the SFRC residual flexural tensile strength, the average values of f_{R1} and f_{R3} are not reported for 39 samples, while for f_{R2} and f_{R4} are not provided for 59 samples. In the DB_s three different prism bending test standards were used, namely the EN 14651 [23], RILEM TC 162-TDF [14], the ASTM C1609 [22] and the UNI 11039[24]. Considering the main differences for the evaluation of the residual flexural strength using the referenced standards, the EN 14651 and RILEM TC 162-TDF adopt a 3-point bending test configuration with notched FRC beams, while the ASTM C1609 and UNI 11039 adopt a 4-point bending test, the former one with un-notched beams, and the later with notched beams.

Due to the use of un-notched SFRC samples on the prisms bending tests performed according to the ASTM C1609, the crack appears at the weakest point between the two loading points, and the measurement of the crack mouth opening displacement (CMOD) is not possible due to the unknown location of the crack. Consequently, in order to calculate the residual flexural strength of the SFRC samples of the database, the CMODs were obtained by correlating the CMOD with the measured mid-span deflection of the SFRC prisms using the methodology proposed in [39] that considers a rigid body kinematic mechanism similar to the one adopted in 3-point bending tests of notched beams [40]. Due to this simplification, the residual flexural strength values that were determined according to the ASTM C1609 can present some error.

Due to the significant number of R/SFRC beams of the DB_s were the residual flexural strength is not reported, there was the need to adopt a model to predict these mechanical properties, which is discussed in the next section.

After implementation of this prediction model, the extreme values of the residual flexural strength of the R/SFRC beams studied in the DB_s are: $f_{R1} = [1.5-8.5]$ MPa, $f_{R2} = [1.8-8.9]$ MPa, $f_{R3} = [1.5-8.9]$ MPa and $f_{R4} = [1.1-7.9]$ MPa.

A first analysis of the DB_s has consisted in the exclusion of test results where a flexural-shear failure mechanism could be feasible. For this purpose, the flexural capacity of each R/SFRC beam was assessed following the guidelines of MC2010. For the samples where the ultimate resisting bending moment was lower than 95% of the acting bending moment (due to shear capacity registered experimentally), the shear test results were not considered in this study. After this analysis, a total of 80 R/SFRC beams were considered for the assessment of the shear resistance prediction models, within which the complete data of the post-cracking residual flexural strength (f_{Ri}, $i = 1,2,3,4$) was available for 42 beams, and partial information (f_{Ri}, $i = 1,3$) was provided for 59 beams. Since some of the beams in the DB_s are casted with the same SFRC mix, the results of the experimental characterization of the post-cracking residual strength for these beams is the same. The 59 beams with experimental data of f_{R1} and f_{R3} were casted from 44 different SFRC mixes, while the 42 beams with f_{R2} and f_{R4} resulted from 35 different SFRC mixes. In Table 2 is presented the different standards considered in the evaluation of the available residual flexural strength values, considering the different SFRC mixes present in the DB_s.

Table 2. Prism bending test standards adopted for the evaluation of the residual flexural strength, f_{Ri}, considering the different SFRC mixes present in the DB_s.

Number of samples	Prism bending standards				
	EN 14651	RILEM TC 16	ASTM C1609	UNI 11039	Total
f_{R1} and f_{R3}	9	9	25	1	44
f_{R2} and f_{R4}	9	0	25	1	35

4. Prediction model of residual flexural strength of SFRC

Based on the analysis of a database (DB_f) formed by 89 samples of hooked end steel fiber reinforced concrete (SFRC) notched beams submitted to 3-point bending according to EN 14651 [23], in [41] is described in detail a strategy for establishing a relationship between the residual flexural strength of SFRC, $f_{Ri}, (i = 1,2,3,4)$, and other material properties capable of being easily obtained in design context.

For the DB_f under consideration, a very good agreement was obtained for the relation between the f_{R1}, the reinforcement index, IR, and the concrete average tensile strength, f_{ctm}. The reinforcement index is defined as:

$$IR = V_f \cdot l_f / d_f \tag{16}$$

The derived relationship between f_{R1}, IR and f_{ctm} is:

$$f_{R1} = k_{1,R1} \cdot f_{ctm} \cdot (IR)^{k_{2,R1}} \tag{17}$$

where $k_{1,R1}$ and $k_{2,R1}$ are model coefficients.

Due to the inherit complexity to perform direct and indirect tensile tests, the tensile strength of concrete can be estimated from the concrete compressive strength by the expression presented in MC2010:

$$f_{ctm} = \begin{cases} 0.3 \cdot (f_{ck})^{2/3} & , f_{ck} \leq 50MPa \\ 2.12 \cdot \ln\left[1 + 0.1 \cdot (f_{ck} + 8)\right] & , f_{ck} > 50MPa \end{cases} \tag{18}$$

Strong correlation between $f_{Ri}\,(i=2,3,4)$ and f_{R1} was also verified in [41], namely:

$$f_{Ri}=k_{1,Ri}\cdot\left(f_{R1}\right)^{k_{2,Ri}} \quad i=2,3,4 \tag{19}$$

Based on the results of the DB$_f$, three intervals of average compressive strength were adopted for best correlate the coefficients $k_{1,Ri}$ and $k_{2,Ri},(i=1,2,3,4)$ with the $f_{cm}[MPa]$, having been obtained the following equations:

$$k_{1,R1}=\begin{cases}3.2 & ,\ f_{cm}<25\\ \dfrac{1.1}{1000}\cdot f_{cm}^{\ 2}-\dfrac{f_{cm}}{10}+5 & ,\ 25\le f_{cm}\le 65\\ 3.2 & ,\ f_{cm}>65\end{cases};\quad k_{2,R1}=\begin{cases}0.5 & ,\ f_{cm}<25\\ \dfrac{8}{1000}\cdot f_{cm}+0.32 & ,\ 25\le f_{cm}\le 65\\ 0.85 & ,\ f_{cm}>65\end{cases} \tag{20}$$

$$k_{1,R2}=\begin{cases}0.9 & ,\ f_{cm}<25\\ \dfrac{-3.8\cdot f_{cm}^{\ 2}}{10000}+\dfrac{3.2\cdot f_{cm}}{100}+0.35 & ,\ 25\le f_{cm}\le 65\\ 0.8 & ,\ f_{cm}>65\end{cases};\quad k_{2,R2}=\begin{cases}1.1 & ,\ f_{cm}<25\\ \dfrac{3\cdot f_{cm}^{\ 2}}{10000}-\dfrac{2.8}{100}\cdot f_{cm}+1.6 & ,\ 25\le f_{cm}\le 65\\ 1.1 & ,\ f_{cm}>65\end{cases} \tag{21}$$

$$k_{1,R3}=\begin{cases}0.88 & ,\ f_{cm}<25\\ \dfrac{-8\cdot f_{cm}^{\ 2}}{10000}+\dfrac{6.05\cdot f_{cm}}{100}-0.14 & ,\ 25\le f_{cm}\le 65\\ 0.4 & ,\ f_{cm}>65\end{cases};\quad k_{2,R3}=\begin{cases}1.12 & ,\ f_{cm}<25\\ \dfrac{8\cdot f_{cm}^{\ 2}}{10000}-\dfrac{6.78\cdot f_{cm}}{100}+2.31 & ,\ 25\le f_{cm}\le 65\\ 1.3 & ,\ f_{cm}>65\end{cases} \tag{22}$$

$$k_{1,R4}=\begin{cases}0.8 & ,\ f_{cm}<25\\ \dfrac{-1.06\cdot f_{cm}^{\ 2}}{1000}+\dfrac{8.4\cdot f_{cm}}{100}-0.66 & ,\ 25\le f_{cm}\le 65\\ 0.3 & ,\ f_{cm}>65\end{cases};\quad k_{2,R4}=\begin{cases}1.12 & ,\ f_{cm}<25\\ \dfrac{1.1\cdot f_{cm}^{\ 2}}{1000}-\dfrac{9.26\cdot f_{cm}}{100}+2.75 & ,\ 25\le f_{cm}\le 65\\ 1.38 & ,\ f_{cm}>65\end{cases} \tag{23}$$

The use of equations (17), (19)-(23) is restricted for $f_{cm}\le 70MPa$, as the formulation tends to overestimate the residual flexural strength of SFRC for $f_{cm}>70MPa$ (few data exist in this interval).

The prediction model of the residual flexural strength presented a very good agreement with the DB$_f$ studied in [41], as the average ratio between the experimental, $f_{Ri,\exp}$, and predicted, $f_{Ri,\mathrm{mod}}$, values of the residual flexural strength ranged between 1.02 to 1.09, while the coefficient of variation varied between 17.8% to 28.94% (Table 3). It is verified that the coefficient of variation has increased with the crack opening at which the f_{Ri} is evaluated, which is a consequence of the increase of the dispersion of the results from f_{R1} to f_{R4}.

The prediction model was also applied to the set of SFRC mixes used in the beams of the DB$_s$ where the residual flexural strength was experimentally characterized (Table 2). In Table 4 is presented the average and coefficient of variation (COV) of the ratio between experimental, $f_{Ri,\exp}$, and estimated values, $f_{Ri,est}$, of the residual flexural strength, considering the prism bending test standard adopted in the evaluation of the experimental values. It is possible to verify that when comparing the estimated and experimental f_{Ri} values determined from the test configuration that adopts notched beams (EN 14651/RILEM TC 162-TDF/UNI 11039), the model exhibits an higher dispersion and average values of the ratio $f_{Ri,\exp}/f_{Ri,est}$ than the observed in [41]. For the samples tested according to the ASTM C1609, the model presents in average a slight underestimation of the SFRC residual flexural strength, which can be related with lower values of $f_{Ri,\exp}$ due to cracking on the weakest section between the loading points of the un-notched beams, and to the simplification adopted in the adopted correlation between the mid-span deflection and CMOD.

Table 3. Statistical results of the residual flexural strength prediction model [41].

Residual flexural strength	$f_{Ri,\exp}/f_{Ri,mod}$		
	Average	SD[a]	COV[b] (%)
f_{R1}	1.02	0.18	17.78
f_{R2}	1.08	0.22	20.61
f_{R3}	1.09	0.26	23.68
f_{R4}	1.05	0.30	28.94

[a] Standard deviation; [b] Coefficient of variation.

Table 4. Statistical analysis of the ratio between the experimental and estimated values of the residual flexural strength, f_{Ri}.

Residual flexural strength	$f_{Ri,\exp}/f_{Ri,est}$					
	Global		EN 14651/RILEM TC 162-TDF/UNI 11039		ASTM C1609	
	Average	COV[a] (%)	Average	COV[a] (%)	Average	COV[a] (%)
f_{R1}	1.06	26.9	1.23	24.4	0.92	19.4
f_{R2}	1.14	28.9	1.49	21.0	0.99	20.2
f_{R3}	1.14	31.9	1.35	27.6	0.99	27.8
f_{R4}	1.09	37.9	1.49	28.3	0.93	30.4

[a] Coefficient of variation.

5. Appraisal of MC2010 shear resistance and residual flexural strength prediction models

The predictive performance of shear models MC2010_EEN and MC2010_MCFT is appraised by using the information collected in the R/SFRC beams shear tests database (DB_s). The shear resistance obtained by both models, $V_{u,\text{model}}$, is compared with the experimentally obtained shear resistance of each R/SFRC beam, $V_{u,\exp}$, and the prediction ratio $\lambda = V_{u,\exp}/V_{u,\text{model}}$ is obtained. For the evaluation of the strain at mid-depth of the effective shear area, ε_x, in the MC2010_MCFT model (equation (8)), the internal forces in the member are determined for a control section at the distance $a/2$ from the beam's support, being a the shear span, where shear failure has occurred, considering that the reaction at the support is equal to the beam's shear resistance. During the analysis, the partial safety factors adopted in the theoretical models are set to 1.0, and average values for the SFRC material properties are adopted, in order to properly compare the experimental and theoretical results, being the predictive performance of the model considered as better as closer to 1.0 is $\lambda = V_{u,\exp}/V_{u,\text{model}}$. In the case of the samples were the residual flexural strength data was not experimentally assessed, it was adopted the prediction model presented in the previous section to estimate the values of this material properties. The characteristic value of the SFRC compressive strength was evaluated from average value present in the DB_s, considering the expression $f_{ck} = f_{cm} - 8MPa$ proposed by MC2010.

In Figure 1 is presented a statistical analysis of the prediction ratio, λ, for each shear resistance model considering the source of the residual flexural strength values (experimental/estimated). When assessing the influence of the use of experimental or estimated values of the residual flexural strength of SFRC for the MC2010_EEN and MC2010_MCFT, it is possible to denote that the average values of the prediction ratio, λ, is slightly higher when using estimated values of f_{Ri}, which is related to the slight underestimation of the estimated values of f_{Ri}, as shown in the column "Global" of Table 4. However, when considering the source of the residual flexural strength values the average values of the prediction ratio λ for the analyzed scenarios have very small differences, including the coefficients of variation. Due to this, it is possible to conclude that the residual flexural strength prediction model presented in the previous section is suitable to be used in the evaluation of the shear resistance of R/SFRC beams, when the values of f_{Ri} are absent in the database. When analysing the global performance of the shear prediction models, both models present a satisfactory approximation to the experimental results (average value of λ equal to 1.26), with acceptable values of dispersion (coefficient of variation lower than 20%). It should be noticed the existence of only one outlier in 80 samples ($\approx 1\%$), which corresponds to a sample of the MC2010_EEN model.

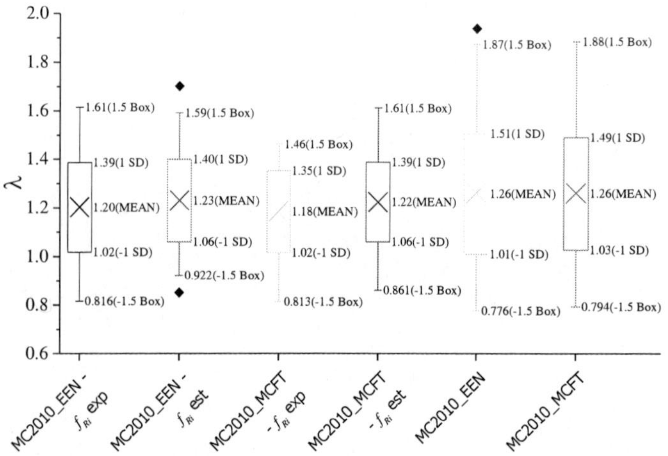

Prediction ratio λ	No. Samples	Average	COV[a] (%)
MC2010_EEN f_{Ri} experimental	59	1.202	15.2
MC2010_EEN f_{Ri} estimated	59	1.230	13.7
MC2010_MCFT f_{Ri} experimental	42	1.184	14.1
MC2010_MCFT f_{Ri} estimated	42	1.224	13.3
MC2010_EEN	80	1.258	19.7
MC2010_MCFT	80	1.259	18.4

[a] Coefficient of variation

Figure 1. Statistical analysis of λ for both MC2010 shear resistance models considering the use of experimental and estimated values of f_{Ri} .

In Figure 2a is presented the comparison between the experimental and prediction values of the shear resistance for each beam. The results are also divided in categories considering the source of the residual flexural strength values considered in the shear resistance model. As can be seen, the predictive performance of the shear models is very similar and a satisfactory agreement with the experimental values of the DB_s is attained. Considering the influence of the source of the residual flexural strength values, it is possible to observe that the use of the f_{Ri} prediction model appears to be related with a more pronounced underestimation of the shear resistance (a relatively high percentage of results is in the interval of $V_{u,\exp} = 1.5 \cdot V_{u,\mathrm{model}}$ to $V_{u,\exp} = 2.0 \cdot V_{u,\mathrm{model}}$) for the R/SFRC beams under analysis. In Figure 2b is presented the overall analysis of both models, regarding that the prediction values are safe for $\lambda \geq 1$ and unsafe for $\lambda < 1$. From these results is possible to verify that the MC2010_MCFT model presents a slightly higher percentage of safe predictions when compared to the MC2010_EEN model. In addition, the comparison between the performance of both MC2010 models was assessed according to an adapted version of the *Demerit Points Classification* proposed in [42]. This classification is based on the determination of the cumulative number of penalties for each value of λ. The penalty points scale is defined in Table 5, and the lower the number of penalties is, the safer is the performance of the model. For this case, both models present a very similar number of total penalties points, with a slight better performance of the MC2010_MCFT model presenting 62 penalty points while the MC2010_EEN model has 68 penalty points.

Table 5. Adapted version of the *Demerit Points Classification.*

$\lambda = V_{u,\exp} / V_{u,\mathrm{model}}$	Classification	Penalty
< 0.50	Extremely Dangerous	10
[0.50-0.85[	Dangerous	5
[0.85-1.15[	Appropriate Safety	0
[1.15-2.00[	Conservative	1
≥ 2.00	Extremely Conservative	2

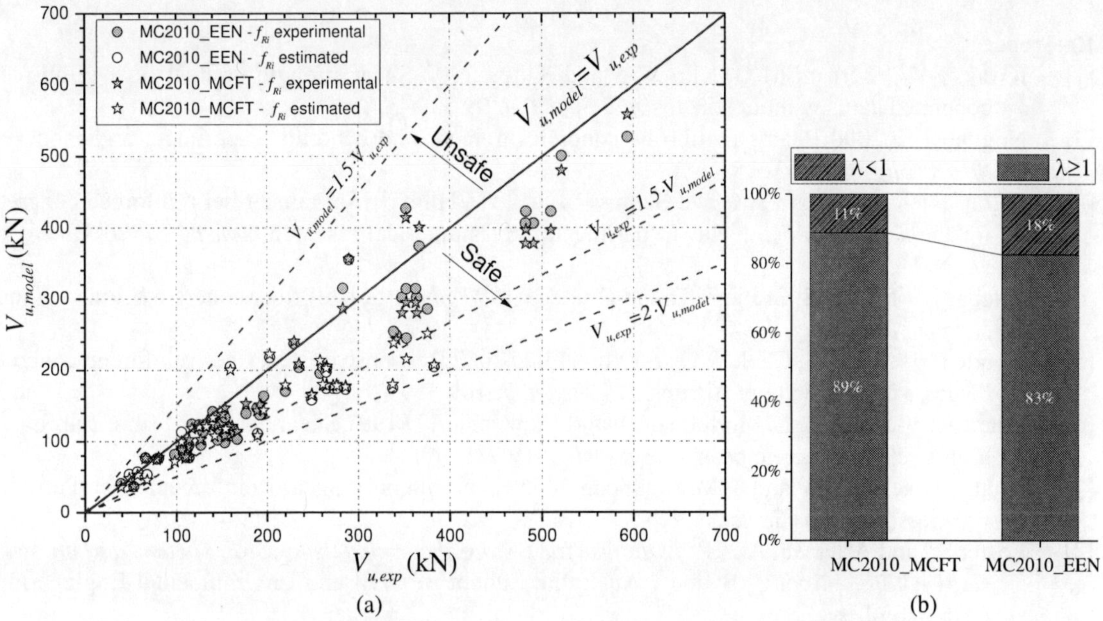

Figure 2. a) Comparison of experimental and shear strength determined from both MC2010 prediction models. b) Predictive performance of theoretical model, safe ($\lambda \geq 1$) and unsafe ($\lambda < 1$).

6. Conclusions

In the present work was assessed the performance of the two shear resistance models available in CEB *fib* Model Code 2010 (MC2010) by comparison of its predictive capability in a database that includes the results of 80 samples of steel fiber reinforced concrete (R/SFRC) beams submitted to shear.

Due to the absence of the characterization of the residual flexural strength of the SFRC used in several tests, a model that estimates these mechanical parameters based on the reinforcement index and concrete strength class was proposed and coupled with the shear resistance models.

For the assessment of the suitability of the residual flexural strength prediction model a comparison of the prediction ratio, λ, was performed, which was computed by considering estimated and experimentally obtained values of the residual flexural strength of the SFRC. The obtained results demonstrated a very small difference between the average and dispersion of the prediction ratio, λ, which validates the use of the residual flexural strength prediction model in the assessment of the shear resistance of FRC elements.

For evaluating the overall performance of the shear models a statistical analysis of the prediction ratio, λ, was conducted, which revealed that the performance of the MC2010_MCFT and MC2010_EEN models is very similar, with the MC2010_MCFT presenting a very slightly minor coefficient of variation for the λ than the MC2010_EEN model (18.4%-19.7%). For the majority of the cases, both MC2010 shear models predict safe values for the shear resistance of the R/SFRC beams, and when the partial safety factor is introduced, the models always return safe predictions.

An additional comparison of the MC2010 shear models was done considering the Demerit Points Classification, having both approaches presented similar performance in this regard (the difference was only 6 penalty points between both models).

Acknowledgements

The first Author would like to acknowledge the grant SFRH/BDE/96381/2013 co-funded by CiviTest - Pesquisa de Novos Materiais para a Engenharia Civil, Lda. and by FCT - Portuguese Foundation for Science and Technology. The authors also acknowledge the support provided by the FCT project

IOP Conf. Series: Materials Science and Engineering **596** (2019) 012004 doi:10.1088/1757-899X/596/1/012004

PTDC/ECM-EST/2635/2014.

References

[1] Kwak Y-K, Eberhard M O, Kim W-S and Kim J 2002 Shear strength of steel fiber-reinforced concrete beams without stirrups *ACI Struct. J.* **99** 530–538

[2] Noghabai K 2000 Beams of fibrous concrete in shear and bending: experiment and model *J. Struct. Eng.* **126** 243–251

[3] Soltanzadeh F, Barros J A O and Santos R F C 2015 High performance fiber reinforced concrete for the shear reinforcement: Experimental and numerical research *Constr. Build. Mater.* **77** 94–109

[4] Adebar P, Mindess S, St Pierre D and Olund B 1997 Shear tests of fiber concrete beams without stirrups *ACI Struct. J.* **94**

[5] Aoude H, Belghiti M, Cook W D and Mitchell D 2012 Response of steel fiber-reinforced concrete beams with and without stirrups *ACI Struct. J.* **109** 359

[6] Imam M, Vandewalle L, Mortelmans F and Van Gemert D 1997 Shear domain of fibre-reinforced high-strength concrete beams *Eng. Struct.* **19** 738–47

[7] Kiang-Hwee Tan P P and K Murugappan 1995 Steel Fibers as Shear Reinforcement in Partially Prestressed Beams *Struct. J.* **92**

[8] Foster S J and Argarwal A 2017 *Database for testing shear models for SFRC beams and one-way slabs without stirrups* (Sydney, Australia: School of Civil and Environmental Engineering, UNSW Sydney)

[9] Foster S J, Agarwal A and Amin A 2017 Design of steel fiber reinforced concrete beams for shear using inverse analysis for determination of residual tensile strength *Struct. Concr.*

[10] Cuenca E, Conforti A, Minelli F, Plizzari G A, Navarro Gregori J and Serna P 2018 A material-performance-based database for FRC and RC elements under shear loading *Mater. Struct.* **51**

[11] Soetens T 2015 *Design models for the shear strength of prestressed precast steel fibre reinforced concrete girders* PhD Thesis (Ghent University)

[12] fib-federation internationale du beton 2013 *fib Model Code for Concrete Structures 2010* (John Wiley & Sons)

[13] Moraes Neto B N, Barros J A O and Melo G S S A 2013 A model for the prediction of the punching resistance of steel fibre reinforced concrete slabs centrically loaded *Constr. Build. Mater.* **46** 211–23

[14] RILEM TC 162-TDF 2002 Recommendations of RILEM TC 162-TDF: Test and design methods for steel fibre reinforced concrete: bending test *Mater. Struct.* **35** 579–82

[15] Bentz E C, Vecchio F J and Collins M P 2006 Simplified modified compression field theory for calculating shear strength of reinforced concrete elements *ACI Struct. J.* **103** 614

[16] Ng T S, Foster S J, Htet M L and Htut T N S 2014 Mixed mode fracture behaviour of steel fibre reinforced concrete *Mater. Struct.* **47** 67–76

[17] European Committee for Standardization 2010 *Eurocode 2: Design of concrete structures Part 1-1: General rules and rules for buildings* vol NP EN 1992-1-1

[18] Marco di Prisco, Giovanni Plizzari and Lucie Vandewalle 2010 MC2010: Overview on the shear provisions for FRC *Shear and punching shear in RC and FRC elements (workshop proceedings)* fib Bulletin (Lausanne: fédération internationale du béton) pp 61–76

[19] Stephen Foster 2010 Design of FRC beams for shear using the VEM and the draft Model Code approach *Shear and punching shear in RC and FRC elements (workshop proceedings)* fib Bulletin (Lausanne: fédération internationale du béton) pp 195–210

[20] Sigrist V, Bentz E, Ruiz M F, Foster S and Muttoni A 2013 Background to the *fib* Model Code 2010 shear provisions - part I: beams and slabs *Struct. Concr.* **14** 195–203

[21] Amin A, Foster S J and Muttoni A 2015 Derivation of the σ- *w* relationship for SFRC from prism bending tests *Struct. Concr.* **16** 93–105

[22] ASTM International 2012 *Standard Test Method for Flexural Performance of Fiber-Reinforced Concrete (Using Beam With Third-Point Loading)* vol ASTM C1609

[23] European Committee for Standardization 2005 *Test method for metallic fibered concrete - Measuring the flexural tensile strength (limit of proportionality (LOP), residual)* vol EN 14651

[24] Italian Board for Standardization 2003 *Steel Fiber Reinforced Concrete—Part I: Definitions, Classification Specification and Conformity—Part II: Test Method for Measuring First Crack Strength and Ductility Indexes* vol UNI 11039

[25] Dinh H H, Parra-Montesinos G J and Wight J K 2011 Shear Strength Model for Steel Fiber Reinforced Concrete Beams without Stirrup Reinforcement *J. Struct. Eng.* **137** 1039–51

[26] Cuenca E 2015 *On Shear Behavior of Structural Elements Made of Steel Fiber Reinforced Concrete* PhD Thesis (Valencia, Spain: Universitat Politècnica de València)

[27] Rosenbusch J and Teutsch M 2003 Shear design with- method *Proc. RILEM TC* 105–117

[28] Minelli F 2005 *Plain and fiber reinforced concrete beams under shear loading: Structural behavior and design aspects* PhD Thesis (Brescia, Italy: Department of Civil Engineering, University of Brescia)

[29] Singh B and Jain K 2014 An Appraisal of Steel Fibers as Minimum Shear Reinforcement in Concrete Beams *ACI Struct. J.* **111**

[30] Conforti A, Minelli F and Plizzari G A 2013 Wide-shallow beams with and without steel fibres: A peculiar behaviour in shear and flexure *Compos. Part B Eng.* **51** 282–90

[31] Jain K and Singh B 2014 Investigation of steel fibres as minimum shear reinforcement *Proc. Inst. Civ. Eng. - Struct. Build.* **167** 285–99

[32] Al-lami K 2000 *Experimental Investigation of Fiber Reinforced Concrete Beams* MSc Thesis (Portland State University)

[33] Bryan Erick Barragán 2002 *Failure and toughness of Steel Fiber Reinforced Concrete under tension and shear* PhD Thesis (Barcelona, Spain: Universitat Politecnica de Catalunya)

[34] Greenough T and Nehdi M 2008 Shear behavior of fiber-reinforced self-consolidating concrete slender beams *Mater. J.* **105** 468–477

[35] Bertozzi A and Reggia A 2006 *Steel fibers for the minimum shear reinforcement of the beams* MSc Thesis (Brescia, Italy: University of Brescia)

[36] Conforti A 2008 *Variable truss model for the shear design of fibre reinforced concrete beams: experimental and analytical study* MSc Thesis (Brescia, Italy: University of Brescia)

[37] Minelli F, Plizzari G A and Vecchio F J 2007 Influence of steel fibers on full-scale RC beams under shear loading *Fracture Mechanics of Concrete Structures – High Performance Concrete, Brick-Masonry and Environmental Aspects* (London: Taylor and Francis) pp 1523–31

[38] Conforti A, Minelli F and Plizzari G A 2017 Influence of Width-to-Effective Depth Ratio on Shear Strength of Reinforced Concrete Elements without Web Reinforcement *ACI Struct. J.* **114**

[39] Stephen J. Foster, Ankit Agarwal and Ali Amin 2017 Design of SFRC Beams for Shear using Inverse Analysis for determination of Residual Tensile Strength *Struct. Concr.* **19** 129–40

[40] Vandewalle L and Dupont D 2003 BENDING TEST AND INTERPRETATION *International RILEM Workshop on Test and Design Methods for Steel Fibre Reinforced Concrete - Background and Experiences* International RILEM Workshop on Test and Design Methods for Steel Fibre Reinforced Concrete - Background and Experiences (RILEM Publications SARL) pp 1–13

[41] Moussa Y H 2017 *Steel fiber efficiency for flexural reinforcement of reinforced concrete beams with minimum reinforcement ratio* MSc thesis (Belém, Brasil: Universidade Federal do Pará)

[42] Collins M 2001 *Evaluation of shear design procedures for concrete structures* (Canada: A Report prepared for the CSA technical committee on reinforced concrete design)

IOP Conf. Series: Materials Science and Engineering **596** (2019) 012005 doi:10.1088/1757-899X/596/1/012005

Nonlinear modeling of SFRC beam using multilinear softening function obtained by inverse analysis

C A Benedetty[1], L H Oliveira[1,2], L C Almeida[1] and L M Trautwein[1]

[1]Departamento de Estruturas, Faculdade de Engenharia Civil, Arquitetura e Urbanismo, Universidade Estadual de Campinas, R Saturnino de Brito, 224, Campinas, SP, Brazil.
[2]Departamento de Engenharia e Tecnologia, Centro Multidisciplinar de Pau dos Ferros, Universidade Federal Rural do Semi-Árido, Pau dos Ferros, RN, Brazil.

E-mail: bendedettiV8@gmail.com

Abstract. The modeling of steel fiber reinforced concrete (SFRC) is a challenging task in comparison with the conventional reinforced concrete structures. Softening functions used to numerically reproduce SFRC fracture need to describe the effects associated with the post-cracking residual strength induced by the fibers in the concrete matrix. In order to do this, multilinear softening functions can be used to consider these effects. The work presents the results of a study in which the behavior of a SFRC beam tested in four-point bending test is compared with the responses obtained in nonlinear simulations using the finite element method. Multilinear softening functions are obtained through an inverse analysis technique, aiming to reproduce the phenomena of appearance and propagation of cracks. The simulations were performed using ATENA/GiD software. The technique adopted to find the softening function of SFRC allowed to reproduce, with a good agreement, the behavior reported experimentally.

1. Introduction

The Steel Fiber Reinforced Concrete (SFRC) is considered a composite material that has an excellent performance in terms of ductility and toughness. These characteristics make the SFRC an interesting material for its application in structural engineering. Structural elements that reach rupture in a fragile way, lead to the collapse of the structure without any indicator (such as opening cracks or high displacements) and, in some cases, put at risk human lives.

Some studies indicate that SFRC can provide improvements in the mechanical behavior of structures subjected to high shear forces, cyclic loads and impact loads [1-3]. The addition of steel fibers in the concrete mixture at specific levels produces changes in mechanical properties demanding a change to describe the concrete cracking mechanism. Properties such as tensile strength (f_t), fracture energy (G_F) and post-cracking capacity are modified. Consequently, the change of these properties, along with the random and discontinuous nature of the fibers present in the concrete matrix, causes the crack propagation mechanism to be a complex fracture phenomenon to reproduce numerically.

The cracking mechanism in quasi-brittle materials such as concrete can be represented through the crack band model [4] approach. In this approach, conventional tensile softening functions (linear, bilinear and exponential) are usually used to numerically reproduce the fracture response of concrete.

However, the use of multilinear softening functions shows a successful alternative to predict the response of the material throughout the fracturing process.

Adoption of a multilinear softening function for the nonlinear analysis of SFRC structures can be done by the inverse analysis technique. Basically, the softening function is found, using the finite element method and an iterative adjustment process of the load - displacement diagrams, obtained in the tests of specimens submitted to three-point bending test. Some studies with nonlinear analysis of SFRC structures shows that the technique has wide applicability and good agreement with experimental tests [5].

This paper presents the results obtained from the implementation of a multilinear softening function in a nonlinear simulation of a four-point bending test of a SFRC beam without stirrups. The finite element analysis was performed using the ATENA v5.4.1/GiD10.0.9 and the inverse analysis technique described by Sajdlová [6].

2. Numerical SFRC simulation using ATENA/GiD

The nonlinear simulation was performed using ATENA/GiD by using the finite element method. The software uses the fracture-plastic constitutive model [7] in order to reproduce numerically the cracking of the concrete. The cracks are represented numerically in the finite element mesh, through the smeared crack approach [8], with the fracture phenomenon of the concrete governed by the crack band model [4].

The simulation of specific phenomena of concrete, due to the characteristics of the material or loading type, can be done through material models in the software [9]. Several material models based on the constitutive model discussed above are available. The CC3DNonLinCementitiousUser material model can be used for the SFRC modeling, basically this allows the user to enter functions that describe numerically the behavior of the material in compression, tension or shear.

The mechanical response of the SFRC is significantly influenced by the properties of the material in tension. Therefore, definition of the softening function is essential for simulating the fracture phenomenon of this composite materials. Several methodologies to establish the softening function can be used. The present work applies the inverse analysis technique using ATENA/GiD.

The purpose of applying the inverse analysis technique in the SFRC simulation is to define a softening multilinear function, reproducing the post-cracking response of the load-displacement diagram obtained from a notched specimen in a three-point bending test (fracture in mode I). The procedures for its execution are purely iterative and the precision of the results can be influenced by parameters such as: the number of SFRC samples tested, the number of the inverse analysis interpolation points and the number of iterations of the analysis.

Figure 1 illustrates the steps for developing such technique. The first step is to simulate a notched specimen subjected to a three-point bending test and define the mechanical properties of the SFRC (e.g. tensile strength, compressive strength and Young's modulus) and a bilinear softening function. The second step consists of comparing the percentage of error (R_i) between the loads of the numerical and experimental response, by setting displacements (δ_i) along the load - displacement relationship. The third step considers the calculation of the fracture strains (ε_f), associated to each displacement δ_i.

Once the strains have been calculated, new points are interpolated in the bilinear softening function, where, consequently, the ordinate value of these points (dimensionless stress) is changed by the R_i value, thus generating the points of the new multilinear softening function. The last step of the analysis is to implement the calibrated softening function to the model of the structure to be simulated.

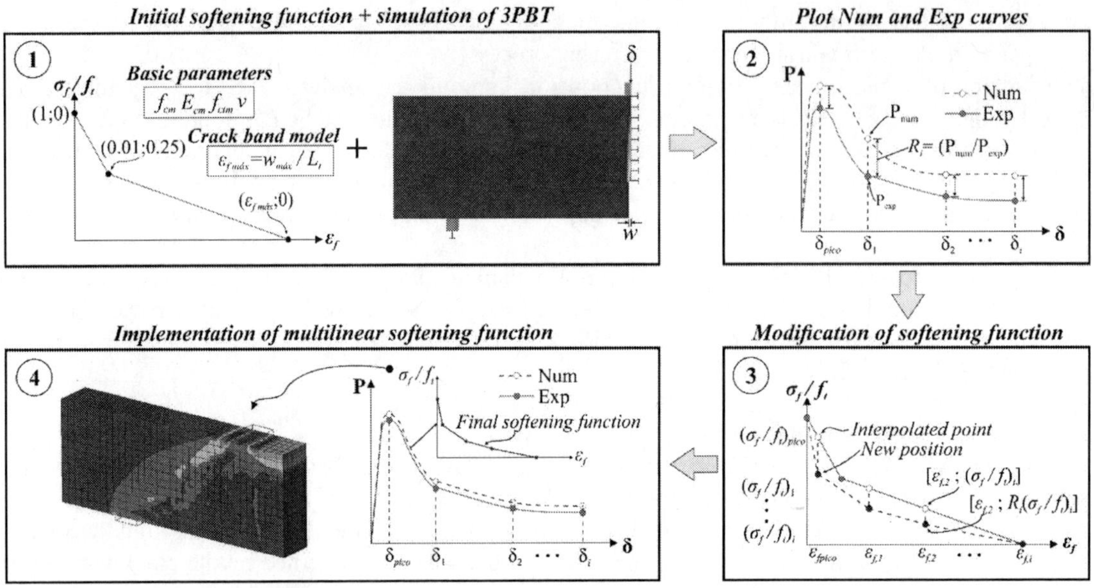

Figure 1. Procedure of the inverse analysis technique.

The inverse analysis technique is an iterative process. The softening function obtained in the third step can be used to simulate the notched specimen again, generating a new load-displacement relationship. If the percentage of error R_i of the diagrams does not reach a maximum permissible by an error criterion, the procedure for obtaining a new function is done again iteratively until an acceptable adjustment of the load-displacement diagram is achieved.

3. Test setup and procedure

The simulated beam is part of an experimental program of a recent research [10]. The test setup submitted the beam to a four-point bending test by load control. Figure 2 shows the test setup, dimensions and the reinforcement used. The load was distributed to the beam by steel plates and rollers positioned at the bottom and at the top of the beam. The displacements in the mid-span were measured by an LVDT. The reinforcements bars strains and concrete strains were measured by electrical strain gauges. Regarding the constrains, both supports restricted the displacement in y-direction, in addition, the right support restricted the displacement in the x-direction.

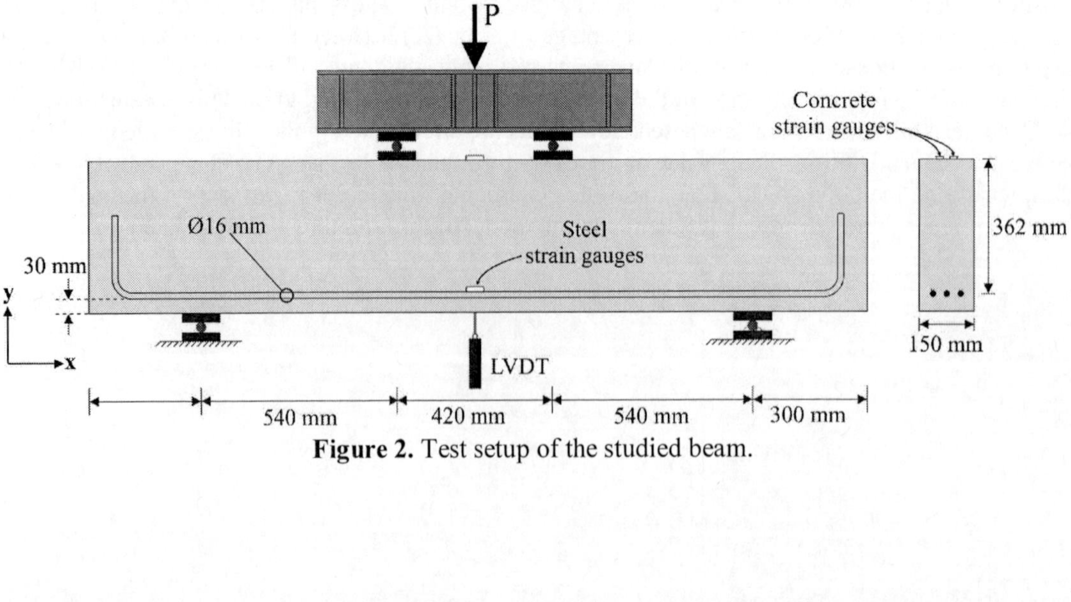

Figure 2. Test setup of the studied beam.

The beam was made with an addition of 60 kg/m^3 (V_f= 0.76%) of steel fibers (DRAMIX® 3D 45/30BL hooked-end). These steel fibers had length L_f= 30 mm and diameter D_f = 0,62 mm. The mechanical properties of the SFRC concrete reported in this study are shown in table 1, values of compressive strength (f_{cm}), split tensile strength ($f_{ct,spm}$) and concrete Young's modulus (E_{cm}) were estimated following the Brazilian standard code. In addition, the mechanical properties of the reinforcement are also presented, the tests performed for the determination of these parameters followed the recommendations of ASTM E8/E8M-16a [11] and ASTM A370-17 [12].

Table 1. Materials Properties

Concrete		**Reinforcement bars**	
Property	Value	Property	Value
Compressive strength, f_{cm} (MPa)	39.64	Yield strength, f_y (MPa)	557.27
Split tensile strength, $f_{ct,spm}$ (MPa)	4.86	Yield strain, ε_y (mm/m)	3.15
Young's modulus, E_{cm} (GPa)	30.19	Young's modulus, E_s (MPa)	190450

Three notched specimens were fabricated from SFRC samples of the beam. These were made with a 100 mm square cross section, a length of 400 mm and a notch depth of 30 mm. Load - displacement relationship were obtained through the three-point bending test performed according to recommendations of the JCI-S-002-2003 standard [13].

4. Numerical modeling description

The generation of the finite element mesh, as well as the definition of material properties, boundary conditions and convergence parameters were done using the GiD preprocessor [14]. The notched specimen model was created using 4-node quadrilateral plane finite elements (CCIsoQuad4_2D). The hypothesis of plane stress was adopted for this model. In the numerical analysis, the symmetric geometrical property was considered in order to optimize the processing time, thus the specimen was modeled as a half specimen. The mesh refinement was increased in the region near the notch, adopting square elements with a minimum length of 2 mm. The numerical method adopted to find the solution of the system of nonlinear equations was Newton-Raphson method. Because it was necessary to know the crack opening (w) for determining the strains of the modified softening function, a crack opening monitoring point was positioned in the notch.

A tridimensional numerical model of the beam was created using 20-node hexahedral solid finite elements (CCIsoBrick20_3D). Regarding the reinforcement bars, these were simulated using discrete bar finite elements embedded in the elements of the concrete.

5. Discussion

The final softening function obtained after the application of the inverse analysis technique is presented in figure 3a and the load-displacement experimental diagrams of the tested specimens are shown in figure 3b. In addition, the load-displacement diagram obtained numerically by ATENA is shown as a result of the implementation of the final softening function in the specimen modeling. It was observed that the multilinear softening function simulated the post-cracking response of the specimens with a good agreement.

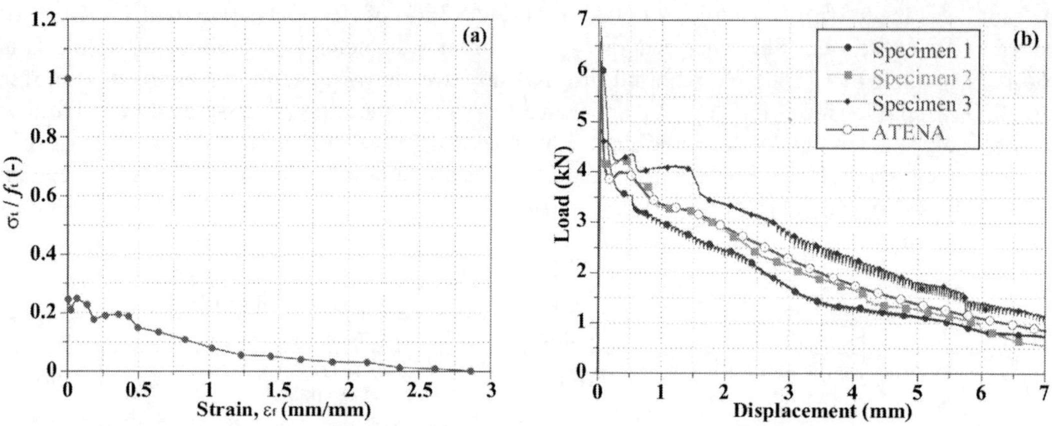

Figure 3. (a) Multilinear softening function and (b) comparison of numerical and experimental response.

Then, in order to evaluate the applicability of the multilinear softening function, the SFRC parameters obtained in the inverse analysis were used as an input for the beam model. The results obtained from modeling were analyzed in terms of the load-displacement response and the crack pattern.

Figure 4 presents the load-displacement experimental and numerical curves. The diagrams presents two representative branches. There is a first branch, where there is a loss of progressive stiffness during the process of loading. After, a peak load is reached, this can be observed in the response of the experimental curve. Once the peak is reached the load-carrying capacity decreases slightly and a second branch begins which describes an apparent ductile plateau.

In general, it was observed that the numerical model was able to reproduce in most of the test the ductile behavior of the SFRC beam. Note that in terms of stiffness, the numerical response of the first branch presents a behavior similar to that recorded experimentally. On the other hand, the model was able to estimate the value of the maximum load with a percentage error of 3.74%.

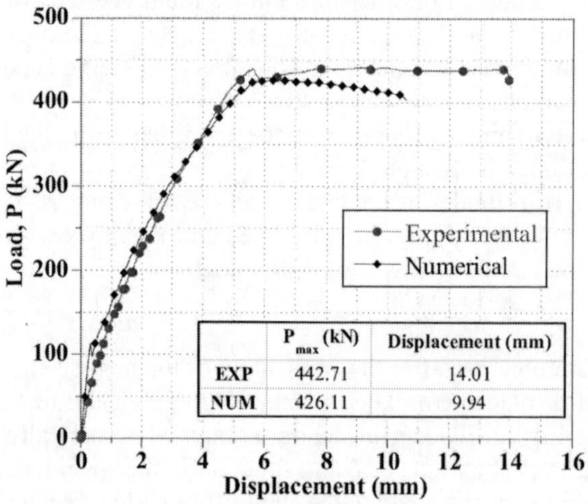

Figure 4. Experimental and numerical load-displacement relationship.

The experimental and numerical cracking pattern of the beam are shown in figure 5. In figure 5a, it can be observed that this beam has a vertical crack in the middle of the span with significant crack width. In addition, there is another diagonal crack with smaller width positioned in the shear span can be visualized. Although the beam does not have stirrups, the beam collapsed due to flexural failure. This is directly related to the effects and alterations produced by the steel fibers in the concrete cracking mechanisms.

By analyzing the numerical cracks pattern shown in the figure 5b, in which the cracks are represented together with the principal tensile strain, it was verified that the numerical model was able to reproduce the propagation of the vertical crack of greater opening.

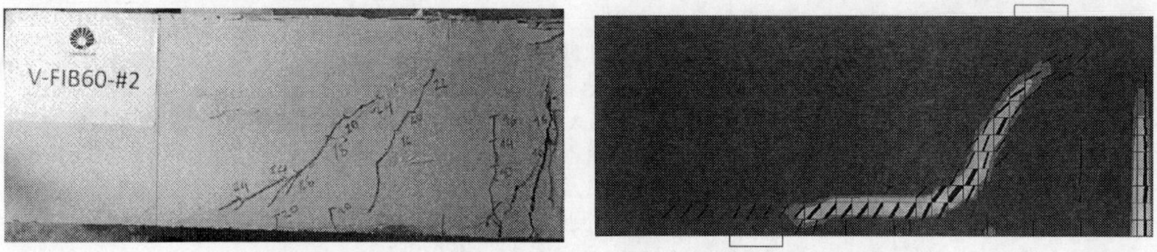

(a) (b)

Figure 5. (a) Experimental and (b) numerical crack pattern.

6. Conclusions

The simulation of a SFRC beam without stirrups was performed using a multilinear softening function in order to describe the fracture phenomena. The inverse analysis technique was used to define the calibrated softening function to simulate the post-cracking behavior of the material.

The results obtained in the notched specimen using the finite element method and the inverse analysis allowed the methodology as a successful alternative to reproduce the post-crack residual strength of the SFRC.

The implementation of the softening function obtained, in the modeling of the beam without stirrups, was able to predict with a good agreement the opening and propagation of the cracks. In addition, other parameters with the prediction of the peak load and the stiffness of the beam along the pre-peak branch of the load-displacement curve could be estimated with excellent accuracy.

Acknowledgments

The authors would like to express their gratitude for the Coordenação de Aperfeiçoamento de Pessoal de Nível Superior (CAPES) (Process 01-P-01879-2016) and the Conselho Nacional de Desenvolvimento Ciêntífico e Tecnologico (CNPq) (process 140135 / 2019-7) for the financial support for this research.

References

[1] Germano F, Tiberti G and Plizzari G 2016 Post-peak fatigue performance of steel fiber reinforced concrete under flexure *Materials and Structures.* **49** (10) 4229-4245.

[2] Ruiz G, de la Rosa A, Almeida L C, Poveda E, Zhang X X, Tarifa M, Wu Z M and Yu R C 2019 Dynamic mixed-mode fracture in SCC reinforced with steel fibers: an experimental study. *International Journal of Impact Engineering.* **129** 101-111.

[3] Ulzurrun G and Zanuy C 2017 Flexural response of SFRC under impact loading *Construction and Building Materials* **134** 397-411.

[4] Bažant Z P and Oh B H 1983 Crack band theory for fracture of concrete *Matériaux et construction* **16** (3) 155-177.

[5] Sajdlová T and Pukl R 2014 Optimization of input parameters for material model of fibre reinforced concrete and application on the numerical simulation of tunnel lining. In ACI-fib International Workshop FRC Montreal Canada.

[6] Sajdlová T 2015 ATENA Program Documentation Part 4-7.

[7] Červenka J and Papanikolaou V K 2008 Three dimensional combined fracture–plastic material model for concrete *International journal of plasticity.* **24** (12) 2192-2220.

[8] Rashid Y R 1968 Ultimate strength analysis of prestressed concrete pressure vessels *Nuclear engineering and design.* **7** (4) 334-344.

[9] Červenka V, Jendele L and Červenka J 2007 ATENA program documentation, Part 1: Theory. Cervenka Consulting Prague 231.

[10] Benedetty C A 2018 Experimental and numerical analysis of the behavior of reinforced concrete beams with steel fibers Master's thesis UNICAMP Campinas Brazil.

[11] ASTM I. ASTM E8/E8M-16a: Standard Test Methods for Tension Testing of Metallic Materials. West Conshohocken, PA, USA: ASTM International, 2016.

[12] ASTM I. ASTM A370 – 17: Standard Test Methods and Definitions for Mechanical Testing of Steel Products. West Conshohocken, PA, USA: ASTM International, 2017.

[13] JCI-S-002-2003. Method of test for load-displacement curve of fiber reinforced concrete by use of notched beam. Japan Concrete Institute Standard; 2003.

[14] GID. The Personal Pre and Post Processor. International Center for Numerical Methods in Engineering Barcelona 2009.

IOP Conf. Series: Materials Science and Engineering **596** (2019) 012006 doi:10.1088/1757-899X/596/1/012006

Numerical validation of a simplified inverse analysis method to characterize the tensile properties in strain-softening UHPFRC

E J Mezquida-Alcaraz*, J Navarro-Gregori and P Serna-Ros

Institute of Science and Concrete Technology, ICITECH, Universitat Politècnica de València, València, 46022, Spain.

E-mail: edmezal@alumni.upv.es

Abstract. In previous research a simplified closed-form non-linear hinge model based on the Third Point Bending Test (TPBT) developed by the authors to derive the tensile material properties of Ultra-High-Performance Fiber-Reinforced Concrete (UHPFRC) was numerically validated. This simplified non-linear hinge model was calibrated for UHPFRC that exhibits strain-hardening constitutive behavior. The aim of this work is the numerical validation of the simplified inverse analysis method to characterize the tensile properties of UHPFRC even with those UHPFRCs which exhibit strain-softening behavior. To get this objective a Finite Element Model (FEM) is carried out. The parameters to characterize the concrete properties from the simplified inverse analysis method by means of TPBT are used in the numerical modelling. The constitutive model for UHPFRC is modelled using a discrete crack approach for the macrocrack position. Different parameters are analysed to quantify the accuracy of the FEM. As a result, the model shows conservative load-deflection response when it is compared to the experimental curves.

1. Introduction

UHPFRC can be considered a special type of high-performance fiber reinforced cement composite (HPFRCC) if the definition in the Preface of the 6th and 7th Symposium on HPFRCC [1,2] and JSCE recommendations [3] is used. It states that HPFRCC can be considered all those concretes that exhibit a strain-hardening tensile stress-strain response accompanied by multiple cracking and a relatively large energy absorption capacity. Reaching strain-hardening behavior in a UHPC matrix depends on fiber type and amount, the matrix strength, bond between matrix and fibers, and also on specimen size and geometry, the pouring system, support conditions, structural redundancy, etc.

The characterization of UHPFRC tensile behavior is still a challenge. Four-point bending test (4PBT) arises as one of the best tests to get this result because of its simplicity. Nevertheless, it requires the use of an inverse analysis methodology to derive the tensile properties from the results obtained from it. Different inverse analysis methods have already been developed to obtain the parameters that define the UHPFRC behavior in tension from 4PBT [4–12].

To determine the validity and accuracy of inverse analysis methods, it is important to develop numerical models. The numerical model presented in [13,14] with the objective of validating the non-linear hinge model and its derived simplified five-point inverse analysis method presented in [15] to derive the tensile UHPFRC's properties from load-deflection response obtained from Third Point

Bending experimental tests (TPBT) was developed for UHPFRCs which exhibits strain-hardening behavior. The aim of this work is the numerical validation of the simplified inverse analysis method to characterize the tensile properties of UHPFRC even with UHPFRCs exhibiting strain-softening behavior.

2. Simplified Four-Point Inverse Analysis Method (4P-IA)

The Simplified Four-Point Inverse Analysis Method (4P-IA) developed and explained in [16,17] is a simplification of the Simplified Five-Point Inverse Analysis Method (5P-IA) fully described in [15,17]. This new simplified methodology is based on the closed-form non-linear hinge model developed in [15,17]. It entails having to select four specific key points extracted from the experimental equivalent bending strength-displacement at mid-span curve (Figure 1). Using these points, parameters that define the assumed quadrilinear stress-strain law inside the hinge can be determined by a back-of-the-envelope calculation. This law is used to determine the constitutive tensile behavior of UHPFRC.

Constitutive model for UHPFRC proposed is depicted in Figure 1 which is defined as a function of six parameters: elastic modulus (E) cracking strength (f_t); ultimate cracking strength (f_{tu}) and its associated strain (ε_{tu}); crack opening at the intersection of the line that defines the initial slope to the w axis (w_0); and the characteristic crack opening ($l_f/4$) defined as one fourth the fiber length. Only the first five are determined using this procedure.

The proposed method requires six input parameters, which derive from only four key points extracted from the experimental flexural strength versus displacement at mid-span curve obtained from a TPBT. The method's output generates the constitutive parameters that define the constitutive behavior of UHPFRC (Figure 1)

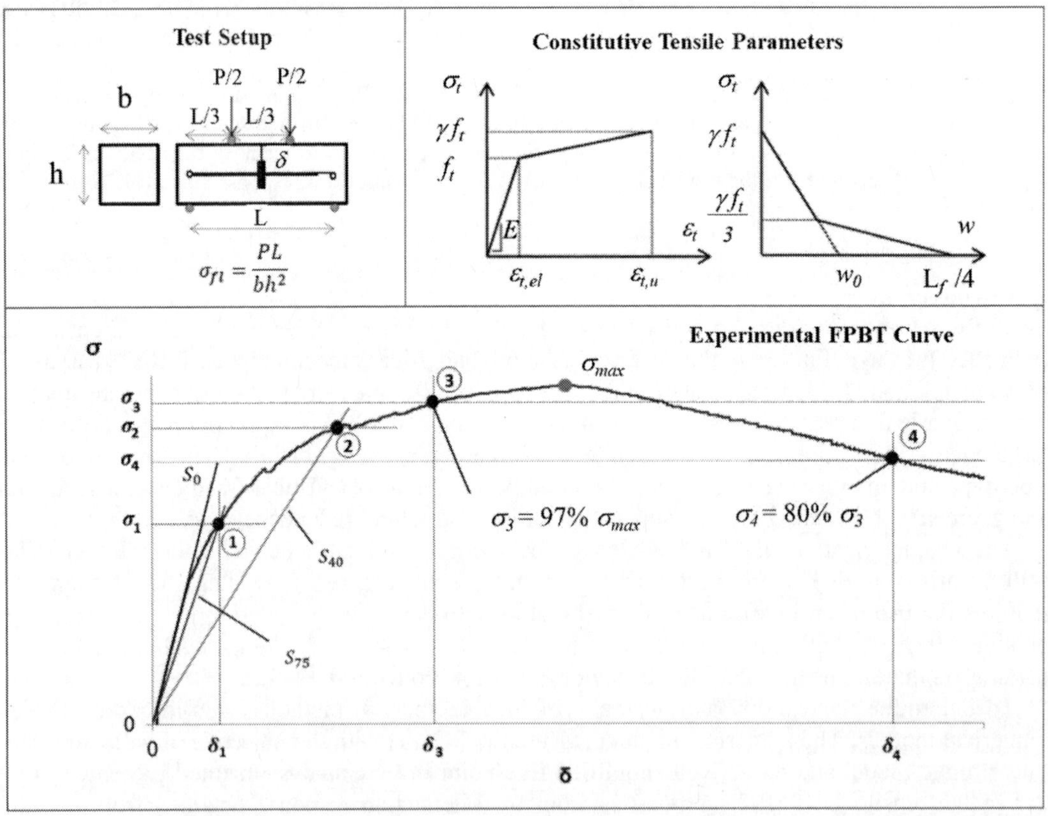

Figure 1. Simplified 4P-IA [16,17].

This 4P-IA (the same with the 5P-IA) is defined as follows: From the non-linear hinge model, analytical load-deflection curves from a TPBT with a specific slenderness ratio can be obtained from a hypothetical stress-strain response of the hinge. By varying the stress-strain parameters, different analytical strain-hardening behavior concretes can be modelled. For each set of tensile parameters, the analytical σ_{fl}-δ curve can be obtained according to the non-linear hinge model developed. From these curves, a set of points $(\delta_i, \sigma_{fl,i})$ can be extracted and statistically analyzed to establish relationships between these points and the tensile parameters.

Taking these into account, what happens if the UHPFRC does not exhibit strain-hardening behavior? Is the 4P-IA reliable enough to obtain the tensile parameters of UHFRC that exhibits strain-softening behavior in tension?

To distinguish between UHPFRCs that exhibit strain-hardening behavior and those that exhibit strain-softening behavior in tension, a hardening ratio (γ), has been defined (1). If $\gamma \geq 1$ the UHPFRC exhibits strain-hardening behavior. If $\gamma < 1$ the UHPFRC exhibits strain-softening behavior.

$$\gamma = \frac{f_{tu}}{f_t} \tag{1}$$

3. Numerical model

In this work, a numerical model is developed using the FE software DIANA [18]. To model the tensile UHPFRC constitutive behavior a discrete cracking approach is used. In this approach the constitutive model for UHPFRC is based on the discrete cracking model as an interface behavior. The constitutive law for discrete cracking in DIANA is based on a total deformation theory [18]. This behavior is only forced at the central section of the beam (Figure 2). To get this objective the tensile strength (f_t) and the ultimate tensile strength (f_{tu}) are reduced in a 2%. The rest of the beam is modelled using a smeared cracking approach based on a fixed total strain crack model expressed as function of crack opening curve (the "fibre reinforced concrete model, FRCCON" from fédération internationale du béton/International Federation for Structural Concrete (fib) working groups).

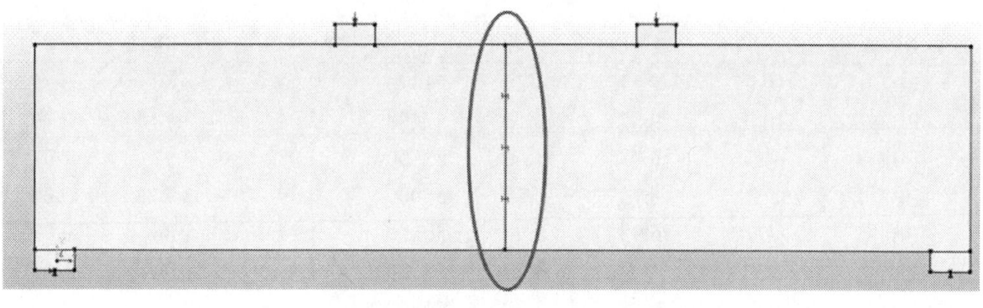

Figure 2. Discrete cracking approach.

The TPBT has been modelled with a 2D quadratic plane stress eight node quadrilateral element. Moreover, a quadratic 2D 3+3 nodes line interface element is situated in the central section of the beam. The load is applied on the steel load plates by a gradual increasing displacement.

A Nonlinear Analysis has been carried out using an incremental-iterative solution procedure. The theoretical formulation of the model, the adopted numerical procedures and solution strategy used are described in detail in DIANA user's manual [18].

4. Experimental program

For the purpose of using the constitutive tensile parameters obtained as a result of the 4P-IA in the FEM developed, eighteen square cross-section specimens with 100mm depth and 500mm length have been tested in a FPBT (Figure 3). A 2% in volume of smooth-straight (13/0.20) steel fibers were used in an Ultra-High-Performance cementitious matrix with an average concrete compressive strength of 145.46MPa, obtained from 32 100mm long cubes with a 1.63% coefficient of variation.

According to Figure 3, two displacement transducers were used to record the displacement at mid-span on the front side and in the back side each one.

Figure 3. Third Point Bending Test (TPBT).

5. Numerical model application

Table 1 shows the constitutive UHPFRC tensile parameters obtained when the 4P-IA is applied for each specimen and the hardening ratio defined in (1).

Table 1. Constitutive UHPFRC tensile parameters.

	constitutive tensile law						
Specimen id.	f_t (MPa)	f_{tu} (MPa)	ε_{tu} (‰)	E (MPa)	w_d (mm)	w_o (mm)	γ
4PBT-1	10.05	7.53	4.25	51300	2.53	3.80	0.75
4PBT-2	10.25	9.87	4.15	52400	2.09	3.14	0.96
4PBT-3	9.52	9.83	3.66	53600	1.98	2.97	1.03
4PBT-4	8.30	6.73	2.25	55000	2.49	3.74	0.81
4PBT-5	8.20	6.99	2.97	50400	1.24	1.86	0.85
4PBT-6	9.71	10.16	4.50	53300	1.29	1.94	1.05
4PBT-7	9.91	8.88	3.39	51400	1.49	2.24	0.90
4PBT-8	10.96	9.63	2.34	49800	1.22	1.83	0.88
4PBT-9	10.52	8.33	5.98	49800	2.19	3.28	0.79
4PBT-10	8.84	8.88	3.83	50000	2.14	3.21	1.00
4PBT-11	10.28	8.22	3.25	48000	1.77	2.66	0.80
4PBT-12	9.78	10.01	7.26	51600	2.13	3.19	1.02
4PBT-13	9.66	9.22	6.73	54100	2.01	3.02	0.95
4PBT-14	10.42	9.95	3.57	51600	1.75	2.62	0.95
4PBT-15	10.45	10.42	4.71	48100	2.04	3.06	1.00
4PBT-16	8.73	5.17	4.92	51259	2.11	3.17	0.59
4PBT-17	9.07	5.74	6.26	48500	1.94	2.91	0.63
4PBT-18	8.79	7.97	1.54	52900	1.24	1.85	0.91

The material properties obtained from the application of the simplified 4P-IA in Table 1 are implemented into the numerical model described in Section 3 and compared to the experimental program results (Figure 4).

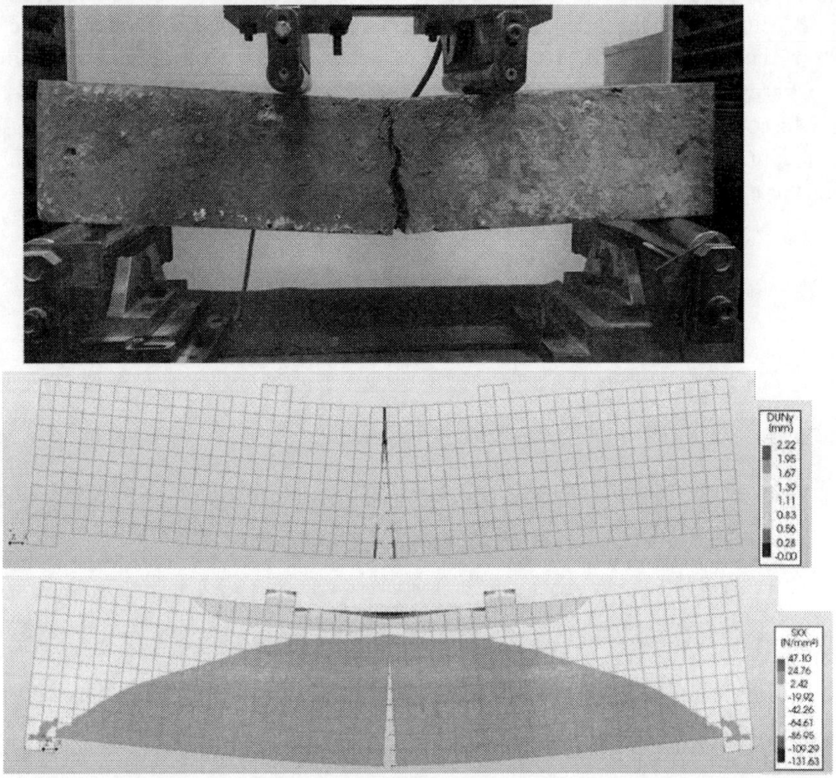

Figure 4. FEM of the Third Point Bending Test specimen.

In Figure 5, a comparison between the experimental and numerical tension-deflection (σ-δ) curve for a specimen that exhibits strain-hardening (Figure 5a) and a specimen that exhibits strain-softening (Figure 5b) has been depicted.

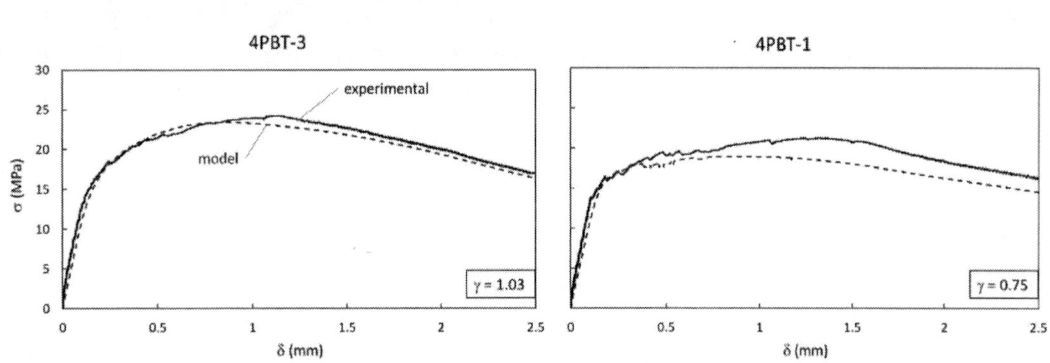

Figure 5. Specimen with strain-hardening **(a)**, specimen with strain-softening **(b)**.

As it can be seen in Figure 5, the σ-δ curve from the finite element model for specimens that exhibit strain-hardening tensile response ($\gamma \geq 1$) is more accurate than for specimens that exhibit strain-softening tensile behavior ($\gamma < 1$). Even though, despite this fact, the inaccuracy observed in the model's curve for the specimens that show strain-softening behavior is on the safe side.

To quantify the accuracy of the finite element model in comparison to the experimental response, a coefficient of security (CS) has been defined as the ratio between the experimental stress and the stress obtained from the model (2). This coefficient of security has been obtained at four levels of the maximum experimental deflection (δ_{maxexp}): $0.25 \cdot \delta_{maxexp}$, $0.50 \cdot \delta_{maxexp}$, δ_{maxexp} and $1.25 \cdot \delta_{maxexp}$ (see Figure 6). Moreover, the energy of the curve delimitated by the δ_{maxexp} level (A1) and by the $1.25 \cdot \delta_{maxexp}$ level (A2) for the experimental and model curves have been obtained. Table 2 shows the values of the parameters above defined.

$$CS = \frac{\sigma_{exp}}{\sigma_{model}} \tag{2}$$

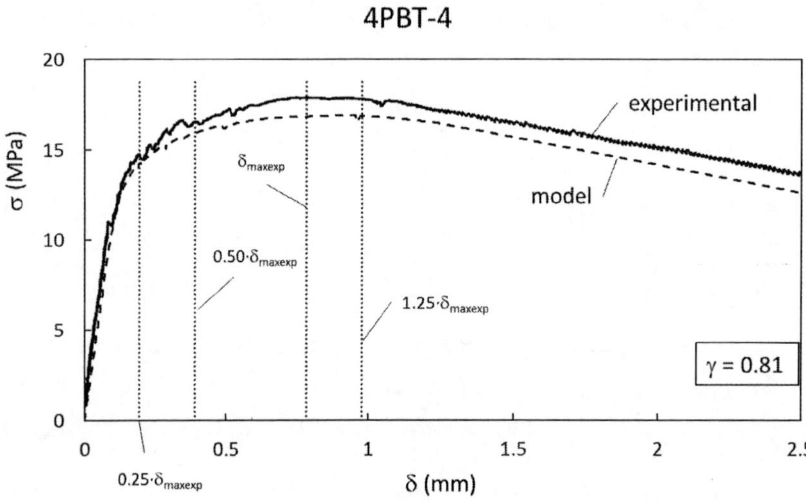

Figure 6. Delimitation of the CS levels.

Figure 7 shows the relation between the hardening ratio (γ) and the Security Coefficient (CS) at the four levels of the maximum experimental deflection (δ_{maxexp}): $0.25 \cdot \delta_{maxexp}$ (Figure 7a), $0.50 \cdot \delta_{maxexp}$ (Figure 7b), δ_{maxexp} (Figure 7c) and $1.25 \cdot \delta_{maxexp}$ (Figure 7d) for the 18 specimens analyzed in this work.

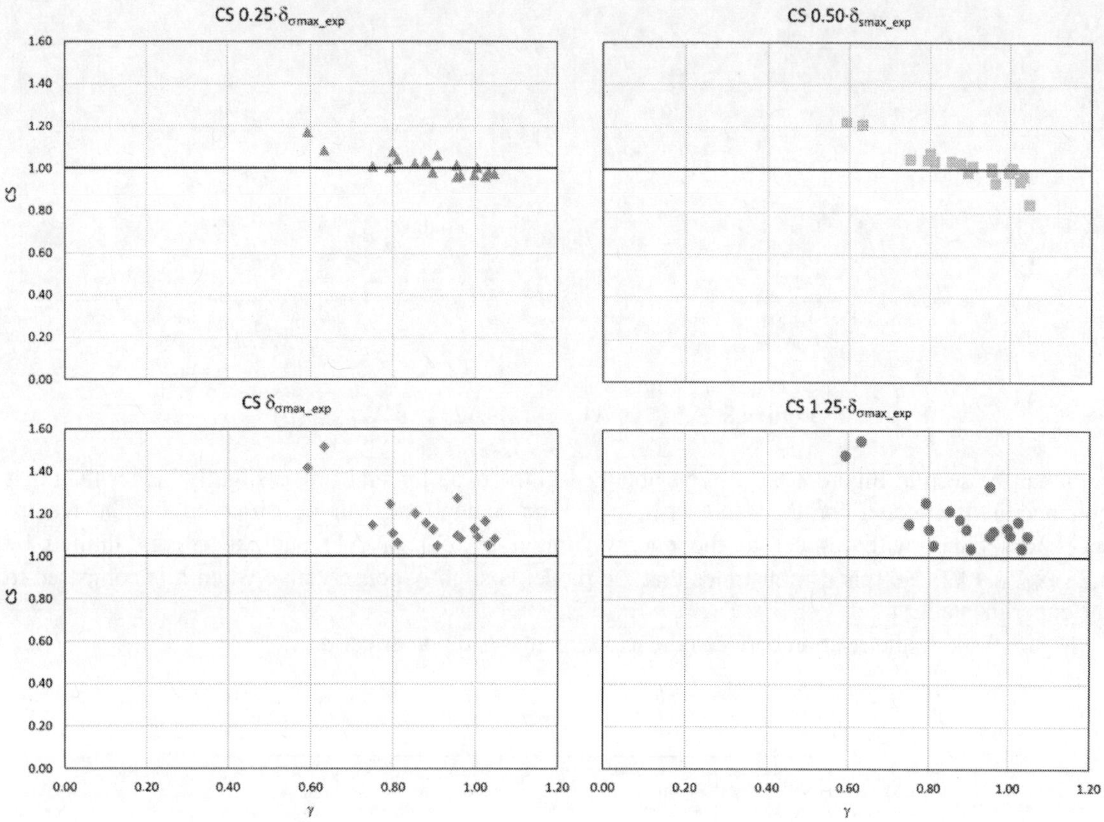

Figure 7. CS vs γ at $0.25\cdot\delta_{maxexp}$ **(a),** $0.50\cdot\delta_{maxexp}$ **(b),** δ_{maxexp} **(c)** and $1.25\cdot\delta_{maxexp}$ **(d).**

As it can be observed in Figure 7, the CS is near 1 in the early stages of the $\sigma-\delta$ curve (at $0.25\cdot\delta_{maxexp}$ and $0.50\cdot\delta_{maxexp}$) and clearly over 1 at the δ_{maxexp} and $1.25\cdot\delta_{maxexp}$ levels. It means that at the firs stages of the $\sigma-\delta$ curve the model adjusts accurately the experimental curve. When the highest level of the curve is reached and the descending branch takes part the accuracy of the model is not as good as in the elastic branch but it seems to be on the safe side because for the same level of δ, the σ value of the model is lower than the experimental one. So the model is reliable enough. If more attention is paid in Figure 7, it can be observed that there is certain inaccuracy in the first two levels, $0.25\cdot\delta_{maxexp}$ and $0.50\cdot\delta_{maxexp}$ (elastic branch of the $\sigma-\delta$ curve), in the unsafety part with the specimens made of UHPFRC that exhibits strain-hardening ($\gamma\geq 1$). Nevertheless, this inaccuracy is negligible. If the other last levels are observed, although they are on the safe side, the stress experimental values are higher than the numerical ones especially in the strain-softening ($\gamma< 1$) case. It seems that the model is more accurate at these stages for specimens made of UHPFRC with strain-hardening behavior.

Figure 8 shows the relation between the energy calculated for the experimental ($A1_{exp}$) and numerical ($A1_{model}$) curves for the δ_{maxexp} level (Figure 8a) and the energy calculated for the experimental ($A2_{exp}$) and numerical ($A2_{model}$) curves for the $1.25\cdot\delta_{maxexp}$ level (Figure 8b).

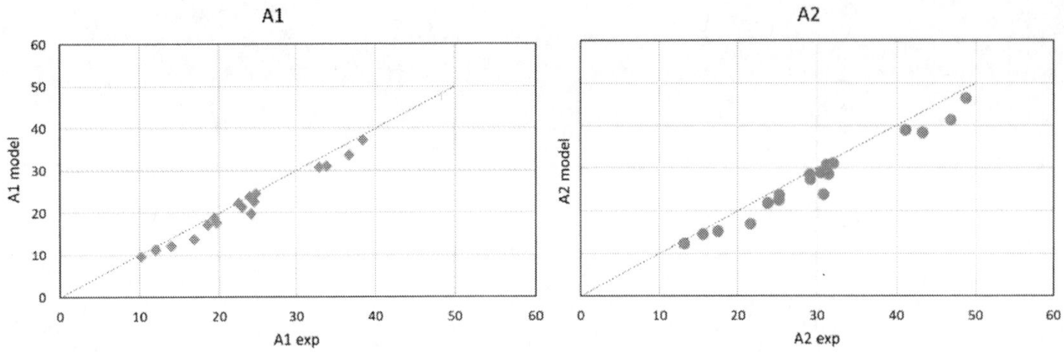

Figure 8. $A1_{exp}$ vs $A1_{model}$ **(a)**, $A2_{exp}$ vs $A2_{model}$ **(b)**.

As it can be seen in Figure 8 the energy obtained in the experimental test is slightly higher than that obtained in the model for the same specimen. The same trend can be observed for the rest of specimens either with respect to the energy limited by δ_{maxexp} (A1) and the energy limited by $1.25 \cdot \delta_{maxexp}$ (A2). So this demonstrates that the model is slightly conservative when it is compared to the experimental test.

Figure 9 shows the relation between the experimental and numerical σ_{max}.

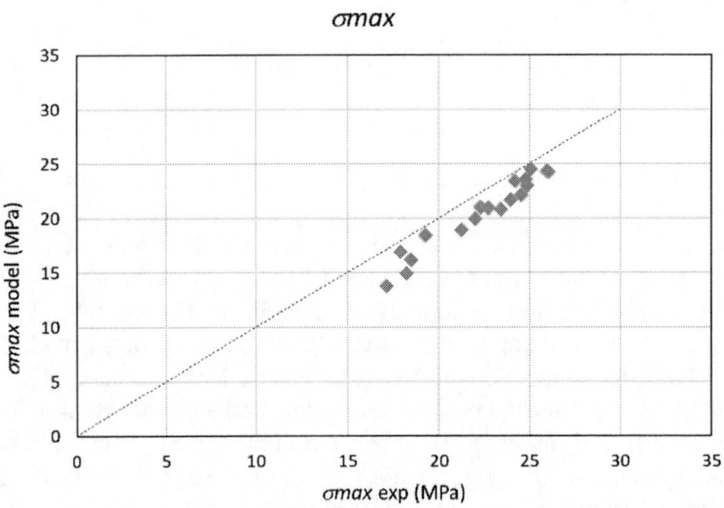

Figure 9. Experimental vs model σ_{max}.

As it is observed in Figure 9, if the maximum stress (σ_{max}) obtained in the experimental test and that in the numerical model are compared for each specimen, it can be seen that the experimental value of σ_{max} is higher than the numerical one for each specimen. So, as it happens with the energy, this value demonstrates that the model is conservative with respect to the experimental results. Moreover, as it can be observed in Figures 7, 8 and 9, the level of accuracy is high.

6. Concluding remarks

The numerical findings presented in the present paper allows to draw the main concluding remarks listed in the following:

1. The Simplified Four-Point Inverse Analysis Method (4P-IA) to obtain the tensile constitutive behavior of UHPFRC from the Third-Point Bending Test, has been validated resorting to a robust non-linear finite element modelling and a set of Third-Point Bending tests.
2. When using the tensile parameters resulting from the application of the 4P-IA, the numerical model shows a slightly conservative response with a considerable level of accuracy for modelling the tensile behavior of UHPFRC that exhibits strain-softening constitutive stress-strain response in a Third-Point Bending test.
3. As a result, inverse analysis methodologies based on the closed-form non-linear hinge model and the derived Simplified Four-Point Inverse Analysis Method proposed can be recommended to derive UHPFRC's tensile properties in Third-Point Bending Tests in both cases: when the UHPFRC exhibits strain-hardening constitutive stress-strain behavior and when exhibits strain-softening behavior.

Acknowledgements
This work forms part of project "BIA2016-78460-C3-1-R" supported by the State Research Agency of Spain

References

[1] Parra-Montesinos G J, Reinhardt H W and Naaman A E High Performance Fiber Reinforced Cement Composites 6 HPFRCC 6
[2] Reinhardt H W, Parra-Montesinos G J and Garrecht H 2015 *HPFRCC-7: Proceedings of the 7th RILEM Workshop on High Performance Fiber Reinforced Cement Composites* (Fraunhofer IRB Verlag)
[3] Yokota H, Rokugo K and Sakata N 2008 (JSCE-2008) Recommendations for Design and Construction of High Performance Fiber Reinforced Cement Composites with Multiple Fine Cracks (HPFRCC)
[4] Rigaud S, Chanvillard G and Chen J 2012 Characterization of Bending and Tensile Behavior of Ultra-High Performance Concrete Containing Glass Fibers *High Performance Fiber Reinforced Cement Composites 6: HPFRCC 6* ed G J Parra-Montesinos, H W Reinhardt and A E Naaman (Dordrecht: Springer Netherlands) pp 373–80
[5] Baby F, Graybeal B, Marchand P and Toutlemonde F 2013 UHPFRC tensile behavior characterization: Inverse analysis of four-point bending test results *Mater. Struct. Constr.* **46** 1337–54
[6] Baby F, Graybeal B, Marchand P and Toutlemonde F 2012 Proposed flexural test method and associated inverse analysis for ultra-high-performance fiber-reinforced concrete *ACI Mater. J.* **109** 545–55
[7] Maalej M and Li V C 1994 Flexural Strength of Fiber Cementitious Composites *ASCE J. Mater. Civ. Eng.* **6** 390–406
[8] Qian S and Li V C 2007 Simplified Inverse Method for Determining the Tensile Strain Capacity of Strain Hardening Cementitious Composites *J. Adv. Concr. Technol.* **5** 235–46
[9] Gröger, Johannes, Viet tue, Nguyen, Wille K 2012 Bending Behaviour and Variation of Flexural Parameters of UHPFRC *3rd International Symposium on UHPC and Nanotechnology for High Performance Construction Materials* (kassel university press GmbH) pp 419–26
[10] Soranakom C and Mobasher B 2007 Closed-form moment-curvature expressions for homogenized fiber-reinforced concrete *ACI Mater. J.* **104** 351–9
[11] Ostergaard L, Walter R and Olesen J F 2005 Method for determination of tensile properties of engineered cementitious composites (ECC), *Construction Materials : Proceedings of ConMat'05 and Mindess Symposium* p 74

[12] López J Á, Serna P, Navarro-Gregori J and Camacho E 2015 An inverse analysis method based on deflection to curvature transformation to determine the tensile properties of UHPFRC *Mater. Struct.* **48** 3703–18

[13] Mezquida-Alcaraz E J, Navarro-Gregori J, Angel Lopez J and Serna-Ros P 2019 Validation of a non-linear hinge model for tensile behavior of UHPFRC using a Finite Element Model *Comput. Concr.* **23** 11–23

[14] Mezquida-Alcaraz E J, Navarro-Gregori J, López J A and Serna-Ros P 2018 Numerical Validation of a Simplified Inverse Analysis Method to Characterise the Tensile Behaviour of UHPFRC, *3rd ACI-fib-RILEM Joint Workshop Fibre Reinforced Concrete: from Design to Structural Applications (FRC 2018)*, Desenzano, Lake Garda, Italy.

[15] López J Á, Serna P, Navarro-Gregori J and Coll H 2016 A simplified five-point inverse analysis method to determine the tensile properties of UHPFRC from unnotched four-point bending tests *Compos. Part B Eng.* **91** 189–204

[16] López J A, Serna P and Navarro-Gregori J 2017 Advances in the development of the first UHPFRC Recommendations in Spain: Material classification, design and characterization *UHPFRC 2017 Designing and Building with UHPFRC: New large-scale implementations, recent technical advances, experience and standards* ed F. Toutlemonde & J. ResplendinoT.Ch. (RILEM Publications SARL) pp 565–74

[17] López J Á 2017 *CHARACTERISATION OF THE TENSILE BEHAVIOUR OF UHPFRC BY MEANS OF FOUR-POINT BENDING TESTS* (Valencia (Spain): Universitat Politècnica de València)

[18] DIANA (Software) 2017 User's Manual -- Release 10.2 *TNO DIANA, Netherlands*

IOP Conf. Series: Materials Science and Engineering **596** (2019) 012007 doi:10.1088/1757-899X/596/1/012007

Simulation of mixed-mode fracture (I-II) on PFRC specimens with various fibre proportions using an embedded cohesive crack model

F Suárez[1] , JC Gálvez[2] , A Enfedaque[2] and MG Alberti[2]

[1]Departamento de Ingeniería Mecánica y Minera. Universidad de Jaén. Campus Científico-Tecnológico de Linares. Avda. de la Universidad (Cinturón Sur) – Linares, Spain

[2]Departamento de Ingeniería Civil-Construcción. Universidad Politécnica de Madrid. E.T.S.I. Caminos, Canales y Puertos, C/ Profesor Aranguren s/n - Madrid, Spain

E-mail: fsuarez@ujaen.es, jaime.galvez@upm.es, alejandro.enfedaque@upm.es, marcos.garcia@upm.es

Abstract. The study of fibre-reinforced concrete (FRC) has become of increasing interest in the last decades. Although it is not a new technology, it has experienced a remarkable progress with the appearance of some recommendations in the standards. More specifically, the use of polyolefin fibres has proved to increase the tensile strength of concrete without the problems usually found with steel fibres, especially those related to corrosion. This type of fibres have been studied in depth and its fracture behaviour has been successfully simulated in the past by means of an embedded crack model using a trilinear softening function. Nevertheless, these simulations have been always focused on cases where fracture took place under pure mode I conditions, namely using the classical three-point bending test on notched specimens. In this study, such embedded crack model is used to reproduce the fracture behaviour on notched specimens subjected to a modified three-point bending test that induces fracture under a combination of modes I and II. Three PFRC mixes are analysed, all of them with the same proportions of concrete components but different proportions of polyolefin fibres. The experimental and numerical diagrams properly agree and allow identifying how the increasing proportion of fibres can be reflected in the trilinear softening function that numerically drives the damage evolution.

1. Introduction

Fibre reinforced concrete is not a new idea. It has been used in practice for decades, especially with steel fibres. Nevertheless, many researchers have lately focused their interest in this technology, partially boosted by the appearance of recommendations in the late standards, such as the Spanish Code for concrete structures [1] and the Model Code [2]. These recommendations help to design fibre-reinforced concrete (FRC) structures and are encouraging engineers to consider it as an alternative to traditional reinforced concrete with steel rebars. They are especially focused on steel-fiber-reinforced concrete, but in the last years some interesting alternative materials have been proposed with success. More specifically, polyolefin fibres have proved to perform remarkably well from the mechanical point of view, overcoming some important drawbacks of the steel fibres, especially related to corrosion [3–6], which not only may reduce the durability of the structure, but also affect its aesthetical aspect, that can be of paramount importance in some cases.

The fracture behaviour of polyolefin fibre reinforced concrete (PFRC) specimens has been numerically reproduced in the past with success by using a well known embedded crack model based on the cohesive crack approach [7,8]. In those simulations, the cohesive model was defined by using a trilinear softening function that allowed reproducing the load recovery due to the fibre contribution after the concrete matrix failure and the final load decay that takes place when fibre is pulled out from concrete. These numerical studies, however, only used the classical three-point bending test for comparison, which develops fracture under pure mode-I conditions. Although recent research studies have analysed the performance of PFRC specimens under mode II conditions [9], the combination of modes I and II, that is a common situation in many real fracture problems, is not well understood and is still under study [10].

This contribution analyses the use of an embedded crack element formulation based on the cohesive zone approach for modelling fracture in PFRC specimens under fracture initiated under a combination of modes I and II. To do this, a modified version of the classical three point bending test described by the RILEM [11], more especifically type 1 tests proposed by Gálvez et al. in [12], is used on PFRC specimens. Three sets of specimens were tested, all of them with the same concrete proportioning but different fibre dosages: 3, 6 and 10 kg/m^3 . These tests were numerically reproduced by using the above-mentioned embedded crack formulation, showing good agreement with the experimental results.

2. Experimental results

In this section the experimental tests are briefly described and then the specimens and the three PFRC mixes analysed in this study presented.

2.1. The modified three-point bending test

The classical three-point bending test on notched specimens can be performed following the recommendations by the RILEM [11], which induces fracture under pure mode I conditions. This test describes a three point bending test where supports and the loading point are symmetrically disposed with respect to the notched specimen. A variation of this test was proposed in [12] to induce fracture under a combination of modes I and II. In that work, the authors proposed two variations of the model, with one of them (type 1) being a true three point bending test and with the other one (type 2) being in practice a four point bending test, since an additional support was added to modify the crack path and, thus, the interaction of modes I and II. In this study type 1 test is considered. The boundary conditions and the geometrical definition of the experimental setup can be seen in Figure 1. Figure 2 shows two specimens, before and after the test is finished.

2.2. Specimens

Three sets of specimens were tested, each of them with different proportions of polyolefin fibres: 3, 6 and 10 kg/m^3. In all cases the concrete proportioning was the same, with the fibre dosage being the only variation. Table 1 shows the composition of each set of specimens.

All specimens were prismatic of dimensions 600 mm × 150 mm × 150 mm and the fibre length was 60 mm.

Table 1. Mix proportioning.

	Cement (kg/m^3)	Limestone powder (kg/m^3)	Water (kg/m^3)	Sand (kg/m^3)	Gravel (kg/m^3)	Grit (kg/m^3)	Superplasticizer (% cem. weight)	Polyolefin fibres (kg/m^3)
PFRC-3	375	100	187.5	916	300	450	1.25	3
PFRC-6	375	100	187.5	916	300	450	1.25	6
PFRC-10	375	100	187.5	916	300	450	1.25	10

IOP Conf. Series: Materials Science and Engineering **596** (2019) 012007 doi:10.1088/1757-899X/596/1/012007

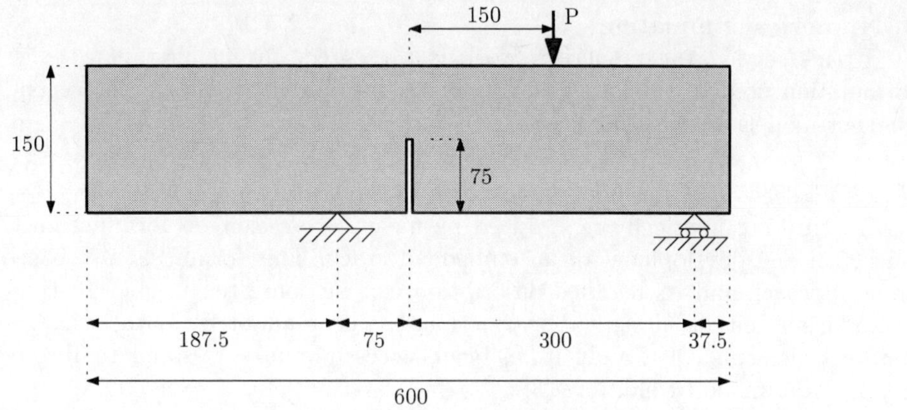

Figure 1. Set-up of the type 1 three point bending test described in [12] (dimensions in mm).

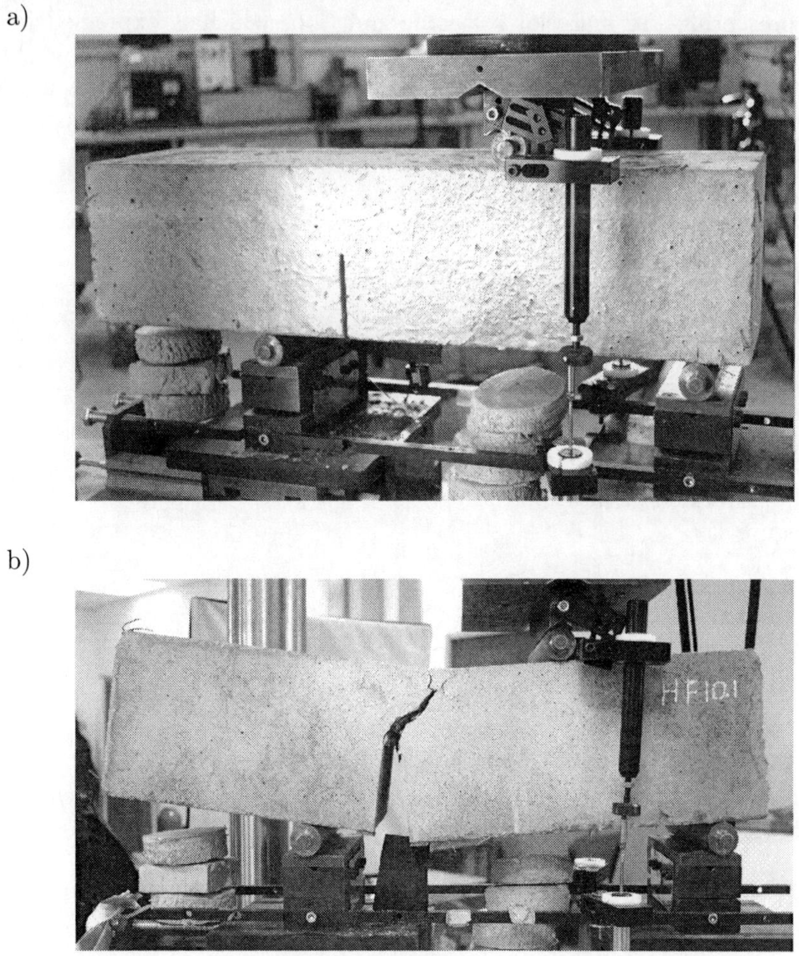

Figure 2. a) One of the tested specimens before testing, b) A specimen at the end of the test.

IOP Conf. Series: Materials Science and Engineering **596** (2019) 012007 doi:10.1088/1757-899X/596/1/012007

3. Numerical simulation

In this section the numerical work is presented. In the first place, the embedded crack formulation used to simulate cracking is briefly described and, in the second place, the finite element models are presented.

3.1. Embedded crack formulation based on the cohesive zone approach

Fracture is numerically reproduced by means of an element formulation that simulates the initiation and development of an embedded crack. This formulation is based on the cohesive zone approach and its detailed description can be consulted in [13]. In the past, the original formulation, initially designed for concrete, has been adapted to other cases, such as brickwork masonry elements [14]. Lately, it has been successfully used to simulate fibre-reinforced concrete mixes under mode I conditions [8].

This formulation is based on the cohesive zone approach proposed by Hillerborg [15] (for more information on this topic, refer to [16]). In this case, it is a central forces model, that is to say, it assumes that the cohesive stress vector $\boldsymbol{t}$ is perpendicular to the crack opening and parallel to the crack displacement vector $\boldsymbol{w}$:

$$\boldsymbol{t} = \frac{f(\tilde{w})}{\tilde{w}}\boldsymbol{w} \qquad \text{with } \tilde{w} = \max(|\boldsymbol{w}|) \tag{1}$$

with $f(|\tilde{w}|)$ representing the material softening function, which is expressed in terms of an equivalent crack opening $\tilde{w}$ that stores the maximum historical crack opening and allows accounting for possible unloading processes. When simulating PFRC specimens, this softening function presents three linear branches [7] as shown in Figure 3, and can be expressed by:

$$\sigma = \begin{cases} f_{ct} + \left(\dfrac{\sigma_k - f_{ct}}{w_k}\right) \cdot w & \text{if } 0 < w \leq w_k \\ \sigma_k + \left(\dfrac{\sigma_r - \sigma_k}{w_r - w_k}\right) \cdot (w - w_k) & \text{if } w_k < w \leq w_r \\ \sigma_r + \left(\dfrac{-\sigma_r}{w_f - w_r}\right) \cdot (w - w_r) & \text{if } w_r < w \leq w_f \\ 0 & \text{if } w > w_f \end{cases} \tag{2}$$

In the models that are presented later, constant strain triangular elements are used; in this case crack can only develop in directions parallel to one of the triangle sides and at mid height, which guarantees local and global equilibria and reduces to three the possible crack paths in an element (see Figure 4).

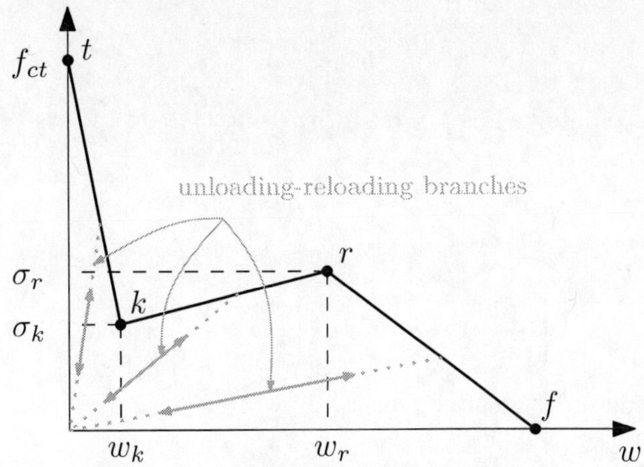

Figure 3. Trilinear softening function.

Once the cracking process starts, the element is divided into two parts, A^+ and A^-. If A stands for the element area, h for the height of the triangle perpendicular to the opposite side to the solitary node, L for the crack length and $\boldsymbol{n}$ for the unit vector perpendicular to the crack, vector $\boldsymbol{t}$ is constant along the crack and can be obtained as:

$$\boldsymbol{t} = \frac{A}{hL}\boldsymbol{\sigma} \cdot \boldsymbol{n} \tag{3}$$

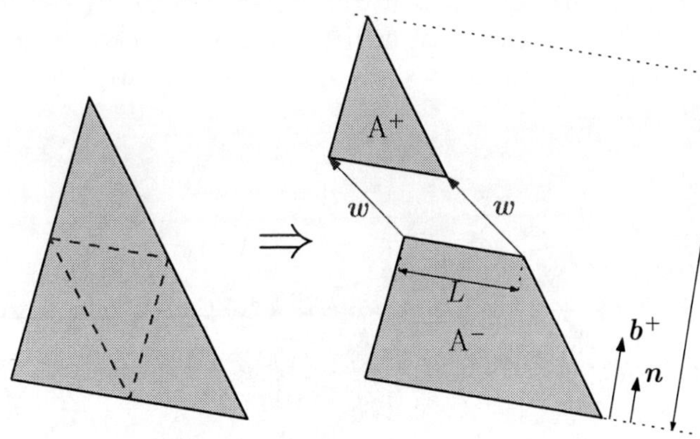

Figure 4. Potential crack paths (left) and geometrical definitions of $\boldsymbol{w}$, $\boldsymbol{n}$ and $\boldsymbol{b}^+$ (right).

The part of the element outside the crack behaves as linear-elastic and the crack displacement vector $\boldsymbol{w}$ can be obtained subtracting the contribution of the crack to the elastic prediction of the stress:

$$\boldsymbol{\sigma} = \boldsymbol{E} : \left[\epsilon^a - \left(\boldsymbol{b}^+ \otimes \boldsymbol{w}\right)^S\right] \tag{4}$$

where $\boldsymbol{E}$ stands for the elastic tangent tensor, $\boldsymbol{b}^+$ for the gradient vector of the solitary node shape function, expressed by (5), ϵ^a for the apparent strain vector. In this expression superscript S indicates the symmetric part of the tensor, : the usual double-dot product ($(\boldsymbol{A} : \boldsymbol{b})_{ij} = A_{ijkl}b_{kl}$) and $\otimes$ the usual direct product ($(\boldsymbol{a} \otimes \boldsymbol{b})_{ij} = a_i b_j$).

IOP Conf. Series: Materials Science and Engineering **596** (2019) 012007 doi:10.1088/1757-899X/596/1/012007

$$\boldsymbol{b}^+ = \frac{1}{h}\boldsymbol{n} \tag{5}$$

In addition, $\boldsymbol{t}$ can be obtained as $\boldsymbol{t} = \boldsymbol{\sigma} \cdot \boldsymbol{n}$, by using (1) and (4):

$$\frac{f(\tilde{w})}{\tilde{w}}\boldsymbol{w} = [\boldsymbol{E} : \boldsymbol{\epsilon}^a] \cdot \boldsymbol{n} - \left[\boldsymbol{E} : \left(\boldsymbol{b}^+ \otimes \boldsymbol{w}\right)^S\right] \cdot \boldsymbol{n}$$

which can be rewritten as:

$$\left[\frac{f(\tilde{w})}{\tilde{w}}\mathbf{1} + \boldsymbol{n} \cdot \boldsymbol{E} \cdot \boldsymbol{b}^+\right] \cdot \boldsymbol{w} = [\boldsymbol{E} : \boldsymbol{\epsilon}^a] \cdot \boldsymbol{n} \tag{6}$$

with $\mathbf{1}$ being the second-order identity tensor.

3.2. FEM models

The embedded crack formulation described above was implemented by means of a UMAT subroutine to be used in Abaqus® and the specimens were simulated with bi-dimensional models using reduced integration triangular elements. In these models, all elements are of the same material, which follows the formulation presented above and are characterised by the same material parameters.

To compare the experimental and the numerical results, the evolution of load with respect to two displacement measurements are analysed: the crack mouth opening displacement (CMOD) and the vertical displacement of the loading point. From the experimental point of view, the CMOD is measured by means of a strain gauge and the loading point displacement by means of a linear variable differential transducer (LVDT). It must be noted that the CMOD is not only a horizontal displacement in this case (it is in the classical symmetrical disposition described in [11], but not in the present case); this must be taken into account when obtaining the diagrams from the numerical results. Figure 5 shows one of the meshes used and the position of the supports and the loading point.

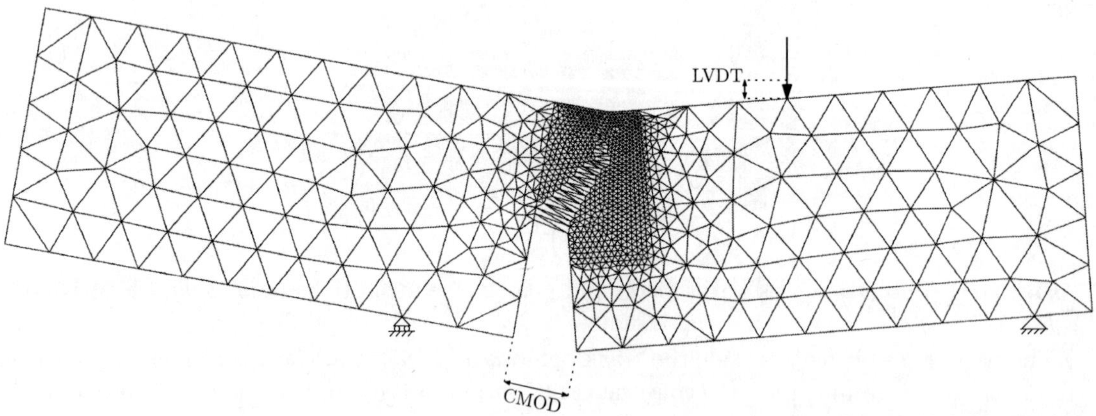

Figure 5. Deformed mesh of one of the models used in this study. The loading and boundary conditions are displayed, as well as the LVDT and CMOD measurements.

The softening function has been defined as trilinear and their characteristic points can be consulted in Table 2 and observed in Figure 6. To calibrate these points the following criteria has been followed: initial and final points, t and f , are equal in all three diagrams, only points k and r are different in order to account for different fibre proportions.

Table 2. Characteristic points of the trilinear softening functions used with each fibre dosage.

	PFRC-3				PFRC-6				PFRC-10			
	t	k	r	f	t	k	r	f	t	k	r	f
w (mm)	0.00	0.15	2.250	7.50	0.00	0.14	2.250	7.50	0.00	0.13	2.250	7.50
σ (MPa)	2.30	0.12	0.245	0.00	2.30	0.33	0.700	0.00	2.30	0.75	1.700	0.00

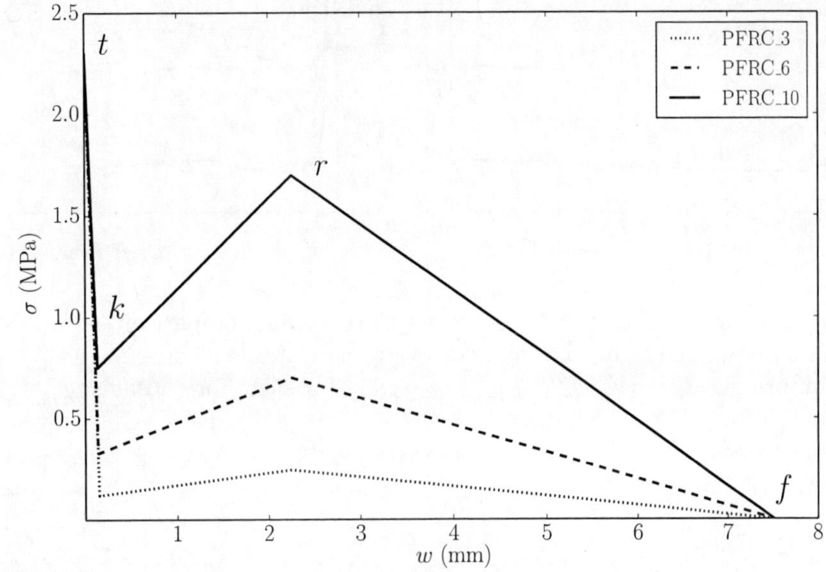

Figure 6. Trilinear softening diagrams used to simulate each fibre proportion.

4. Results

Figures 7 and 8 compare the load-CMOD and load-LVDT diagrams obtained experimentally and numerically. These figures show that the softening functions of Figure 6 provide an adequate approximation to the experimental behaviour of the three analysed concrete mixes. The peak load is properly reproduced in all cases, as well as the load decrease after the concrete matrix initial failure. Regarding the later behaviour, all numerical simulations remain inside the experimental envelope, no matter if the CMOD or the LVDT displacement is used for comparison. The numerical results reproduce adequately the load increase due to the performance of the polyolefin fibres as well as the later load decay that takes place as a result of the pulling out of the fibres.

Finally, Figure 9 shows the crack path obtained numerically (damaged elements are shaded in blue) and the experimental envelope of the crack paths (overlapped in red) for each PFRC mix. The initial branches of these paths obtained numerically are in all cases reasonably accurate when compared with the experimental envelopes, although the path deviates considerably for later stages of the tests. It must be born in mind that in PFRC specimens the fracture path is highly influenced by the local presence of fibres around the crack tip during the test, which can explain these differences.

5. Final remarks

The results presented in this contribution confirm the validity of trilinear softening functions with PFRC specimens subjected to a modified three-point bending test that induces fracture

IOP Conf. Series: Materials Science and Engineering **596** (2019) 012007 doi:10.1088/1757-899X/596/1/012007

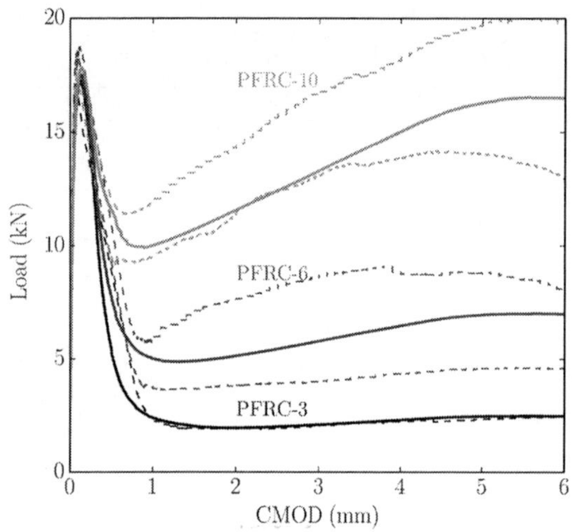

Figure 7. Comparison of Load-CMOD diagrams obtained experimentally and numerically for each fibre dosage.

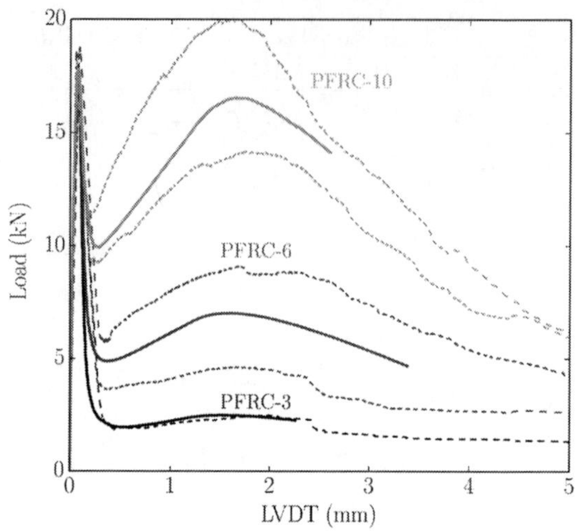

Figure 8. Comparison of Load-LVDT diagrams obtained experimentally and numerically for each fibre dosage.

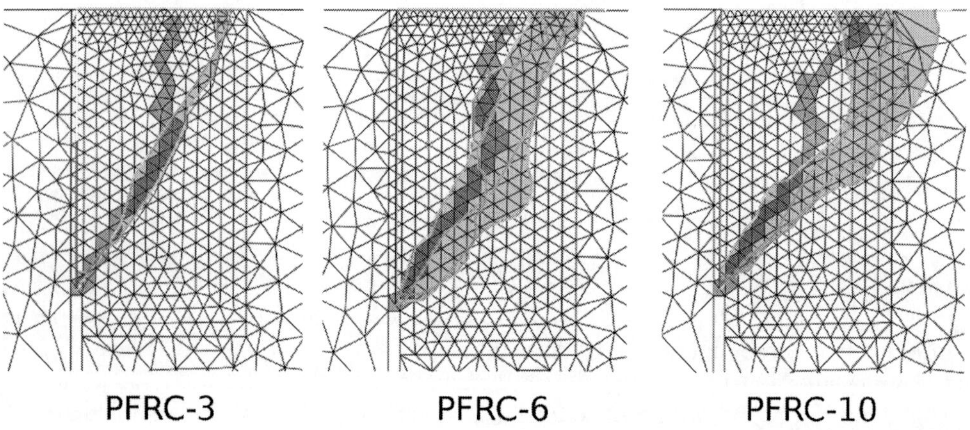

Figure 9. Comparison of the crack paths obtained experimentally and numerically for each of the analysed PFRC mixes.

under mixed-mode (I+II) conditions. This observation has been confirmed with PFRC mixes of the same concrete proportioning but various fibre dosages: 3, 6 and 10 kg/m^3 . Only modifying the two intermediate points of the softening function is enough to adjust the fibre behaviour of each fibre proportion.

Acknowledgements

The authors gratefully acknowledge the financial contribution for this Research provided by the Ministry of Economy and Competitiveness of Spain by means of the Research Fund Project BIA2016-78742-C2-2-R. They also thank SIKA SAU for supplying the polyolefin fibres used in the experimental work.

References

[1] EHE-08. Instrucción de hormigón estructural. *Ministerio de Fomento, Madrid, España*, 2008.

[2] Paul Beverly. *fib model code for concrete structures 2010*. Ernst & Sohn, 2013.

[3] M.G. Alberti, A. Enfedaque, and J.C. Gálvez. On the mechanical properties and fracture behavior of polyolefin fiber-reinforced self-compacting concrete. *Construction and Building Materials*, 55:274 – 288, 2014.

[4] M.G. Alberti, A. Enfedaque, and J.C. Gálvez. Comparison between polyolefin fibre reinforced vibrated conventional concrete and self-compacting concrete. *Construction and Building Materials*, 85:182 – 194, 2015.

[5] A. Enfedaque, M.G. Alberti, J.A. Paredes, and J.C. Gálvez. Interface properties of polyolefin fibres embedded in self-compacting concrete with a bond improver admixture. *Theoretical and Applied Fracture Mechanics*, 90:287 – 293, 2017.

[6] A. Enfedaque, M.G. Alberti, J. Gálvez, M. Rivera, and J. Simón-Talero. Can polyolefin fibre reinforced concrete improve the sustainability of a flyover bridge? *Sustainability*, 10(12):4583, 2018.

[7] M.G. Alberti, A. Enfedaque, J.C. Gálvez, and E. Reyes. Numerical modelling of the fracture of polyolefin fibre reinforced concrete by using a cohesive fracture approach. *Composites Part B: Engineering*, 111:200– 210, 2017.

[8] A. Enfedaque, M.G. Alberti, J.C. Gálvez, and J. Domingo. Numerical simulation of the fracture behaviour of glass fibre reinforced cement. *Construction and Building Materials*, 136:108 – 117, 2017.

[9] A. Picazo, J.C. Gálvez, M.G. Alberti, and A. Enfedaque. Assessment of the shear behaviour of polyolefin fibre reinforced concrete and verification by means of digital image correlation. *Construction and Building Materials*, 181:565 – 578, 2018.

[10] F. Suárez, J.C. Gálvez, A. Enfedaque, and M.G. Alberti. Modelling fracture on polyolefin fibre reinforced concrete specimens subjected to mixed-mode loading. *Engineering Fracture Mechanics*, 211:244 – 253, 2019.

[11] RILEM. Determination of the fracture energy of mortar and concrete by means of three-point bend tests on notched beams. *Mater Struct*, 18(106):285–290, 1985.

[12] J.C. Gálvez, M. Elices, G. Guinea, and J. Planas. Mixed mode fracture of concrete under proportional and nonproportional loading. *Int J Fract*, 94(3):267–284, 1998.

[13] J.M. Sancho, J. Planas, D.A. Cendón, E. Reyes, and J.C. Gálvez. An embedded crack model for finite element analysis of concrete fracture. *Engineering Fracture Mechanics*, 74(1):75 – 86, 2007. Fracture of Concrete Materials and Structures.

[14] E. Reyes, J.C. Gálvez, M.J. Casati, D.A. Cendón, J.M. Sancho, and J. Planas. An embedded cohesive crack model for finite element analysis of brickwork masonry fracture. *Engineering Fracture Mechanics*, 76(12):1930 – 1944, 2009.

[15] A. Hillerborg, M. Modéer, and P.-E. Petersson. Analysis of crack formation and crack growth in concrete by means of fracture mechanics and finite elements. *Cem Concr Res*, 6(6):773 – 781, 1976.

[16] Zdeněk P Bažant and Jaime Planas. *Fracture and size effect in concrete and other quasibrittle materials*. CRC Press, 1997.

IOP Conf. Series: Materials Science and Engineering **596** (2019) 012008 doi:10.1088/1757-899X/596/1/012008

Material parameters for computer analysis of fibre reinforced concrete structures

R Pukl[1], D Lehký[2], M Lipowczan[3] and D Novák[4]

[1] Project manager, Červenka Consulting s.r.o., Prague, Czech Republic
[2] Assoc. Prof., Institute of Structural Mechanics, Faculty of Civil Engineering, Brno University of Technology, Brno, Czech Republic
[3] PhD student, ISM FCE BUT, Brno, Czech Republic
[4] Professor, ISM FCE BUT, Brno, Czech Republic

E-mail: radomir.pukl@cervenka.cz

Abstract. Methods and software tools for material parameter identification of high-performance cementitious composites are presented. The aim is to provide techniques for the purpose of advanced assessment of their fracture–mechanical properties and subsequent numerical simulations of components/structures made of these materials. The paper describes development of computational and material models utilized for efficient material parameter determination of studied composite. Such a determination is performed with the help of experimental data from four-point bending test used in inverse analysis based on artificial neural networks. It is a part of complex methodology for statistical and reliability analyses of fibre reinforced concrete structures.

1. Introduction

Advanced assessment of fracture–mechanical properties is primarily important for subsequent numerical simulations of components/structures made of fibre reinforced concrete (FRC). The variability of experimental results using specimens made of quasi-brittle materials such as FRC is high due to the natural heterogeneity of this materials. Then an assessment of fracture–mechanical parameters is much more difficult and problematic. To remain at deterministic level is therefore impossible and without virtual statistical approach, simulation and probabilistic result assessment the consequent practical design of quasi-brittle material-based structures can be risky. The aim of the performed research is to deepen knowledge about the complexity of the behaviour of this advanced and progressive composite material especially in relation to its resistance to crack propagation. The obtained knowledge is a prerequisite for efficient design of this composite and the consequent expansion of its applicability for increase of sustainability of constructed elements, structures and buildings.

The research is focused on several topics: First, development of suitable constitutive law for FRC cementitious composites under the framework of nonlinear fracture mechanics software ATENA [1] is of primary importance. The computational model of unnotched beam subjected to four-point bending has been created and verified using data from Dura Technology Sdn Bhd., Malaysia. Sensitivity analysis which shows importance of parameters of computational model have been performed. Finally, material parameters identification will be performed based on the combination of fracture tests, numerical simulation and artificial neural network-based (ANN) inverse analysis [2, 3].

IOP Conf. Series: Materials Science and Engineering **596** (2019) 012008 doi:10.1088/1757-899X/596/1/012008

2. Computational model

The nonlinear numerical simulation of the four-point bending beam tests (Figure 1) has been performed using FEM software ATENA [5]. It has been developed by the Červenka Consulting company for realistic simulation of concrete and reinforced concrete structures at deterministic level. ATENA software enables to calculate response of the structural member including material damage and is used here as an advanced failure function.

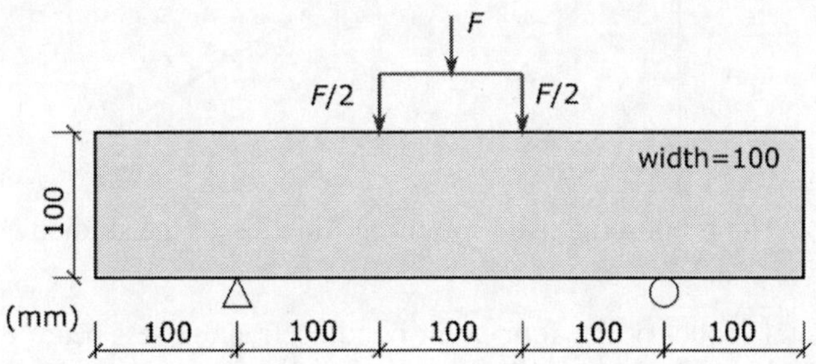

Figure 1. Diagram of the unnotched specimen tested in the four-point bending configuration.

The crucial role by the nonlinear numerical analysis plays the constitutive law at material point level. The realistic structural response, damage and failure calculated by the computer model depends strongly on the material model quality, which decides about the exactness and accuracy of the achieved results. Since the fibre reinforced concrete is rather complex heterogeneous material with strongly nonlinear response the "3DNonlinerCementitious2User" material model has been utilized for the realistic computation of this composite. This material model can capture all the important aspect of the FRC material behavior and response under tensile as well us compressive loading.

Concrete in tension is described within the smeared cracks concept by the nonlinear fracture mechanics with crack band approach. The main material model parameters are tensile strength and form of the softening function which characterizes the crack mouth opening related to the remaining tensile stress. The real discrete crack is in the model represented as a band of localized strains where the strain corresponding to the crack width is related to the size, form a orientation of the finite element. The softening function for the material law used in the smeared crack concept has to be determined for every material point (or finite element) in such a way that the objective crack opening law will be preserved. Only such an approach based on energetical formulation assures objective solution independent on the finite element mesh [1]. In case of the fibre reinforced concrete an appropriate softening function should be developed and applied.

The proposed function for optimal reproduction of the tensile softening of the investigated composite material is shown in Figure 2. Its form attempts to capture all the particular stages of the FRC material damage such as reaching of strength of the cement matrix, delayed activation of the fibers within the composite, and consequent stiffening of the composite with activated fibers after exhausting of the matrix bearing capacity. The form of the function is described by tensile strength and four additional parameters C_1 to C_4 as shown in Figure 2.

Concrete in compression exhibits strong confinement effect – i.e. increase of the compressive strength under concentrated three-dimensional compressive stress state. This effect is covered by the special plasticity theory with a non-associated plastic flow law in the combined fracture-plastic models in ATENA. The compressive ductility of FRC should be appropriately accounted in the material model as well.

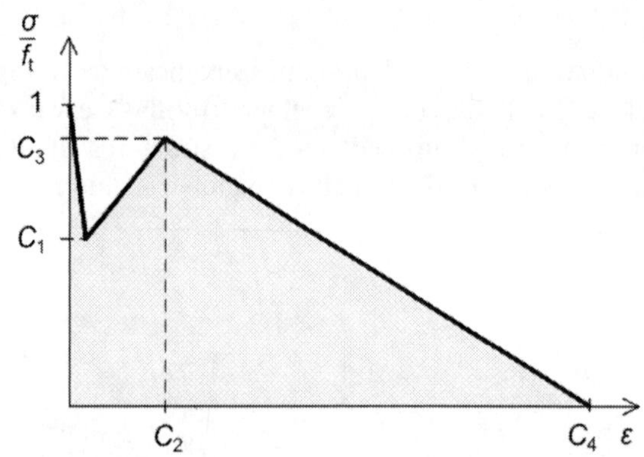

Figure 2. The proposed tensile softening function of the studied composite.

The created computer model of the four-point bending fracture test beam is shown in Figure 3. It consists of 500 four-node iso-parametric finite elements. It is loaded in 120 load steps by prescribed displacement at the marked loading point locations (see Figure 3, top). The nonlinear solution is performed by modified Newton-Raphson iterative method. The resulting crack pattern from the deterministic analysis when reaching the ultimate capacity of the cement matrix is shown in Figure 3, bottom.

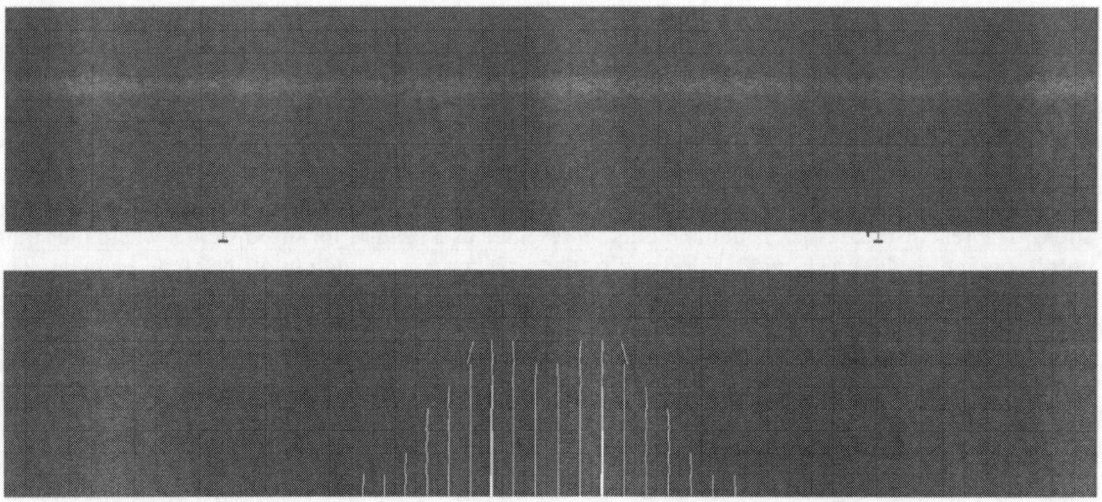

Figure 3. FEM model of beam tested in four-point bending configuration (top) and crack pattern when reaching the ultimate capacity of cement matrix (bottom).

3. Material parameter identification and its software implementation

Mechanical fracture parameter values are determined using the results of fracture tests in suitable configurations. In case of FRC, four-point bending (4PB) tests on unnotched prism specimens is widely used. The outcome of each test is a force–deflection diagram (*F–d* diagram), which is subsequently used for mechanical fracture parameter determination.

An artificial intelligence-based inverse procedure developed by Novák and Lehký [2] transforms fracture test response data into the desired mechanical fracture parameters. This approach is based on matching laboratory measurements with the results gained by reproducing the same test numerically.

The ANN is used here as a surrogate model of an unknown inverse function between input mechanical fracture parameters and corresponding response parameters.

The set for training the ANN is prepared numerically via the utilization of an FEM model (Figure 3) which simulates 4PB test with random realizations of material parameters. These are generated with the help of the stratified sampling method and by performing an inverse transformation of the distribution function in order to reflect the probability distribution of the parameter.

The random responses from the computational model and the corresponding random realizations of parameters serve as input–output elements for the ANN training set. After training, the ANN is ready to solve the main task, which is to provide the best material parameters in order for the numerical simulation to achieve the best agreement with the experiment. This is performed by simulating a network using the previously measured responses as an input. This results in a set of identified material parameters. The last step is result verification – the calculation of the computational model using the identified parameters. Comparison with the experiment will show the extent to which the inverse analysis was successful.

Note that the importance of the training sample preparation has be emphasized and tested by Tong and Liu [4], including LHS scheme. In spite of the fact that these authors concluded that the number-theoretic methods appear as the most efficient, LHS scheme also provided very good results. Moreover, our focus to LHS is also determined by the general applicability of this small-sample simulation technique for practical statistical, sensitivity and reliability analyses in many fields of engineering.

The methodology described above for parameter identification combining non-linear simulations with the training of an artificial neural network is relatively time consuming and of high complexity. Therefore, the whole procedure has been implemented in the software FraMePID-3PB [3], successfully used for material parameters identification of plain concrete. Now, the software is being modified and updated for fibre reinforced concrete including constitutive law as described in Section 2. Screenshot of the first version of the software named FRCID-4PB is shown in Figure 4.

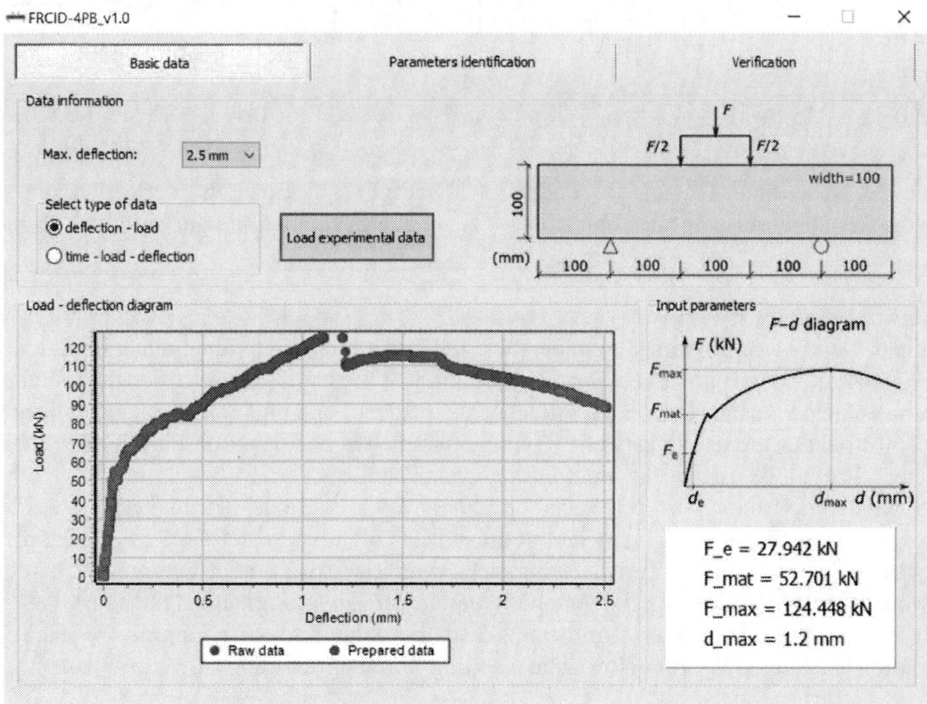

Figure 4. FRCID-4PB software – screenshot.

IOP Conf. Series: Materials Science and Engineering **596** (2019) 012008 doi:10.1088/1757-899X/596/1/012008

4. Results

The approach has been applied for fibre reinforced concrete produced by DURA Technology company, Malaysia. The presented results show outcomes from a pilot study serving for determination of the model feasibility and obtaining a range and sensitivity of the particular model parameters for the purpose of the consequent identification of FRC material parameters using the neural network technology.

4.1. Deterministic simulation of experiment

A set of five reference experimental load-deflection diagrams obtained from Dura Technology Sdn Bhd's laboratories was used to create and set up a four-point bending test computational model (Figure 5). The manufacturer's recommendations have been used to set the default parameters of 3DNonlinerCementitious2User material model. The four specific parameters C_1-C_4 of the tensile softening model were set based on experience and engineering estimation. The numerical simulation with the initial set of parameters results in a *F–d* diagram shown in Figure 6. Note that the range of experimental curves is plotted using grey dashed lines.

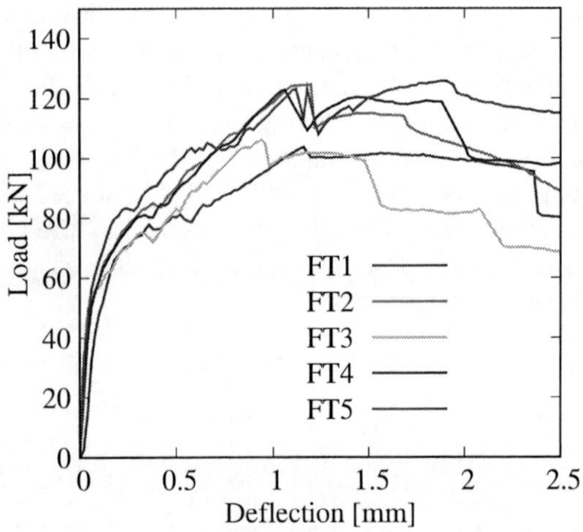

Figure 5. Experimental load vs. deflection diagrams from the four-point bending test.

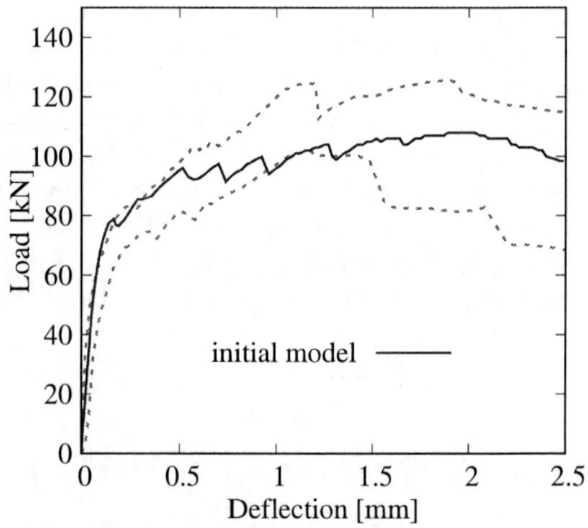

Figure 6. Load vs. deflection diagram from model simulation with initial set of parameters.

4.2. Sensitivity analysis

A parametric study was performed to understand the effect of the individual material parameters on the beam response in the four-point bending test. The aim was to identify the so-called dominant parameters, to determine the parameters that do not influence the response and thus can be excluded from the inverse analysis, to find out the real ranges of individual parameters and to propose a set of suitable response parameters that will be used as the input of the inverse analysis.

The parametric study was performed gradually for six model parameters – tensile strength, compressive strength and four parameters of the tensile softening model C_1-C_4. For each parameter, its value was gradually changed, always in four steps (with the exception of compressive strength, where only two steps were considered), the remaining parameters always remained unchanged at their default values. One of the cases always corresponded to the default set of parameters – marked with the abbreviation "ini" in the figures below. The resulting *F–d* diagrams are shown in Figure 7.

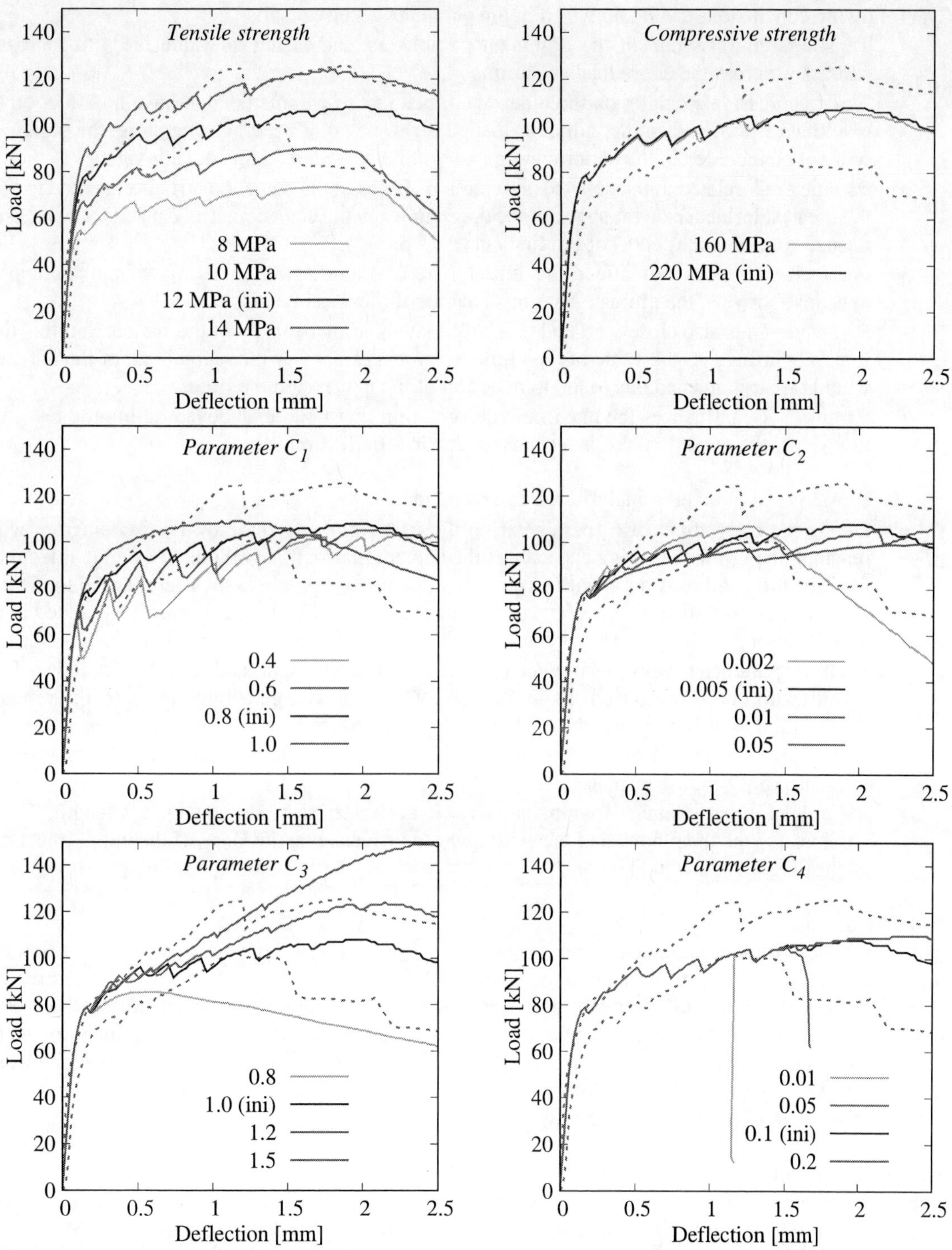

Figure 7. Four-point bending test simulation for various values of tensile strength (top left), compressive strength (top right), parameter C_1 (middle left), parameter C_2 (middle right), parameter C_3 (bottom left), and parameter C_4 (bottom right).

The following conclusions can be drawn from the parametric study results:

- Tensile strength is one of the dominant parameters. Increasing its value leads to increased resistance across the entire load spectrum.
- The compressive strength, on the other hand, belongs to parameters that have no effect on the resulting response when the sample is loaded in a four-point bending. Therefore, the parameter will not be included in the identification and will be considered at its default value.
- Parameter C_1 related to the delayed activation of fibres in the cement matrix is expected to have the greatest influence on the shape of the diagram when the cement matrix capacity is exhausted. Its size controls the "depth" of the first curve drop.
- Parameter C_2 does not affect the initial part of the diagram, but affects the amount of deformation when the ultimate bearing capacity of the specimen is reached.
- Parameter C_3 is also closely related to the ultimate bearing capacity of the sample, but this time it mainly influences the value of the ultimate force and slope of the second part of the diagram, when it is strengthened due to the joint action of the matrix and the fibres.
- Parameter C_4 influences the maximum deformation that will be achieved during the test. At a low value, the numerical test is terminated at a low deflection value.

For the purposes of inverse analysis, we can conclude:

- Due to the clear effect of C_1 parameter on the first visible decrease of the *F–d* diagram when reaching the ultimate bearing capacity of the cement matrix it is possible to exclude it from the identification and consider it with a constant value of $C_1 = 0.8$. In case of future need to identify samples with significantly “deeper” first drop in diagram, C_1 parameter may be included in the identification.
- Another parameter whose value can be considered constant and can be excluded from identification is parameter C_4. Its initial value $C_4 = 0.1$ is large enough to reach the required deformation, which corresponds to the ductility of the material.
- The last parameter that falls out of the identification is the compressive strength. Its effect on the four-point bending response is negligible.
- Identification set contains four parameters of the material model – tensile strength, C_2, C_3 parameters, and the modulus of elasticity, whose influence on the slope of the initial linear part of the diagram are well known.

5. Conclusion

The paper describes methodology and software tool which can be routinely used for indirect determination of the fracture-mechanical parameters of fibre reinforced concrete based on an artificial neural network. User-friendly software has been developed for application based on the three-point bending testing of notched specimens, and a neural network has been trained and predefined for such a configuration for the purpose of future inverse analysis. Material parameters identification technique is based on the combination of the statistical simulation method. Methodology and software tool have been verified using data from DURA technology company, Malaysia.

Acknowledgments

The results presented in this paper were achieved within projects FV30244 "IdeMaS" by MPO TRIO3 and No. LO1408 "AdMaS UP - Advanced Materials, Structures and Technologies" by Ministry of Education, Youth and Sports of the Czech Republic, National Sustainability Programme.

References

[1] Červenka V, Červenka J and Pukl R 2002 ATENA – A tool for engineering analysis of fracture in concrete *Sadhana* **27** (4) 485-492

[2] Novák D and Lehký D 2006 ANN Inverse Analysis Based on Stochastic Small-Sample Training Set Simulation *Engineering Applications of Artificial Intelligence* **19** 731-740

[3] Lehký D, Keršner Z and Novák D 2014 FraMePID-3PB Software for Material Parameters Identification Using Fracture Test and Inverse Analysis *Advances in Engineering Software* **72** 147–154
[4] Tong F and Liu X L 2004 On training sample selection for artificial neural networks using number-theoretic methods *Proc. of the 4th Int. conf. on Enginering Computation Technology* (Civil-Comp Press, Stirling, Scotland) 1-15
[5] Červenka V, Jendele L and Červenka J 2019 ATENA program documentation – Part 1: theory (Červenka Consulting, Prague, Czech Republic)
[6] Novák D, Vořechovský M and Teplý B 2014 FReET: Software for the statistical and reliability analysis of engineering problems and FReET-D: Degradation module *Advances in Engineering Software* **72** 179-192

IOP Conf. Series: Materials Science and Engineering **596** (2019) 012009 doi:10.1088/1757-899X/596/1/012009

Application of the design code for steel fibre reinforced concrete into finite element software

L Dlouhý [1] and S Pouillon [2]

[1] SCIA CZ, Slavíčkova 1A, Brno. Czech Republic
[2] N.V. Bekaert S.A., Bekaertstraat, B-8550 Zwevegem, Belgium

Email: l.dlouhy@scia.net

Abstract. Steel fibre reinforced concrete (SFRC) is more and more used as construction material. Well-known are applications for industrial floors. Nevertheless, this material is also widely used in structural concrete elements where it is especially effective to cover crack widths in combination with standard conventional reinforcement. Currently there is no general part in Eurocode 2 [1] for buildings regarding steel fibre concrete. There are various standards and recommendations for the calculation of the steel fibre contribution. One of the most applied publications is the German guideline DAfStb [2] which was used as a basis for the numerical analysis outlined in this paper. Thus, company SCIA N.V. together with Bekaert N.V. as producer of DRAMIX® fibres developed a special tool for designing and checking SFRC structures. This tool was built in into finite element software SCIA Engineer and based on the German guideline. To gain a fundamental understanding of the structural behaviour, we have also developed non-linear FEM models, which contain a built-in module for SFRC structures. Numerical simulation results are compared to experimental performance of SFRC slab on piles of the Nanhui project in Shanghai in terms of ultimate load and crack pattern.

1. Introduction

An expansion of new materials brings together with it higher demands on their design. This is valid also in case of concrete structures with steel fibers. When steel fibers are added to concrete matrices this leads to significant tension stiffening and higher toughness of the resulting material. The main question remains how to effectively verify such kind of structures. There is no general part in Eurocode 2 [1] for buildings regarding steel fibre concrete. There exist only recommendations like RILEM [3], Model code 2010 [4] from 2014 and mainly the German guideline DAfStb [2] which was used as a basis for the numerical analysis outlined in this paper. This guideline also extends the German national annex DIN EN 1992-1-1/NA. Currently preparation is already ongoing of the new draft of Eurocode where a special chapter about steel fibre reinforced concrete is presented in part L of this code.

The first complex solution on the market is resulting from the cooperation of the top worldwide producer of steel fibers DRAMIX®, Bekaert N.V. and SCIA N.V., a developer of software SCIA Engineer for finite element analysis of structures. The proposed solution is integrated into the program combining (i) a material library for a selection of predefined DRAMIX® steel fibres and (ii) a calculation tool providing design and check features of steel fibre concrete structures based on the German guideline DAfStb [2]. The design approach implemented in SCIA Engineer in relation to the DAfStb from 2012 is described in this paper.

IOP Conf. Series: Materials Science and Engineering **596** (2019) 012009 doi:10.1088/1757-899X/596/1/012009

2. Structural steel fibre concrete structures in SCIA Engineer

SCIA N.V. together with Bekaert N.V. developed a unique solution for designing and checking of steel fibre reinforced concrete slabs. This is a one of the first known implementation of such a feature into finite element software (in this case SCIA Engineer). Bekaert N.V. provided their long-term knowledge in development and production of DRAMIX® steel fibres. A special library for steel fibre reinforced concrete is built-in for different concrete classes and dosages based on the selected DRAMIX fibre type. There are 3 basic types of DRAMIX® steel fibres included: 5D65/60BG; 5D65/60GG and 4D65/60BG [5].

The verification of the structure is performed based on ultimate limit state (ULS) checks:

- check capacity (normal forces + bending moment),
- shear check,

and serviceability limit state (SLS) checks:

- crack width check,
- stress limitation.

Moreover, it is possible to verify local effects of column-punching on the slab. All the checks can be run on a fixed dosage or an optimized dosage based on the ultimate limit state. An equilibrium system for this state can be achieved if at least one of the following conditions are met (i) analysis with redistribution of moments within statically indeterminate systems, (ii) combination with reinforcement or (iii) normal compression due to external actions.

2.1. Material characteristics

2.1.1. Characteristic tensile strength on material level. The calculation of material characteristics is dependent on the fibre type and dosage and based on DAfStb. The mean residual strength values $f_{R1,m}$ and $f_{R4,m}$ are the basis for the calculation of the material behaviour (σ–ε). They are determined by means of beam tests and are mainly influenced by fibre type, dosage and concrete performance. The characteristic residual flexural strength values (($f^{f}_{cflk,L1}$; $f^{f}_{cflk,L2}$) are derived from the mean values. As these mean values are based on a high amount of tests, it can be assumed that the threshold ($0.51f_R$) is reached in all cases – see formula O.3 from the DAfStb Guideline [2]. The calculation of the characteristic residual tensile strength values ($f^{f}_{ct0,L1}$; $f^{f}_{ct0,L2}$; $f^{f}_{ct0,u}$; $f^{f}_{ct0,s}$) is the next step of the process. In principle, the characteristic residual flexural strength values are multiplied with a beta factor. Determination of beta factors (βL1 and βL2) is done based on figures P.1 and P.2 from DAfStb [2].

$$f^{f}_{ct0,L1} = f^{f}_{cflk,L1} \cdot \beta_{L1} \tag{1}$$

$$f^{f}_{ct0,L2} = f^{f}_{cflk,L2} \cdot \beta_{L2} \tag{2}$$

$$f^{f}_{ct0,u} = f^{f}_{cflk,L2} \cdot \beta_{u} \tag{3}$$

$$f^{f}_{ct0,s} = f^{f}_{cflk,L2} \cdot \beta_{s} \; for \; L2/L1 \leq 1{,}0 \tag{4}$$

$$f^{f}_{ct0,s} = f^{f}_{cflk,L1} \cdot \beta_{s} \; for \; L2/L1 > 1{,}0 \tag{5}$$

In SCIA Engineer, these characteristic residual tensile strength values are listed in the material library in function of the fibre type, dosage and concrete class.

2.1.2. Characteristic tensile strength on structural element level. The second step in the calculation of the material characteristics is the calculation of residual values from characteristic ones by considering the member size and the fibre orientation effect in the calculation. By default, considering a large, horizontally casted slab, following values are taken into account:

- κ^{f}_{G} =1,7; member size effect; with one factor per material and per slab

- κ_F^f fibre orientation effect (1,0 for bending and tensile loads; 0,5 for shear)

$$f_{ctR,L1}^f = \kappa_F^f \cdot \kappa_G^f \cdot f_{ct0,L1}^f \quad (6)$$

$$f_{ctR,L2}^f = \kappa_F^f \cdot \kappa_G^f \cdot f_{ct0,L2}^f \quad (7)$$

$$f_{ctR,u}^f = \kappa_F^f \cdot \kappa_G^f \cdot f_{ct0,u}^f \quad (8)$$

$$f_{ctR,s}^f = \kappa_F^f \cdot \kappa_G^f \cdot f_{ct0,s}^f \quad (9)$$

2.1.3. Design values

The last step in the calculation of material characteristic concerns the calculation of design values from characteristic ones by applying the reduction factor (α^f_c=0,85) for long-term effects and the partial safety factor (γ^f_{ct}=1,25) according to Table R.1 from [2] on the residual tensile strength of the steel fibre reinforced concrete.

$$f_{ctd,L1}^f = \alpha_c^f \cdot f_{ctR,L1}^f / \gamma_{ct}^f \quad (10)$$

$$f_{ctd,L2}^f = \alpha_c^f \cdot f_{ctR,L2}^f / \gamma_{ct}^f \quad (11)$$

$$f_{ctd,u}^f = \alpha_c^f \cdot f_{ctR,u}^f / \gamma_{ct}^f \quad (12)$$

$$f_{ctd,s}^f = \alpha_c^f \cdot f_{ctR,s}^f / \gamma_{ct}^f \quad (13)$$

2.1.4. Material diagrams. The material diagrams are different than those of standard concrete material. The effect of fibres is considered by the tensile branch. Here only the behaviour in tension is described. There are two types of material diagrams used in the analysis, dependent on the analysis type, see Figure 1.

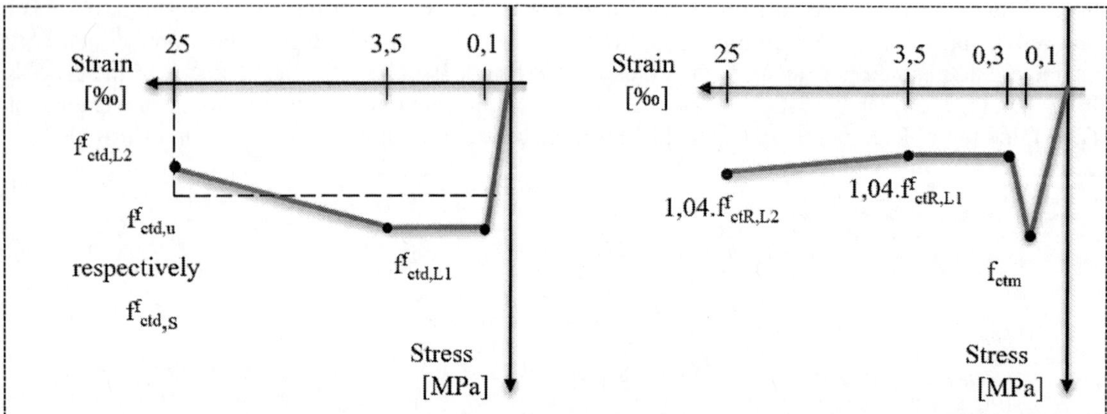

Figure 1. Stress-strain diagram in tension for a) linear analysis and cross-section verification in ULS b) non-linear analysis.

2.2. Dosage design and check capacity in ultimate limit state (ULS)

The determination of the optimal dosage design is based on an iterative calculation of the ULS capacity loaded by a combination of normal forces and bending moment. The procedure searches the optimal fibre dosage in each finite element of the structure to reach maximal exploitation 100% and obtain equilibrium. The effect of conventional reinforcement is also included in the calculation in case this reinforcement is defined. The equilibrium of the cross-section is demonstrated on Figure 2. This figure clearly indicates the main advantage in the usage of steel fibre concrete in tension due to the existence of the steel fibres. The basic assumptions are also valid for steel fibre analysis:

- Steel fibre reinforced concrete is active in compression and in tension
- Plain section remains plain after load application
- Bending moment and normal force equilibrium on cross-section
- Parabolic rectangular stress-strain diagram in compression is used, maximal strain in compression is dependent on concrete class
- Stress-strain diagram in tension is visible from figure and maximal strain in tension is 25‰

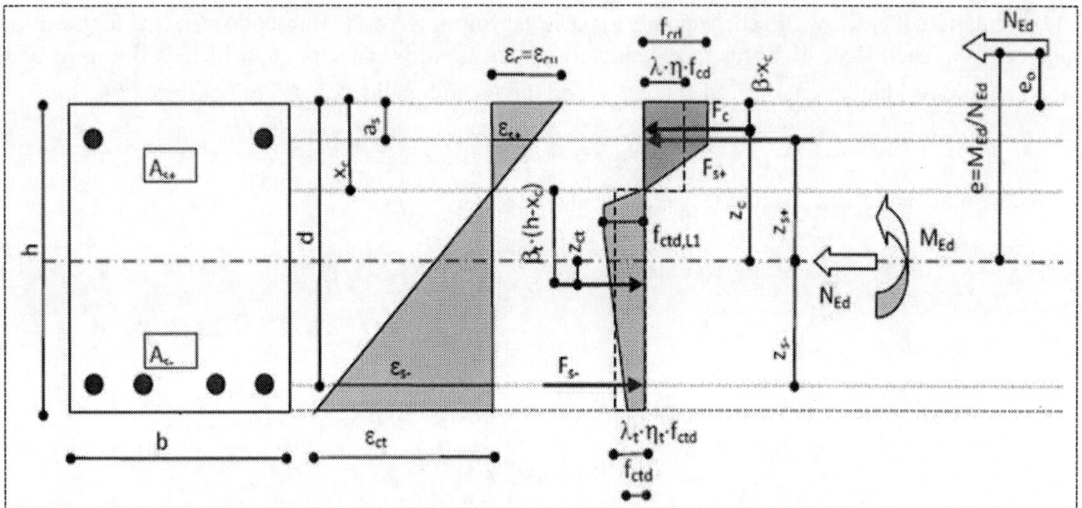

Figure 2. Stress-strain distribution in the cross-section.

The equations for static equilibrium of normal forces and bending moments are indicated in the following formulas.

$$F_c + F_{ct} + F_{s+} + F_{s-} + N_{Ed} = 0 \quad (14)$$

$$F_c z_c + F_{ct} . z_{ct} + F_{s+} z_{s+} + F_{s-} . z_{s-} + M_{Ed} = 0 \quad (15)$$

2.3. Shear capacity and punching

The effect of steel fibres ($v_{Rd,cf}$) is considered in the calculation of the total shear resistance ($v^f_{Rd,c}$) in addition to the regular shear capacity of concrete ($v_{Rd,c}$). The steel fibres behave as spread reinforcement and participate on the transfer of the shear loading. This can lead to significant savings on the traditional shear reinforcement, or even neglect it completely.

$$v^f_{Rd,c} = v_{Rd,c} + v_{Rd,cf} \quad (16)$$

The shear concrete capacity $v_{Rd,c}$ is calculated based on chapter 6.2 from the code [1]. The incrementation of the capacity caused by the steel fibres is expressed according to 6.2.2 [2] and it is dependent on the material characteristic of steel fibre concrete as follows.

$$v_{Rd,cf} = \alpha^f_c \, f^f_{ctR,u} \, b_w \, \frac{h}{\gamma^f_{ct}} \quad (17)$$

Where $f^f_{ctR,u}$ – characteristic residual concrete strength in ULS [MPa].

Additionally, the final value of the shear capacity of steel fibre concrete is limited by 140% of the shear capacity of traditional concrete.

$$v_{Rd,c}^{f} = min(v_{Rd,c} + v_{Rd,cf};\ 1{,}4\ v_{Rd,c}) \tag{18}$$

The steel fibre contribution is considered in the same way in case of a punching verification. Nevertheless, when the required area of conventional reinforcement for punching is designed then based on Re 6.4.5 [2] the effect of fibres cannot be taken not account and only concrete without fibers is analyzed.

2.4. Crack width and stress limitation

The calculation of crack width corresponds to basic assumptions of the Eurocode standard for traditional concrete [1]. The effect of fibres is considered by an additional factor α_f, which is the ratio of the characteristic residual strength in SLS ($f_{ctR,s}^{f}$) and the mean tensile strength of concrete (f_{ctm}).

$$w_k = s_{r,max}\left(\varepsilon_{sm}^{f} - \varepsilon_{cm}\right) \tag{19}$$

Where maximal crack spacing ($s_{r,max}$) is calculated as

$$s_{r,max} = \frac{(1-\alpha_f)\phi_s}{(3{,}6\ \rho_{p,eff})} \leq \frac{(1-\alpha_f)(\sigma_s\ \phi_s)}{3{,}6 f_{ct,eff}} \tag{20}$$

$$\alpha_f = \frac{f_{ctR,s}^{f}}{f_{ctm}} \tag{21}$$

And the difference of mean strain value of concrete and reinforcement $\left(\varepsilon_{sm}^{f} - \varepsilon_{cm}\right)$

$$\left(\varepsilon_{sm}^{f} - \varepsilon_{cm}\right) = \frac{(1-\alpha_f)\left(\sigma_s - 0{,}4\frac{f_{ct,eff}}{\rho_{p,eff}}\right)}{E_s} \geq 0{,}6\ \frac{(1-\alpha_f)\sigma_s}{E_s} \tag{22}$$

where:

- ε_{sm}^{f} – mean strain in reinforcement
- ε_{cm} – mean strain in concrete between cracks
- α_f – the ratio of characteristic residual strength in SLS ($f_{ctR,s}^{f}$) and mean tensile strength of concrete (f_{ctm}).
- ϕ_s – diameter of reinforcement
- $\rho_{p,eff}$ – effective ratio of reinforcement
- $f_{ct,eff}$ – effective tensile concrete strength
- E_s – modulus of elasticity of reinforcement
- σ_s – stress in the reinforcement

In case of steel fibre concrete without conventional reinforcement it is necessary to apply a different approach where the crack width is calculated based on the following formula according to [2].

$$w_k = s_w^{f}\ \varepsilon_{ct}^{f} \tag{23}$$

where:

- s_w^{f} – 140 mm
- ε_{ct}^{f} – strain in steel fibre concrete

When the crack width is presented in the cross-section then the check of stress limitation usually doesn't satisfy. The basic assumptions are the same as for the crack width calculation. The following list of stresses in particular components are checked in this verification:

- Stress in concrete from characteristic combination based on 7.2 [1]
- Stress in concrete from quasi-permanent combination based on 7.2.(3) [1]
- Stress in conventional reinforcement from characteristic combination based on 7.2.(5) [1]

The stress in conventional reinforcement is calculated without fibres being present which leads to fictitious values. As such this value can be higher than the yield strength. The effective steel stress, taking into account the presence of the steel fibres, should be calculated and compared to the reinforcement strength limit. The effective steel strength is calculated as follows:

$$\sigma_{s,eff} = \sigma_s \cdot (1 - \alpha_f) \tag{24}$$

3. Physical non-linear analysis

As steel reinforced fibre concrete is being used more often, the real behaviour and investigation of the structure is essential. Especially for steel fibre reinforced concrete this non-linear analysis can provide the real behaviour of the structure with different results compared to the linear analysis. This behaviour is heavily influenced by the crack formation which is typical for fibre concrete. The toughness of fibre concrete is significantly increased by the existence of steel fibres up to values very close to standard reinforcement.

Physical non-linear analysis represents a very powerful tool for analysing any kind of structure in civil engineering created not only from steel fibre reinforced concrete but also from other materials. Generally, there could be significant differences of results compared to a linear analysis, especially in case of hyper-static structures. In case of linear analysis, only the E-modulus of the material is considered for the preparation of the stiffness matrix. There is no stress distribution based on increasing the strain in the component. On the other side the non-linear analysis provides a stress distribution and increase of the bearing capacity after reaching the ultimate strain until the collapse mechanism. This mechanism is based on different values of stress and strain, dependent on a predefined non-linear stress-strain relationship in the material diagram as defined on Figure 1.

The damage Mazarz material model is used for the non-linear calculation. This model efficiently enables the considered stress-strain diagrams with descending branches which is typical for steel fibre concrete in tension. The practical application of the non-linear analysis is demonstrated on a real test project of a foundation slab on piles in Shanghai [6].

3.1. Foundation slab on piles (Nanhui project in Shanghai)

In order to investigate the real behaviour of SFRC slabs on piles, a campaign of tests has been conducted at the Tongji University of Shanghai, China. The load test was performed on SFRC slabs from C40/50 and dimensions 4,3m x 4,3m with thickness 187,5mm supported by rectangular supports representing the piles, see Figure 3. There were used slabs with DRAMIX 5D fibres with 30kg/m^3 and 35kg/m^3. The load was imposed gradually by a hydraulic jack from the bottom to the top. At first the material characteristics were established by laboratory tests on 6 specimens based on EN14651 where especially their residual strength L1 and L2 were determined.

The presented non-linear model provides a good match with the experimental test. The comparison of the results with the load-displacement diagram coming from the test is visible on Figure 4 for Dramix fibres 5D65/60BG with dosage 30kg/m^3 and 35kg/m^3. This capacity is at the load level 283,9kN (299,0kN for 35kg/m3 respectively) where the ultimate tensile strain in steel fibre reinforced concrete (25‰) is reached.

The tight peak in the tension part of the diagram works as a "crack localizator". This is perfectly visible in Figure 5 where the load is continuously increased. The red colour represents the stress close to the tensile strength and continuous cracking of the slab from the centre of the slab to the edges. The extending middle part to the edges uses just the residual tensile strength of the steel fibre reinforced

IOP Conf. Series: Materials Science and Engineering **596** (2019) 012009 doi:10.1088/1757-899X/596/1/012009

concrete after cracking. Additionally, these stresses show the places with crack formation in the middle of the slab and their spreading of cracks during the increasing of the load.

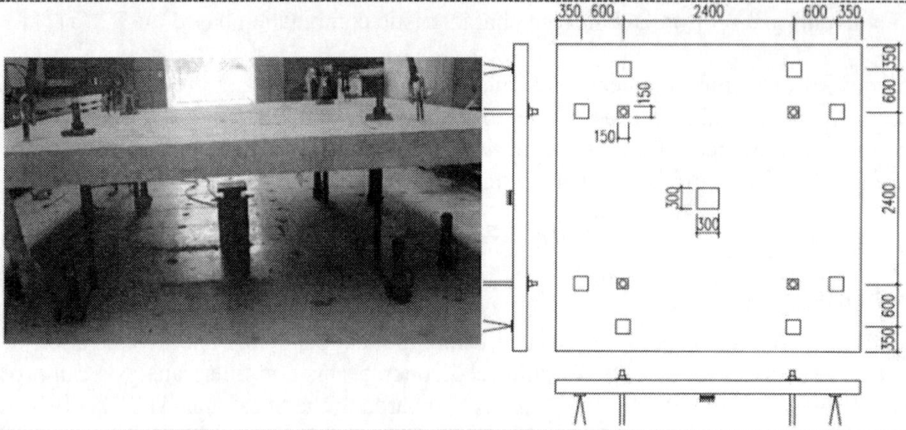

Figure 3. Preparation of testing device a) and b) dimensions of the slab.

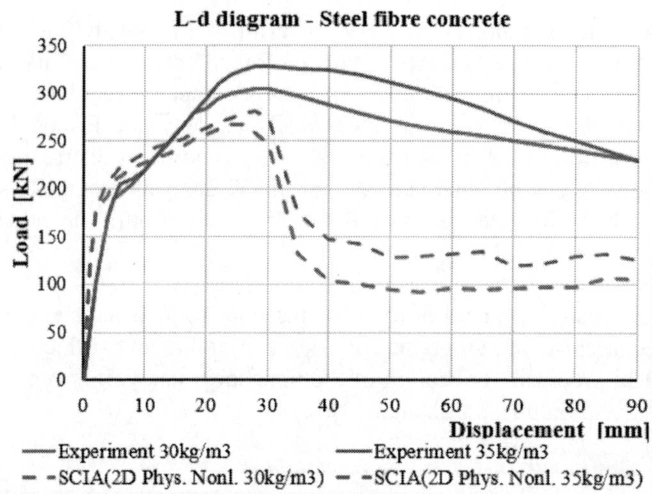

Figure 4. L-d diagrams for dosage 30 a 35kg/m^3 respectively.

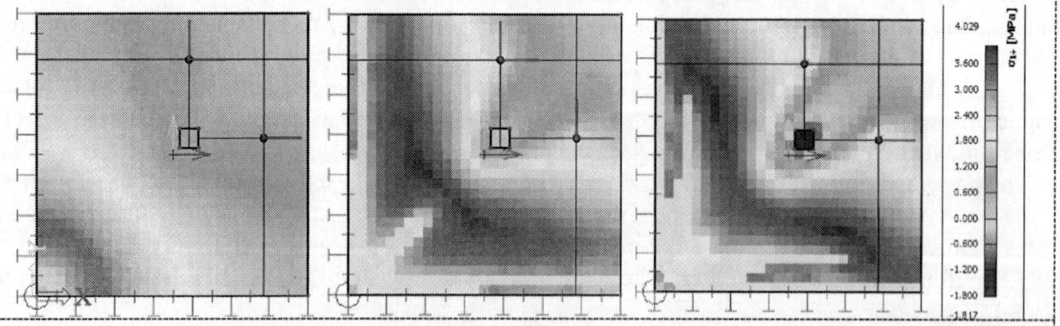

Figure 5. Evaluation of principal stresses in upper surface of 30kg/m^3 specimen for load level 1mm; 12mm and 28 mm respectively.

Figure 6 represents a graph of the dependency between the load level and the stress-strain for the most critical point in the middle of the slab and upper fibres. When looking at the Load-Stress distribution, it can be deducted that the behaviour of the structure is linear till load level 55kN. The structure is approaching the top of the stress strain diagram peak. A strain very close to 0.1‰ is reached. After that there is a phase of crack formation and the stress is coming down significantly based on the stress-strain diagram. The residual strength in tension (1,09MPa) is reached nearly at load level 100kN corresponding to a strain of 0,3‰. A load higher than 100kN means the spreading of the cracks with smooth hardening of the material done by the increasing branch of the stress-strain diagram in tension till maximal value of 1,13MPa and the plastic load is reached by the load of 283,9kN where the strain reaches the limit of 25‰.

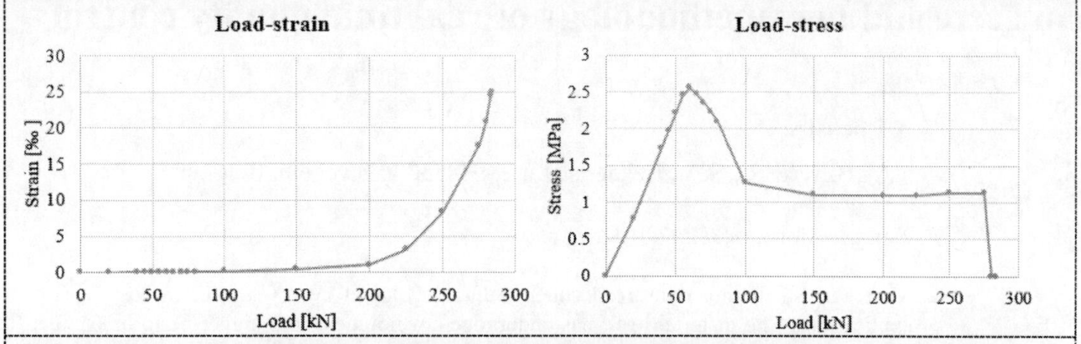

Figure 6. a) Load - stress b) load strain distribution for centre slab point for dosage 30kg/m^3.

Table 1. The comparison of the ultimate load capacity.

	Ultimate load capacity F (kN)	
	DRAMIX 5D 65/60BG 30kg/m^3	DRAMIX 5D 65/60BG 35kg/m^3
Experiment	302,7	330,8
SCIA	283,9	299,0

4. Conclusions

SCIA N.V. together with Bekaert N.V. developed a unique solution for designing and checking of steel fibre reinforced concrete slabs based on the latest standard DAfStb [2]. This tool brings new possibilities for a very fast and easy design and check of steel fibre reinforced concrete.

In addition, a non-linear analysis method has been developed for simulating the real behaviour of the structure based on the non-linear stress-strain diagram with respect to cracks. This analysis method was proven to provide a very good comparison with real-life test results. A comparison was done for the deflection of two slabs made from C40/50 with fibres DRAMIX 5D 65/60BG and dosage 30kg/m^3 and 35kg/m^3 respectively. The comparison of the maximal loading capacity mentioned in Table 1. clearly indicates the higher maximal load is achieved by the test. This indicates the non-linear FE calculation provided in SCIA Engineer is on the safe side.

References

[1] EN1992-1-1, Eurocode 2: Design of concrete structures - Part 1-1: General rules and rules for buildings
[2] DAfStb-Richtlinie, Stahlfaserbeton, Ausgabe November 2012
[3] RILEM TC 162 TDF, Design of Steel fibre reinforced concrete – Method, recommendations, Material and Structures, March 2001
[4] ModelCode2010, The fib Model Code for Concrete Structures 2010
[5] https://www.bekaert.com/en/product-catalog/construction/dramix/all-round-concrete-reinforcement-solution"
[6] Dlouhý L and Roland R 2018 *Proc. Fibre Reinforced Concrete: from Design to Structural Applications Joint ACI-fib-RILEM International Workshop* (Desenzeano de Garda Italy) p 74

IOP Conf. Series: Materials Science and Engineering **596** (2019) 012010 doi:10.1088/1757-899X/596/1/012010

Advanced design methodologies for steel fibre reinforced concrete and new methodology of real time quality control

A Hoekstra

N.V. BEKAERT S.A., Bekaertstraat 2, B-8550 Zwevegem, Belgium.

E-mail: Anne.hoekstra@bekaert.com

Abstract. Steel fibre reinforced concrete has developed over years for the use in structural applications. The material itself has undergone development into higher performing steel fibre types and thus into higher performing steel fibre concretes. Whilst in regards to standards the development went from recommendations into building codes the way of designing – linear elastic design methods or yield line theory – and the way of quality control seemed to remain unchanged from the past. This paper shall emphasize the recent development in a design tool that enables a non-linear design methodology and a new advanced way of real time quality control.

1. Introduction

Steel fibre reinforced concrete (SFRC) is a composite material that provides post crack strength to concrete. With the publication of Model Code 2010 and various national standards, advanced guidelines for the design of SFRC for structural applications are available. During the last decade steel fibres are increasingly being used in structural applications, e.g. rafts, segmental tunnel linings, pre-stressed beams or even elevated slabs. Depending on the actual application, steel fibres may act as the sole reinforcement or may be combined with traditional concrete reinforcement. Typically the plasticity theory is used to work out the capacity of steel fibre reinforced structures. Meanwhile non-linear design methodologies are integrated into software that allows for a very accurate calculation of the material contribution. Due to more optimized design calculations and the increasing use of SFRC in structural applications, quality control has gained additional importance. Beam testing is one possibility for performance verification but is not always preferred due to effort and variation of results. Therefore alternative quality control methods have been developed, based on the combination of initial beam testing and focus on process control. An ideal solution to measure fibre type, amount and distribution over a full truck load with a measuring device will be one main focus of this paper.

2. Material development

Steel fibres increase the ductility of the concrete by bridging cracks and transferring tensile stresses over these cracked zones. Knowing that national codes are under development and the intention to integrate SFRC as structural reinforcing material into Eurocode 2 were trigger to increase physical performance of steel fibres. Many different parameters influence the performance of the steel fibre concrete: the tensile strength of the steel wire, fibre shape, ductility of the wire, the dosage of steel fibres and the concrete quality. The newly developed Dramix® steel fibres, the families 4D and 5D, are especially designed to reinforce structural concrete elements.

IOP Conf. Series: Materials Science and Engineering **596** (2019) 012010 doi:10.1088/1757-899X/596/1/012010

The main difference between the 4D/5D family and the 3D family is the significantly increased performance in serviceability limit state (4D) and ultimate limit state (5D). The combination of type of anchorage (more end hooks), stronger wires (up to 2200 MPa) and more ductile wires (up to 7% of elongation) are key in this increased performance. The tensile strength of a steel fibre has to increase in parallel with the strength of its anchorage. Only in this way can the fibre resist the forces acting upon it.

This development of steel fibres has been applied some years back and high performing fibres are used for structural applications.

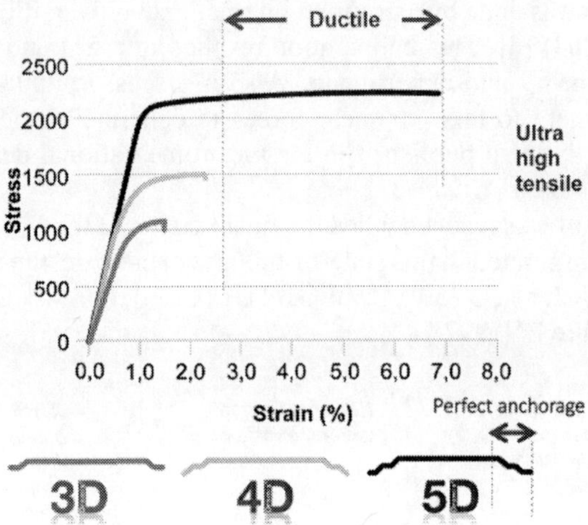

Figure 1. Yield strength and strain capacity of the wire used for the different steel fibre types.

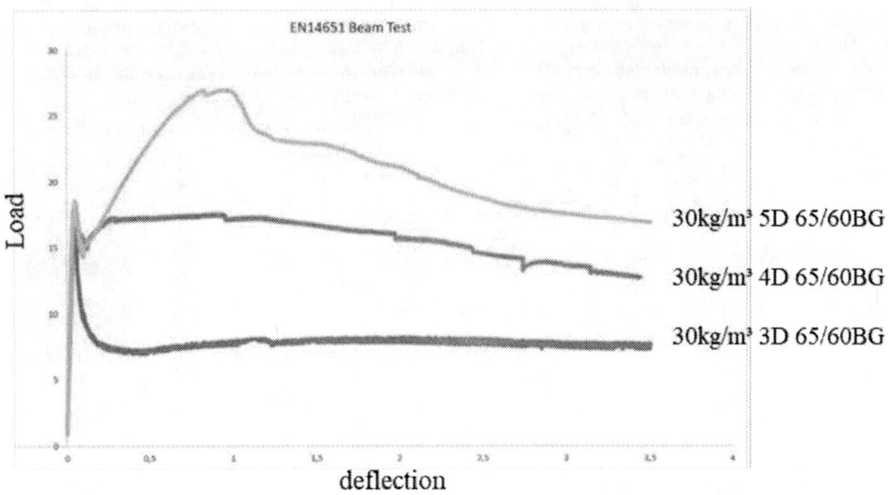

Figure 2. Beam test curves with each 30kg/m³ dosage rate. 3D and 4D steel fibre type with typical strain softening behaviour, 5D steel fibre type with bending hardening behaviour.

3. Development of codes and standards

Steel fibre reinforced concrete has been subject of much research over the last 40 years. In line with the knowledge built up, guidelines and standards developed gradually.

IOP Conf. Series: Materials Science and Engineering **596** (2019) 012010 doi:10.1088/1757-899X/596/1/012010

In the early 1980's, the SFRC test standards – JSCE-SF4 and ASTM C1018 – were in place and widely used. The performance was described as "flexural toughness" and represented the energy dissipation. Actually, the EN14651 is the test standard mostly used in Europe, but also widely accepted outside Europe. The performance of the SFRC expressed directly in flexural post crack stresses at specific crack mouth openings.

The first design guideline was published in 1995: "Dramix Design Guideline". This document contained the first basic design rules in order to design concrete structures with steel fibres. This document was a result of a cooperation between N.V. Bekaert S.A., the Belgian universities KULeuven and UGent and the Belgian building research institute WTCB.

A deeper investigation was done by a group of international universities brought together by RILEM (RILEM TC162-TDF, 2003) [5]. This cooperation resulted in a test and a design method worked out and fully backed-up by theory and experiments. Also overseas, in 2008, ACI 318 incorporated steel fibres as a structural element to take up shear forces in concrete elements. Different countries (e.g. Germany, Sweden and Italy) also published in the meantime national design guidelines and/or codes, often as an add-on to EN1992-1-1.

A new milestone in 2014 was reached when the final version of the Model Code 2010 was published. Steel fibres were fully incorporated in this state-of-the-art concrete design guideline. This document was also the trigger for the launch of TC 250/SC2/WG1/TG2 (CEN); this task group works on the integration of steel fibres into the future EN1992-1-1.

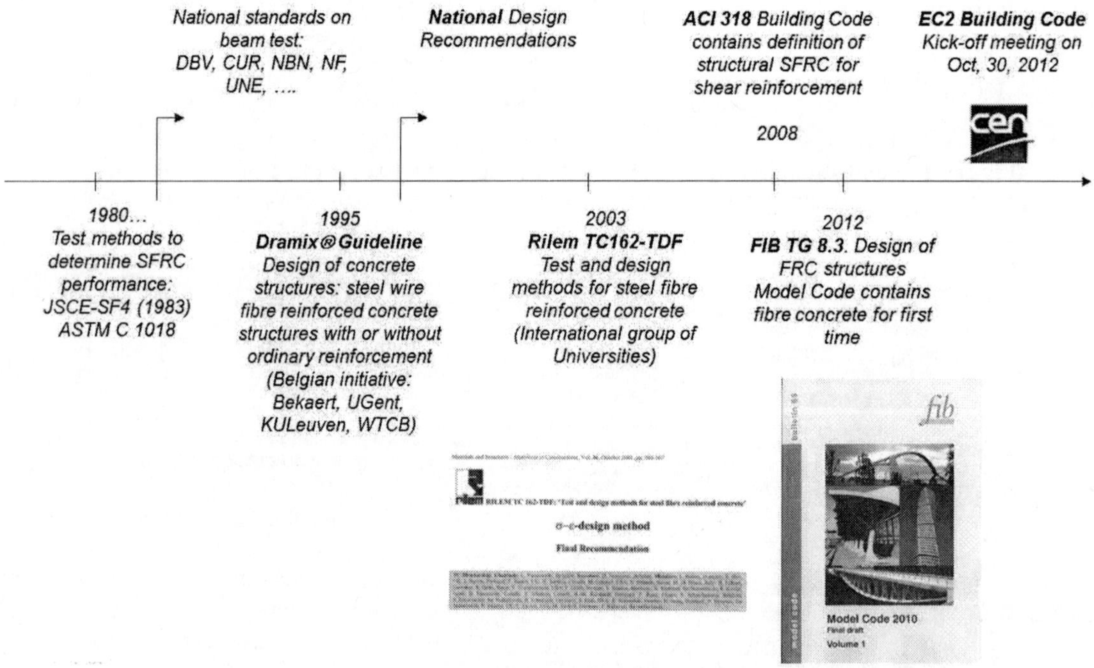

Figure 3. Evolution of SFRC test and design standards from 1980 to present.

4. Design approach and software development

In the available standards, the constitutive laws of SFRC are available as basis for the design. The results of the initial type testing are used for deriving the section capacity according to the sigma-epsilon diagram. Based on this, all section capacities are worked out. Until then, software tools to make use of the section capacity under bending/shear/punching and for a crack width calculation in a design were still in backlog. Meanwhile some relevant development has been undergone to provide design tools.

IOP Conf. Series: Materials Science and Engineering **596** (2019) 012010 doi:10.1088/1757-899X/596/1/012010

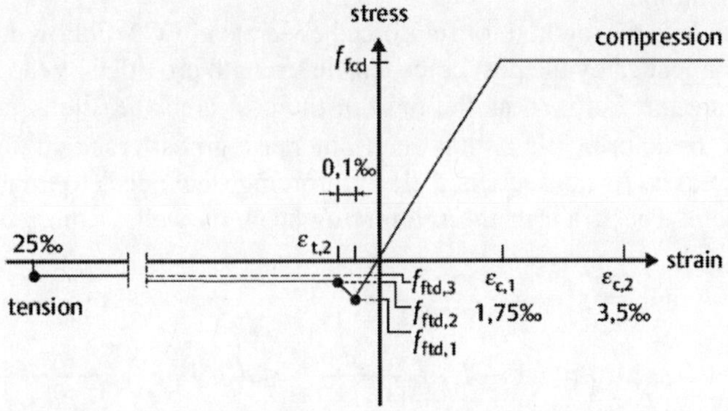

Figure 4. Stress-strain diagram of steel fibre reinforced concrete (design values for strength).

4.1 Bending resistance

The bending moment capacity of steel fibre-reinforced section can be calculated for a cross-section with or without axial force based on the following assumptions

- Flat cross-sections remain flat.
- The strain distribution is aligned with the strain distribution of reinforced concrete.

In order to calculate the cross-section bending moment capacity, the static equilibrium needs to be determined. Figure 4 schematically represents the relation between stresses and strains.

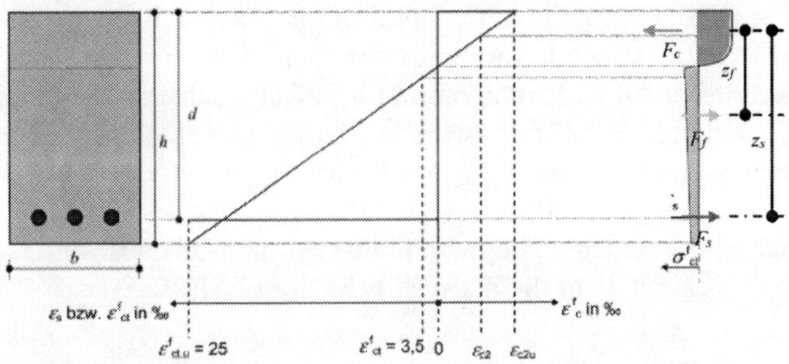

Figure 5*:* Static equilibrium of the cross section under bending.

With the calculation tool: Dramix® Pro Moment Capacity, the bending moment capacity of SFRC only or combined reinforcement can be calculated.

4.2 Shear and punching resistance

The effect of the steel fibres in shear and punching resistance is taken into account by an additional component in the equations. Steel fibres act like a shear reinforcement over the entire cross section of the structure. The shear capacity of the element increases as a function of the performance of the steel fibre concrete used. This can lead to a significant reduction or elimination of conventional shear reinforcement. [3] takes the increased shear resistance due to steel fibre concrete into account by introducing an additional element $V_{Rd,cf}$ into the equations for conventional shear design.

$$V_{Rd,c}^{f} = V_{Rd,c} + V_{Rd,cf} \text{ (without conventional shear reinforcement)} \quad (1)$$

$$V_{Rd,s}^{f} = V_{Rd,s} + V_{Rd,cf} \leq V_{Rd,max} \text{ (with conventional shear reinforcement)} \quad (2)$$

Dramix® Pro Shear Capacity is able to calculate the capacity of the section for shear or punching shear.

4.3 Crack width calculation

The design corresponds to the method for reinforced concrete in EC2. Following the DAfStb-Richtlinie, the rules of EC2 are amended by the post crack tensile strength provided by the steel fibre concrete. This is done by introducing the factor α_f as the ratio of the post crack tensile strength over the first crack tensile strength. The basic principle is that due to increasing post crack strength the released force at crack formation decreases. As consequence, the reinforcing steel needs to transfer only a reduced force back into the concrete. The strain in the reinforcing steel, as well as the required transfer length, is reduced.

$$w_k = s_{r,max} \cdot \left(\varepsilon_{sm}^f - \varepsilon_{cm}\right) \quad (3)$$

$$s_{r,max} = \left(1 - \alpha_f\right) \cdot \frac{d_s}{3.6 \cdot \rho_{p,eff}} \leq \left(1 - \alpha_f\right) \cdot \frac{\sigma_s \cdot d_s}{3.6 \cdot f_{ct,eff}} \quad (4)$$

$$\varepsilon_{sm}^f - \varepsilon_{cm} = \frac{(1-\alpha_f) \cdot \left(\sigma_s - 0.4 \cdot \frac{f_{ct,eff}}{\rho_{p,eff}}\right)}{E_s} \geq 0.6 \cdot \left(1 - \alpha_f\right) \cdot \frac{\sigma_s}{E_s} \quad (5)$$

The use of steel fibres can significantly decrease the required amount of reinforcing steel. By defining the normalized ratio of α_f based on the 28-day strengths and by subsequently multiplying it with$f_{ct,eff}$, the fibre effect is aligned to the concrete age at the time considered for the design. The tool Dramix® Pro Combi Slab calculates the mean value of the crack width under the chosen condition.

4.4 Software development

Software tools based on Excel programs for applications like slab on ground, slab on piles, raft foundations are available and meanwhile further developed.

A new milestone in terms of software integration for SFRC is achieved in cooperation between SCIA and Bekaert. SFRC is integrated in a Finite Element Method tool that enables the design for SFRC or combined reinforcement in both methods, linear-elastic and non-linear. This development gives the engineer the opportunity to design quickly with steel fibre concrete or with a combination of steel fibre concrete and traditional reinforcement. The methodology of non-linear design is regulated in EC2 and applies also to SFRC as it is suitable to the material behavior of SFRC.

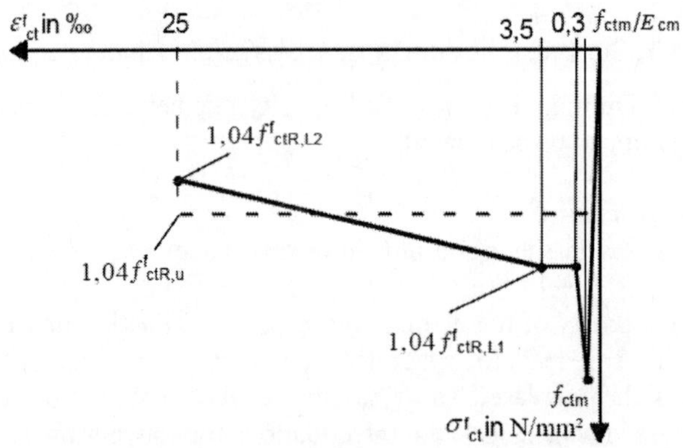

Figure 6. Stress-strain diagram for non-linear design methodology proposed in [3].

Non-linear design method is a very complex and detailed design methodology that engages the entire concrete member behaviour considering the stiffness change in each deformation state and for each crack opening in every face of the concrete member. That enables a simulation of real material behaviour at different crack phases. Those design methods need intensive verifying in order to ensure the rightness of results for such a design methodology. The following picture shows the comparison between a large scale test (real case), an elevated slab that was accordingly loaded and stresses and cracks were monitored and evaluated and between a result of a design using the non-linear methodology of that software solution. The result difference with approx. only 2.38% proofs the extremely high conformity between full scale test and design outcome and thus the optimisation and the projection of the actual material behaviour.

$$\delta = \frac{22820 - 22290}{22290} = +2{,}38\%$$

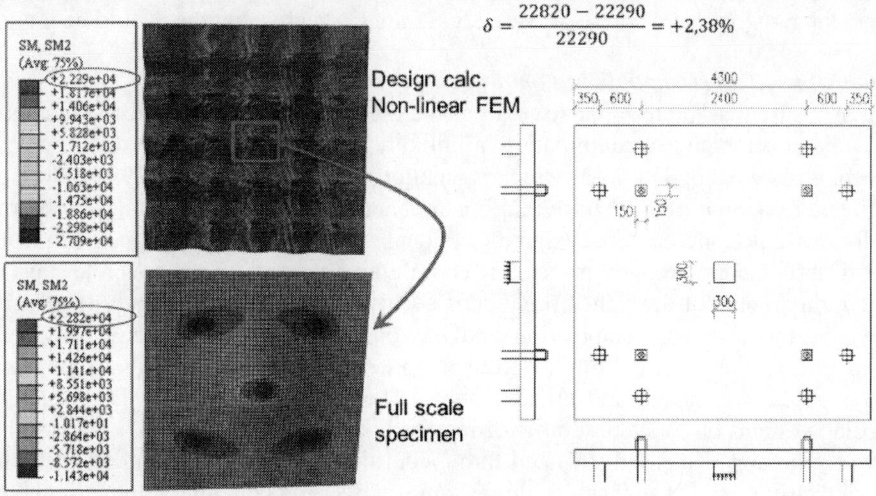

Figure 7. Comparison between full scale test and SCIA non-linear design methodology for SFRC.

5. Quality control system for SFRC

5.1 Introduction

Two main approaches for quality control (QC) of steel fibre concrete can be distinguished:

- Direct verification of the residual strength by testing
- Indirect verification of the residual strength by process control

5.2 Verification of residual strength based on beam testing

One way to check the residual strength of the delivered steel fibre concrete is beam testing. The defined number of specimens is cast on site at a defined sampling frequency. The residual strength is determined in the laboratory later on, typically at the age of 28 days. The test results can then be compared to the specified limits and tolerances.

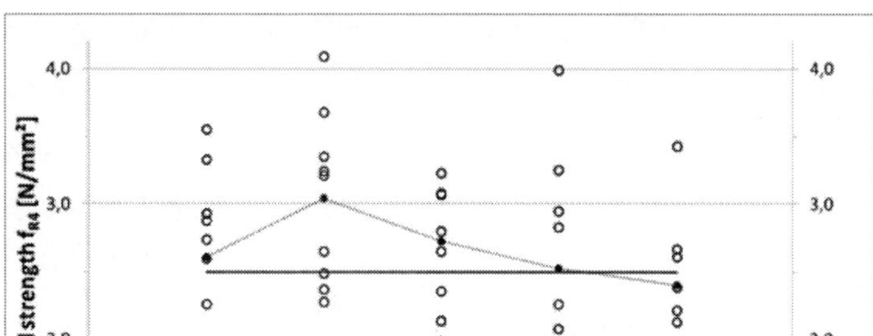

Figure 8. Results from in situ sampling.

In this example for each sampling series 9 beams were randomly taken over the volume of the truck mixer. The residual flexural strength was determined in accordance with EN 14651.

When comparing individual f_{R4} values (circle) to a specified minimum values (dashed line), all samples pass the specific requirement of (in this case) 1.5 N/mm. Alternatively the average values of each series (full dot) could be compared to a specified average value (full line). In that case, series 5 would not pass the requirement of (in this case) 2.5 N/mm. Corrective actions would be required.

5.3 Verification of residual strength based on process control

Another way to ensure that the residual strength of the delivered steel fibre concrete is conforming to the specified residual strength is a combination of initial type testing (ITT) and process control.

This concept is already applied in [3] which is a national amendment to EN 1992-1-1 on code level for the design and execution of steel fibre reinforced concrete structures. The process control concept accepts that for the same concrete mix design the residual strength will not change, given that the same constituents are used and are properly mixed, the compressive strength of the concrete stays within the limits and the right amount of the right fibre type is evenly distributed throughout the mix. From that moment onwards, the concrete composition and its constituents are not changed anymore and full focus is on controlling the process from constituents to delivery. The correct fibre type can be easily identified by the delivery papers and quality labels. The correct dosage can be ensured by using automatic dosing systems or counted number of required packaging units. Finally it has to be verified that the fibres are homogeneously distributed throughout the concrete. Wash-out tests can be used to verify the fibre distribution. [3] suggests taking 3 samples over the concrete volume of about 10 litres each. Individual samples should not contain less than 80% of the specified dosage. The average should not be less than 85% of the specified values. In Table 1 wash-out results taken from 6 different truck loads are presented. All results comply with the limit on the average value (-15%). However, the third sample of truck 1 misses out on the individual limit (-20%). In this case, corrective actions could be initiated as soon as the results for truck 1 were available.

Table 1. Results from in situ wash-out tests.

Truck	$mf,1$	$mf,2$	$mf,3$	$mf,average$
truck 1	13%	4%	-26%	-3%
truck 2	3%	-17%	26%	4%
truck 3	9%	-11%	0%	-1%
truck 4	-12%	-4%	1%	-5%
truck 5	9%	-2%	10%	6%
truck 6	16%	-10%	6%	4%

5.4 Assessment of the quality control principles

Both quality control approaches are suitable methods to ensure the quality of SFRC. Practical experience is available as well as national standards [3]. The advantage of the material property based approach described in section 5.1 is the direct verification of the residual strength which is delivered on site. One major disadvantage is the time which has to pass until the beam test results are available. The process control approach described in section 5.3 allows for much shorter reaction time. If required, results from wash-out test can be available within a couple of hours. Although there is no confirmation of the residual strength, the mix design in combination with the compressive strength testing closes that gap. However, this approach requires detailed procedures and relies on well-controlled constituents.

IOP Conf. Series: Materials Science and Engineering **596** (2019) 012010 doi:10.1088/1757-899X/596/1/012010

5.5 Methods to determine homogeneity in fresh concrete

Whatever quality control approach is applied, a homogeneous fibre distribution is crucial. Wash-out test methods are available to verify this in fresh concrete and are an established and feasible method to determine the fibre amount and fibre distribution. Results can be available within a couple of hours. They are already implemented in standards like EN 206-1 / EN 14721 and [3]. As this methodology is regulated and applied since long, further introduction into the principles are not within the scope of this paper.

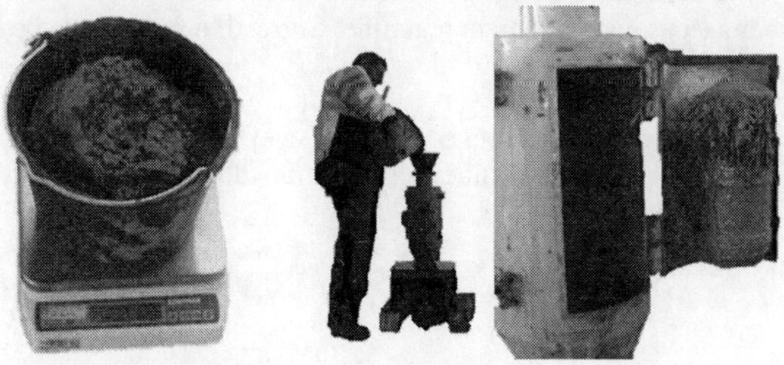

Figure 9. Wash-out test using pipe magnet.

Despite the fact that wash-out tests are rather simple to execute, a certain effort is required and only a small volume can be assessed. Still quite some time passes before a possible need of corrective actions is identified. It is obvious that wash-out can only provide a snapshot of the real situation.

6. New quality control system for SFRC: "Real time" quality control: eyeD®

The construction industry is advancing. Production processes are optimized and smart solutions are now being used to monitor both production and execution. In that context, beam tests or wash-out tests can hardly be integrated in a process driven, smart environment. The resolution of these test methods is small and immediate results cannot be obtained. An automated real time quality control system would be the ideal solution, both in terms of process integration as well as quality assurance.

6.1 EyeD® measuring device

NV Bekaert SA has developed the eyeD® measuring device – which

- allows for the quality control of steel fibre concrete in real time
- can be configured to determine fibre type, amount and distribution
- can be linked into the production and quality control processes
- operates mostly automatically
- allows remote access of results

Figure 10. eyeD® system connected to the chute of a truck mixer.

The eyeD® is connected to the truck mixer (Figure 10) by an adjustable interface and should fit all major truck mixer types. The concrete flow is continuously determined through regular ultrasound height sensors and especially developed speed sensors. The variation of the fibre's electromagnetic response is measured and plotted as a deviation from the mean value. Results can be accessed as soon as the unloading process has finished. This can be done with any WiFi-capable device right on site. Furthermore, the eyeD® is equipped with a GSM module which transfers the data to a central server. It is accessible from any internet connected device as long as there is a GSM connection on the jobsite. The eyeD® provides instantaneous quality data in real time. Corrective actions can be taken immediately.

6.2 Measurement results

More than 100 truckloads have been examined with the eyeD®. Figure 6 shows a typical example of a homogeneous mix, expressed in relative values for both fibre dosage and concrete volume.

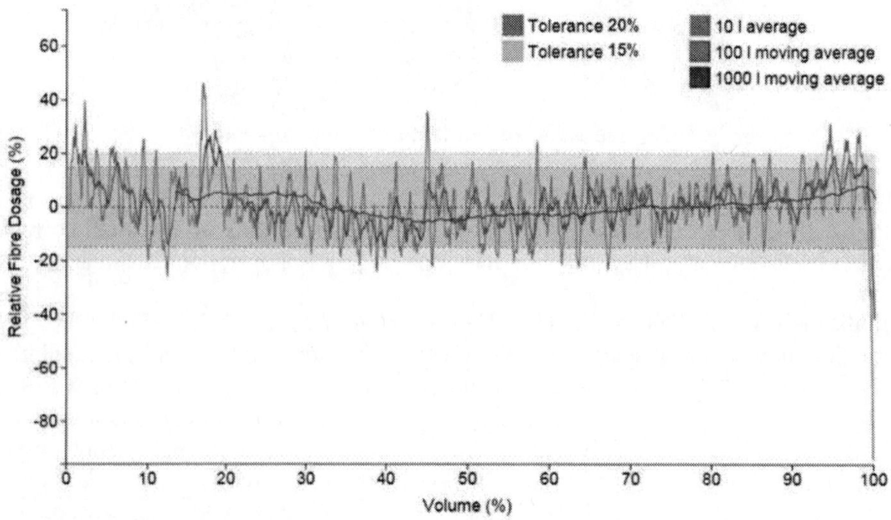

Figure 11. eyeD® measurement result: homogeneous fibre distribution.

The measured data can be processed to match different sample sizes. Moving averages of e.g. 100 litres and 1000 litres allow for an improved interpretation of the fibre distribution. In Figure 12 an insufficient fibre distribution is shown. The fibres were dosed right into the truck mixer using a conveyor belt. However, the required additional mixing time after fibre addition was not met. Additional concrete remixing before pouring did not take place. The result on fibre distribution could be figured out immediately with the eyeD® device.

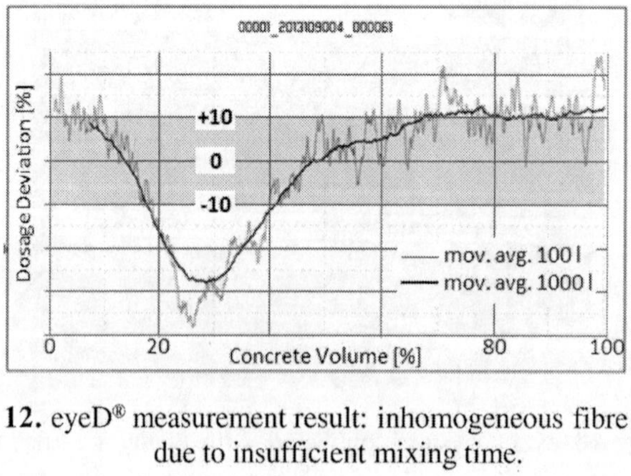

Figure 12. eyeD® measurement result: inhomogeneous fibre distribution due to insufficient mixing time.

IOP Conf. Series: Materials Science and Engineering **596** (2019) 012010 doi:10.1088/1757-899X/596/1/012010

6.3 Interpretation of results
The eyeD® system allows for full and detailed monitoring of the steel fibre distribution for the whole truck load. Therefore it is possible to judge the homogeneity of the complete concrete volume. Peaks can be identified as peaks and trends can be identified as trends. Even if individual data points plotted with a 10 litre resolution should not meet the tolerance limit defined for 10 litre wash-out samples (e.g. –20%), the whole truck load can be judged and the result can be put in perspective. Based on the measurements done with the eyeD® system so far, it appears that a moving average of 100 litres and a moving average of 1000 litres allow for a more appropriate judgement on the homogeneity.

6.4 Summary
The eyeD® system allows for a full assessment of type, amount and distribution of steel fibres in steel fibre reinforced concrete. The measuring device is attached to the chute of a truck mixer so that the full truck load can be assessed. Results are delivered in almost real time, as soon as unloading is completed. Access via local WiFi or remotely via the internet is possible so that immediate actions can be triggered in case the measurements should not comply with the specifications.

7. Conclusion
Steel fibre concrete is a material that has developed from use in minor applications to the use for structural applications, steel fibre only or combined reinforcement. Guidelines and recommendations have evolved into national standards and codes. Next version of EC 2 intends to subject SFRC for structural concrete too. Design methods, design tools are further established and also the quality control is going to developed more and more in a smart way.

References
[1] International Federation for Structural Concrete (*fib*) 2013 Model Code for Concrete Structures 2010 (Lausanne, Switzerland)
[2] European Committee for Standardization (CEN) 2005 EN 14651: Test method for metallic fibered concrete – Measuring the flexural tensile strength (limit of proportionality (LOP), residual
[3] German Committee for Structural Concrete (DAfStb) 2012 DAfStb Richtlinie Stahlfaserbeton (Berlin, Germany)
[4] European Committee for Standardization (CEN) 2003 EN 1992-1-1 Design of concrete structures – Part 1-1: General rules and rules for buildings
[5] Rilem TC162-TDF 2003: Test and design methods for steel fibre reinforced concrete
[6] Losberg A 1961 *Design Methods for Structurally Reinforced Concrete Pavements* (Chalmers University)
[7] Kennedy G, Goodchild C. H. 2004 *Practical Yield Line Design* (The Concrete Centre 09/2004)
[8] Vitt G *"Real time" Quality control system for steel fibre concrete*

IOP Conf. Series: Materials Science and Engineering **596** (2019) 012011 doi:10.1088/1757-899X/596/1/012011

New UHPFRC bridges in the Czech Republic

J Marek[1], J Kolísko[2], P Tej[2], D Čítek[2], J Komanec[3], M Kalný[3] and L Vráblík[4]

[1]KŠ PREFA s.r.o., Jinonická 805/57, 150 00 Praha 5-Košíře, Czech Republic
[2]Klokner institute, Czech Technical University in Prague, Šolínova 7, 160 00 Praha 6-Dejvice, Czech Republic
[3]Pontex Consulting Engineers, Ltd., Bezová 1658, 147 00 Praha 4-Braník, Czech Republic
[4]NOVÁK & PARTNER, V Olšinách 2300/75, 100 00 Praha-Strašnice, Czech Republic

E-mail: j.marek@ksprefa.cz

Abstract. This paper deals with new UHPFRC (Ultra High Performance Fibre Reinforced Concrete) bridges produced or constructed in 2018 in the Czech Republic. The paper includes a short description of 3 unique pedestrian bridges with load bearing structure made of UHPFRC and 2 steel bridges over the railway track with lost formwork for bridge decks and cornices made of thin UHPFRC slabs. Focus is given on the testing of the material, comparison of test specimens and summary of results from UHPFRC production in real scale. The size effect of control specimen and the results analysis is given in the last chapter. The 3 mentioned unique pedestrian bridges are Footbridge over the Lubina river in city Příbor, Pedestrian and cycle bridge in the vicinity of Black Bridges in city Tábor and Footbridge over Dřetovice stream in Vrapice - a district of Kladno city. Thin panels as permanent lost formwork for two bridges over the railway tracks are in city Přerov. All new UHPFRC structures should be worthy of attention of bridge experts and community of specialists in the field of concrete engineering.

1. About UHPC

UHPC, Ultra High Performance Concrete, usually reinforced with steel (usually high strength) or polyolefin fibres (PP, PA or PVA), is therefore better named as UHPFRC, (Ultra High Performance Fibre Reinforced Concrete). It is a modern material based on principles similar to normal strength concrete (NSC), but applied in the smaller scale, characterized by its material optimization: fine-grained sand, silica fume, fibres, Portland cement with no large aggregate.

High compressive strength, strain hardening, and extremely low porosity leads to the high durability and very long serviceability of UHPFRC structures.

Many lab researchers are currently working with UHPC [7,8], therefore no more detailed information will be written here about UHPC/UHPFRC composition, its preparation, casting and curing in laboratory conditions with small-scale results.

Used mixture was finally developed in 2017 after several years of research and applied tests in cooperation of KŠ PREFA company and Klokner institute of CTU. The exact design of the mixture cannot be published, but it is quite similar to standard UHPC mixtures.

The details of dosing, batching and curing are business secrets of KŠ PREFA Ltd.

2. Standards

There are no valid (or validated by legislative) normative standards in the Czech Republic concerning UHPC, except three methodical manuals published by Klokner institute of CTU in Prague in 2015 [9-11]. The real scale application of UHPC into the structures of bridges or footbridges brings out many questions about the material properties and production quality controlling and testing, e.g.:

- What is the meaning of usually required compressive strength above 150 MPa?
- Which test specimens should be used?
- How to set up quality control schedule?
- How to describe design mechanical parameters?
- Which test specimens are the best for which structural elements?
- Is the test equipment for standard concrete reliable enough for UHPC?
- In the field of concrete structures, an essential characteristic of UHPFRC structure – size effect – should be taken into account.

3. UHPC production in real industrial scale

For application of UHPC in the real size structures, the very important question is the way of dosing and batching the fresh mixture in volume of thousands of litres and necessary technical equipment. Standard industrial concrete batching plants are not ready for so many components in the mixture, required accuracy of dosing and very tricky mixing with fibres.

After successfully managed mixing, the next step is how to cast the fresh mixture into the mould, keeping in mind that moulds have to be used repeatedly, and the element must be cured all the time from casting to demoulding. Moulding technology from common producers is not prepared for hydrostatic pressure and non-Newtonian fluid behaviour of fresh UHPC with microparticles eager to leak through even the smallest gap.

Lifting and transportation of UHPC elements is very different compared to NSC, because UHPC allows to create thin walls and slabs which are not tested by common industry producers of anchors and devices, and their standards, certification and limitations are out of range for UHPC.

One can think of higher strength of UHPFRC as advantage compared to NSC, but other parts of standardized lifting and transport systems (parts not directly connected to concrete) are designed for NSC, which limits the whole system.

As mentioned above, the transformation of technology from laboratory scale UHPC preparation and mixing to real industrial production is not a fast and easy process. It is a part of technology evolution, involving long period of applied research and testing.

3.1. History of real scale use of UHPC in the Czech Republic

Industrial use of UHPC in bridge structures in the Czech Republic started in 2012 in facility Skanska Štětí with first application of ribbed thin slabs as permanent lost formwork for bridge deck 272-008 near city Benátky nad Jizerou, where the volume of the UHPFRC slab was about 50 litres. Another smaller slabs each about 80 litres were produced at the same place as façade tiles for building in Malmö in Sweden.

Some small elements were casted for the family house façade cladding in village Mokrá-Horákov and a few pieces of safety barriers were created for Prague airport.

One of the largest UHPC precast elements till 2018 were segments for footbridge Čelákovice from 2013, where company Metrostav with its fresh concrete producer TBG casted 3.8 m^3 in one step for each segment. Similar volume, ca 4.0 m^3 was casted in 2014 in Skanska Štětí, one Pi - beam for footbridge over Opatovický channel in village Čeperka. [1-6, 12-16]

3.2. Real scale use of UHPC in the Czech Republic in 2018

Since 2015, company KŠ PREFA Ltd. have become the successor of Skanska company in the facility Štětí and continued with applied research of UHPC. In cooperation with Klokner institute of CTU in Prague and company Pontex Consulting Engineers the research project supported by Epsilon programme of Technical Agency of Czech Republic called Service life enhancement and construction speed-up of elements of traffic infrastructure using UHPC is being carried out.

The main goal of the project is the application of UHPC into the real structures and projects of traffic infrastructure.

In first year of the project, 2017, were made necessary adjustments of batching plants and preparation of technology, simultaneously with designs of bridges based on the UHPC technology.

Simultaneously were made many laboratory tests of mixture itself, the results were validated by mixing in the batching plants in factory Štětí, and the mixture was optimized for industrial scale.

Year 2018 became the milestone for industrial use of UHPFRC in bridge engineering in the Czech Republic.

UHPFRC load bearing structures of 3 footbridges were designed and built and thin slabs as lost formwork for the bridge deck and cornices of 2 steel bridges were used.

The importance is given by the fact, that all these projects were part of the open competition, and the solution with structures made of UHPFRC was winner of the designing and bidding process.

In this way the UHPFRC came into the standard business world in structural engineering. In total number more than 70 m^3, 175 tons of UHPFRC, were commercially produced in the Czech Republic in 2018.

4. Footbridge over the Lubina river in city Příbor

The load bearing structure of the footbridge is a single span prismatic rectangular beam. Span of almost 36 meters and the height of cross section of 800 mm declares boldness of a structural engineer working with concrete and the ratio 1:45 fulfils the requirement of an architect.

The structure itself is made of 5 elements connected by post tensioned monostrand cables, injected in three channels with parabolic shape.

Connections of the elements were sealed by epoxy mortar, for channels were used special waterproof connectors.

The casting of the elements was preceded by tests of 1.0 and 2.0 m^3 for checking and testing of heat of hydration, next pre step was 2 test elements together in volume of one typical element casted simultaneously as a test of batching and casting and test of the mould. In that moment it was the largest UHPFRC casting in CR ever. These elements were used for sealing and cable tensioning test, for testing of the handrail anchoring.

Final 5 elements were created in 4 next weeks, in volumes of 2 x 7.8 + 3 x 5.4 m^3.

Architectural design of Petr Tej and Marek Blank was statically proofed by Lukáš Vráblík from Novák & Partner, KŠ PREFA cooperated with Klokner institute of CTU on optimization of the mixture, the creation and assembly of the load bearing structural elements were made by KŠ PREFA. Post tensioning was made by Freyssinet, general constructor was Strabag. Finished footbridge is shown in Fig. 1. Quality control tests were performed, and the results are given in Table 1. The analysis of obtained data is given in the last chapter.

Table 1. Average results of 28 day control material tests - Footbridge Příbor.

Test / Set (Element)	1	2	3	4	5	Avg.	Unit
Young´s Modulus - cylinder 150x300mm	49.3	49.1	49.6	47.5	47.8	48.7	GPa
compressive strength - cylinder 150x300mm	140.2	139.1	145.3	140.5	137.6	140.5	MPa
compressive strength - cube 100x100mm	157.3	150.0	149.6	140.6	144.1	148.32	MPa
compressive strength - cube 150x150mm	144.5	146.8	139.2	139.1	141.9	142.3	MPa
compressive strength - beam 40x40x160mm	166.0	183.0	186.0	168.0	171.0	174.8	MPa
flexural strength - beam 100x100x400mm	11.1	15.5	15.2	14.6	18.5	15.0	MPa
flexural strength - beam 40x40x160mm	31.7	37.7	36.2	33.5	29.2	33.7	MPa

IOP Conf. Series: Materials Science and Engineering **596** (2019) 012011 doi:10.1088/1757-899X/596/1/012011

Figure 1. Footbridge over the Lubina river in city Příbor.

5. Pedestrian and cycle bridge in the vicinity of Black Bridges in city Tábor

Presented footbridge creates an essential part of complex of pavements and bike paths over the Budějovická street and its access roads in densely built-up part of the city Tábor.

Footbridge is parallel to the steel railway bridge, in the distance of ca 1.1 m.

Load bearing structure is single span Pi - beam, with height of 0.94 m, width 3.0 m and length of 27.0 m; prefabricated as one prestressed element consisting of 12 m^3 of UHPFRC. Bridge deck with thickness of 60 mm is reinforced only by ribs in the distance of 1000 mm. Steel rebars are not used in the deck itself.

Structure is designed according to EN 1991-2 and methodical manual for designing of UHPC structures by Klokner institute of CTU [9-11]. Uniform load caused by crowd of people of 5 kN/m^2, single load of service vehicle of total weight 3.5 t with dynamic coefficient $\delta = 1.10$ and special vehicle of 12 t are considered.

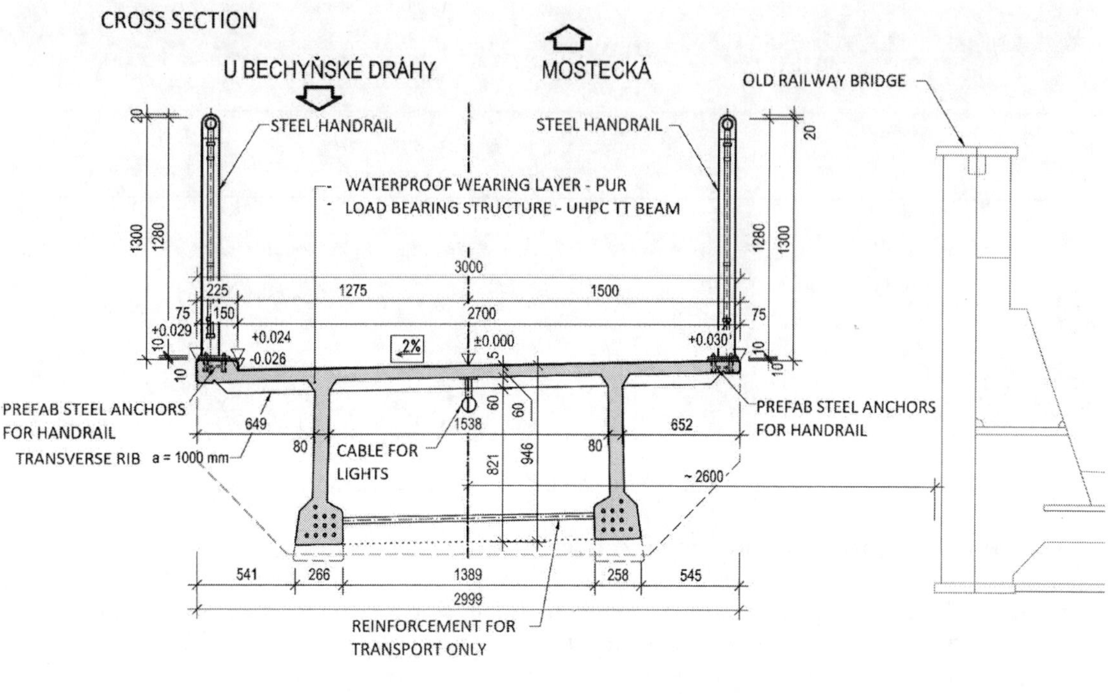

Figure 2. Pedestrian and cycle bridge in city Tábor - cross section.

IOP Conf. Series: Materials Science and Engineering **596** (2019) 012011 doi:10.1088/1757-899X/596/1/012011

Casting of UHPFRC in one step in one element with volume of 12 m^3 is unique in Europe.

After casting elements for footbridge in Příbor, gained experience allowed to cast whole load bearing beam for footbridge Tábor in one 2-hour long work.

Detailed preparation, comprehensive planning and on-site testing enabled the casting process without any pause, which was essential to avoid cold joints and to achieve optimal scatter of steel fibres, flawless surface and unexceptionable serviceability and durability of the finished UHPFRC element.

One of the largest UHPFRC elements poured in one take in Europe was successfully created, assembled and put in use of public.

Design of load-bearing structure was made by Jan Komanec and it was statically proofed by his team of bridge engineers from company Pontex Engineers Ltd. KŠ PREFA cooperated with Klokner institute of CTU on optimization of the mixture, the creation, prestressing and assembly of the beam were made by KŠ PREFA in factory Štětí. General constructor was Firesta Ltd. Finished footbridge is shown in Fig. 3. Quality control tests during the production were performed, and the results are given in Table 2. The analysis of obtained data is given in the last chapter.

Table 2. Average results of 28 day control material tests - Pedestrian and cycle bridge Tábor.

Test	Avg.	Unit
Young´s Modulus - cylinder 150x300mm	45.6	GPa
compressive strength - cylinder 150x300mm	138.0	MPa
compressive strength - cube 100x100mm	135.1	MPa
compressive strength - cube 150x150mm	125.9	MPa
compressive strength - beam 40x40x160mm	160.5	MPa
flexural strength - beam 100x100x400mm	17.9	MPa
flexural strength - beam 40x40x160mm	29.0	MPa

Figure 3. Pedestrian and cycle bridge Tábor.

6. Footbridge over Dřetovice stream in Vrapice – a district of Kladno city

Single span footbridge with unique U-shaped cross section leaps over the stream in two arches, both vertically and horizontally. Beam with its length of 10.5 m, thickness of the walls 40 mm, thickness of the deck from 45 to 55 mm, without any steel rebars or prestressing cables, made this experimental structure a unique bridge in the world scale.

IOP Conf. Series: Materials Science and Engineering **596** (2019) 012011 doi:10.1088/1757-899X/596/1/012011

Volume of used UHPFRC ca 1.6 m^3 was casted in specially designed mould twice. The first attempt was success in casting but unsuccessful in load testing of whole bridge element. The second attempt was casted with thicker walls and upgraded geometry of wall-deck connection. The second bridge beam successfully accomplished load testing and was assembled in place over the stream. Due to its experimental nature, it was given to the city government as a gift from KŠ PREFA and Klokner institute of CTU. Finished footbridge is shown in Fig. 4.

Figure 4. Footbridge over Dřetovice stream in Vrapice – a district of Kladno city.

IOP Conf. Series: Materials Science and Engineering **596** (2019) 012011 doi:10.1088/1757-899X/596/1/012011

7. Bridges over the railway tracks in city Přerov

Both steel bridges are over the railway track Přerov – Bohumín and Česká Třebová – Přerov. General constructor, Strabag, was replacing old steel load-bearing structures of both bridges with new ones. Support structures were repaired.

Bridge deck was casted over the steel crossbeams with NSC, but as the lost formwork were used thin UHPFRC panels, both inside between the main steel girders and outside as sidewalk decks. Thickness of the NSC deck varied from 220 to 400 mm. UHPFRC slabs with thickness of 60 mm covered the distance between steel crossbeams with span ca 1.7 m. For the sidewalk part the NSC deck was 190 mm thick and the UHPFRC slab was 50 mm thick. Assembled slabs before placing the reinforcement of the main deck are shown in Figure 5.

Lost formwork panels for bridge cornices were 1900x600 mm and 50 mm thick.

Before casting of the thin panels, the initial testing of the early age properties of the elements had to be performed. The early loading of the panels is the reason for different control testing schedule, as sufficient safety of 7-day old slabs had to be proven. Load bearing parameters together with material parameters were tested in different stages, from 4 days to 14 days, and after that in 28 days.

Table 3. Average results of control material tests - lost formwork panels for Přerov bridges

Test	Avg.	Unit
compressive strength - cube 100x100mm, 4 days	92.0	MPa
compressive strength - cube 100x100mm, 28 days	156.6	MPa
compressive strength - beam 40x40x160mm, 4 days	105.0	MPa
compressive strength - beam 40x40x160mm, 28 days	170.0	MPa
flexural strength - beam 40x40x160mm, 4 days	19.2	MPa
flexural strength - beam 40x40x160mm, 28 days	29.4	MPa
flexural strength - beam 100x100x400mm, without notch, 28 days	15.3	MPa
flexural strength - beam 100x100x400mm, with notch, 28 days	19.7	MPa

Figure 5. Lost formwork panels for Bridges over the railway tracks in city Přerov.

8. Analysis of control testing results

All the precast elements mentioned above were made of UHPC, class C 110/130 XC4+XD3+XF4, with declared parameters:

- Tensile strength in the time of crack initiation – average minimum 18 MPa, total minimum 15 MPa
- Residual strength class min: $0.7 < f_{R3K}/f_{R1K}$

Initial and quality control tests proved all the declared parameters, but more can be said from the results.

Used mixture could have been declared as higher strength class concrete, but due to the lack of valid standards, it must have been declared as C110/130, the highest class defined in EN 206.

Size effect can be easily seen in comparison of higher results reached on smaller specimens, and lower results reached on larger ones. (the highest compressive strength in total was 186 MPa on beams 40x40x160 mm).

The effect of the grinding of cylinder (150x300 mm) specimens before tests might be probably responsible for higher results of cylinders, in average comparable to the results of cubes (150 mm). The cubes were not grinded, so it is not confirmed at the moment. Another possibility is that the smooth round surface of the cylinders causes better scatter of the fibres. This hypothesis will be verified in this year by ongoing experimental research.

Another effect was observed. Comparing the compressive test results on new equipment with computer aided testing with older ones with handmade test driving led up to the necessity of proper setting up of the new equipment. The handmade test driving led to higher results, because the operator could continue in stressing of the specimen after the first crack.

With larger specimen (cube 150 mm) it was very common to get higher results after first crack and after activation of steel fibres. This effect cannot be seen with new computer aided test equipment. Also, the smaller specimen (cube 100mm, beam 40 mm) has not shown this difference.

Acknowledgement
The financial support of research connected to this work by the Technology Agency of the Czech Republic, EPSILON programme, Project No. TH02020373, is gratefully acknowledged.

References

[1] Kalný M, Kvasnička V and Komanec J 2016 First practical applications of the UHPC in the Czech Republic *HiPerMat 2016* (Kassel, Germany)

[2] Kalný M, Komanec J, Kvasnička V and Fiala C 2017 Experience with UHPFRC applications in the Czech Republic *UHPFRC 2017* (Montpellier, France)

[3] Kolísko J, Tichý J, Kalný M, Huňka P, Hájek P and Trefil V 2012 Vývoj ultravysokohodnotného betonu (UHPC) na bázi surovin dostupných v ČR *BETON TKS 6/2012, annex Betonové konstrukce 21. století, betony s přidanou hodnotou,* pp 51-56

[4] Rydval M, Kolísko J, Huňka P and Tichý J 2013 Závislost únosnosti prvků vyrobených z UHPFRC na distribuci vláken *20. Betonářské dny 27.-28.11.2013* (Hradec Králové, Czech Republic)

[5] Kolísko J, Rydval M and Huňka P. 2013 UHPC – Assessing the Distribution of the Steel Fibre and Homogeneity of the Matrix *fib Symposium 2013* (Tel Aviv, Israel)

[6] Vítek J L, Coufal R and Čítek D 2013 UHPC – Development and Testing on Structural Elements *Concrete and Concrete Structures 2013* (Žilina, Slovakia: University of Žilina) pp 218-223

[7] Abbas S, Nehdi M L and Saleem M A 2016 Ultra-High Performance Concrete: Mechanical Performance, Durability, Sustainability and Implementation Challenges, *International Journal of Concrete Structures and Materials* 10(3) pp 271-295

[8] Duque L F M, Varga I and Graybeal B A 2016 Fiber Reinforcement Influence on the Tensile Response of UHPFRC *First International Interactive Symposium on UHPC* (Des Moines, Iowa, USA)

[9] Metodika 1 – Metodika pro návrh UHPC a materiálové zkoušky *Result of project TAČR TA010110269* (Prague, Czech Republic: Klokner institute, 2014)

[10] Metodika 2 – Metodika pro navrhování prvků z UHPC *Result of project TAČR TA010110269* (Prague, Czech Republic: Klokner institute, 2014)

[11] Metodika 3 – Metodika pro výrobu prvků z UHPC a pro kontrolu jejich provedení *Result of project TAČR TA010110269* (Prague, Czech Republic: Klokner institute, 2014)

[12] Kolísko J, Čítek D and Tej P 2017 Technologie výroby tenkostěnné obloukové dvojitě zakřivené lávky z UHPFRC *14. konference TECHNOLOGIE 2017* (Jihlava, Czech Republic: Česká betonářská společnost ČSSI)

[13] Kolísko J, Čítek D, Tej P and Rydval M 2017 Production of Footbridge with Double Curvature Made of UHPC *IOP Conference Series: Materials Science and Engineering* **246**

[14] Kněž P, Tej P, Čítek D and Kolísko J 2017 Design of Footbridge with Double Curvature Made of UHPC *IOP Conference Series: Materials Science and Engineering* **246**

[15] Kabele P, Sajdlová T, Rydval M and Kolísko J 2015 Modeling of High-Strength FRC Structural Elements with Spatially Non-Uniform Fiber Volume Fraction *Journal of Advanced Concrete Technology* **13**(6) pp 311-324

[16] Blank M., Tej P, Kolísko J and Vráblík L 2016 Design of Experimental Suspended Footbridge with Deck Made of UHPC *3rd International Conference on Mechanics and Mechatronics Research ICMMR 2016* (Chongqing, China)

IOP Conf. Series: Materials Science and Engineering **596** (2019) 012012 doi:10.1088/1757-899X/596/1/012012

Development of bridge precast beams from high performance concrete in EUROVIA CS

Michal Sýkora[1], Petr Jedlinský[1], Jan Komanec[2] and Václav Kvasnička[2]

[1] EUROVIA CS, a.s., Národní 138/10, 110 00 Praha 1, Czech Republic
[2] Pontex spol. s r.o., Bezová 1658, 147 14 Praha 4

E-mail: jkm@pontex.cz

Abstract. The paper presents a summary of the development of bridge precast beams in EUROVIA CS company in the last years. Several types of beams are described. The first beam that has already been applied in real construction is an Omega beam which consists of shell outside envelope enabling construction of a bridge without intermediate supports in midspan. The second one is an inverted T beam for railway and road bridges. At present, the development of UHPC beams for footbridges is going on. The plan is to create a set of shapes and solutions usable for various configurations with spans up to 15 m, fully exploiting the UHPC advantages. Another element for bridge construction developed by Eurovia and its partners is the lost formwork panel used for bridge superstructures built from both concrete and steel beams.

1. Introduction

The development of precast beams from high performance concrete since 2016 is described in this paper. Participants in the development are consultants Pontex s.r.o., precast plant PrefaPro, a.s. and Stachema. The aim was to develop precast beams for small and medium span bridges and footbridges and panels for lost formwork.

The basic idea was using high performance concrete to obtain lightweight products. The weight of the precast members is often the limiting factor in construction of bridges in complicated approach conditions on site and in the cases when the internal span has to be installed with crane standing behind the abutment. The reduced weight of members can have favorable effect on the economy, reducing both transport and assembly costs in the order of tens of percentage points.

We have decided to use UHPC for the elements of bridges and footbridges in combination with prestressing. The lost formwork elements are designed from UHPC only.

2. Omega beam

The first type of such beams was the "OMEGA" beam. Its development started in 2016 (for more details, please refer to [1, 2]) and it was used on site for the first time in 2017 on the bridge in the quarry Třebnůška (Fig. 4). The beam voided shape is based on a typical "T" beam shape normally used in the PrefaPro production site in Brandýs nad Labem.

The "OMEGA" beam consists of shell outside envelope with the shape of inverted letter Ω. The thickness of the shell is 70 – 100 mm, the height of the beam is 550 – 950 mm, depending on the span. Concrete SCC C60/75-XF2 with consistency F7 was used together with prestressing by strands Y1860

IOP Conf. Series: Materials Science and Engineering **596** (2019) 012012 doi:10.1088/1757-899X/596/1/012012

S7. The key requirement was that the beam withstands the construction loads without intermediate support in midspan. The prestressing tendons with preinstalled anchors are placed in the casting yard (Fig. 1). The tendons are later prestressed on site after the composite reinforced concrete deck was cast.

The inner form for this beam had to be designed taking into account the uplift forces from the concrete mixture. The company Česká DOKA was involved in the design of the form and the demolding elements (Fig. 2).

Figure 1. Complete Ω beam

Figure 2. The form of the void

As a part of the development, several tests were carried out.:

1. The pull-out test of the wave transport anchor
2. Test of bond between core and shell
3. Load test (Fig. 3)

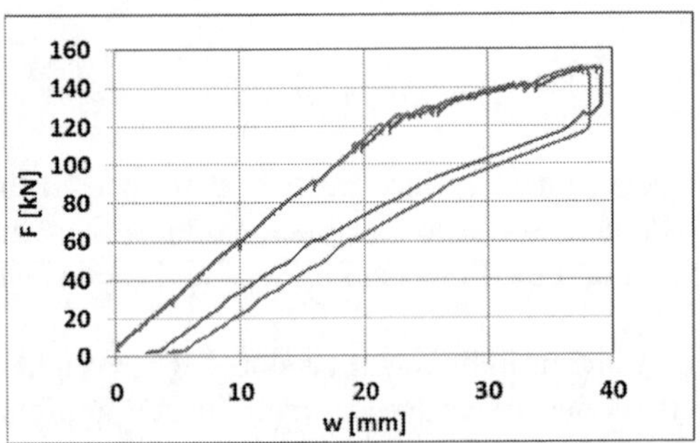

Figure 3a. Force-deflection diagram from the load test

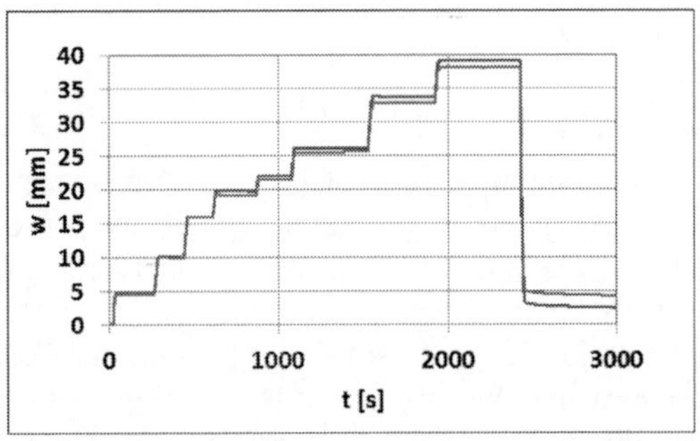

Figure 3b. Measured deflections during test

An economic comparison of assembly costs was performed comparing the typical "T" prestressed beams and the OMEGA beams (Table 1).

IOP Conf. Series: Materials Science and Engineering **596** (2019) 012012 doi:10.1088/1757-899X/596/1/012012

Table 1. Economic comparison

Single span	Ω15	T15	Ω21	T21	Ω24	T24
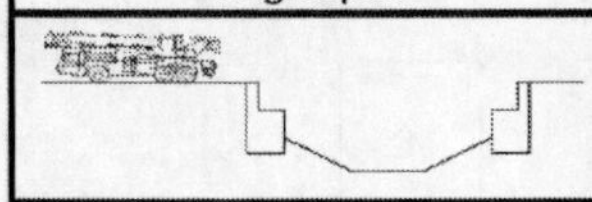	LTM 1055	LTM 1200	LTM 1100	LTM 1200	LTM 1200	LTM 1300
	810 EUR	3890 EUR	1670 EUR	3890 EUR	3890 EUR	5470 EUR
2 span	2 x Ω21	2 x T21	2 x Ω27	2 x T27		
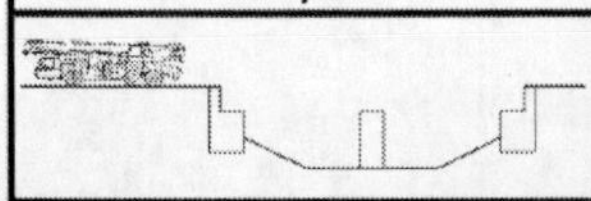	LTM 1200	LTM 1500	LTM 1500	LTM 1500*	* Including 2 assembly positions	
	5520 EUR	7470 EUR	9260 EUR	11260 EUR		

Figure 4a. Beams assembly and the completed bridge

Figure 4b. Beams assembly and the completed bridge

A bid for construction of 3-span bridges was prepared in 2018, where, due to the terrain and approach limitations, the beams assembly will have to be done by crane standing behind the abutment. Again, the classical "T" beams and OMEGA beams were compared. The overall cost savings were above 10 %.

3. Inverted T beams

In 2017, a beam from UHPC for railway and road bridges was developed. The aim was to create a UHPC beam with the potential of replacing the steel "I" or "HEB" beams in these bridge structures. In the case of railway bridges, the beams are embedded in concrete deck. In road bridges, beams are supporting a composite reinforced concrete deck (Fig. 5).

The beam cross section is an inverted "T". For the length of 8.6 m (as used in the trial structure in Chrtníky bridge) the depth of beams is 350 mm (Fig. 6).

In the case of railway bridges, the transversal forces have to be carried. In the case of embedded rigid beams these are covered by transverse reinforcement running through the holes in the webs of steel beams. In the case of concrete beams with web thickness of 100 mm, the void is hard to be filled properly with the concrete of the deck which is necessary for proper functioning of the reinforcement and corrosion protection. This problem was easily solved by using transverse prestressing with MONOSTRAND strands (Fig. 7).

IOP Conf. Series: Materials Science and Engineering **596** (2019) 012012 doi:10.1088/1757-899X/596/1/012012

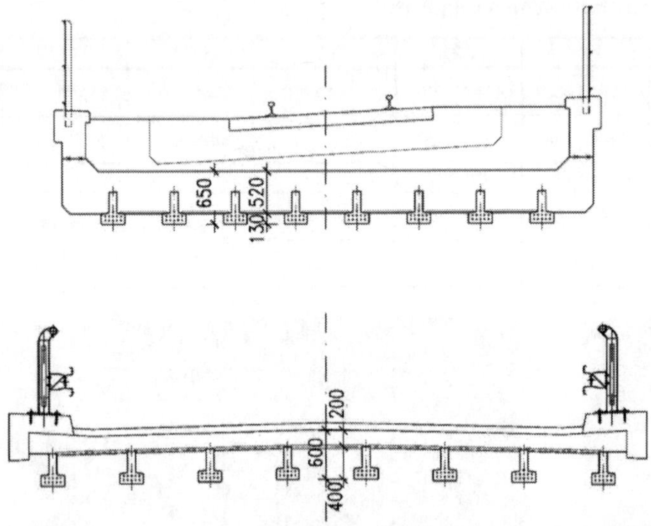

Figure 5. Bridges cross sections

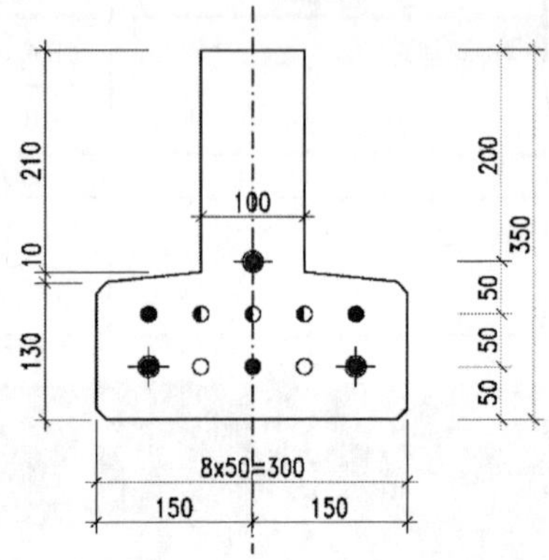

Figure 6. Beam cross section

Figure 7. Transverse prestressing

The new beams were used for the first time on the bridge in Chrtníky quarry (fig. 9) near the town of Přelouč. Due to the requirement of testing the more complicated layout designed mainly for the railway bridges and to confirm the results of the structural analysis, the alternative with embedded beams was used here for testing purposes. Plates CETRIS were used as the lost formwork. For a bigger distance between the beams, it is necessary to use lost formwork panels from UHPC to span this gap without any supports.

As a part of the development of this beam, a load test was carried out. It confirmed the results of the structural analysis. At load level of 2x242 kN the deflection of beams was 60 mm with no visible cracks. The test was finished at load level of 2x376 kN, still without a visible destruction of the specimen (Fig. 8).

More measurements are done during the long-term inspection of the bridge using built-in strain gauges and recording devices in service. These measurements were commenced in April 2018, repeated in March 2019.

Figure 8. Deflected beam during load test

Figure 9. Assembled beams on site

After the bridge construction, a full scale load test was carried out by an accredited testing laboratory using a truck Tatra Jamal with the weight of 33.5 t (Fig. 10). Measured deflections were smaller than those calculated. The load test confirmed appropriate functioning of the structure.

Figure 10. Load test of the bridge

4. UHPC Beams for Footbridges

At present, the development of UHPC beams for footbridges is going on. The plan is to create a set of shapes and solutions usable for various configurations with spans up to 15 m, fully exploiting the UHPC advantages, durability, resistance to environmental effects etc. These beams are planned to be assembled in precast yard completely, i.e. including railing (Fig. 11).

For this project, a form is being prepared currently. Testing casts are planned for June 2019 followed by a load test in August 2019.

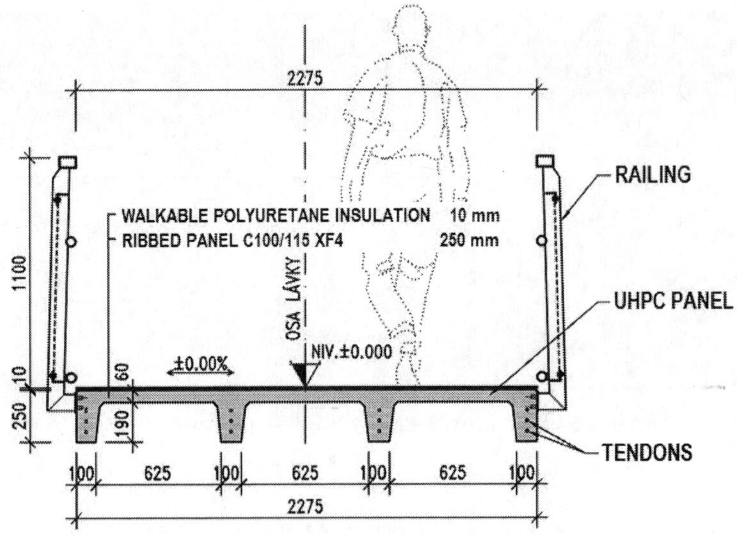

Figure 11. Footbridge precast UHPC beam

5. Lost formwork panels

Another element for bridge construction, developed by Eurovia and its partners, is the lost formwork used for bridge superstructures built from both concrete and steel beams.

The development of these elements is based on economic reasons, when the classical formwork from CETRIS plates is economically feasible up to spans 500 mm with technical limit of 800 mm.

In order to cover larger spans, decks from UHPC are being developed. For smaller spans, plain plates are used. For longer spans, decks with edge ribs are used. (fig. 12).

Figure 12. Ribbed deck of lost formwork

A load test of ribbed decks of lost formwork was done at the Klokner Institute in Prague on 1st April 2019, leading to 2 sets of rather different results (Fig. 13). The load was applied by gradual increase of the loading weight.

Due to the found differences, further investigation of the sources of differences will be done. The potential reasons are different cement used and age of tested elements.

The tested elements will be inspected in detail on the fracture areas. Some additional tests are planned to be done on each of the remaining parts of specimen to confirm the trends of previous tests on these elements. Also, the residual strength will be measured.

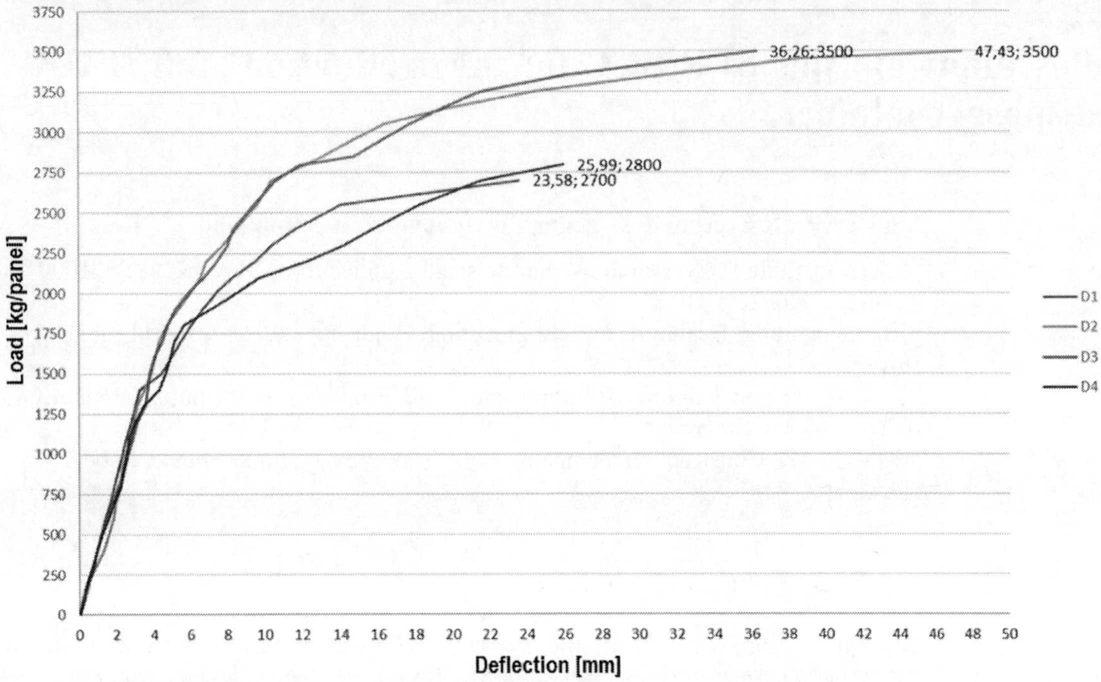

Figure 13. Load test diagrams – lost formwork elements (20 mm slabs + 50 m ribs)

6. Conclusions

The outcome of the nearly 4 year development is a set of structural members for bridge construction using high performance concrete with advantages not only in quality and durability but also in economics of the construction.

Acknowledgements

This paper was prepared thanks to the support of the Technology Agency of the Czech Republic (TAČR), project "Centre for Effective and Sustainable Transport Infrastructure (CESTI)" (no. TE01020168).

References

[1] Bílý P, Kohoutková A and Jedlinský P 2017 FEM simulation of static loading test of the Omega beam *IOP Conference Series: Materials Science and Engineering (MSE)* **236**(1), paper no. 012066.

[2] Sýkora M, Jedlinský P and Komanec J 2017 Lightweight HPC beam OMEGA *IOP Conference Series: Materials Science and Engineering (MSE)* **236**(1), paper no. 012055.

IOP Conf. Series: Materials Science and Engineering **596** (2019) 012013 doi:10.1088/1757-899X/596/1/012013

Possibility of using SFR-SCC for the production of the LILW disposal container

J Šušteršič[1], R Ercegovič[1], F Sinur[2], B Duhovnik[2], A Šajna[3] and T T Resnik[4]

[1] IRMA Institute for Research in Materials and Applications, Slovenčeva 95, 1000 Ljubljana, Slovenia
[2] IBE Consulting, Designing and Engineering, Hajdrihova ulica 4, 1000 Ljubljana, Slovenia
[3] ZAG Slovenian National Building and Civil Engineering Institute, Dimičeva 12, 1000 Ljubljana, Slovenia
[4] POMGRAD, Construction Company, Bakovska ulica 31, 9000 Murska Sobota, Slovenia

E-mail: jakob.sustersic@irma.si

Abstract. The disposal container with low and intermediate level radioactive waste (LILW) was developed using self-consolidating concrete (SCC), which provides long term mechanical properties and extremely low permeability for liquids. In the paper, some results of additional investigations are presented, which complement the assessment of the behaviour of SCC during use. The main focus is the presentation and discussion of the results of SFR-SCC (steel fibre reinforced - self-consolidating concrete) tests in fresh and harden state. From these results it can be seen that there is a possibility of further improving the properties of the SCC and thus the container.

1. Introduction

SCC was developed in the framework of the development project for the construction of a low and intermediate-level radioactive waste (LILW) disposal container. It is used for the production of reinforced concrete walls, bottom and container cover. In addition to the SCC, the following two composites were developed: 1. filling mortar to fill the space between the inserted barrels with LILW and the inner surface of the container and 2. sealing mortar to seal the unfilled space between the filling mortar and the inner surface of the cover and to seal the contact between the cover and the top surface of the walls.

Within the project, the requirements for the properties of the basic materials used for the preparation of composites and for the properties of all three composites (SCC, filling and sealing mortar) have been determined. These requirements are specified in the Program of testing of containers, measurements and tests [1]. They are also given in the paper by Sinur and Duhovnik [2]. In order to obtain the Slovenian Technical Approval (STA), it was necessary to develop composites with properties that met all the requirements. The results of the tests given in the report Results of container testing, measurements and tests [3] showed that all requirements were met. Containers with such composites were produced [4] and tested at the test facility at the Pomgrad production plant in Lipovci [5].

This paper provides only brief information about the container and the results of SCC investigations. The following section discusses results of investigations some of the key characteristics of SFR-SCC, based on which it would be able to make assessment of the possibility of using SFR-SCC for the production of the disposal container for LILW.

2. Short information on the disposal container and the properties of the SCC

A large team of experts from IBE (LILW disposal site and container designer), Pomgrad (manufacturer of container prototype), ZAG and IRMA (co-operating Institutes in the project) participated in the development project of the disposal container designated N2d.

2.1. Disposal container

Concrete container as one of the most important elements of the engineering part of the multi-tier system for preventing the passage of radioactive substances from the disposal site to the environment acts as: (a) biological shield in the time before the disposal, (b) mechanical protection of LILW during storage and disposal, (c) the basic safety element during the transport and internal transport (displacement) of LILW in the container, (d) the basic criterion of the profile in the process of preparing waste for disposal and (e) the central object of handling LILW in the area of the storage silo.

The most important and most specific container requirement is, in addition to the resistance and stability of all anticipated loads in the filling phase and the transport before the final disposal, also the durability in the expected life span of 300 years.

In view of the known environmental conditions to which containers will be exposed after disposal in the silo, the containers will be exposed to carbonization in the first phase (in the period 2020 – 2061), after the filling and closing the silo, and the abandonment of water pumping from 2062 onwards, the silo with containers exposed to underground water.

Storage containers ready for disposal will be transported to the disposal site in accordance with the provisions of the requirements for the transport of dangerous goods by road (ADR)[6]. Slovenia has implemented the ADR provisions [6] in national legislation through the Law on the Transport of Dangerous Goods (ZPNB)[7]. This precisely prescribes that containers for the transport of dangerous goods must be tested with a "drop test".

In total, 950 containers of type N2d will be deposited, with a maximum permissible mass of 40 t. Containers deposited in the silo with the help of a portal crane and special grips. Empty spaces between containers and containers and the silo wall will be filled with filling concrete.

2.1.1. Geometric characteristic of the N2d container. The basic geometry of the container was determined based on the installation of 4 TTC barrels. The N2d container is an upgrade of the N2b container, where it differs substantially only in the design of the cover and anchorage of the cover in the reinforced concrete container. The specific solutions of the container are given below and are result of the development of the container through the acquisition of the Slovenian Technical Approval: (1) External dimensions (Width = 1,95 m, Length = 1,95 m, Height = 3,30 m), (2) Thickness of the bottom slab = 23 cm, (3) Wall thickness at the top = 20 cm, (4) Wall thickness at the bottom = 23 cm, (5) Cover thickness = 20 cm, (6) Mass of empty container with the cover = 14,92 t, (7) Maximum permissible mass of the full container = 40 t. *3D display of the container with reinforcement is shown in Figure 1.*

2.2. Some properties of the SCC

In order to achieve a complete filling of densely reinforced elements of the container (walls, bottom and cover) [2], the mix proportion of SCC developed as part of the development project. SCC was placed into the elements by the continuous casting through the tube [4]. In preparing the mix proportion of SCC, the rules for SCC, which are set out in SIST EN 206:2013, have been followed. A large number of SCC mixtures were prepared in the laboratory, changing the proportions and ratios of the individual components in such a way that a mixture was obtained that was suitable for the selected method of placing.

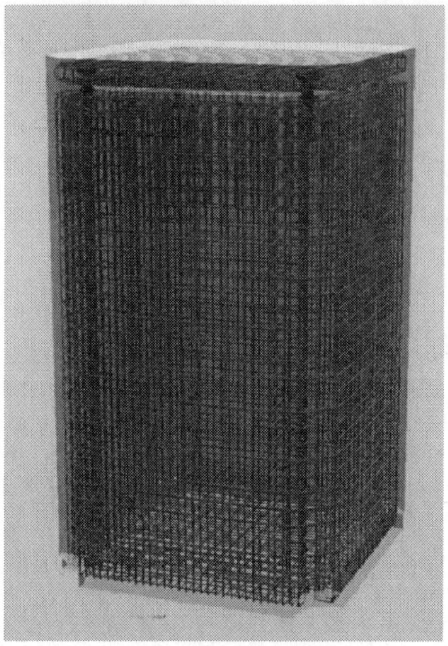

Figure 1. 3D display of the container with reinforcement.

The average results of measurements of fresh and hardened SCC are given in Tables 1 and 2. The results are within the limits of the required values [2, 3].

Table 1. The average results of the tests of the properties of fresh SCC.

Property (measured by standard)	Average value
Slump – flow (SIST EN 12350-8:2010)	680/710 mm
Density (SIST EN 12350-6:2009)	2393 kg/m^3
Air content (SIST EN 12350-7:2009, Chapter.5)	2,0 %
w/c ratio (SIST 1026:2016, Appendix NC)	0,37

Table 2. The average results of the tests of the properties of hardened SCC.

Property (measured by standard)	Average value
Compressive strength at 28 days (SIST EN 12390-3:2009)	86,0 MPa
Water penetration at 28 days (SIST EN 12390-8:2009)	3,0 mm
Shrinkage at 182 days (DIN 4227-Part 1)	0,374 mm/m
Resistance to freeze/thaw up to 200 cycles (SIST 1026:2016, Appendix ND)	101,0 %
Modulus of elasticity at 28 days (DIN 1048)	42500 MPa
Total porosity (EN 1936: 2006)	10,85 %
Chloride content (SIST EN 206: 2013, Chapter 5.2.8)	0,070 %

2.2.1. Increasing the compressive strength of SCC. The increase in the compressive strength of the SCC, the filling and sealing mortar depending on the time can be seen from Figure 2.

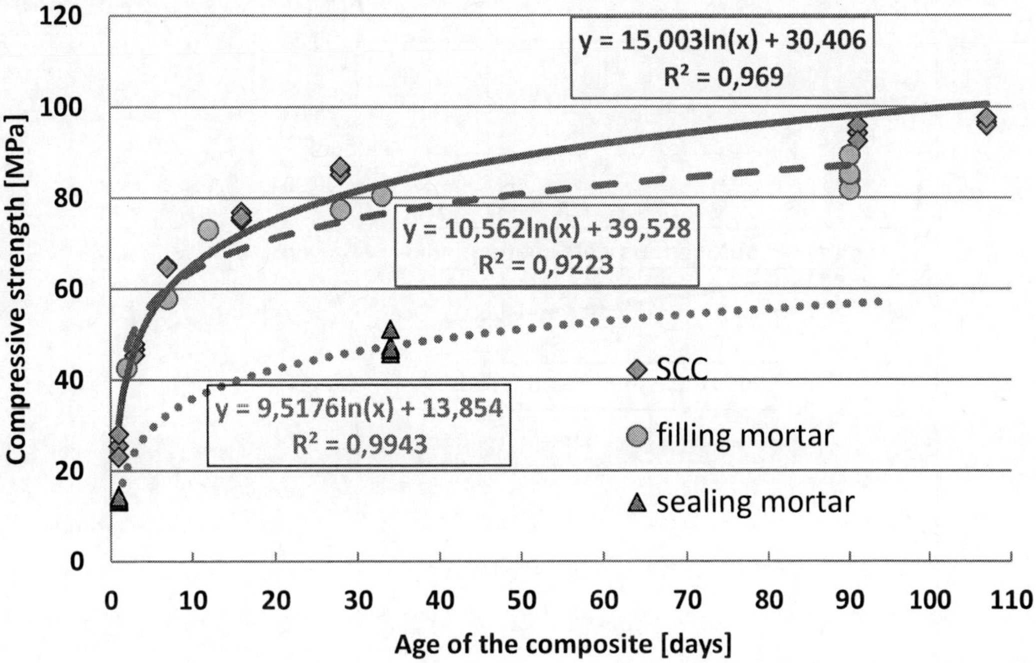

Figure 2. Compressive strength of SCC, filling and sealing mortar depending on their age.

The increase in compressive strength at an earlier age of SCC and a filling mortar up to 28 days is much more intense than in the case of an age above 28 days. A similar increase in the SCC modulus of elasticity was observed over time.

2.2.2. Measurement of autogenous shrinkage of SCC. Autogenous shrinkage is greater in high-strength concrete, or concrete, with a lower w/c ratio [8]. Autogenous shrinkage of concrete, also called hydration shrinkage, is the result of self-drying in the pores of cement stone that is water consumption in the process of hydration of cement. Thus, autogenous shrinkage occurs as soon as the process of hydration of the cement in the concrete begins. Depending on the type of concrete mixture, the shrinkage process begins about 2 to 24 hours after the mixing. Since concrete curing is already being established during this time and the concrete is placed in the formwork, evaporation of the water is significantly prevented. During this time, concrete deformations due to autogenous shrinkage and temperature deformations of concrete occur. After the curing of concrete is finished, the concrete is deformed - it is shrinking due to the evaporation of water from the surfaces of the concrete element.

Measurements of autogenous shrinkage were carried out at the laboratory of the Faculty of Civil Engineering and Geodesy in Ljubljana. Measurements were carried out in accordance with the Japanese standard JIS A 1129.

In Figure 3, the thin line shows the progress of the entire measured early shrinkage of SCC. This is the average value of the measurement results on three prisms with dimensions of 10 × 10 × 40 cm. The development of the temperature of the SCC as a function of time is given by a red line (upper diagram). The course of the estimated autogenous deformation of SCC is presented with a thick line, which is determined by deducting the temperature deformation of concrete from the total measured shrinkage. The results show a relatively small autogenous shrinkage of SCC. Also, a smaller shrinkage of SCC due to drying was determined. Average shrinkage of SCC at the age of 182 days is 0,374 mm/m.

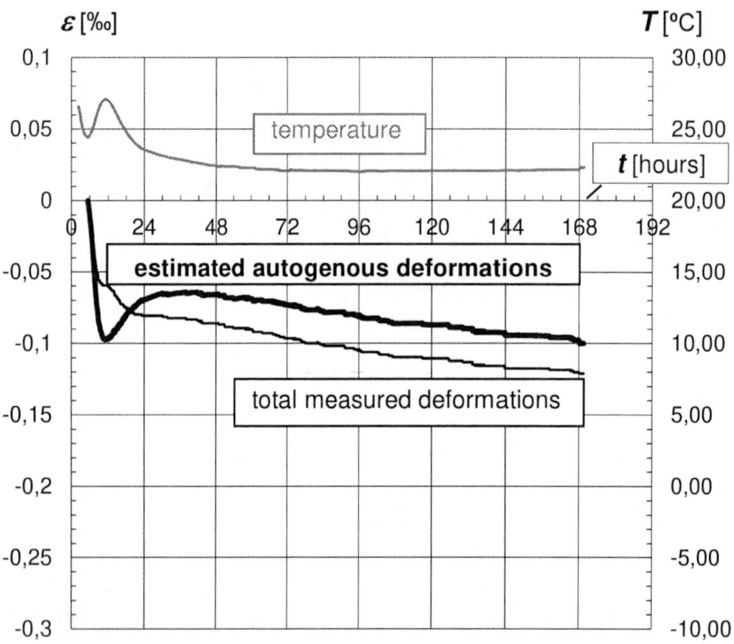

Figure 3. Results of SCC autogenous shrinkage measurement up to 7 days after mixing.

Within one of the previous research projects [9], laboratory tests were carried out into the time history of the shrinkage of fibre reinforced high-performance concrete with volumetric contents of 0.25%, 0.50% and 0.75% containing longer (l = 32 mm, d = 0,5 mm) or shorter (l = 16 mm, d = 0,5 mm) steel fibres or polypropylene fibres.

Based on the results, and discussion of results, of the performed experimental investigation into the shrinkage of high performance concrete containing polypropylene or shorter / longer steel fibres, the following conclusions can be reached: (1) The total shrinkage of the fibre reinforced concrete was, in the case of all of the investigated volumetric contents and types of fibres used, less than that of the comparable concrete without fibres. At the end of the period of measurement, the total shrinkage of the investigated concrete was, depending on the volumetric content and type of fibres used, approximately 17% to 29% less than that of the comparable concrete without fibres. The differences in the total shrinkage of the composites, with different volumetric contents and containing different types of fibres, were relatively small. (2) For the reduction of the early autogenous shrinkage of concrete, the use of steel fibres is more effective than the use of dry polypropylene fibres. In the case of using of steel fibres with volumetric contents of 0,25% or 0,50%, the longer fibres are more effective, whereas in the case of a volumetric content of 0,75% shorter fibres are more effective. (3) For reducing the total autogenous shrinkage, at a volumetric fibre content of 0,25% longer steel fibres are the most, whereas polypropylene fibres are the least effective. In the case of volumetric contents of fibres of 0,50% and 0,75%, however, there is relatively little difference in the effectiveness of steel and polypropylene fibres in the reduction of total autogenous shrinkage. (4) • The drying shrinkage of concrete is, in the case of all the investigated volumetric contents and types of fibres used, significantly less than the corresponding drying shrinkage of the comparable plain concrete without fibres.

3. SFR-SCC

In the continuation of the laboratory investigations SFR-SCC (Steel Fibre Reinforced Self-Consolidating Concrete) was prepared and investigated. Different quantities (0,25, 0,5, 0,75 and 1 vol.%) of steel fibres with anchors were added, with an equal length of 16 mm and a different thickness (diameter of cross-section) of 0,50, 0,40 and 0,35 mm, and aspect ratios (the ratio between length and diameter of the fibre) l/d = 32, 40 and 46. This paper provides only information and some of the results of the investigations of SFR-SCC. The aim was to identify possible potential for further improvement in the behaviour of the container during use, in particular, ensuring its long life.

3.1 Investigation of fresh SFR-SCC

The same tests of fresh SFR-SCC were performed as for fresh SCC. It was searched how the amount of fibres and their aspect ratios affect the workability and place-ability of SFR-SCC. In this paper, only the results of the slump flow test are given. The mean spread of fresh SFR-SCC according to the standard SIST EN 12350-8 was measured. The obtained results are given in Figure 4 depending on the amount of added fibres and their aspect ratio (l/d = 32, 40 and 46).

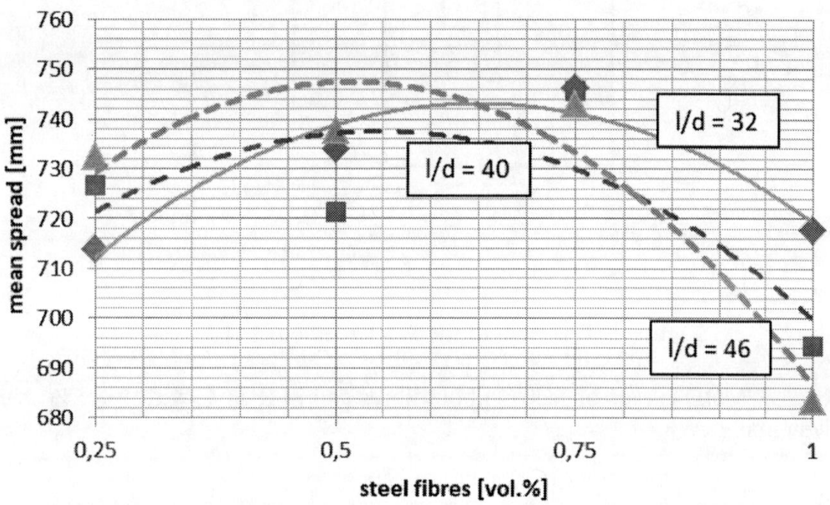

Figure 4. Mean spread of fresh SFR-SCC depends on the amount of fibres and their aspect ratio.

The results show that the fibres in amount of up to 0,75 vol.% do not affect workability and place-ability of fresh SFR-SCC. Some results suggest that the presence of fibres even improves workability, as if the fibres increase the internal "sliding" of the fresh SFR-SCC mass and do not hinder its workability - the flow-ability, as happens with a larger amount of longer fibres. The workability of fresh SFR-SCC is slightly reduced when 1 vol.% of fibres are added. The reduction in workability is greater when fibres with a larger aspect ratio are used.

3.2. Investigation of hardened SFR-SCC

The main purpose of adding fibres to concrete is to increase its ductility, toughness and resistance to crack propagation. Several types of tests were performed. This paper will only provide some of the results of the wedge spit test (WST).

WST is a test method to perform stable fracture mechanics tests on concrete and concrete-like materials. It was proposed by Brühwiler and Wittman [10], and Linsbauer and Tschegg [11]. The method proposed by last authors [11] was used in discussed and previous [12-14] investigations in order to obtain load – CMOD curves. Equivalent strengths up to selected crack width (CW = 0,1, 0,2,

0,3 and 0,4 mm) are calculated by equation, in which parameters, derived from load-CMOD curve are used.

Tests in accordance with the WST method were carried out on cubes with an edge length of 150 mm and with initial notch depth of 50 mm, at ages of 28 days. The principle of the testing method is shown in Figure 5 [15].

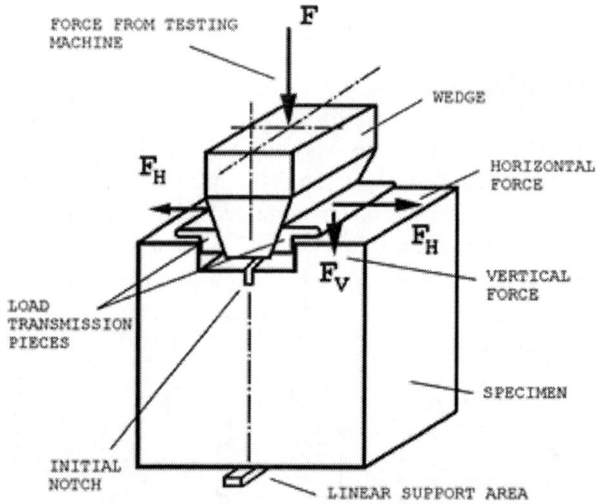

Figure 5. Principle of the WST method [15].

Characteristic load – CMOD curves of SCC without fibres and SFR-SCC with 1 vol.% steel fibres with l/d = 30 are given in Figure 6.

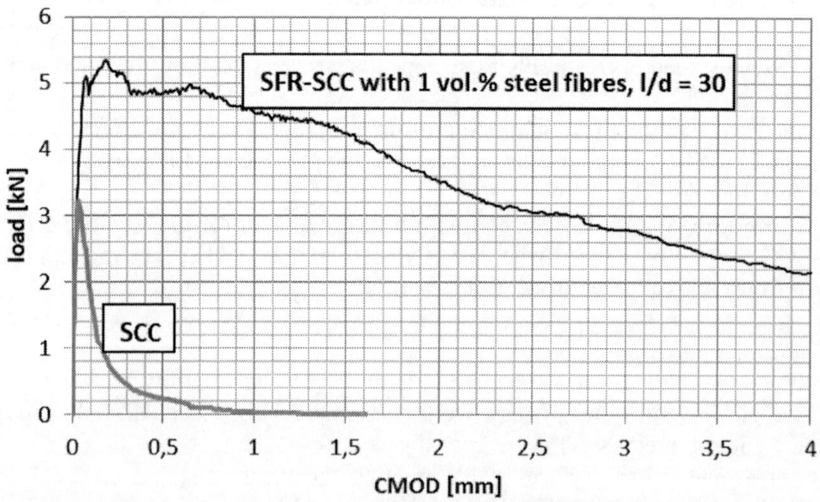

Figure 6. Characteristic load – CMOD curves of SCC without fibres and SFR-SCC with 1 vol.% steel fibres with l/d = 30.

There is a very large difference between the areas under the curves, which represent the amount of absorbed energy SCC and SFR-SCC during the test. From such diagrams the following strengths are calculated: ultimate strength f_{ct}, strength at first crack f_{FC} and equivalent strengths up to selected crack width (CW = 0,1, 0,2, 0,3 and 0,4 mm). For SCC, SFR-SCC with 1 vol.% steel fibres with l/d = 30 and 40, the average values of these strengths in Table 3 are given.

Table 3. The average results of strengths of SCC and SFR-SCC obtained by WST.

Type of concrete	f_{FC} (MPa)	f_{ct} (MPa)	$f_{0,1}$ (MPa)	$f_{0,2}$ (MPa)	$f_{0,3}$ (MPa)	$f_{0,4}$ (MPa)
SCC	2,8	3,2	2,7	2,2	1,8	1,5
SFR-SCC with 1vol.% fibers with l/d = 30	4,9	5,7	4,5	4,9	4,9	4,8
SFR-SCC with 1vol.% fibers with l/d = 40	5,2	6,1	5,0	5,2	5,4	5,5

The average ultimate strengths f_{ct} are given in Figure 7 depending on the aspect ratios l/d (aspect ratio l/d = 0 has SCC because it has no fibres).

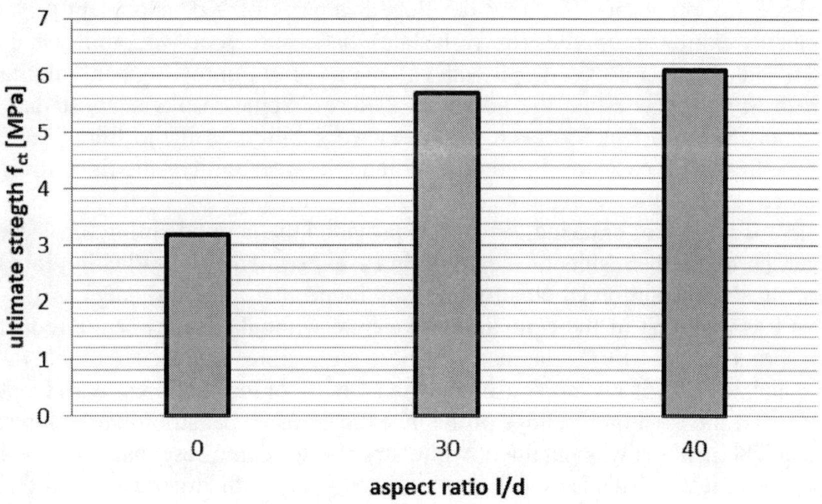

Figure 7. The average ultimate strengths f_{ct} depending on the aspect ratios l/d.

The average strength at first crack f_{FC} and equivalent strengths up to selected crack width f_{CW} are given in Figure 8 depending on the crack width (CW = 0, 0,1, 0,2, 0,3 and 0,4 mm). For the crack width CW = 0 mm, the first cracking strengths f_{FC} are applied.

From the average strength results (Table 3, and Figures 7 and 8), it is evident that they are much higher in SFR-SCC than in SCC. Fibres, of course have significant influence on post crack behaviour of concrete. Strain – softening response was obtained, when SCC without fibres were tested (Figure 8). On the other hand, strain - hardening response of SFR-SCC were gained up to the crack width of 0,4 mm (Figure 8). This is especially true when fibres with larger aspect ratio (l/d = 40) are used.

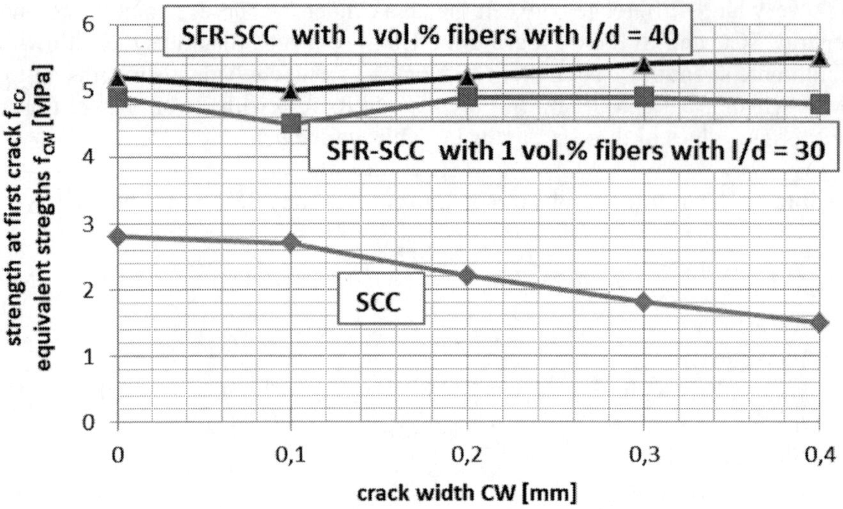

Figure 8. The average strength at first crack f_{FC} and equivalent strengths up to selected crack width f_{CW} depending on the crack width (CW = 0, 0,1, 0,2, 0,3 and 0,4 mm).

4. Conclusion

The SCC whose mix proportion has been developed as part of the development project for the LILW disposal container exhibits properties (in fresh state) that ensure good workability and denseness in the walls, bottom and container cover, despite the large amount of reinforcing bars. Similar behaviour was found in fresh SFR-SCC, despite the addition of fibres. Some results suggest that the presence of fibres in quantity up to 0,75vol.% even improves workability, as if the fibres increase the internal "sliding" of the fresh SFR-SCC mass and do not hinder its workability - the flow-ability.

The obtained results of measurements of autogenous shrinkage and shrinkage due to drying of SCC show that there is no risk of cracking, of course, provided that a good curing of placed SCC is carried out. However, based on the results of the preliminary experimental investigation into the shrinkage of high performance fibre reinforced concrete, it was found that total shrinkage of the fibre reinforced concrete was less than that of the comparable concrete without fibres. For the reduction of the early autogenous shrinkage of SFRC, the use of short steel fibres (with length of 16 mm) and with volumetric contents of 0,75% is more effective than the use of longer fibres (with length of 32 mm).

It is also one of the essential findings of the investigations is that strain-hardening response of SFR-SCC with 1 vol.% of fibres was obtained, while strain – softening response was obtained, when SCC without fibres were tested. This is especially true when fibres with larger aspect ratio are used.

On the basis of all the research results so far obtained, it can be concluded that there is a possibility to increase the resistance of SCC (for the construction of a LILW disposal container) to the crack propagation by adding a certain amount of steel fibres, without worsening the workability and place-ability of SCC.

References

[1] Duhovnik B, Kavnik T, Šušteršič J and Šajna A 2017 Program of testing of containers, measurements and tests necessary for the certification of the container, No NRVB-POM_Program (in the Slovenian language), Murska Sobota, Pomgrad.

[2] Sinur F and Duhovnik B 2018 Concrete container for disposal of low and intermediate level radioactive waste. (in the Slovenian language) In: *Proc. 25th Slo. Colloq. on Concrete* (pp. 53-60). Ljubljana, IRMA.

[3] Török Resnik T, Šušteršič J., Šajna A., Kavnik T. and Sinur F 2017 Results of testing of containers, measurements and tests necessary for the certification of the container, No NRVB-POM_Results (in the Slovenian language), Murska Sobota, Pomgrad.

[4] Török Resnik T and Kavnik T 2018 Disposal container with low and intermediate level radioactive waste (LILW), Production of container prototypes and drop testing. (in the Slovenian language) In: *Proc. 25th Slo. Colloq. on Concrete* (pp. 61-68). Ljubljana, IRMA.

[5] Bohinc U, Robič S and Šajna A 2018 LILW concrete container drop-test and damage assessment. (in the Slovenian language) In: *Proc. 25th Slo. Colloq. on Concrete* (pp. 79-87). Ljubljana, IRMA.

[6] European Agreement concerning the International Carriage of Dangerous Goods by Road (ADR); http://www.unece.org/trans/danger/publi/adr/adr2017/17contentse0.htlm.

[7] Zakon o prevozu nevarnega blaga (ZPNB), Ur.l. UPB1 33/06, 41/09, 97/10.

[8] Saje D 2005 Shrinkage of high strength concrete in the first days after placement. (in the Slovenian language) In*: Proc. 12th Slo. Colloq. on Concrete* (pp. 41-48). Ljubljana, IRMA.

[9] Saje D, Bandelj B, Šušteršič J, Lopatič J and Saje F 2012 Autogenous and Drying Shrinkage of Fibre Reinforced High-Performance Concrete. *Journal of advanced concrete technology*, Feb. 2012, Vol. **10**, No. 2, pp. 59-73

[10] Brühwiler E and Wittman F H 1988 The Wedge Splitting Test, a New Method of Performing Stable Fracture Mechanics Tests. *Fracture and Damage of Concrete and Rock*. Pergamon Press. 1988. Ed. H.P. Rossmanith. (pp. 117 – 125).

[11] Linsbauer H and Tschegg E K 1986 'Die Bestimmung der Bruchenegie an Würfelproben' (Fracture Energy Determination of Concrete with Cube-Shaped Specimens). *Zement und Beton*, **31** 1, (pp. 38 – 40).

[12] Šušteršič J, Kolenc M, Zajc A, Riček F and Zajc P M 1999 High-Performance Fibre Reinforced Concrete for Mine Roadway Support Panels. *Proceedings, Second CANMET/ACI International Conference Gramado*, RS, Brazil,. SP-**186**. (pp. 101 – 112).

[13] Šušteršič J, Ukrainczyk V, Zajc A and Šajna A 2001 Evaluation of Crack Opening Resistance of SFRC. *Concrete Under Severe Conditions*. Vol. **2**. Vancouver.. Eds. N. Banthia, K. Sakai, O.E. Gjørv. (pp. 1594 – 1601).

[14] Šušteršič J, Zajc A, Leskovar I, Dobnikar V 2003 Improvement in the crack opening resistance of FRC with low content of short fibres. Ed.: Dhir, Ravindra K. *Role of concrete in sustainable development : proceedings of the International Symposium dedicated to professor Surendra Shah, Northwestern University, USA* held on 3-4 September 2003 at the University of Dundee, Scotland, UK. London: "Thomas Telford", pp. 167-174.

[15] Tschegg E K 1991 'New Equipments for Fracture Tests on Concrete'; *Materialprüfung* **33** 11 - 12, München, pp. 338 - 342.

IOP Conf. Series: Materials Science and Engineering **596** (2019) 012014 doi:10.1088/1757-899X/596/1/012014

Steel microfibres in fly ash geopolymer for multifunctional conductive composites

C Mizerová, I Kusák, P Rovnaník and P Bayer

Brno University of Technology, Faculty of Civil Engineering, Brno, Czech Republic

E-mail: mizerova.c@fce.vutbr.cz

Abstract. Steel fibres in construction materials not only enhance their mechanical and fracture properties but also increase their capability of electrical conductivity. This feature thus enables the material to obtain advanced multifunctional properties that can be utilized in self-sensing and other smart applications. Fly ash geopolymers with steel microfibres in 5–30% wt. were prepared to assess the influence of the fibres on selected electrical properties of the composite (resistance, capacitance and relative permittivity) with respect to mechanical performance. Microstructure and interfacial bonding of matrix and steel fibres was observed by electron microscopy imaging. Steel fibres caused an improvement in all assessed electrical properties but only above a certain frequency values. Furthermore, steel reinforced samples exhibited increase in both compressive and flexural strength.

1. Introduction

Development of smart conductive composites could support the growing efforts to produce multifunctional materials. Since the electrical properties of construction materials reflect the changes of external conditions, they can be implemented in self-sensing solutions for real time health monitoring of civil infrastructures, i.e. sense strain, detect cracks and other defects or observe the changes of temperature and humidity. Such applications eliminate or reduce the need for embedded or attached devices, which are expensive and limited in durability [1]. Moreover, this idea is adopted for design of advanced traffic monitoring systems to track the weight and speed of vehicles and traffic density [2]. Other applications include self-heating concretes for both indoor and outdoor structures, especially for de-icing and winter maintenance of roads, car parks or airport areas [3]. Electrically conductive materials with carbon filler also exhibit increased ability of electromagnetic shielding needed in protective structures for strategic communication and control systems [4].

Geopolymers are novel cementitious materials synthetized by the alkaline activation of solid aluminosilicate precursors, frequently industrial waste materials or by-products. Such raw materials with preferably high amorphous glassy content with reactive Si and Al are dissolved in alkaline activator solution, usually mixture of alkali hydroxides and alkali silicates [5]. The following geopolymerization process is described as a formation of highly coordinated polymeric structure of SiO_4 and AlO_4 tetrahedra balanced by Na^+/K^+ cations supplied by the alkaline activator [6]. Geopolymer materials produced from low Ca fly ash, metakaolin or other calcined clays have been widely investigated regarding their good strength, chemical resistance and exceptional performance in exposure to elevated temperatures [7]. Considerable savings of nonrenewable resources, energy and emissions make geopolymer binders more environmentally friendly alternative to Portland cement [8].

Alkali activated materials may represent promising materials for smart composites as they are considered to possess lower electrical resistivity compared with cement-based binders [9]. Direct piezoelectric effect in geopolymeric mortars was explained by a charge imbalance and the formation of local dipoles generated by the migration of the hydrated Na^+ cations [10]. According to Hanjitsuwan [9,11], the electrical conductivity of fly ash geopolymer matrix is affected by the NaOH concentration in activator solution, liquid activator/ash ratio and frequency spectrum.

The application of steel fibres in geopolymeric materials was mostly discussed in studies concerning their mechanical performance and fracture properties and they confirm significant improvement of reinforced geopolymer in flexural strength, fracture toughness and transition of failure mode from brittle to ductile [12]. The effect of steel fibres on the electrical properties is scarcely mentioned. Chung [13] compared steel fibre and carbon fibre (CF) reinforced cement pastes stating that CF paste is a better piezoressistive strain sensor than stainless steel fibre paste at a similar fibre volume fraction. Conductivity of fly ash geopolymers could be enhanced by a very low dosage of CNTs (up to 0.5%) but only in case of effective distribution within the matrix [14]. Recently, exceptional improvement of performance in both mechanical and electrical properties was observed in studies of fly ash geopolymer doped with graphene [15].

2. Materials and methods

Main source materials for the geopolymer binder was fly ash (FA) from black coal combustion with the chemical composition given in **Table 1** and commercially sold water glass (sodium silicate solution with molar ratio $Na_2O/SiO_2 = 1.6$). Brass coated steel microfibers DM 6/0,175 were used as a conductive admixture in 5–30 % wt. of the fly ash. We used standardized quartz sand as fine aggregate and a defoaming agent Lukosan S to diminish excessive pore generation in fresh mortars.

Table 1. Chemical composition of fly ash.

	SiO_2	Al_2O_3	Fe_2O_3	CaO	MgO	S_{total}	Na_2O	K_2O
(%)	49.82	24.67	7.50	3.91	2.68	0.91	0.70	2.78

Each of the mixes was prepared according to mix compositions shown in **Table 2**, following these steps: first, mixing the sodium silicate solution with fly ash and water. After proper binder homogenization, quartz sand was added from fine to coarse fractions. Then the steel microfibers were added into the fresh mortar and the defoaming agent was added at the very end of the mixing procedure.

Fresh mortars were poured into prismatic moulds (40 × 40 × 160 mm) and covered with plastic sealant to protect the binder from undesirable moisture loss from the surface. After 2 hours in laboratory conditions (22 ± 2 °C, $\varphi = 45 \pm 5$ %), the samples were thermally treated at 40 °C for 24 h. Demoulded specimens were stored in laboratory till the age of testing.

Table 2. Mix design of geopolymer mortars.

	REF	SF 5	SF 10	SF 15	SF 20	SF 25	SF 30
Fly ash (g)	350	350	350	350	350	350	350
Sodium silicate (g)	280	280	280	280	280	280	280
Steel fibres (g)	-	17.5	35	52.5	70	87.5	105
Sand (g)	1050	1050	1050	1050	1050	1050	1050
Lukosan S 1% (ml)	-	2	2	2	3	3	3
Water (g)	35	35	35	35	35	35	35

Before the measurements of electrical properties, the prepared prismatic samples were dried at 105 °C to reduce the influence of moisture on assessed properties. The specimens were placed between parallel brass electrodes (30 × 100 mm) so that a distance between electrodes was 40 mm and characterized by impedance spectroscopy in the range of 40 Hz to 1 MHz using an Agilent 33220A sinusoidal signal generator and an Agilent 54645A dual-channel oscilloscope. The output voltage of the signal generator was 5.5 V, the input values for the electrical capacity and the resistance of the oscilloscope were 13 pF and 1 MΩ, respectively. Relative permitivity measurements were performed in the range of 100 to 3000 MHz using the vector analyser and DAK probe. Geopolymers were tested for their mechanical properties (compressive and flexural strength) and scanned by electron microscope (SEM) Tescan MIRA3 XMU to obtain the images of microstructure of the geopolymers.

3. Results and discussion

3.1. Electrical properties

The main purpose of conductive filler in smart materials is to enhance their conductivity and reduce electrical resistance. Electrical resistance of geopolymer mortars is depicted in **Figure 1** and decreased with higher frequency applied. Up to 600 Hz, the lowest resistance was unexpectedly attained by sample SF 5. Curve shapes of samples with 5–25% steel fibres seem to be somehow “inverted” from this certain point and increasing steel fibres content resulted in drop of resistance only above this frequency.

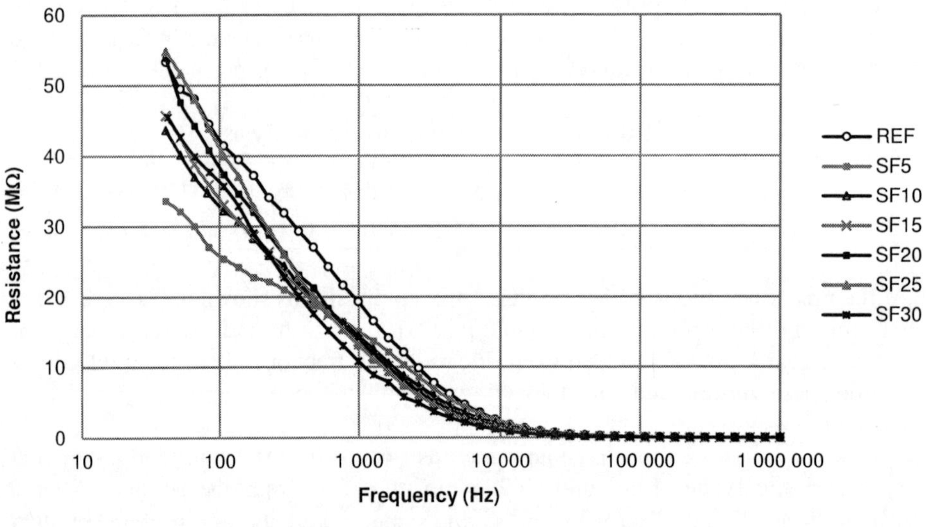

Figure 1. Electrical resistance of FA geopolymer mortars with different steel fibres content.

Similar behaviour was observed regarding the electrical capacitance (**Figure 2**). The capacitance decreased with frequency and increased with steel fibres content but this was noted only after the critical value, this time at frequency equalling to approximately 200 Hz. Capacitance of all samples exhibited minor differences at higher frequency. **Table 3** summarizes the values of resistance and capacitance at chosen reference frequency 1 kHz. Resistance and capacitance values above the points of frequency previously mentioned follow the expectations of reduced electrical resistance and growing capacitance with high steel fibres content.

As far as the output of electrical conductivity measurements was not applicable (due to a number of points with zero conductivity values measured across the frequency spectrum), the electrical properties at higher frequency up to 3000 MHz are represented by the relative permittivity (**Figure 3**). Relative permittivity increased with the steel fibres content across the spectrum and slightly declined with higher frequency. Initial outlying values related to the transition between the electrode and the samples were omitted and the curves could be replaced by trend lines as the fluctuations are caused by the measuring device. Significant increment of relative permittivity in case of maximum steel fibres dose (SF 30) is probably attributed to the surface conductivity as the transmitting and receiving parts of probe were located on the same surface of the sample.

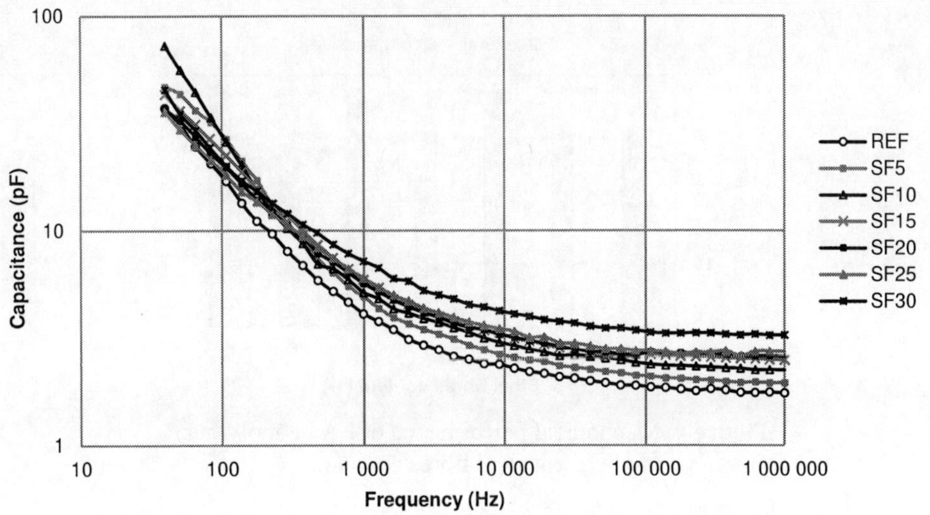

Figure 2. Capacitance of FA geopolymer mortars with different steel fibres content.

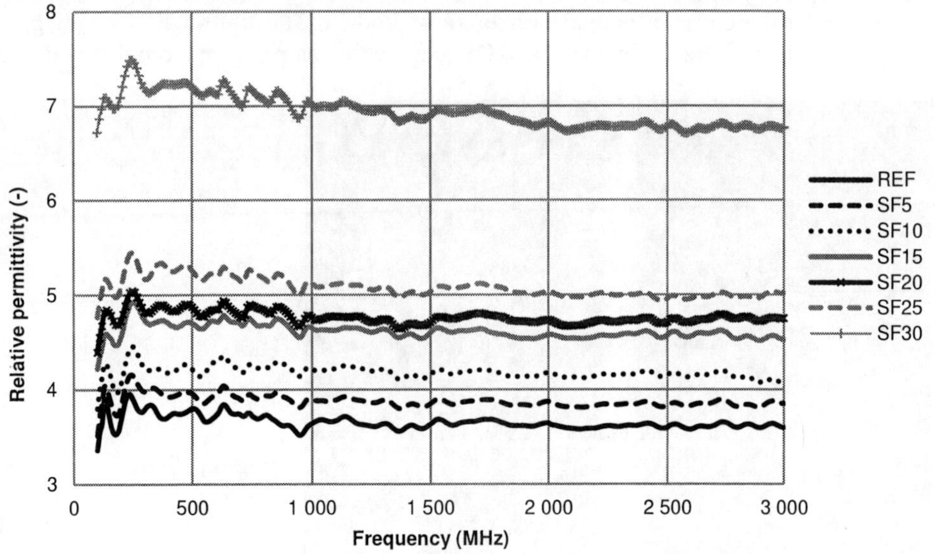

Figure 3. Relative permittivity of FA geopolymer mortars with different steel fibres content.

Table 3. Electrical resistance and capacitance at chosen frequency 1 kHz.

	REF	SF 5	SF 10	SF 15	SF 20	SF 25	SF 30
Electrical resistance (MΩ)	19.29	15.05	14.40	13.60	14.06	13.13	10.83
Electrical capacitance (pF)	4.11	4.92	5.17	5.90	5.68	6.02	7.22

3.2. *Mechanical properties*

Mechanical performance of steel fibre doped geopolymers is shown in **Figure 4**. Geopolymer with 5% of steel fibres exhibited similar strength values as the reference sample, slight reduction in compressive strength might be associated with possible increase in porosity of the material and insufficient bonding ability of the matrix and still relatively low dosage of fibres. Higher steel fibres content resulted in improvement in both flexural strength (by up to 80%) and compressive strength (by up to 20%).

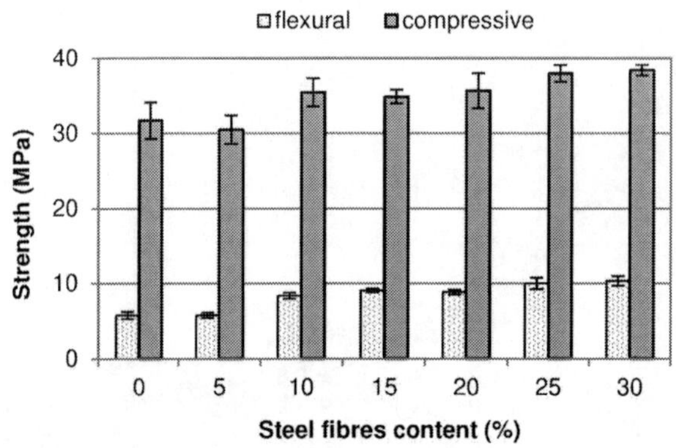

Figure 4. Mechanical performance of FA geopolymer mortars with different steel fibres content.

3.3. *Microstructure*

The nature of the interfacial zone between steel fibre and geopolymer matrix is shown in **Figure 5**. Steel fibre as received is depicted in picture A and as one can see in pictures B and C, the fibre in matrix is coated by a fine layer of geopolymer binding product. The fibre-matrix bonding is ensured by local geopolymer bond bridges (image C and D) because the matrix is not completely dense.

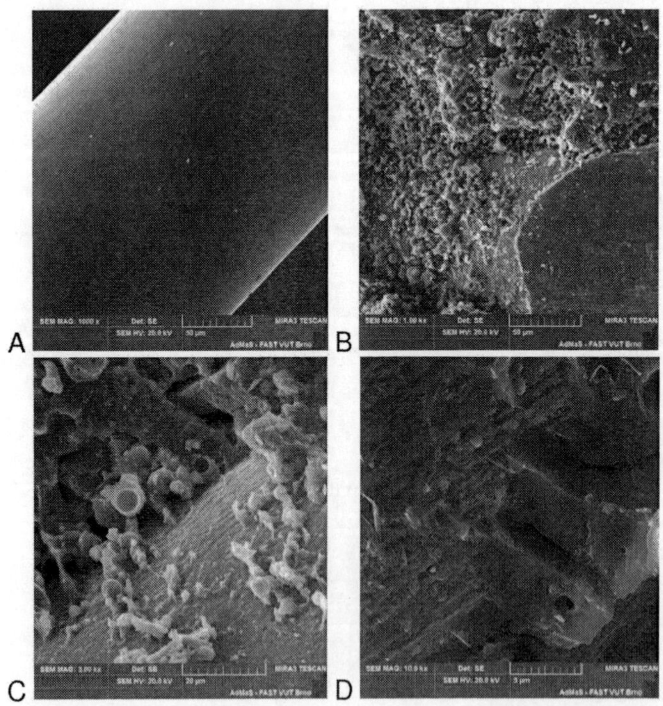

Figure 5. SEM images of steel fibres in geopolymer matrix.
A: steel fibre as received,
B: cross-section of the steel fibre,
C: spot bonds of the matrix and binding product coated surface of the fibre,
D: bonding in detail.

4. Conclusions

Steel fibres are usually applied in concrete to reduce shrinkage and increase flexural strength or fracture toughness. This research aimed to assess their contribution to the electrical properties of fly ash geopolymer intended to design advanced conductive composites. Geopolymer mortars with 5-30% of steel microfibers were analysed by means of impedance spectroscopy. Reduced electrical resistance and increase in capacitance with higher steel fibres content were registered only above certain points in frequency spectrum, approximately 600 Hz or 200 Hz, respectively. Samples with 15-25% of steel fibres exhibited similar behaviour in terms of electrical properties, electrical resistance and capacitance across the frequency spectrum did not considerably differ. Relative permittivity at 15% of steel fibres clearly increased and it was comparable with 20% dosage. As far as the differences among these samples were small, this concentration range might refer to percolation threshold or other change in microstructure that affects the transfer of electric charge. Concerning the mechanical performance, the steel fibres addition of 10% and higher resulted in improvement of both flexural and compressive strength. From the results obtained, 15% steel fibres concentration seems to meet the requirements for enhanced electrical and mechanical performance of fly ash geopolymer.

Acknowledgements
This study has been financially supported by the Czech Science Foundation under the project No. 19-11516S and the BUT student research project FAST-J-19-6016.

References

[1] Han B, Yu X and Ou J 2014 *Self-sensing concrete in smart structures* (Oxford: Butterworth-Heinemann)
[2] Shi Z Q and Chung D D L 1999 Carbon fiber-reinforced concrete for traffic monitoring and weighing in motion. *Cement and Concrete Research* **29(3)** pp 435-439
[3] Chang C, Ho M, Song G, Mo Y L and Li H 2009 A feasibility study of self-heating concrete utilizing carbon nanofiber heating elements. *Smart Materials and Structures* **18(12)** pp 127001
[4] Chung D D L 2012 Carbon materials for structural self-sensing, electromagnetic shielding and thermal interfacing. *Carbon* **50.9** pp 3342-3353
[5] Provis L, Van Deventer J S J 2014 Alkali Activated Materials (Springer)
[6] Davidovits J 1994 Properties of geopolymer cements. *First international conference on alkaline cements and concretes* (Kiev: Kiev State Technical University, Scientific Research Institute on Binders and Materials) pp 131-149
[7] Duxson P, Fernández-Jiménez A, Provis J L, Lukey G C, Palomo A and Van Deventer J S J 2007 Geopolymer technology: the current state of the art. *Journal of materials science* **42(9)** pp 2917-2933
[8] Van Deventer J S J, Provis L, Duxson P and Brice D G 2010 Chemical research and climate change as drivers in the commercial adoption of alkali activated materials. Waste and Biomass Valorization 1.1 pp 145-155 K
[9] Lamuta C, Candamano S, Crea F, and Pagnotta L 2016 Direct piezoelectric effect in geopolymeric mortars. *Materials & Design* **107** pp 57-64
[10] Hanjitsuwan S, Hunpratub S, Thongbai P, Maensiri S, Sata V and Chindaprasirt P 2014 Effects of NaOH concentrations on physical and electrical properties of high calcium fly ash geopolymer paste. *Cement and Concrete Composites* **45** pp 9-14
[11] Hanjitsuwan S, Chindaprasirt P and Pimraksa K 2011 Electrical conductivity and dielectric property of fly ash geopolymer pastes. *International Journal of Minerals, Metallurgy, and Materials* **18.1** pp 94-99
[12] Ranjbar N, Mehrali M, Mehrali M, Alengaram U J and Jumaat M Z 2016 High tensile strength fly ash based geopolymer composite using copper coated micro steel fiber. *Construction and building materials* **112** pp 629-638

[13] Wen S and Chung D D L 2003 A comparative study of steel-and carbon-fibre cement as piezoresistive strain sensors. *Advances in cement research* **15(3)** pp 119-128

[14] Saafi M, Andrew K., Tang P L, McGhon D, Taylor S, Rahman M, ... and Zhou X 2013 Multifunctional properties of carbon nanotube/fly ash geopolymeric nanocomposites. *Construction and Building Materials* **49** pp 46-55.

[15] Saafi M, Tang L, Fung J, Rahman M, Sillars F, Liggat J and Zhou X 2014 Graphene/fly ash geopolymeric composites as self-sensing structural materials. *Smart materials and structures* **23(6)** pp 065006.

IOP Conf. Series: Materials Science and Engineering **596** (2019) 012015 doi:10.1088/1757-899X/596/1/012015

Properties of cement-lime render containing perlite as lightweight aggregate

J Pokorný, M Pavlíková and Z Pavlík

Czech technical university in Prague, Faculty of Civil Engineering, Department of Material Engineering and Chemistry, Czech Republic

E-mail: jaroslav.pokorny@fsv.cvut.cz

Abstract. Lightweight renders with improved thermal performance, resistant against moisture and respecting specific behaviour of original plasters and substrates became attractive materials for surface finishing of both historical and old wall structures. In this respect, structural, mechanical and thermal properties of cement-lime based renders containing a different amount of perlite as fine silica sand substitution are researched in the presented paper. Experimental tests conducted for 28 days samples showed significant lightening of hardened renders with increased perlite content in their fresh mix. Gained strength parameters were, because of higher total open porosity of perlite enriched samples, lower compared with reference material. On the other hand, newly developed lightweight renders exhibited improved thermal insulation performance due to the highly porous perlite particles. In summary, cement-lime renders with incorporated lightweight aggregate based on expanded perlite were considered as alternative and promising materials for application in repair and restoration of historical masonry.

1. Introduction

Lime-based mortars and renders have been the most widespread used building composites not only in the historical monuments but also in buildings constructed at the beginning of 20th century [1]. As the basement of such buildings has not yet been provided with horizontal waterproof insulation or it has been already damaged there might occur a rising damp from the ground upward through permeable wall structure [2]. Advantageous property of lime-based composites, given by their pore structure with the network of capillaries, is being mentioned water vapour permeability helping to keep masonry structures breathable. However, the use of aerial lime as the main binder component provides only low mechanical strength and poor water resistance of renders in the wet or water environment respectively [3]. The incorporation of Portland cement into plaster mix helps to create denser and more durable porous structure of hardened material better resisting to water induced action due to the presence of highly stable C-S-H phases.

The compatibility between the applied repair/retrofitting render and the original substrate and all the masonry elements is very important. Gulbe et al. [4] recommend add lime binder into hydraulic based mortars in order to get conformity and comparable properties with historical mortars.

Rising damp in many cases brings difficulties with water soluble salts coming from subsoil or from agents applied for winter maintenance.

Salts nucleation and subsequent crystallization is accompanied by the creation of high volume products that fulfil porous structure of renders and cause their gradual decay [2], [5]. Consequently, renders used for restoration should dispose high porosity exceeding 40 % according to the WTA-2-9-04 manual recommendation [6]. Their composition is often spread about the addition of light filling materials, such as perlite, polystyrene, vermiculite, etc., which are capable to store a great amount of salts [5]. In addition, highly porous aggregates enhance thermal properties of produced plasters which, like a part of building envelope, help to reduce energy demands for heating and cooling required in EU directives [7].

The effect of perlite incorporation on behaviour of cement-lime renders was examined in the paper. The analysis of experimental results revealed the reduction in unit weight and strength parameters of tested lightweight renders compared to reference material without perlite use. High open porosity, however, rapidly decreased thermal conductivity of materials with perlite-based aggregate.

2. Experimental

The experimental investigation was performed on the set of cement-lime based renders with improved thermal performance. The obtained data was compared with results assessed for reference render with silica sand aggregate. In all prepared render mixes, ordinary Portland cement (PC), type 42.5 R and hydrated lime CL 90-S were used. For the determination of chemical composition of both binders, X-ray fluorescence device ED-XRF Quant-X (Thermo Scientific) was employed. During the measurement taking place in argon atmosphere, the basic elements of tested powders were determined and subsequently by the software Uni Quant ED 6.28 recalculated to oxides. Particle size distribution of lime-hydrate and PC was accessed by the apparatus Analysette 22 NanoTec (Fritsch) equipped with the red and green lasers that enable to identify particles with diameter in the range of 0.08–2 000 μm. Specific surface was measured using Blaine device model 62 (Controls, Spa, Italy) according to the standard EN 196-6 [8]. Specific densities of both bonders were determined by automatic helium pycnometer Pycnomatic ATC (Thermo Scientific).

Oxides composition (only main oxides are given) and physical properties including particle size distribution of lime hydrate and PC are summarized in Tables 1 and 2. In particular, obtained material characteristics indicated approx. 6 times higher specific surface of lime hydrate contrary to PC. That observation was also confirmed by measured particle size distribution when 50 % of hydrated lime particles were lower than 4.2 μm, 22.8 μm in the case of used cement type respectively.

Table 1. Composition of used cement and hydrated lime.

Oxides composition (mass %)	PC	LH
SiO_2	19.00	0.15
Al_2O_3	4.31	0.10
Fe_2O_3	2.40	0.12
TiO_2	0.28	-
CaO	62.90	98.70
MgO	1.80	0.41
K_2O	0.82	-
Na_2O	0.14	-
SO_3	3.24	0.12

Table 2. Material characteristics of lime hydrate and PC used.

Material	Specific surface ($m^2 \cdot kg$)	Powder density ($kg \cdot m^{-3}$)	Specific density ($kg \cdot m^{-3}$)	d_{10} (μm)	d_{50} (μm)	d_{90} (μm)
Hydrated lime	2 211	232.8	2 210	0.8	4.2	50.3
Portland cement	360	970	3 129	6.0	22.8	32.4

As a reference material, cement-lime render (REF) was produced. It consisted of PC and lime hydrate dosed in 1/1 wt. ratio, three kinds of silica sand providing a smooth particle size distribution curve with particles size up to 2 mm and batch water. In other renders, dense silica aggregate was substituted in the portion of 25–100 vol.% by lightweight aggregate – a mix of commercially available perlite EP 100 (Perlite 0.0/0.5) and EP 150 PB (Perlite 0.0/2.0) mixed together in the volumetric ratio ¼. Finely particulate perlite served as a supplementary material increasing the amount of fine grains in perlite mix whose particle size distribution was thus more resembling to that of silica aggregate. Before mixing, perlite aggregate was stored in closed plastic box filled with tap water for 24 h. The amount of added batch water kept similar workability of all developed render mixtures with the value of spreading of 160 × 160 ± 5 mm. Mix proportions of reference as well as of lightweight renders are given in Table 3.

Table 3. Mix proportions of reference and lightweight renders.

Substance	Content ($kg \cdot m^{-3}$)				
	REF	LR 25	LR 50	LR 75	LR 100
Lime hydrate	241.9	239.9	250.4	262.2	274.3
Portland cement	241.9	239.9	250.4	262.2	274.3
Sand 0.0/0.5	451.6	335.9	233.8	122.3	-
Sand 0.5/1.0	451.6	335.9	233.8	122.3	-
Sand 1.0/2.0	451.6	335.9	233.8	122.3	-
Perlite 0.0/0.5	-	2.4	5.0	7.3	9.5
Perlite 0.0/2.0	-	40.8	85.6	123.8	173.4
Water	348.3	263.9	210.3	157.3	98.8

Raw materials were mixed by mixer Spar D 200-B, from fresh renders prisms having dimensions of 40 × 40 × 160 mm and 70 mm cubes were casted. Filled metal moulds were than compacted by 5 hits on laboratory table. After 48 h of initial hardening in laboratory conditions at temperature (20 ± 1) °C and relative humidity (45 ± 5) %, samples were demoulded and stored in conditions with high relative humidity (~ 98 %) for overall time of 26 days.

Particle size distribution of used aggregates was assessed by sieve analysis according to the standard EN 933-1 [9]. At first, a representative sample of the particular tested aggregate was dried to constant mass at (105 ± 2) °C and naturally cooled. Then, it was placed into series of sieves with mesh size 2.0; 1.0; 0.5; 0.25; 0.125, and 0.063 mm, and mechanically shaken by vibratory sieve shaker Retsch AS 200.

Value of spreading of fresh mixtures was determined according to the standard EN 1015-3 [10] in two perpendicular directions after 15 hits on the shook table.

For the hardened 28 days cured samples, basic physical parameters, mechanical properties and thermo-physical properties were determined. The bulk density measurements were conducted according to the standard EN 1015-10 [11]. The specific densities were accessed by helium pycnometer Pycnomatic ATC (Thermo-Scientific). The total open porosity was calculated based on both the bulk and matrix density values. The relative expanded uncertainty of the applied testing method was 5 %. The mechanical properties, namely flexural and compressive strength were assessed according to the standard EN 1015 – 11 [12]. Three point bending test was carried out on prismatic specimens having size 40 × 40 × 160 mm. On the rest of broken prisms, compressive strength was tested [13]. The expanded measuring uncertainty of the both strength tests was 1.6 %. Among tested mechanical parameters, determination of dynamic Young's modulus was done using apparatus Dio 562 NLF (Starmans Electronics) [14]. The relative expanded uncertainty of the dynamic Young's modulus measurement was 2 %.

Thermo-physical properties, represented by thermal conductivity λ ($W \cdot m^{-1} \cdot K^{-1}$), thermal diffusivity a ($m^2 \cdot s^{-1}$) and volumetric heat capacity c ($J \cdot m^{-3} \cdot K^{-1}$), were measured on 70 mm dried cubes by the portable device ISOMET 2114 (Applied Precision) equipped with circular plate probe. The measuring range of thermal conductivity was from 0.04 to 3 $W \cdot m^{-1} \cdot K^{-1}$ with the measurement accuracy 5 % of reading + 0.001 $W \cdot m^{-1} \cdot K^{-1}$. For volumetric heat capacity was the accuracy 10 % of reading and the reproducibility was 3 % of reading + $1 \cdot 10^3$ J $m^{-3}K^{-1}$ [15]. Tests were performed under laboratory conditions at temperature (21 ± 2) °C.

3. Results and discussion

Basic material properties of perlite and sand fractions are introduced in Table 4. From the powder density as well as matrix density results, one can observe significantly lower unit weight of perlite compared to silica sand that led to above indicated requirement of volumetric substitutions of dense silica aggregate. Accordingly, high porosity of perlite particles expectedly help to mitigate thermal conductivity of hardened specimens. Particle size distribution of silica and perlite aggregates is shown in Figure 1. Both obtained distribution curves had practically similar shape. However, perlite aggregate contained slightly higher representation of particles with diameter size above 0.5 mm.

Table 4. Material parameters of perlite and sand fractions.

Material	Perlite 0.0/0.5	Perlite 0.0/2.0	Sand 0.0/0.5	Sand 0.5/1.0	Sand 1.0/2.0
Powder density (kg·m^{-3})	55	233	1 478	1 505	1 568
Matrix density (kg·m^{-3})	651	572	2 654	2 648	2 640

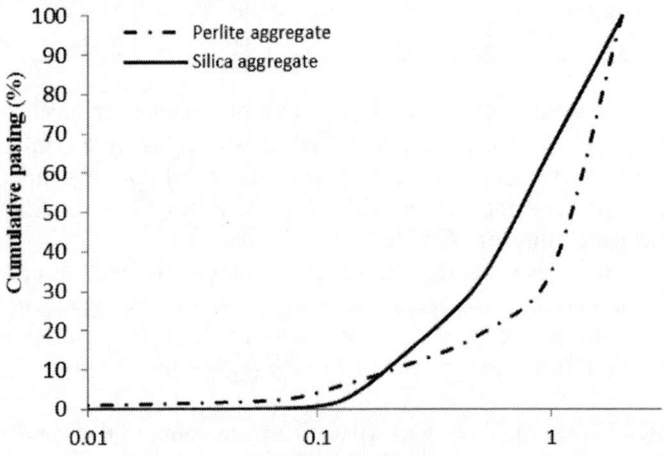

Figure 1. Sieve analysis of perlite and fine silica aggregates.

The effect of perlite dosage on basic material properties of examined renders is apparent from Table 5. It is evident that an increase of perlite content resulted in a gradual lightening of hardened renders when bulk density values dropped from initial 1 814 kg m^{-3} determined for reference sand-based material to 635 kg·m^{-3} obtained in the case of sand-free render. Total open porosity data shoved corresponding trend. Moreover, tested hardened specimens containing lightweight aggregate equal or above 50 vol. % reached open porosity exceeding 40 % thereby these met requirements according to the WTA-2-2-04 [6] manual for restoration renders. According to the standard EN 998-1 [16] these materials could be also sorted as lightweight renders with dry bulk density below 1 300 kg·m^{-3}.

Table 5. Basic material properties.

Render	Bulk density (kg·m^{-3})	Matrix density (kg·m^{-3})	Total open porosity (%)
REF	1 814	2 525	28.1
LR 25	1 536	2 330	34.1
LR 50	1 163	2 090	44.4
LR 75	908	2 034	55.3
LR 100	635	1 618	60.8

Strength properties and dynamic Young's modulus are summarized in Table 6. Obtained experimental results are in agreement with reported basic material properties. There was recorded strength as well as Young's modulus decline proportional to the growth of perlite content. In this respect, the lowest compressive strength of 2.9 MPa was measured for LR 100. According to the EN 998-1 [16], recorded compressive strength fit in the class CS II (strength values from 1.5 $N \cdot mm^{-2}$ to 5.0 $N \cdot mm^{-2}$) and thus can be used as lightweight render for renovation purposes as well.

Table 6. Mechanical properties.

Render	Flexural strength (MPa)	Compressive strength (MPa)	Young's modulus (GPa)
REF	2.7	8.1	10.9
LR 25	2.3	7.8	8.0
LR 50	2.0	5.2	3.7
LR 75	1.4	3.9	2.5
LR 100	1.2	2.9	1.6

The development of thermo-physical properties in dependence on perlite content is given in Table 7. Due to low weight and high porosity of perlite aggregate, a steep decrease in the thermal conductivity and thermal diffusivity was recorded. In porous structure entrapped air acts as an excellent thermal insulator and improves thermal insulation function of composite materials [17]. On this account, render with full silica sand substitution by perlite reached approx. 11.5 times lower thermal conductivity compared with the reference samples. Very low thermal conductivity (0.12 $W \cdot m^{-1} \cdot K^{-1}$) suggests its considerable thermal insulation function and its possible classification as a thermal insulating render meeting thermal performance demand $T_2 \leq 0.2$ $W \cdot m^{-1} \cdot K^{-1}$ (EN 998-1) [16].

Table 7. Thermo-physical characteristics.

Render	λ ($W \cdot m^{-1} \cdot K^{-1}$)	$a \times 10^{-6}$ ($m^2 \cdot s^{-1}$)	$c \times 10^6$ ($J \cdot m^{-3} \cdot K^{-1}$)
REF	1.39	0.82	1.70
LR 25	0.91	0.55	1.65
LR 50	0.42	0.26	1.56
LR 75	0.22	0,18	1.24
LR 100	0.12	0.16	0.78

4. Conclusion

The paper was focused on the effect of a partial and full substitution of commonly used dense silica fine aggregate by perlite mix on properties of developed lightweight renders. Practical experiments revealed very low weight of perlite contrary to silica sand that resulted in significant decrease of bulk density and lightening of hardened render specimens. Higher representation of lightweight aggregate, however, led to mechanical resistance reduction. On the other hand, highly porous perlite particles helped to improve thermal insulation function of developed renders.

Designed rendes, in particular these without silica fine aggregate, dispose of low unit weight, high open porosity, sufficient mechanical resistance, and thus meet the requirements according to the standard EN 998-1 for lightweight thermal insulating repair renders suitable for renewal purposes.

Acknowledgments

The authors gratefully acknowledge the financial support received from the Czech Science Foundation under project No 18-07322S, and by the CTU in Prague under project No SGS17/166/OHK1/3T/11.

References

[1] del Mar Barbero-Barrera M, Maldonado-Ramos L, Van Balen K, García-Santos A and Neila-González F J 2014 *J. Cult. Herit.* **15** 326
[2] Brachaczek W 2018 *Constr. Build. Mater.* **182** 418
[3] Stefanidou M, Kesikidou F and Antoniadis K 2017 *Procedia Environ. Sci.* **38** 242
[4] Gulbe L, Vitina I and Setina J 2017 *Procedia Eng.* **172** 325
[5] Hunek B D, Siddique R, Klimek B and Franus M 2018 *Constr. Build. Mater.* **142** 162
[6] WTA Merkblatt 2-9-04 Sanierputzsysteme
[7] Gomes G M, Flores-Colen I, Melo H and Soares A 2019 *Constr. Build. Mater.* **198** 786
[8] EN 196-6, Methods of testing cement – Part 6: Determination of fineness, CEN, Brussels, 2010
[9] EN 933-1, Test of geometrical properties of aggregate – Part 1: Determination of particle size distribution – Sieving method, CEN, Brussels, 2012
[10] EN 1015-3, Methods of test for mortar for masonry – Part 3: Determination of consistence of fresh mortar, CEN, Brussels, 2000
[11] EN 1015-10, Methods of test for mortar for masonry - Part 10: Determination of dry bulk density of hardened mortar, CEN, Brussels, 1999
[12] EN 1015-11, Methods of test for mortar for masonry - Part 11: Determination of flexural and compressive strength of hardened mortar, CEN, 1999
[13] Kulovaná T, Pokorný J, Keppert M, Pavlík Z and Černý R 2016 *Mater. Sci.* **22** 82
[14] Záleská M, Pavlíková M, Pokorný J, Jankovský O, Pavlík Z and Černý R 2018 *Constr. Build. Mater.* **180** 1
[15] Záleská M, Pavlíková M, Jankovský O, Pokorný J and Pavlík Z 2018 *Ceram.-Silik.* **66** 221
[16] EN 998-1, Specification for masonry – Part 1: Rendering and plastering mortar, CEN, Brussels, 2016
[17] Abili S, Nait-Ali B, Joliff Y and Favotto C 2015 *Compos. Part B* **68** 392

IOP Conf. Series: Materials Science and Engineering **596** (2019) 012016 doi:10.1088/1757-899X/596/1/012016

Improved methodology for determining tensile strength of fibre reinforced concrete

V Křístek, J Vodička, L Musil and H Hanzlová

Department of Concrete and Masonry Structures, Czech Technical University in Prague, Faculty of Civil Engineering, Thákurova 7/2077, Prague, Czech Republic

E-mail: vladimirkristek@seznam.cz

Abstract. Two configurations are currently used to determine the tensile strength of fibre reinforced concrete: simply supported beams of square cross section loaded either by two forces at thirds of spans (four-point bending test) or by one force at the mid-span where the specified cross section is weakened on the tensile side by a notch (three-point bending test). Based on elementary formulas, the tensile strength of fibre reinforced concrete is then determined. Considering the unacceptability of these formulas to capture the 3D problem and, in particular, for the three-point bending test, also because of significant singularities at the notch and the degree of statistical uncertainty in determining the actual position of the weakest area of the test specimen, such a test does not respect the real stress distribution in the tensile area which is changing in the process of load increase. The proposed adjustment of the test specimen eliminates these fundamental drawbacks by removing the singular areas of the specimen and leaving only the cross-sectional areas where only uniaxial stress state may be assumed. The results obtained by this methodology are presented, statistically evaluated and compared with the currently used procedures.

1. Introduction

Tensile strength is the basic material characteristics of fibre reinforced concrete. Several tests are used to determine the tensile strength in practice. The tests are different in their layout, technical complexity and credibility of the results.

In practice, the most common test is the bending test. The three-point bending test performed on a beam with notch or the four-point bending of the beam without notch are commonly used. In particular, the fundamental deficiency of the three-point bending test with notch is the predetermined failure mode and location. The correct properties throughout the whole test beam are not detected at all. The state of stress above the notch is very complicated that cannot be described by the elementary technical formulas. A significant non-linear behavior occurs in the bending test prior to macrocracking. As a result of all influences, the value of the tensile strength determined by the bending test is always higher than the real axial tensile strength of the material.

The following relationships are recommended to calculate the axial tensile strength from the bending strength provided by the bending test:

- three-point bending: $$f_{fct,ax} = 0.6 * f_{fct,fl} \quad (1)$$

- four-point bending: $$f_{fct,ax} = f_{fct,fl}/1.45 \quad (2)$$

where $f_{fct,ax}$ is the axial tensile strength of fibre reinforced concrete and $f_{fct,fl}$ is the value of the tensile bending strength from the bending test.

In particular, it should be noted that the source of the correction coefficients of 0.6 and 1.45 in these formulas cannot be confirmed. The principal drawback of formulas (1), (2) is that they use the same coefficients for all varieties of concretes. The formulas are exactly the same for different types and material characteristics of the concrete, type and content of fibres. This is, of course, an essential point to the unreliability of the formulas.

The only presently used test trying to provide the axial tensile strength value is the axial tension test: cylindrical test samples with a notch in the middle of the cylinder height - the crack location is thus unambiguously defined by the notch. Only one spot of the test specimen is examined, i.e. with all of the principal drawbacks mentioned above.

An alternative to the described arrangement is to use a test specimen with a variable cross section along the length ("a dogbone") [2]. Both test configurations make it possible to detect the strength of the material directly in the axial tension - but the first test provides tensile strength in a defined cross section only. However, the test arrangements are technically demanding since the axial tension test is very sensitive to the inevitable imperfections and the load control thus places high demands on the testing technique. For this reason, such a test technique is currently used only exceptionally, primarily as a calibration test for deriving strength conversion relationships.

The wedge test is currently a less widely used method for determining the axial tensile strength of fibre reinforced concrete composites. In this case, the test body forms a cube with a notch (on one or both sides) into which a steel wedge is pressed through the cylindrical bearings. However, even in this case it is not possible to obtain the value of the axial tensile strength of the material directly, because the 3D state of stress results in very complex distribution of stress in the critical region.

An alternative method for determination of the tensile strength of a concrete composite, used today rather marginally, is the Barcelona Test (Tensile splitting strength of fibre reinforced concrete).

2. New methodology

The proposed methodology meets the main demands for usability for standard tests:

- It is easy to perform and is very close to its arrangement and execution at standard beam bending tests.
- It does not require any additional equipment and tasks in the tests, except for the drilling of a hole in the specimen.
- Provides a distribution of tensile stresses that is very close to uniform at the examined region.

The principle of the new test is that the test beam of standard dimensions 150x150x700 mm is modified by drilling a circular hole in the middle of the span (Figure 1., Figure 7.). The test is arranged as a four-point bending, i.e. in the inner third of the span is bending only, without shear effects. This creates a system consisting of two narrow areas in the middle of the span - upper compression zone and lower tensile zone (Figure 3.).

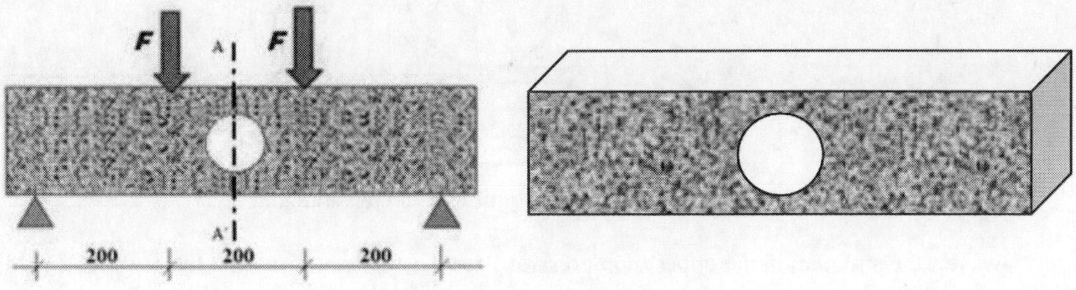

Figure 1. Test Layout Scheme.

The main advantage of the proposed methodology is the fact that the circular opening with its smooth round contour ensures elimination of singularities. Such singularities would arise from sharp interventions in the test specimen, such as a notch in a three-point bending test. An even more important principle of the proposed methodology is that the trajectories of the stress circulate smoothly around the curved periphery of the hole. This results in stress increase at the aperture surface, which compensates for the tension reduction that would correspond to the elementary assumption of the plane stress distribution (Figure 2a). Thus, in the lower tensile zone, a practically uniform stress distribution is obtained over the full region depth. This results in real axial tension.

In Figure 2a, a false assumption of the stress distribution corresponding to the elementary Navier assumption is indicated. Figure 2b and Figure 3 then show the actual stress distribution affected by its concentration at the curved surface of the hole.

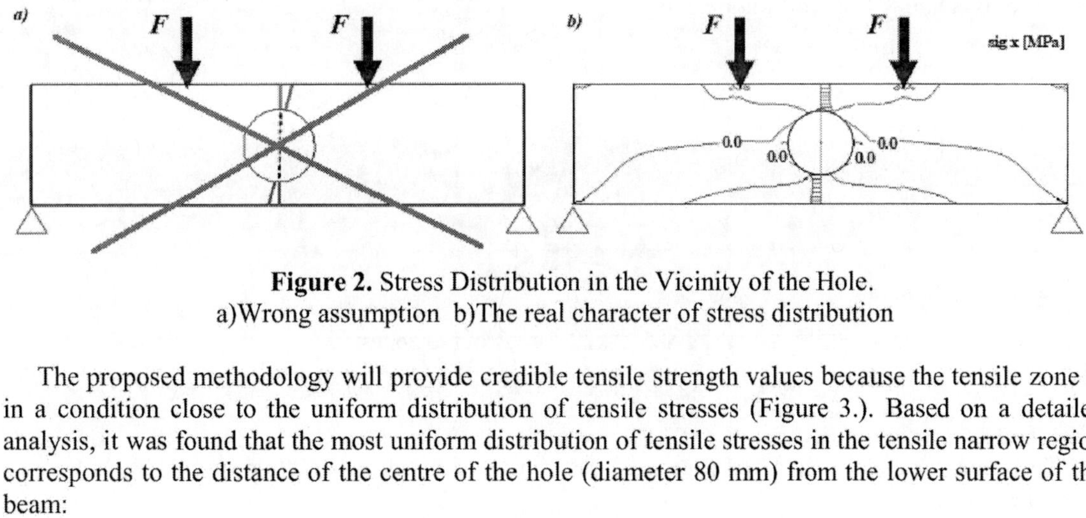

Figure 2. Stress Distribution in the Vicinity of the Hole.
a)Wrong assumption b)The real character of stress distribution

The proposed methodology will provide credible tensile strength values because the tensile zone is in a condition close to the uniform distribution of tensile stresses (Figure 3.). Based on a detailed analysis, it was found that the most uniform distribution of tensile stresses in the tensile narrow region corresponds to the distance of the centre of the hole (diameter 80 mm) from the lower surface of the beam:

$$v = \pi^2 h/20 \tag{3}$$

where h is the depth of the test beam cross-section

For the depth h = 150 mm, v= $7.5\pi^2$ = 74 mm, the corresponding tensile stress distribution is shown in Figure 3. It appears that the deviations from the average stress (indicated by the dashed line in Figure 3.) are negligible.

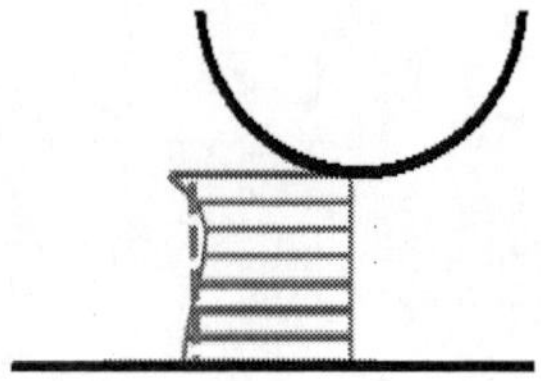

Figure 3. Stress distribution in the tension zone.

However, the situation in the upper (compression) zone is different from the lower (tensile) region; the stress distribution is uneven here due to the proximity of the loading forces (Figure 4.).

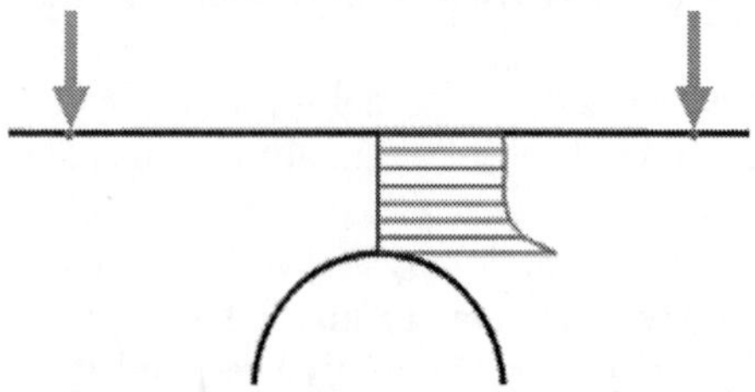

Figure 4. Stress distribution in the compression zone.

This is to be taken into account when determining the magnitude of the internal force lever arm; the internal forces N are the resultants of the stresses (Figure 5.). Considering also the insignificant unevenness of the stress distribution in the upper (compression) zone, this lever arm is r = 111 mm for the beam height h = 150 mm (Figure 5.).

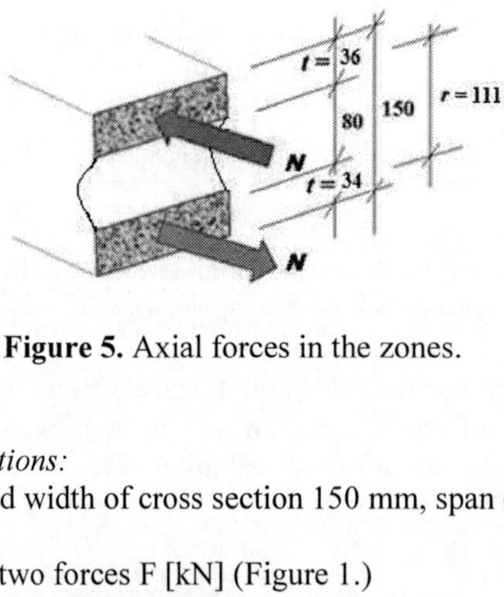

Figure 5. Axial forces in the zones.

2.1. Test Evaluation –Final Relations:

Test beam geometry: depth and width of cross section 150 mm, span 600 mm
Hole diameter: 80 mm
The load action is realized by two forces F [kN] (Figure 1.)
Bending moment in centre region of beam [kNm]
The thickness of the tensile zone resisting to axial tension t = 34 mm = 0.034 m
Tensile zone area of the lower region:

$$A = bt = 150 * 34 = 5100\ mm^2 = 0.0051\ m^2 \tag{4}$$

Lever arm of internal forces r = 111mm = 0.111 m
Force in tension area:

$$N = M/r = (0.2F)/0.111 = 1.797F\ [kN] \tag{5}$$

The tension stress in the tensile region:

$$\sigma = N/A = (1.797F)/0.0051 = 352.35F\ [kN/m^2 \ldots kPa] \tag{6}$$

Obviously, this methodology is completely transparent. No fault can occur here, because the process is based solely on the equilibrium of the external load forces and the resisting internal forces. The results do not need to be refined by any corrective coefficients of doubtful origin, as in relations (1) and (2). This methodology uses only the assumption that tensile stress distribution in the lower region is close to the uniform (Figure 3.).

2.2. Course of the test and its evaluation:

a) Initially, the system operates according to Figure 2b: The lower region is in tension mode, the upper region is in compression.
b) Lower region rupture means reaching the axial tensile strength, which is the aim of the test. The strength obtained corresponds to the cracking force and can be taken from the load F versus deflection diagram (Figure 8b).
c) The highest load force value F_{crack} is reached. (This is the force at which the crack develops.)
d) Calculation of tensile strength of fibre reinforced concrete according to the mentioned relations (6).
e) If the test process continues, the deformation continues to increase. The lower element is already broken and thus almost falls out of action. The system changes completely with a dramatic decrease in the load force: already only the stretched fibres are active in the lower area and the upper area has to take almost the entire flexure. Therefore, the upper region behaves like a low bent "beam", which has its lower surface in tensile regime (Figure 6, Figure 7), which means the formation of cracks at the lower edge of the upper, initially only compression part (Figure 7.).

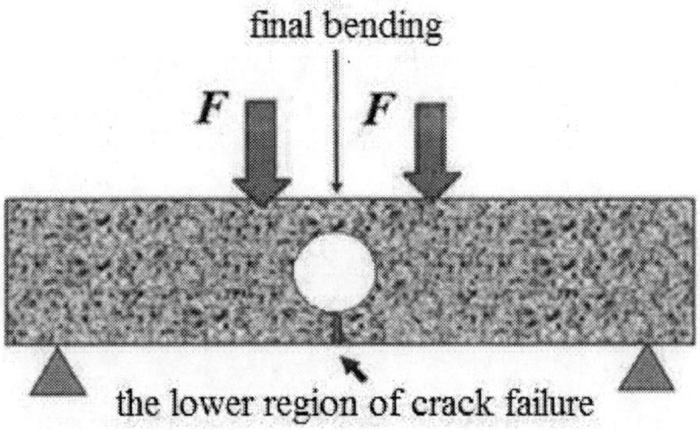

Figure 6. The final state.

Figure 7. Cracked Test Specimen with Hole.

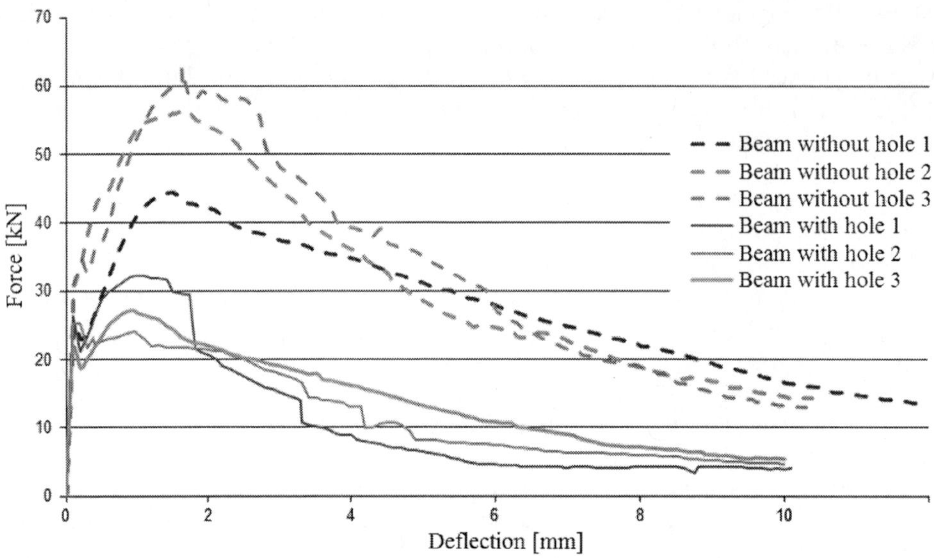

Figure 8a). Load-deflection diagrams with and without holes.

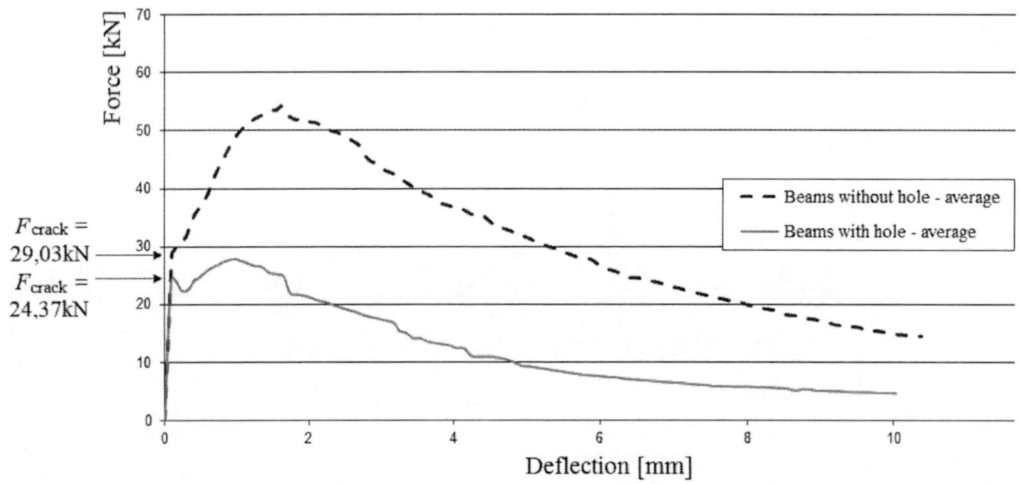

Figure 8b). Average Load-deflection diagrams with and without holes.

An experimental program was performed as a pilot project to demonstrate the advantages of the proposed methodology. Four-point bending tests were carried out on the classical arrangement of beam specimens (i.e., with no holes) and on the beam specimens prepared according to the proposed methodology with 80 mm diameter holes. The aim was to determine the axial tensile strength of fibre reinforced concrete experimentally.

In the beam tests, with 80 mm drilled holes (Figure 1), the total ultimate loading of 24.37 kN (two forces F = 12.18 kN) was reached (Figure 8b). The corresponding tensile strength in the lower element (according to formula (5)) is $\sigma = 4.29$ MPa.

According to the classic four-point bending test (without a hole), from the graph shown in Figure 8 band when formula (2) is used, the average tensile strength value of 5.16 MPa is obtained, which is 1.2 times higher than the tensile strength determined by the proposed methodology. The significant deviation from the calculated results according to the proposed methodology according to formula (5) is clearly evident.

Also, the thorough experimental study 2, performed using the "dog bone" method on samples with different fibre contents, showed that classical approach (Eq.(2)) provides

higher, unrealistic, tensile strength values compared to reality. The "dog bone" method, which can be considered as a pure tensile test, thus shows the same tendency as the proposed methodology of specimens with holes.

Figure 8a Load-deflection diagrams of specimens with and without holes also demonstrate a suitability of the recipe of the fibre reinforced concrete used and the uniformity of the wire distribution in its structure.

In the frame of the experimental program, supplementary strength tests of the used fibre reinforced concrete were carried out; the following results were obtained:

- compressive strength - cube ... average of 6 measurements: 83.76 MPa
- tensile splitting strength ... average of 6 measurements: 7.94 MPa

3. Summary

The presented new test of concrete axial tensile strength (or of fibre reinforced concrete) proves to be much more suitable than standard tests used so far, as it is easy to perform them, it does not rely on any inexact assumptions and in particular it is able to fully respect the type and all material characteristics of the used concrete and the type and the fibre content, which is not attempted by standard tests of formulas (1) and (2).

Acknowledgment

The presented results were obtained with the support of the Technology Agency of the Czech Republic within project TH02010375 and Grant Agency of the Czech Technical University in Prague, grant number SGS18/115/OHK1/2T/11.

References

[1] Podstawka A, Kovář M, Foglar M, Křístek V 2015 Analytical and statistical Evaluationof FRC bending tests layouts. *Advanced Materials Research*. 2015, ISSN 1022-6680.

[2] Tipka M 2017 Tensile properties of structural fibre reinforced concrete. *Ph.D. thesis*. CTU in Prague, Faculty of Civil Engineering. Prague 2017

IOP Conf. Series: Materials Science and Engineering **596** (2019) 012017 doi:10.1088/1757-899X/596/1/012017

Comparison of two UHPFRC premixes blast resistance with the commonly used concrete based on the failure mode prediction

O Janota[1], E Bonnet[2] and M Foglar[1]

[1] Faculty of Civil Engineering, Czech Technical University in Prague, Thákurova 7/2077, 166 29 Prague 6, Czech Republic
[2] R&D Lafargeholcim, 95 rue du Montmurier, Saint Quentin Fallavier, France

E mail: ondrej.janota@fsv.cvut.cz

Abstract. This paper presents results of extensive experimental programme which took place in 2019. Two proprietary ultra-high performance fibre reinforced composite materials were tested for their blast (contact and close-in) resistance. In total, twenty-eight specimens were tested. Specimens were loaded with explosive charge of weight of 100 - 1000 g in various distances. All specimens were visually evaluated for the damage extend, failure mode and crack pattern. Results from the experiment were compared to the results available in literature for normal strength reinforced concrete. Results showed that both tested premixes performed better than ordinary mixtures in terms of blast resistance. Difference between tested materials and commonly used mixture are described as well. Finally, both materials were compared from the material properties point of view as well as by their blast resistance performance.

1. Experimental program

Experimental program took place at the fields of the University of Pardubice in 2018. Twenty–eight specimens with dimensions 1000 x 1000 mm and three thicknesses (100; 150 and 200 mm) were tested. For manipulation with the specimen, four handlings were placed at the contact side in each corner and two handlings were placed at the side edge. For the selection of the dimensions two criteria were considered [1]. Primary, the dimensions were set by the numerical simulation which confirmed that the pressure wave reached the bottom surface before the reflection from the side edges occurred. Secondly, the manipulation with the specimen was considered.

Blast load was created by the explosive charge of SEMTEX 1A. The weight of the explosive (0 – 1000 g) as well as the distance (0 – 100 mm) between the explosive and top surface varied. Therefore, both types of blast load, contact and close-in blast, were performed. Shape of the explosive charges was cylinder with length/diameter ration equal to one. Detonation point was set approximately 20 mm under the top edge of each explosive charge.

Each specimen was put on steel frame (Fig. 1). Therefore, the specimen can be considered as a slab simply supported along all four edges. Height of the steel frame was set 720 mm and it served as a place for a mirror which enabled the recording of the bottom surface needed to be placed under the specimen. Reflection of the pressure wave from the ground was prevented by the elevation of the specimen [2] [3]. Each shot was recorded by two high-speed cameras. For experiment scheme see Fig. 2.

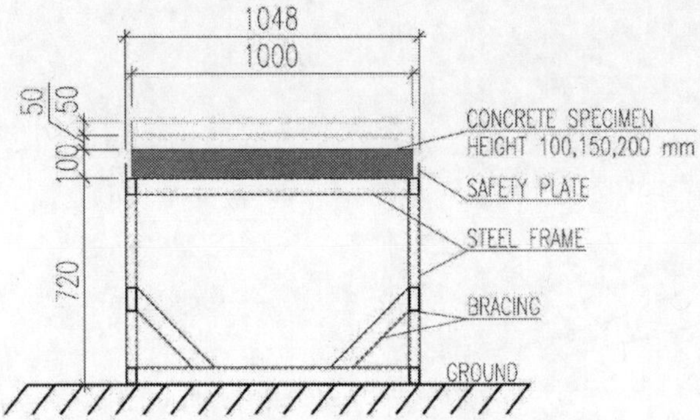

Figure 1. Left – cross section of the specimen supported by the steel frame; Right – specimen with the explosive charge.

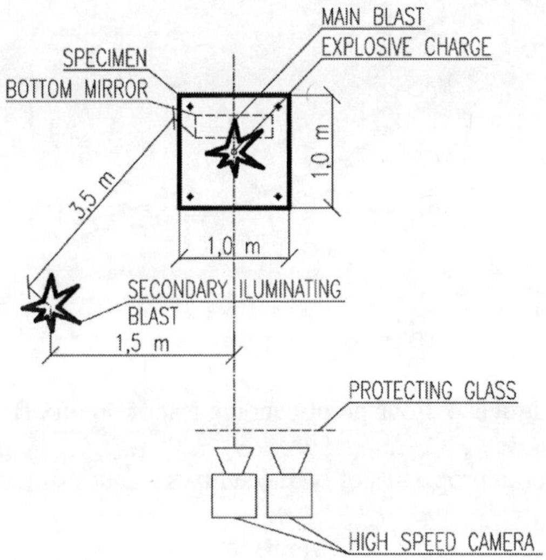

Figure 2. Scheme of the experiment site.

2. Material properties

Two proprietary UHPFRC materials with compressive strength higher than 150 MPa were tested during this experimental program. First material was Premix A with compressive strength 180 MPa, second material was Premix B with compressive strength 150 MPa. Amount of fibres in each material, their length and yield strength were similar in both premixes. Both materials were tested on three specimens for their compressive and tensile strength and flexural energy. Tensile strength and flexural energy were calculated from the results of the four-point bending tests (Tab. 1). Compressive strength was calculated from the compressive test (Tab. 2). Results of the tests are presented in the Fig. 3 and 4.

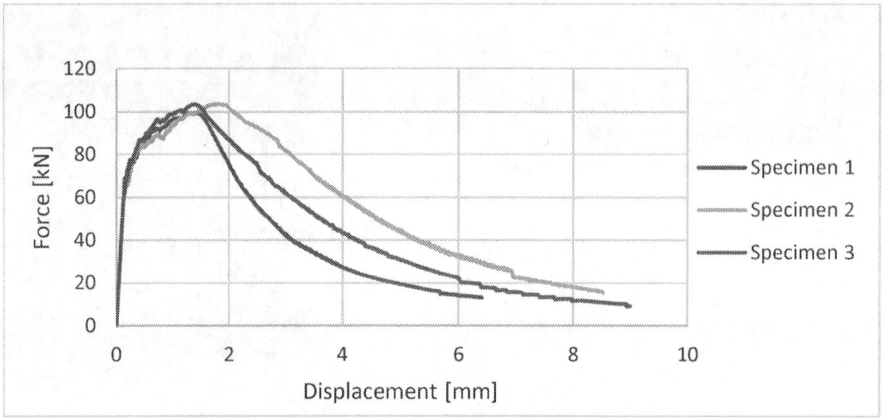

Figure 3. Four-point bending test of premix A.

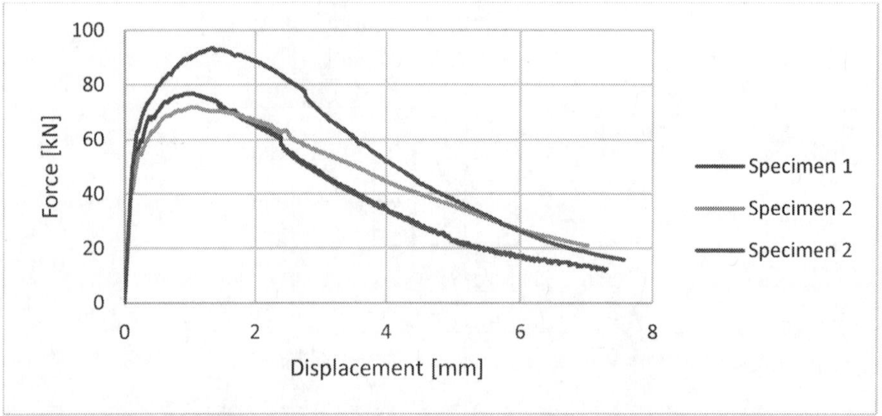

Figure. 4. Four-point bending test of premix B.

Table 1. Material properties of both premixes – four-point bending test.

				Premix A				
Spec. Nr [-]	**Spec. length [mm]**	**Spec. width [mm]**	**Spec. height [mm]**	**Spec. weight [kg]**	**Density [kg/m³]**	**Max. force [kN]**	**Ten. Strength [MPa]**	**Flex. energy [MJ]**
1	700	151.4	149.9	39.1	2460	99.48	17.4	13.7
2	700	151.1	148.2	39.0	2488	103.581	18.4	21.7
3	700	150.7	150.5	39.1	2464	103.341	18.1	18.2
Average value	-	-	-	-	**2471**	**102.13**	**18.0**	**17.9**
				Premix A				
Spec. Nr [-]	**Spec. Nr [mm]**	**Spec. Nr [mm]**	**Spec. Nr [mm]**	**Spec. Nr [kg]**	**Spec. Nr [kg/m³]**	**Spec. Nr [kN]**	**Spec. Nr [MPa]**	**Spec. Nr [MJ]**
1	700	151.4	151.0	39.5	2483	76.76	13.5	13.4
2	700	151.3	148.3	39.4	2524	71.781	12.9	14.9
3	700	150.9	148.2	39.9	2551	93.581	16.6	18.4
Average value	-	-	-	-	**2519**	**80.71**	**14.3**	**15.5**

IOP Conf. Series: Materials Science and Engineering **596** (2019) 012017 doi:10.1088/1757-899X/596/1/012017

Table 2. Material properties of both premixes – compressive test.

Spec. Nr	Spec. length	Spec. width	Max. force	Comp. strength
[-]	**[mm]**	**[mm]**	**[kN]**	**[Mpa]**
1-1	150	150	3476.0	154.5
1-2	150	150	3875.0	172.2
2-1	150	150	4108.0	182.6
2-2	150	150	3934.0	174.8
3-1	150	150	3630.0	161.3
3-2	150	150	3613.0	160.6
Average value	-	-	**3773.0**	**166.7-**
Spec. Nr	**Spec. Nr**	**Spec. Nr**	**Spec. Nr**	**Spec. Nr**
[-]	**[mm]**	**[mm]**	**[kN]**	**[Mpa]**
1-1	150	150	2985.0	132.7
1-2	150	150	3426.0	152.3
2-1	150	150	3552.0	157.9
2-2	150	150	3535.0	157.1
3-1	150	150	3577.0	159.0
3-2	150	150	3671.0	163.1
Average value	-	-	**3458.0**	**153.7**

3. Experimental results

The evaluation of the results presented in this paper is based mainly on the visual inspection and damage measurements. Fundamental damage parameters (damage depth, damage size, failure mode and shape) were measured for the evaluation of the empirical prediction which were derived for concrete specimens by McVay [4] and Morishita [5], [6]. Both approaches are based on the following parameters:

- Scaled thickness ($T/W^{1/3}$) – thickness of the specimen (T) divided by the weight of the explosive (W).
- Scaled distance ($R/W^{1/3}$) – distance between detonation point and top surface of specimen (R) divided by the weight of the explosive (W).
- Scaled damage ($(C_D + S_D)/T$) – Summation of damage depth of soffit and contact side ($C_D + S_D$) divided by the thickness of the specimen (T).
- Failure mode – type of the final damage; generally, three types are considered in literature (Fig. 5).

Based on huge amount of experimental results and above-mentioned parameters, McVay derived empirical approach for the prediction of the result. However, these prediction curves were derived for the normal strength and reinforced concrete. As there are the differences in the material properties between normal strength concrete and UHPFRC, the validation of this approach needed to be done. Moreover, the superior blast resistance of UHPFRC to normal strength concrete can be proven with this approach. Morishita's [5], [6] approach, which is more recent, was based on the calculation of the scaled damage based on the scaled thickness. This approach was also derived for the normal strength reinforced concrete.

IOP Conf. Series: Materials Science and Engineering **596** (2019) 012017 doi:10.1088/1757-899X/596/1/012017

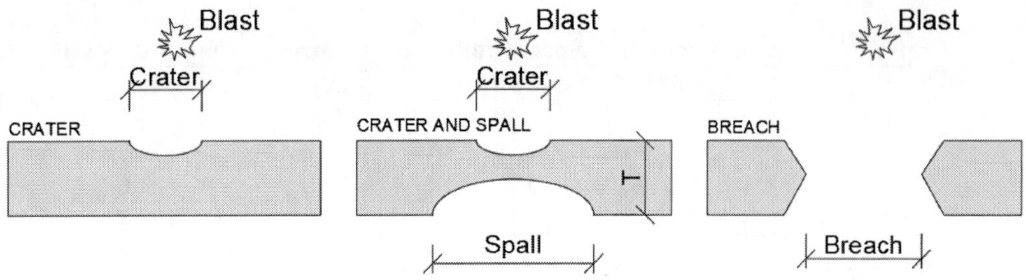

Figure. 5. Failure modes – Left – Crater; Middle – Crater and Spall; Right – Breach.

One of the experiment targets was to describe the boundaries between different failure modes. Seventeen of the specimens ended with crater and spall failure mode. Three of the specimens ended with only spall damage. Five specimens ended with only crater or no damage failure mode and six were breached. Overview of the parameters of each shot and reached failure mode is presented in Tab. 3.

Table 3. List of experiment dispositions and final failure mode for each specimen.

Shot number	Premix	Spec. thick.	Exp. weight	Exp. length	Distance	Failure mode	Scaled distance
[-]	[-]	[mm]	[g]	[mm]	[mm]	[-]	[m/kg$^{1/3}$]
1	A	100	400	61	0	Breach	0.051
2	A	100	200	58	0	Crater and spall	0.060
3	A	150	400	66	0	Crater and spall	0.057
4	A	100	400	74	100	No damage	0.192
5	A	100	1000	90	60	Crater and spall	0.119
6	A	100	400	60	10	Crater and spall	0.062
7	A	100	200	58	0	Crater and spall	0.060
8	A	200	1000	95	20	Crater and spall	0.087
9	B	100	400	700	100	Spall	0.187
10	A	100	400	700	100	Spall	0.187
11	A	100	400	700	100	No damage	0.187
12	A	100	400	700	0	Breach	0.062
13	A	100	400	60	0	Breach	0.050
14	A	150	1000	96	0	Breach	0.070
15	A	200	1000	90	30	Crater and spall	0.092
16	A	100	200	55	25	Crater and spall	0.094
17	B	100	300	70	100	Crater and spall	0.205
18	B	100	300	70	0	Crater and spall	0.068
19	B	200	500	75	0	Crater and spall	0.063
20	B	100	1000	80	60	Breach	0.110
21	B	100	200	50	100	Spall	0.204
22	B	100	150	45	25	Crater and spall	0.086
23	B	150	500	75	40	Crater and spall	0.110
24	B	100	500	75	20	Breach	0.087
25	B	100	100	47	25	Crater	0.103
26	B	150	400	64	30	Crater	0.092
27	B	100	800	76	60	Crater and spall	0.114
28	B	200	300	50	0	Crater	0.041

3.1. Crack pattern

Two types of crack systems were observed. Firstly, the radial cracks appeared on most of the specimens. These cracks were wider; mostly starting on the border of damage zone and propagating in all directions. However, mostly in the direction towards the edge with side handlings. Moreover, on some of these specimens one main radial crack penetrated fully to the opposite surface. These cracks occurred on all of the specimens damaged by the breach and on the most of the specimens damaged by the crater and spall.

IOP Conf. Series: Materials Science and Engineering **596** (2019) 012017 doi:10.1088/1757-899X/596/1/012017

Second type of cracks was thin peripheral cracks appearing mostly on contact surface and in some cases on soffit. These peripheral cracks were observed on specimens with all three failure modes, mostly on the contact side. On some specimens these cracks were continuously joined into a circle (Fig. 6). However, peripheral cracks were thin and did not propagate deeper into the specimen which was in contrary to some of the radial cracks that propagated through the specimen thickness.

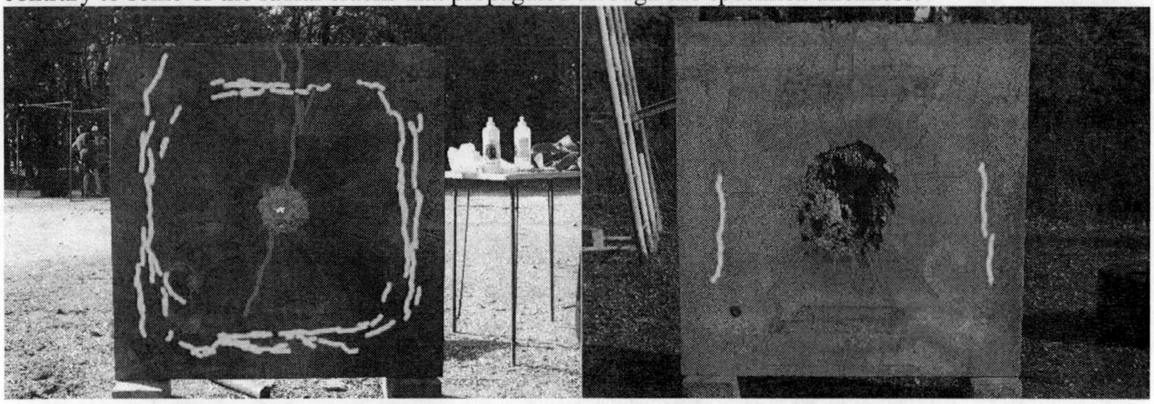

Figure 6. Cracks patterns, red – radial cracks, green – peripheral cracks.

3.2. Multiple wave reflection

Spalling of the debris on the soffit revealed multiple wave reflections as the pressure wave reached the interface between concrete and the air. Concrete spall created new interface for the still propagating pressure wave. This multiple wave reflection shaped probably spall contours. If there was only one reflection, the damaged surface was smooth. Once the multiple reflection occurred, surface was more rugged and not so continuous as in the case with only one reflection, see Fig.7.

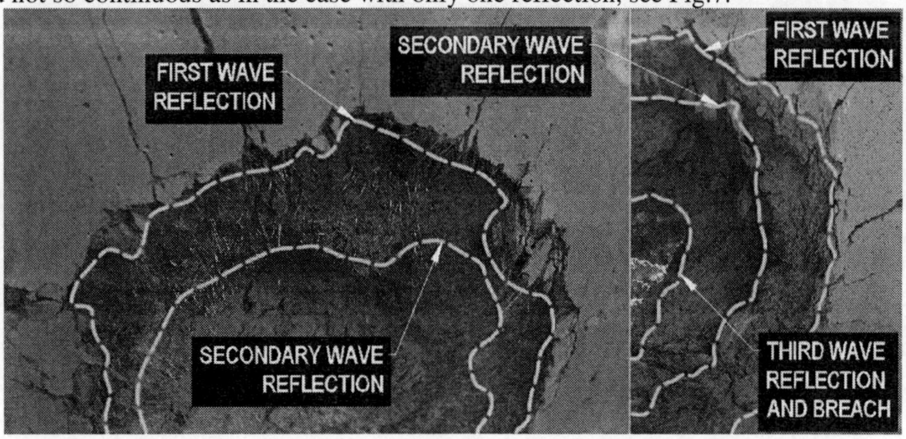

Figure 7. Damage areas with traces obvious for multiple reflection of pressure wave.

4. Empirical approach to the failure mode estimation for the UHPFRC

4.1. McVay approach to failure mode prediction

Although, the experiment covered only a part of the McVay's graph and for better specification of the boundary curves and their mathematical description more experiments are needed, twenty-eight specimens and their failure modes provided good opportunity for the comparison of the initially derived results with the results of UHPFRC material. Results indicates (Fig. 8) that originally derived boundaries are not fully valid for the UHPFRC material. Crater or no damage and breach failure modes clearly defined both boundaries. Even some results which were classified as crater and spall and lay near the boundary between no damage and crater and spall were threshold spall. Both curves defining boundary between the failure modes were shifted downward. Inclination of both curves stayed similar. This indicates that UHPFRC material is more blast resistant than the normal strength concrete and more energy needs to be produced to reach spall or breach failure mode.

Both materials behaved very similar but slight difference was observed. Specimens made of Premix B were more likely breached with the lower scaled distance. This difference was probably caused by the difference of the material properties. As the results were located only in part of the McVay's graph, the difference between both materials would be covered by the statistical dispersion with increase of the results.

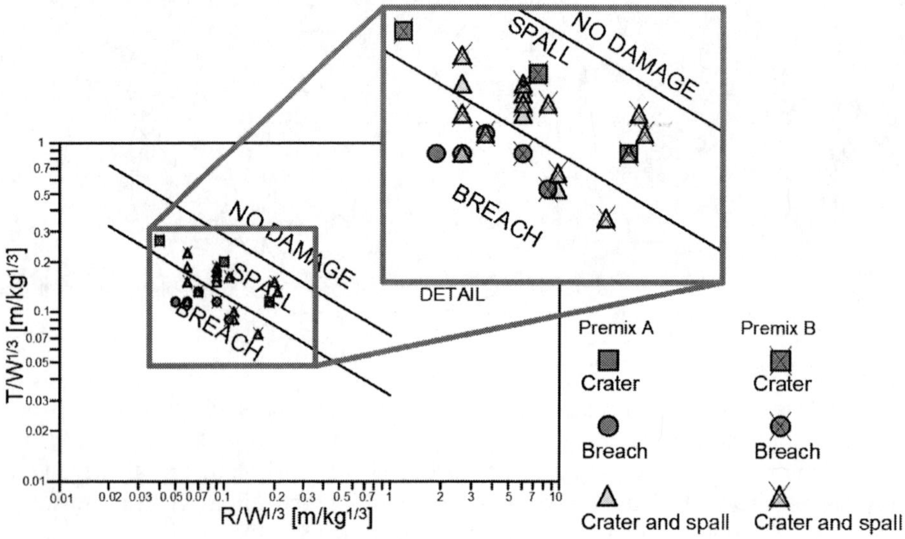

Figure 8. Comparison of the experimental results with the failure mode prediction derived by McVay.

4.2. Morishita approach to failure mode prediction

Morishita's approach was initially derived for the contact blast. Therefore, only scaled thickness was considered. It is important to emphasize that not all specimens were loaded by contact blast. Therefore, the results compared to Morishita's approach can be distorted by this fact. When applying the Morishita's [2] [3] equations for different failure modes on the tested UHPFRC material, it is obvious that limits (2.0 for breach, 3.6 for crater and spall) changed.

Results presented in Fig. 9 clearly showed new limits for failure modes. Limit value of scaled thickness for boundary between crater/no damage and crater and spall was set 2.1 cm/g$^{1/3}$. Boundary between crater and spall and breach failure mode was set 1.3 cm/g$^{1/3}$. Moreover, these limits also indicate that inclination of the crater and spall damage depth prediction curve is steeper than the original Morishita's curve and boundaries for this type of failure are closer to each other. This phenomenon is in agreement with the results from the McVay's approach.

As the previously described approach did not consider influence of the explosive distance from the specimen top surface, dependency of the scaled damage on the scaled distance is presented in Fig. 10. Results indicates that if the scaled distance is 1.0 cm/g$^{1/3}$ or less the damage depth is 40% of the specimen thickness or more. On the other side, specimens with scaled distance 1.5 – 2.0 cm/g$^{1/3}$ ended with damage depth lesser than 40% of the specimen thickness. Range of scaled distance between 1.1 and 1.5 cm/g$^{1/3}$ cannot be described for lack of experiments. Results of specimens with crater and spall indicates direct ration between damage depth and scaled distance.

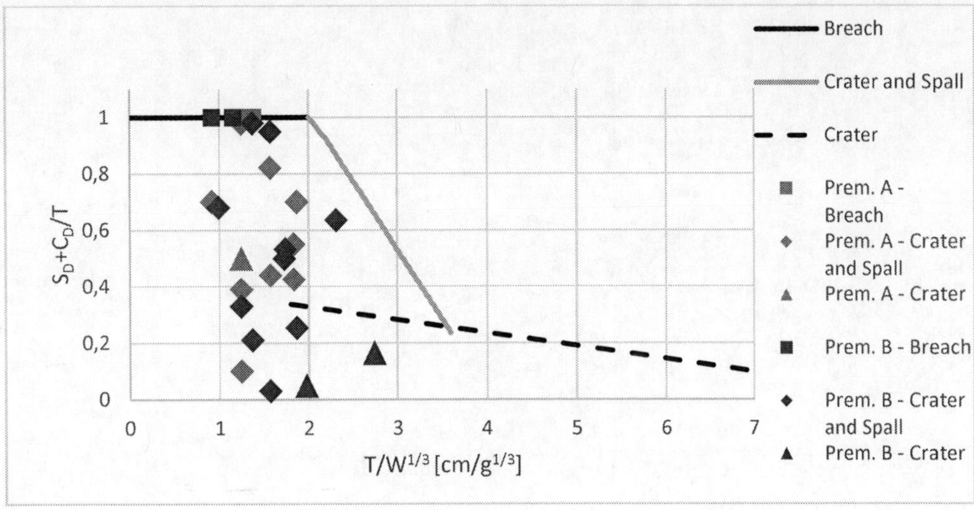

Figure 9. Comparison of the experimental results with the failure mode prediction derived by Morishita.

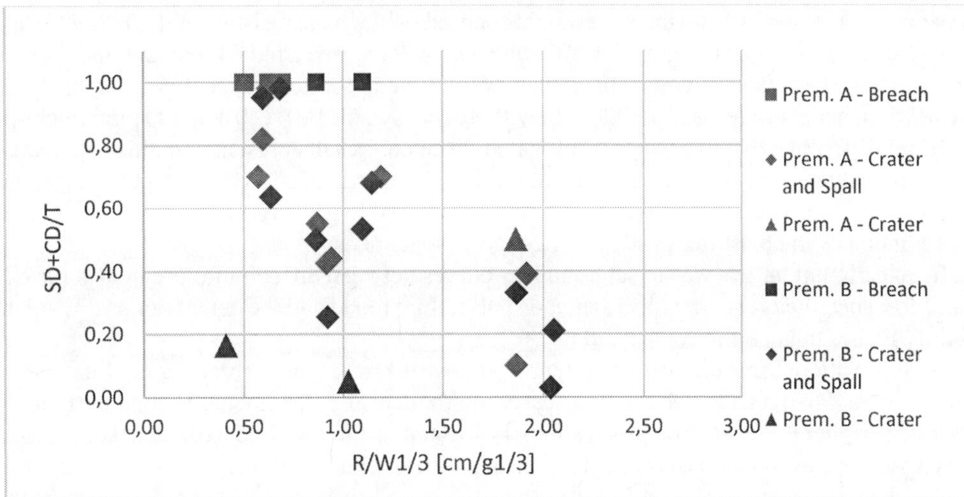

Figure 10. Development of the scaled damage and its dependency on the scalded distance.

As in the McVay's approach both materials were compared to each other (Fig. 11). Premix A with the better material properties performed better. The difference between two materials for the tested scaled distances varied from 10 % to 20 % of the specimen thickness. The difference between both materials was more obvious for lower values of the scaled distance. With the increase of scaled distance, the difference decreased. Regression analysis performed on the results for each material revealed that the dependency of the scaled damage on the scaled distance can be described by the exponential function. However, R-value which describes strength of the linear analysis was relatively low. Results from the Morishita's approach are in agreement with the results from McVay's approach.

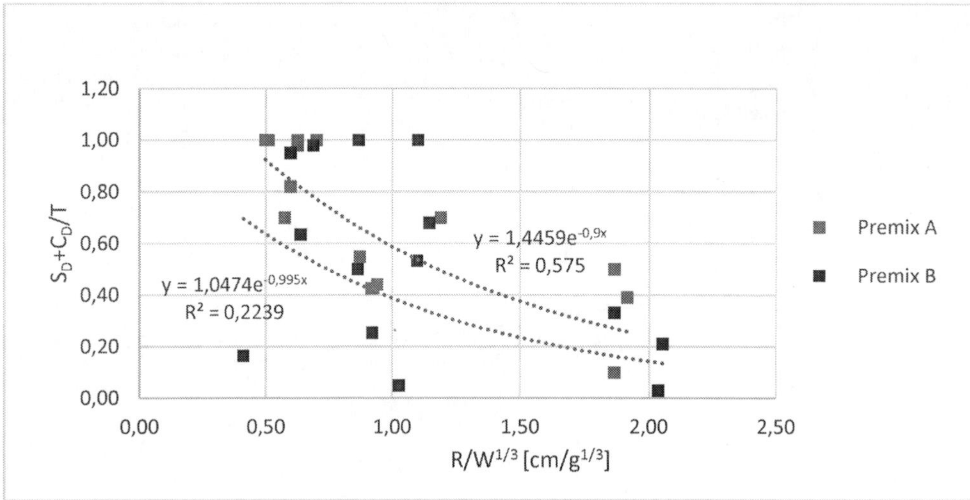

Figure 11. Regression analysis for damage prediction.

5. Conclusion

This paper presented results from the experiment focused on the blast resistance of UHPFRC slabs. Two materials with compressive strength 150 MPa and 180 MPa were tested for contact and close-in blast resistance. Altogether, twenty-eight specimens were tested, fourteen specimens for each material. Specimens' dimensions were 1000 x 1000 mm with thicknesses of 100, 150 and 200 mm. Each specimen was tested for different explosive weight and scaled distance. Results presented in this paper were based on the visual evaluation of the experiment results. Firstly, the damage extend, depth and failure mode were observed and measured. Then the results were compared to McVay's and Morishitas's approaches for the prediction of the blast performance of concrete specimens.

One of the experiment targets was to set boundary curves between different failure modes for UHPFRC material. Most specimens ended with crater and spall failure mode and the extent of damage on the soffit was slightly higher than on the contact side.

Visual observations revealed two different systems of cracks on both surfaces. First system was represented by radial cracks which were observed on most of the specimens. Second system of cracks consisted of peripheral cracks which were mostly located at the specimen edges. Radial cracks were relatively wide, and in some cases propagated through the thickness of the specimen. Contrary to radial cracks, peripheral cracks were located only at the contact and soffit surfaces and did not propagate into the specimens.

Comparison of experiment results to the empirical approaches derived by McVay and Morishita revealed superior behaviour of UHPFRC to concrete with normal strength. Both boundaries between different failure modes were clearly defined and shifted towards lower numbers of scaled thickness/distance. Comparison of both approaches revealed that both approaches were in agreement. Substitution of scaled thickness with scaled distance revealed clear influence of this factor. With the increasing scaled distance by two or four times, the total damage depth decreases up to 20% of the initial value.

Both materials were compared to each other as well. Comparison revealed similar blast resistance of both materials. However, the premix A performed slightly better than premix B. This was caused by the better material properties of Premix A. The difference was not clearly set, and it is probable that if the number of specimens increased, the difference would be dissolved into the statistical dispersion.

Acknowledgements
This work was financially supported by Grant Agency of the Czech Republic, project no. 17-23067S, which is gratefully acknowledged.

References

[1] Kovar M, Foglar M and Hajek R 2014. *The blast performance of real-scale reinforced concrete specimens with varying fiber types and content. Structures Under Shock and Impact XIII,* pp 159-169.

[2] Foglar M, Pachman J, Pelikan V, Hájek R, Künzel M and Kovář M 2014 *The structural response of a reinforced concrete specimen subjected to adjacent blast loading*, pp. 171-179.

[3] Foglar M, Hajek R, Fladr J, Pachman J and Stoller J 2017. *Full-scale experimental testing of the blast resistance of HPFRC and UHPFRC bridge decks. Construction and Building Materials*, 145, pp.588-601.

[4] McVay M 1988. *Spall damage of concrete structures.*

[5] Morishita M, Tanaka H, Ando T and Hagiya H 2004. *Effects of Concrete Strength and Reinforcing Clear Distance on the Damage of Reinforced Concrete Slabs Subjected to Contact Detonations. Concrete Research and Technology*, 15(2), pp.89-98.

[6] Morishita M, Tanaka H, Ito T and Yamaguchi H 2000. *Damage of Reinforced Concrete Slabs Subjected to Contact Detonations. Journal of Structural Engineering*, Vol.46A, pp.1787-1797, (in Japanese)

IOP Conf. Series: Materials Science and Engineering **596** (2019) 012018 doi:10.1088/1757-899X/596/1/012018

The impact of various types of steel fibres on the strength parameters and blast resistance of high – performance concrete

P Bibora, M Drdlová, V Prachař and O Sviták

Research Institute for Building Materials, Hněvkovského 30/65, 617 00 Brno, Czech Republic

E-mail: bibora@vustah.cz

Abstract. The paper describes composition proposals and subsequent optimization and testing of developed fibre-reinforced UHPC mixtures, which seem to be optimal as construction material for the blast and ballistic protective systems, such as mobile blast-resistant walls and covers, which are the expected outcomes of the presented research project. Methods of laboratory testing and optimization of developed UHPC mixtures, with the aim of achieving the best possible values of physical-mechanical parameters, are described in this paper. In connection with laboratory verification, field blast tests of UHPC board samples and consequently of the entire blast-resistant wall system were realized and resistance to dynamic stress was investigated.

1. Introduction

Presented paper summarizes the outcomes of advanced phase of research and development conducted as one of the CAMPT research centre's activities. This facility focuses on increasing safety of citizens, protection of critical infrastructure, defence and national security. It was founded in 2014 to operate for 6 years, associates 9 partners from the field of research and development and commercial sphere and it is supported by Technology Agency of the Czech Republic as one of the "Competence Centres".

Described research works were aimed at verification of the influence of various types of fibre reinforcement on resulting properties of developed fine-grained UHPC, especially the physical-mechanical characteristics - flexural strength, compressive strength, and subsequently blast resistance. Based on testing, the most suitable UHPC mixture was chosen to develop a mobile blast-resistant wall.

Research works depicted in this article build on previous phases of the research published in [1], where the same testing methodology of reinforced concrete blast resistance was used.

Fibre reinforcement is commonly used to provide enhanced toughness and ductility to brittle cementitious matrix. [2]. Owing to its dense microstructure, UHPC has both higher strength and also higher stiffness (expressed by the modulus of elasticity) when compared with normal- and high-strength concretes. It is well known that the behaviour of concrete becomes more brittle as its strength increases. This effect is particularly evident with UHPC. However, this disadvantage can be countered effectively by adding fibres, e.g. high-strength steel fibres, which bring about a considerable improvement in the post-peak behaviour in compression and tension, and that in turn means a more favourable response generally when failure is pending [3].

Fibre reinforcement in combination with high performance matrix is a way to improve blast resistance of concrete. Optimal composition with emphasis on the compatibility of the matrix and fibre is required, especially in the case of relatively thin wall elements [4] [5].

It is well known that the response of elements to dynamic loads depends on the rise time and the applied peak with respect to the natural period of vibration and the resistance of the element. The theory of the high dynamics of concrete structures and failure mechanisms of concrete elements loaded by a blast wave, including simulation methods, are described in [6] and [7]. The propagation of the compression wave during a blast creates strains in the concrete, and where the tensile strength of the concrete is exceeded, fragments are created on the rear side of the concrete element. This failure usually occurs in the surface zone, and conventional steel reinforcement is not able to help in controlling this event. On the other hand, randomly dispersed fibres are able to reinforce the concrete surface layers and thus enhance the tensile strength and prevent the concrete from fragmentation [8].

For the investigation of explosion resistance of concrete elements, the effect of concrete composition and mainly of different types of reinforcement was observed and subsequently the improved explosion resistance was evaluated.

2. Experimental works

All described works were aimed mainly on advanced optimization of developed UHPC composition so that the highest physical-mechanical values were achieved, including higher resistance to dynamic load, i.e. explosion. Such optimization lied in testing the influence of various dispersed steel fibre reinforcement combined with adjustments of water-cement ratio.

Predominantly, the physical-mechanical properties tested in new and optimized UHPC compositions were compression strength, flexural strength and bulk density. After testing, average values were calculated to choose the most suitable UHPC composition for blast tests, depicted in the latter half of this paper.

2.1. UHPC mixture composition

For evaluation and verification of physical-mechanical parameters of UHPC, total of 5 fine-grain UHPCs were proposed and verified. The testing bodies were sized 100 × 100 × 400 mm and their manufacture was guided by methodology corresponding with ČSN EN 12390-2 Testing hardened concrete – Part 2: Making and curing specimens for strength tests. The compositions of individual UHPC mixtures are listed in Table 1.

Table 1. UHPC mixtures for the mechanical performance investigation.

Composition of proposed UHPC mixtures					
UHPC mixture	S2B6% 100	S2B6% 50/50	S2B6% 70/30	S2B8% 25	S2B6% MIX
Cement 52.5 R	955	955	955	955	955
Sand 0-1 mm	1100	1100	1100	1100	1100
Mikrosilika Elkem 940	143	143	143	143	143
Plasticizer	25	25	25	25	25
Water	230	230	230	210	210
Water-cement ratio	0.24	0.24	0.24	0.22	0.22
Steel fibres 25/0.6	_	_	_	628	63
Steel fibres 30/0.6	_	234	312	_	_
Steel fibres 20/0.8 (corrugated)	468	234	156	_	408

Fine-grained mixture S2B with the best results in physical-mechanical and blast testing, based on the results presented in earlier stages of the research [1], was newly modified with new components - various types of steel fibre reinforcement, new plasticizer and new water-cement ratio, so that five variants of fine-grained UHPC mixture S2B were prepared. The individual variants differed mainly in type and amount of fibre reinforcement. The binder-filler ratio and also the plasticizer dose was identical in all five mixtures. The water-cement ratio was 0.22 – 0.24. The volume of fibres used for reinforcement represented 6 – 8 % of dry-mix volume. Mixtures S2B6%100 and S2B8%25 contained only one type of reinforcement while two types of fibres in different ratios were combined in the other samples.

2.2. *Testing of physical-mechanical parameters*

In this development stage, the tests were aimed at verification of influence of various types of dispersed steel reinforcement on UHPC mechanical parameters (see Table 1). Compression strength, flexural strength and bulk density were measured according to standards ČSN EN 12390-7 Testing hardened concrete – Part 7: Density of hardened concrete, ČSN EN 12390-5 Testing hardened concrete – Part 5: Flexural strength of test specimens, ČSN EN 12390-3 Testing hardened concrete – Part 3: Compressive strength of test specimens. The resulting average values calculated from testing six sample bodies are in Figures 1 – 3.

2.3. *Blast resistance testing*

Testing and subsequent evaluation of blast resistance values of designed and optimized UHPCs with various steel fibre reinforcement was conducted. Mixture compositions are listed in Table 1, their physical-mechanical properties are summarized in Table 2. The influence of various parameters of concrete mixture on blast resistance was investigated. The testing was conducted according to the following two methods:

2.3.1. Modified method M-T0-VTUO 10/09. Sample board is firmly attached to the test bench and loaded with a blast wave generated by initiation of the charge located at a defined distance from the test sample. After the test, the damage of the board is visually evaluated, together with determination of weight loss caused by fragmentation, and sheet metal plate for fragment capture placed under the sample. For this purpose, aluminium plates of 0.5 mm thickness were used. The charge initiated was 150 g of Semtex 10 (Eq. 180 g TNT) at a distance of 100 mm from the centre of the board sample. All samples were dried to constant weight before testing. The test configuration is in Figure 1 [9].

Figure 1. Testing bench T0B and test configuration according to modified method M-T0 VTUO 10/09.

2.3.2. The speed changes of ultrasonic wave propagation through the sample. The sample was attached to the test bench T0B and marked with 6 points, where the speed of ultrasonic wave propagation through the sample was measured before and after the blast. For the measurements, Proceq Tico device was used with 54kHz probe. The rate of increase in wave propagation time is directly dependent on the sample damage. The charge used was 100 g Semtex 10 (Eq. 120 g TNT). The charge weight was chosen in order to preserve the sample board integrity and, at the same time, to cause a measurable change in the ultrasonic wave propagation time and deflection so that individual concrete samples could be compared.

For blast test according to methods mentioned above, ten sample boards were prepared, two of each concrete mixture, sized 500 × 500 × 40 mm. The results of both used testing methods are listed Table 3.

3. Results

The following figures and tables show average values of realized laboratory testing and external blast tests. The average strength and bulk density parameters of each UHPC mixture have always been calculated from measurements on at least six test specimens – beams of size 100 x 100 x 400 mm. Blast tests were always carried out on two test boards from each UHPC mixture. The test board dimension was 500 x 500 x 40 mm.

3.1. Bulk density after 28 days

The average bulk densities of the UHPC samples (see Figure 2) ranged 2.600 – 2.700 kg/m3. There were no significant differences between the bulk densities of the individual UHPC samples, as the composition differed only in the type of reinforcement, dosed in the same volume of 6 %. The only mixture containing more reinforcement was S2B8%25 with 8 % fibre, therefore, the bulk density value of this sample was higher and reached 2,691 kg/m^3.

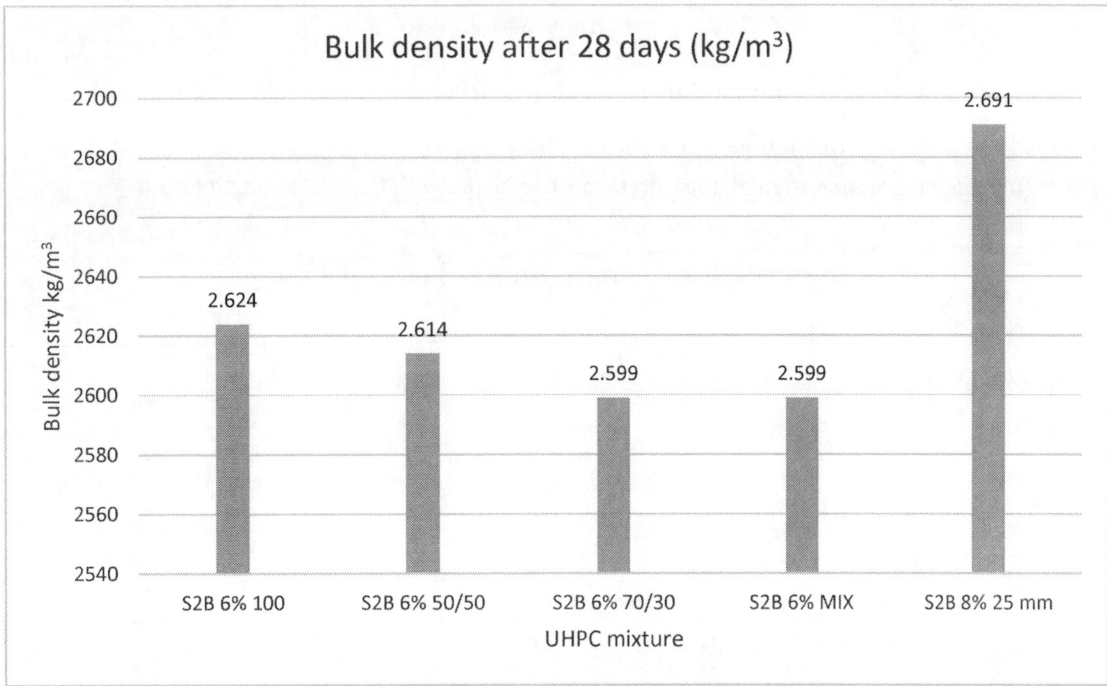

Figure 2. Average values of bulk density of the UHPC samples after 28 days of maturing.

3.2. Flexural strength after 28 days

The best results of flexural strength values were achieved with S2B8%25 UHPC mixture. Thanks to the type and amount of fibre reinforcement, the S2B8%25 mixture reached flexural strength of 28.41 MPa. The S2B6%50/50 mixture also exhibited satisfying result of 21 MPa as it contained 6 % of two types of fibre reinforcement differing in shape and length in 1:1 ratio.

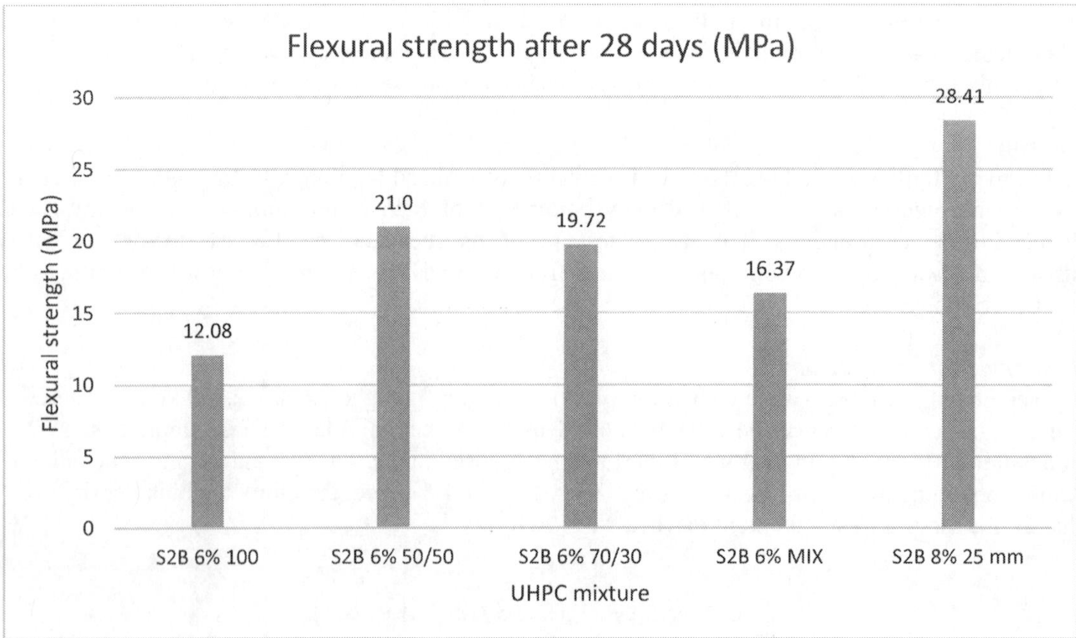

Figure 3. Average flexural strength values of the UHPC samples after 28 days of maturing.

3.3. Compressive strength after 28 days

The results of compressive strength measurements correlated with flexural strength results.

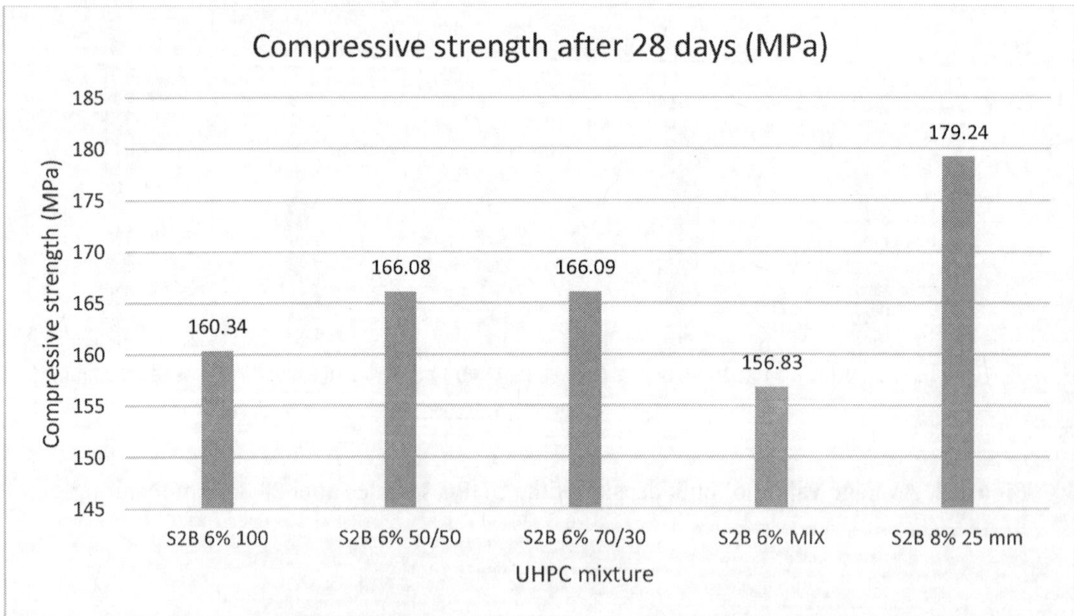

Figure 4. Average compressive strength values of the UHPC samples after 28 days of maturing.

Mixture composition S2B8%25 (containing 8 % of 25 mm fibre reinforcement) exhibiting the best flexural strength values reached also the best compressive strength values, 179.2 MPa in average. All results of physical-mechanical properties from Figures 2 – 4 are listed in Table 2.

Table 2. Summary of average physical-mechanical values of tested UHPC mixtures.

UHPC mixture	Bulk density (kg/m^3)	Flexural strength (MPa)	Compressive strength (MPa)
S2B6mix	2.599	16.40	156.80
S2B8%25	2.691	28.40	179.20
S2B6%75/25	2.599	19.70	166.10
S2B6%100	2.624	12.10	160.30
S2B650/50	2.614	21.00	166.10

3.4. Blast resistance testing

After physical-mechanical testing, blast resistance tests were conducted, according to methods depicted in 2.3. The results of both testing methods are summarized in Table 3. Measured ultrasonic wave propagation time is expressed as percentage increase after the explosion test. Absolute and relative weight loss of UHPC is also listed, together with photographs of sample board rear sides after testing according to the first method, i.e. after explosion of 150 g Semtex 10 charge. Based on results comparison, blast resistance of individual UHPC mixture compositions was evaluated.

Table 3. Summary of blast tests results.

UHPC mixture	Increase of UTS passage time [%]	Weight loss [g]	Weight loss [%]	Back of sample (150 g of Semtex 10 load)
S2B6mix	94.54 217.66 96.98 89.09 86.95 88.83 **112.3**	2240	8.6	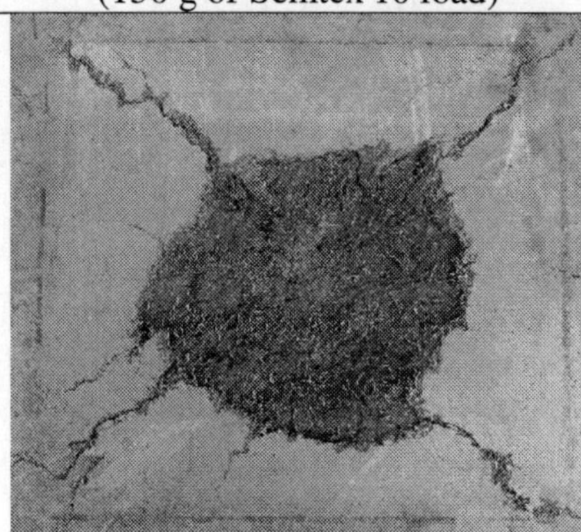
S2B8%25	49.48 42.81 43.82 31.77 31.89 3.31 **39.5**	498	1.9	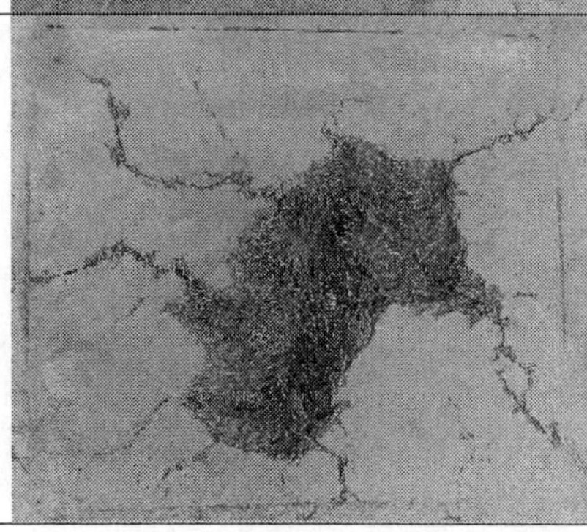

S2B6%7525	55.10 61.78 69.19 79.78 61.35 76.98 **67.4**	1048	4.3	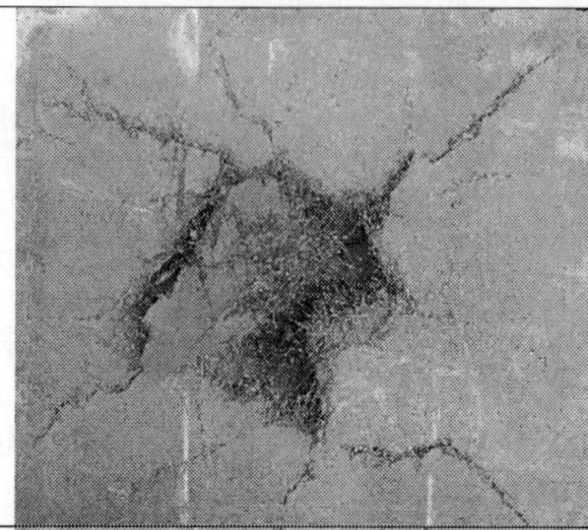
S2B6%100	114.53 351.67 129.27 404.67 575.19 357.80 **322.2**	2316	8.7	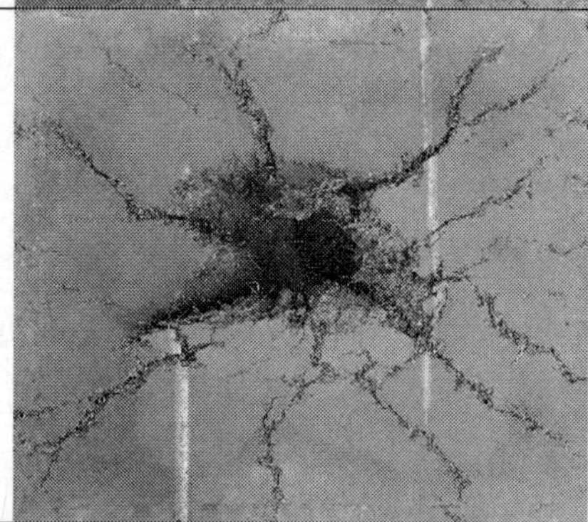
S2B650/50	65.88 62.12 72.03 68.05 64.83 75.02 **68.0**	508	2.0	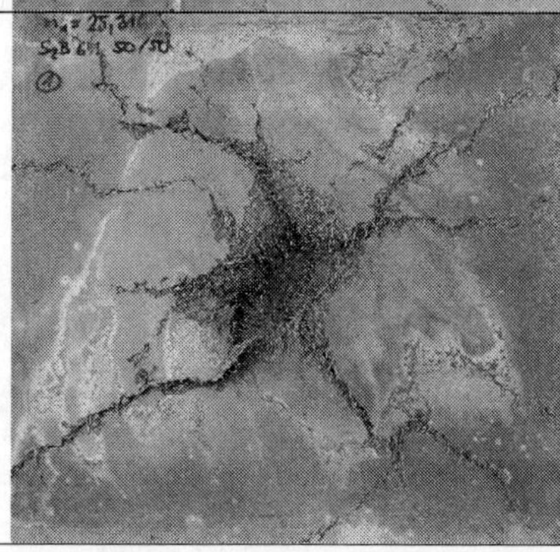

As seen from Table 3, from the point of view of ultrasonic wave propagation method, the best results of blast resistance of UHPC boards sized 500 × 500 × 40 mm were achieved with S2B8% 25 mixture composition. This sample showed also the best results of physical-mechanical testing thanks to the highest fibre content of all samples. Presumably due to the highest bulk density, these boards exhibited the lowest ultrasonic wave propagation increase after the blast. Subsequent weight loss (498 g, 1.9 %) caused by fragmentation was also the lowest at UHPC boards of mixture S2B8% 25. The second lowest weight loss after blast resistance test (508 g, 2 % w/w) occurred at UHPC boards of SB6% 50/50 mixture.

3.5. Application and advance testing of developed UHPC

Next advance stage was blast resistance testing of chosen UHPC mixtures in the shape of real application. The aim of this stage was to design and verify a mobile wall system resistant against highly intensive explosions. A protective wall was designed, composed of steel beams, effective anchoring and sandwich panels consisting of two UHPC layers (mixture S2B8% 25) and one absorption layer made of lightweight artificial aggregates and organic binder.

IOP Conf. Series: Materials Science and Engineering **596** (2019) 012018 doi:10.1088/1757-899X/596/1/012018

The whole developed blast protective system, its shape and dimensions were first designed in a 3D software, including the overall visualization. Subsequently, a cooperating company using LS-DYNA software, performed an explosion simulation of defined parameters in order to preliminarily determine a theoretical resistance of the designed wall and presumed deformations after the blast.

Based on the results of previous analyses, the explosion resistance of the developed protective wall system was verified. The ability of the developed protective wall to withstand the effects of a high intensity explosion was tested. The charge of the Semtex 10 plastic explosive was therefore set to the maximum possible value of 30 kg, the explosive was placed in 3 m distance from the tested wall. The course, intensity of the explosion and its effect on the tested shield system were recorded using several high-speed cameras and pressure sensors. The tested protective wall was made of steel construction and concrete panels sized 1000 × 500 × 40 mm, which were made from optimized UHPC S2B8% 25 composition. Functionality and durability of the entire developed shield system against the effects of high intensity explosion of was then verified.

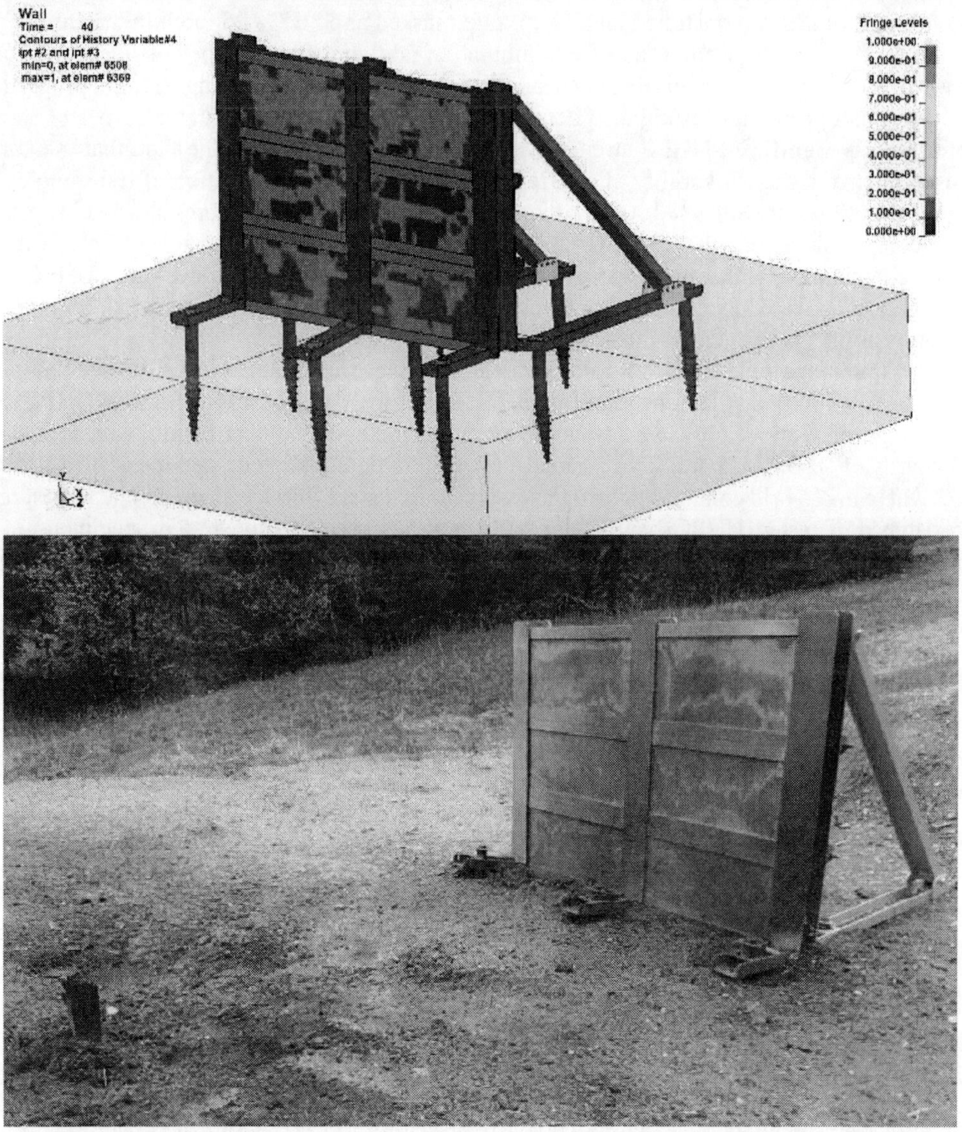

Figure 5: Comparison of results of software simulation and real explosion test.

As seen in Figure 5, the rate of damage, simulated in LS-DYNA software, was very similar to the real state of the shield system after the blast test. Panels made from UHPC mixture S2B8% 25 resisted the effects of explosion without visible damage and undesirable fragmentation. The most intensive load and slight shift were recorded only at the bottom of the shield wall as well as in the simulation, thus it can be said that the software simulations can be successfully used as an attractive alternative to the very expensive field explosive tests, in the early stages of development.

4. Conclusion

Realized research works were mainly focused on the impact of different composition of proposed UHPC mixtures on their properties. The influence of the type and amount of steel fibres content on physical-mechanical parameters and blast resistance of developed UHPC mixtures was evaluated.

On the basis of completed research works, it can be determined, that the utilization of higher volume of steel fibres content improves the strength characteristics of the UHPC samples significantly. The highest strength values demonstrated UHPC mixture marked as S2B8% 25, both in the case of flexural and compressive strength. Rather than the combination od two different types of steel fibres used, the increased dose of one type of steel fibres had significant impact on the resulting strength properties. The results of strength properties testing also demonstrated, that the use of corrugated types of steel fibres did not bring any significant UHPC strength value improvement. Despite the fact, that the corrugated fibres used improved the workability of the fresh mixture during the production of test samples, UHPC test specimens of mixture S2B6% 100 with the content of only corrugated fibres showed lowest flexural and even compressive strength values. On the other hand, if this type of corrugated fibres was used as hybrid, in combination with other type of straight steel fibres, results of developed UHPC mixtures strength parameters were higher, compare to test specimens where only the corrugated fibres were used. This theory was also confirmed in the case of the realized blast tests.

The test specimens of S2B8% 25 mixture exhibited the best integrity of all tested samples after explosion tests, which was demonstrated mainly as the lowest weight loss of tested UHPC samples. From the point of view of ultrasonic wave propagation method, the best results were achieved with UHPC boards of S2B8% 25 mixture. Second best results demonstrated specimens of S2B6% 50/50 mixture, where thy hybrid content of corrugated and straight steel fibres was used. The lowest results of blast resistance of tested UHPC boards, for both methods used, demonstrated specimens of UHPC mixture S2B6% 100 with the content of only corrugated fibres. So as in the case of physical-mechanical properties tests, using the hybrid fibre type, the blast resistance parameters of the tested UHPC mixtures were improved, compared to UHPC mixture where only the type of corrugated fibre was used.

Acknowledgments

The authors wish to express their gratitude and sincere appreciation to the authority of The Technology Agency of the Czech Republic, project No. TE02000162 for financial support.

References

[1] Bibora P, Drdlova M, Prachar V, Svitak O 2017 UHPC for Blast and Ballistic Protection, Explosion Testing and Composition Optimization, *et al IOP Conf. Ser.: Mater. Sci. Eng.* **251** 012004

[2] Drdlova M, Popovic M, Koutny O 2018 Blast resistance of hybrid fibre reinforced concrete containing polyvinyl alcohol, polypropylene and steel fibres with various shape parameters, et al. *European Physical Journal-Special Topics Volume: 227 Issue: 1-2,* pp 111-126

[3] Fehling, E., Schmidt, M., Walraven, J., Leutbecher, T., Fröhlich, S., *Ultra-High Performance Concrete UHPC*, Germany (2014), 23-27

[4] Drdlova M, Cechmanek R and Ridky R 2015 Blast impact behaviour of concrete with different fibre reinforcement, *Proc. DYMAT 2015 – 11th International Conference on the Mechanical*

and Physical Behaviour of Materials under Dynamic Loading, *EPJ Web of Conferences*. **94**. 05006. 10.1051/epjconf/20159405006
[5] Drdlova M, Buchar J and Ridky R 2015 Blast resistance characteristics of concrete with different types of fibre reinforcement, et al. *Structural Concrete Volume: **16** Issue: 4* pp 508-517
[6] Schrader E K 1981 Impact resistance and test procedure for concrete.*ACI Journal **78***, pp 141–146
[7] Gebbeken N and Krauthammer T 2013 Understanding of the dynamic response of concrete to loading: Practical examples. *Understanding of tensile properties of concrete*, pp 338–364
[8] Gebbeken N, Hubner M, Larcher M, Michaloudis G and Pietzsch A 2013 Beton und Stahlbetonkonstruktionen unter Explosion und Impakt, *Beton- und Stahlbetonbau **8***, pp 515–527
[9] Belko J, Calkovsky P and Krestan J 2011Testing of the safety objects M-VTÚO 11/11 *(version 1.0). Methodology of VTÚO Brno*(Brno)

IOP Conf. Series: Materials Science and Engineering **596** (2019) 012019 doi:10.1088/1757-899X/596/1/012019

Material properties of various types of fibre reinforced concrete exposed to elevated temperature

K Horníková and M Foglar

Czech Technical University in Prague, Faculty of Civil Engineering, Prague, Czech Republic

E-mail: katerina.hornikova@fsv.cvut.cz

Abstract. This paper presents experimental program focused on determination of material properties of various types of fiber reinforced concrete exposed to elevated temperatures up to 1000° C. Material properties were measured by several methods which are described and compared. The resulting changes of material properties depending on temperature are presented. The values are compared with values from literature and standards. The aim of this paper is to provide a comprehensive overview of material properties of these types of concrete depending on elevated temperature and possibility of their measurement.

1. Introduction

The requirements on durability and resistance of structures increase, together with requirements on their fire resistance. The fire resistance of concrete structures depends on the properties of the main component. These properties are defined by dimensions of the specimen, method of reinforcement and by material properties of concrete. Therefore, material properties of concrete are measured and evaluated in many experimental programs. However, usually only partial measurements of several characteristics are undertaken.

This paper presents an experimental program which was focused on determining the material properties of three types of concrete. Basic mechanical and thermal properties were measured: bulk density, thermal conductivity, specific heat capacity, porosity, permeability and compressive strength. The description and results of the tests are presented in the following chapters.

As mentioned above, the material properties are studied by many experimental programs and the results may be found in literature. Basic summary of material properties at normal temperature is presented in [1]. Description of the material characteristics at high temperatures are given in [2]. The book [2] describes material properties and their changes, defines basic mathematical relations between all properties. On the other hand, [3] deals with material characteristics and their influence on fire resistance of structure. Reference [4] focuses on the use of material characteristics to determine the behavior of elements at high temperatures. [5] and [6] can serve as an example of an experiment focused on measurement of only one characteristic. The first paper deals with commonly used techniques for measuring thermal conductivity of concrete. The second paper presents the results of measurement of thermal conductivity of eight materials. The paper [7] describes one type of measurement method for compressive strength and its results for some cement materials.

IOP Conf. Series: Materials Science and Engineering **596** (2019) 012019 doi:10.1088/1757-899X/596/1/012019

2. Materials

Material properties were measured for the following types of concrete:

- Common concrete C30/37 – C marked
- Concrete with polypropylene fibres – P marked
- Ultra-high performance concrete with steel fibres – U marked

Common concrete was chosen as a reference material to be used for further comparison.

3. Material properties and their measurement

Some material properties were measured by more than one method to compare and validate the results (or methods). All used methods are described in the following paragraphs.

3.1. Bulk density

The first measured characteristic was the bulk density depending on elevated temperature. Bulk density was measured using two methods. The first measurement (marked TZ) was carried out according to EN 993-1 [8]. For the second time, bulk density and matrix density were measured by the helium pycnometry (Pycnomatic ATC device) combined with gravimetric method (marked MA). The samples dimensions were 50 x 50 x 100 mm.

3.2. Compressive strength

The compressive strength of the concrete depending on elevated temperature was measured using a common hydraulic press. The system was installed in an electric furnace. A sample was placed in the furnace and then heated to the required temperature. After reaching of required temperature, the normal test of compressive strength was performed in the furnace. For the measurement of one temperature level, three cylinder specimens with 200 mm high and 100 mm diameter were used.

3.3. Thermal conductivity

Thermal conductivity depending on elevated temperature was measured by two methods. The first method (marked TZ) uses the heating wire is described in EN 993-15 [9]. For the measurement, three specimens with dimension 150 x 150 x 700 mm were used. The second method (marked MA) uses series of thermocouples, which record increasing temperature in the element. This method is described in [8]. This method is based on an approach taking into account moisture transport, radiative modes of heat transport and other reactions. The dimensions of samples were 50 x 50 x 50 mm and three samples were used.

3.4. Specific heat capacity

Specific heat capacity was determined using a water calorimeter (mixing vessel with water). The method used is described in [9] and [10]. It is a type of non-adiabatic method. The samples with dimensions of 50 x 100 x 100 mm were exposed to elevated temperature and heated to required temperature. After heating, the specimen was placed into mixing vessels and continuously stirred. Temperature of water was measured by system of five thermocouples, which were connected to monitoring systems. By this system specific heat capacity of used materials was defined.

3.5. Porosity

Porosity was measured by two methods. Using the first method, the porosity was calculated as ratio of weight dried samples and saturated samples (marked TZ). Using the second method, the porosity was calculated as ration of mean of matrix density and bulk density (marked MA).

3.6. Permeability

In this case, gas permeability was measured according to EN 993-4 [13]. The measurement was performed for three pressure levels. During the test, amount of gas passing through the specimen, was recorded. Three cylinder specimens with 50 mm high and 50 mm diameter were used for each tested pressure level.

4. Results of material properties measurement

All data obtained from the experimental measurement were compared. Results from literature were added to the comparison. The data from available and valid standards were also used. The results of each test are listed in this chapter.

4.1. Bulk density

Bulk density was measured by two methods. The results of both measurements are shown in Figure 1. The measured values were compared with the assumed values based on the parameters defined by the valid standard.

As expected, the bulk density of all materials decreased with increasing temperature affecting the specimen. Results of the both types of measurements were comparable for all materials up to 600 °C. From 600 °C to 1000 °C bulk density still decreased, but for measurement marked MA decreased much faster. Bulk density at 1000 °C of all materials was one quarter lower than initial value.

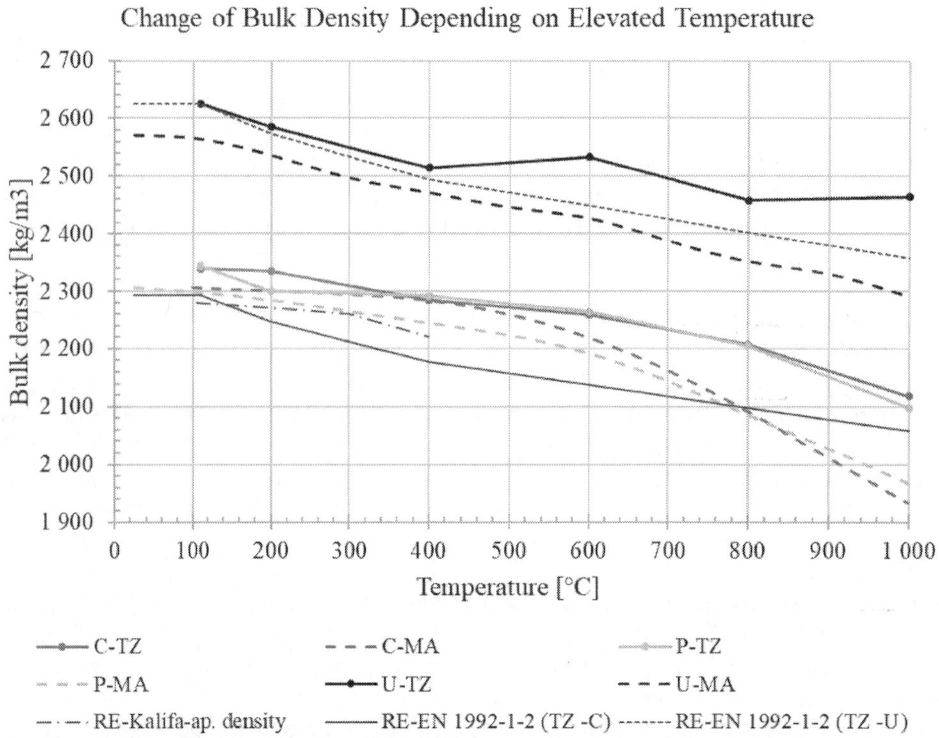

Figure 1. Results of bulk density measurement. [14], [15]

4.2. Compressive strength

The compressive strength of the concrete was measured by one method. The results are shown in Figure 2. The results were compared with the assumed values based on the parameters defined by the valid standard and the available literature. The results were comparable with standards values. The exception is the area between 100 °C and 200 °C where the values were different. Standards indicated no changes in this area, but experimental values were increasing. This may be because the standard conservatively neglects the increase in strength at lower temperatures. However, the resulting compressive strength at the highest temperature reached the expected values. The only exception was high-performance concrete with fibres (marked U), where the measured compressive strength was in all cases lower than expected.

IOP Conf. Series: Materials Science and Engineering **596** (2019) 012019 doi:10.1088/1757-899X/596/1/012019

The resulting compressive strength of all materials used decreased to 20 % of the initial compressive strength.

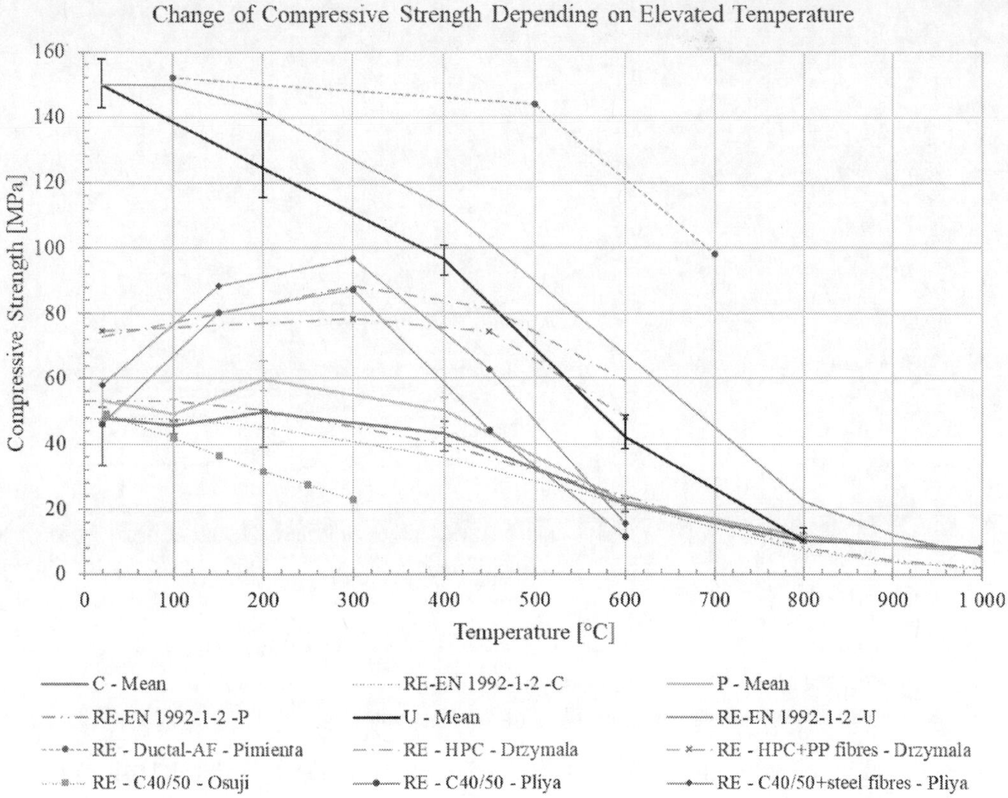

Figure 2. Results of compressive strength measurement. [6], [15], [16], [17], [18], [19]

4.3. Thermal conductivity

The thermal conductivity was measured by two different methods but with an incomparable result as shown in Figure 3.

The thermal conductivity determined according to the method specified in the standard (marked TZ) corresponded to the assumed values based on the parameters defined by the valid standard and the available literature. Thermal conductivity decreased with increasing temperature. However, the values obtained by the second measurement method (marked MA) were completely different. After reaching the temperature approximately 250 °C, the thermal conductivity stopped decreasing and started slowly increasing. For ultra-high performance, concrete with steel fibres it was growing at all measured points. The final thermal conductivity at the highest temperatures significantly exceeded the thermal conductivity of the material at normal temperature.

4.4. Specific heat capacity

Specific heat capacity was measured by one method and the results are shown in Figure 4. The results were compared with the assumed values based on the parameters defined by the valid standard. They were compared with curve for 0% moisture. This corresponds with the measurement because the samples were dried before the test.

The specific heat capacity of all materials used increased to 600 °C, respectively 700 °C, and then decreased. However, the final specific heat capacity was higher than its initial values. The specific heat capacity was affected by the moisture of the material in all its forms. This was reflected in its resulting course.

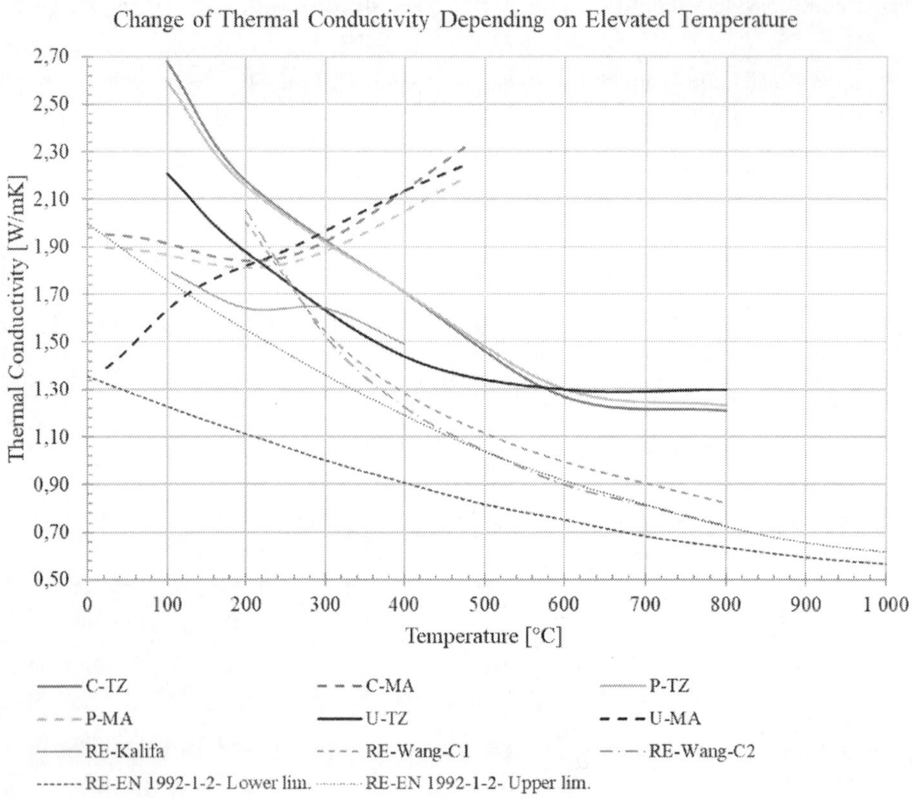

Figure 3. Results of thermal conductivity measurement. [15], [6], [14]

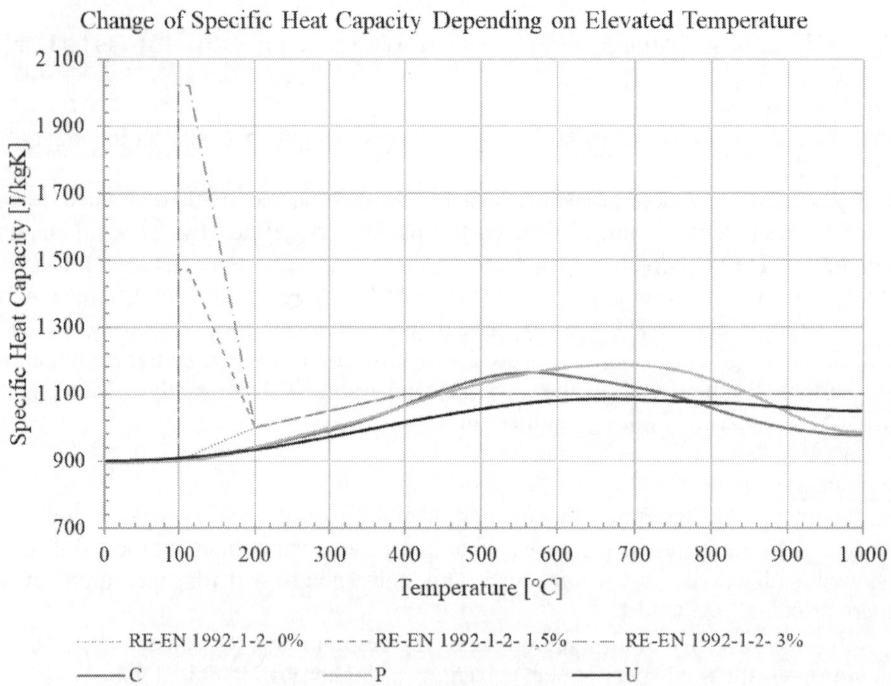

Figure 4. Results of specific heat capacity measurement. [15]

4.5. Porosity

Porosity was measured by two different methods and the results are shown in Figure 5. The results of both measurements were comparable and reached approximately the same values. The only exception was high-performance concrete with steel fibres (marked U), where one measurement showed almost double porosity than the other. The porosity of the material increased to maximum elevated temperature. The final porosity at 1000 °C was more than two times higher than initial value for all materials.

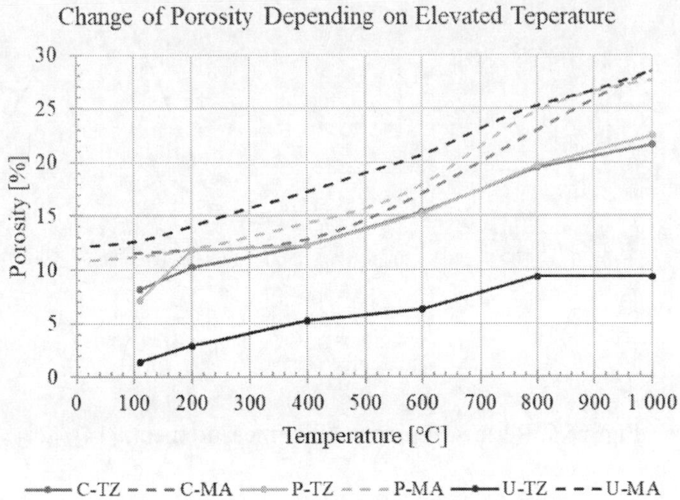

Figure 5. Results of porosity measurement.

4.6. Permeability

Permeability is closely related to porosity. As porosity increases, the amount of open pores also increases, causing increased permeability. In this case, the permeability was measured as gas permeability. The results are shown in Figure 6.

The results were compared with those obtained from the literature. The permeability increased with increasing temperature and increased most at high temperatures above 600 °C. Final permeability at the highest temperature was many times higher than initial value. That indicates significant damage of samples during its heating.

The lowest permeability was measured for ultra-high performance concrete with steel fibres (marked U). This material had zero permeability up to temperature 400 °C.

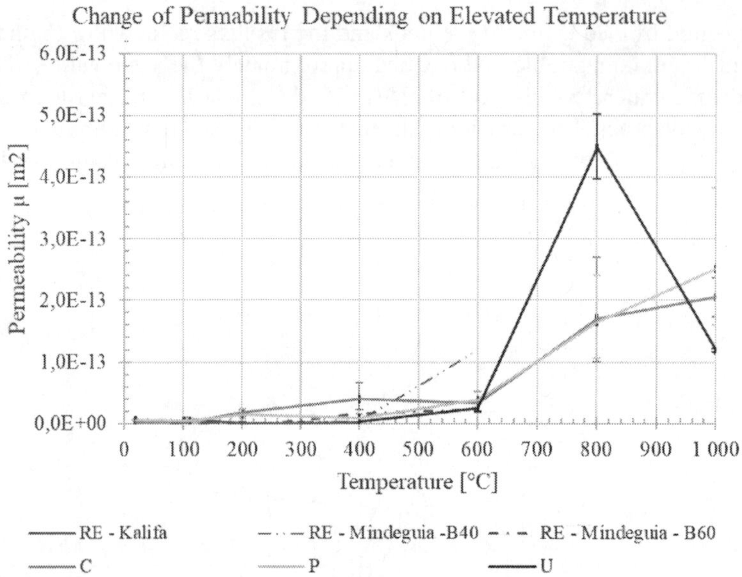

Figure 6. Results of permeability measurement. [14], [20]

5. Conclusion

This paper presents the results of an experimental program focused on experimental evaluation of material properties of various types of (fiber) concrete. Material characteristics of three types of concrete were measured at elevated temperatures up to 1000 °C. The reference material was common concrete, the second material was common concrete with polypropylene fibres and the last material was high-performance concrete with steel fibres. The material properties of common concrete are described in the valid standards and are also measured in many experimental programs. Concrete with polypropylene fibres is a commonly used material with high fire resistance. Therefore, it is important to know its behavior and material properties at high temperatures. The high-performance concrete is used mainly for its high load carrying capacity but with low fire resistance and high probability of spalling.

Methods of measurement and the results are presented in this article. According to assumptions with increasing temperature, bulk density, compressive strength and thermal conductivity of all materials decreased. On the other hand, the specific heat capacity, porosity and permeability of all materials increased. All these changes are mainly influenced by the moisture (and its transport) through the material. Moisture transport affects the pore pressure in the element. High pore pressure can damage the material (small cracks can be formed). Due to the small cracks, the air can pass to material structure and affect all thermal properties. Air and moisture have a major influence on thermal properties. On the other hand, cracks has a major influence on mechanical properties especially the compressive strength of material. The influence on the material properties is also the temperature itself. When the critical temperatures of material are reached, some material components can be changed.

Acknowledgements

Financial support from Grant Agency of the Czech Republic project no. 17-23067S is gratefully acknowledged.

References

[1] Neville A. M. 1963 *Properties of concrete* (Longman)
[2] Bazant Z P and Kaplan M F 1996 *Concrete at high temperatures: material properties and mathematical models* (Addison Wesley)
[3] Kodur V 2014 *ISRN Civil engineering* **2014**.

[4] Guo Z and Shi X 2011 *Experiment and calculation of reinforced concrete at elevated temperatures* (Elsevier)
[5] Asadi I., Shafigh P, Hassan Z F B A, and Mahyuddin N B 2018. *Journal of Building Engineering.*
[6] Wang W C 2017 *Construction and Building Materials* **147** 305-311.
[7] Novak J and Kohoutkova A 2018 *Fire Safety Journal* **95** 66-76.
[8] ČSN EN 993-1 1995 *Methods of test for dense shaped refractory products – Part 1: Determination of bulk density, apparent porosity and true porosity* (CEN) [in Czech]
[9] ČSN EN 993-15 2005 *Methods of test for dense shaped refractory products – Part 15: Determination of thermal conductivity by the hot-wire (parallel) method* (CEN) [in Czech]
[10] Černý R and Vejmelková E 2011 *Measurement*, **44** 1220-1228.
[11] Poděbradská J, Pavlík J, Toman J, and Černý R 2003 Specific heat capacity of cementitious composites in high-temperature range. In *Proceedings of the Thermophysics 2003 meeting of the Thermophysical Society Working Group of the Slovak Physical Society* pp. 18-23
[12] Toman J, and Černy R 1995 *Journal of Thermal Analysis and Calorimetry* **43** 489-496.
[13] ČSN EN 993-4 1997 *Methods of test for dense shaped refractory products – Part 4: Determination of permeability to gases* (CEN) [in Czech]
[14] Kalifa P, Tsimbrovska M, and Baroghel-Bouny V 1998 High performance concrete at elevated temperatures—an extensive experimental investigation on thermal, hygral and microstructure properties. In *Proceedings of the International Symposium on High-Performanceand Reactive Powder Concretes* pp. 259-279
[15] EN 1992-1-2 2004 *Eurocode 2: Design of concrete structures-Part 1-2: General rules-structural fire design* (CEN)
[16] Drzymała T, Jackiewicz-Rek W, Tomaszewski M, Kuś A, Gałaj J and Šukys R 2017 *Procedia Engineering* **172** 256-263.
[17] Pliya P, Beaucour A L and Noumowé A 2009 *Strength and porosity of concrete incorporating polypropylene and steel fibres subjected to high temperature* Tech. Rep.
[18] Pimienta P, Mindeguia J C, Simon A, and Behloul M 2011 *Designing and Building with UHPFRC*, 579-600.
[19] Osuji S O and Ukeme U 2015 *Nigerian Journal of Technology* **34** 472-477.
[20] Mindeguia J C 2009 *Contribution expérimentale à la compréhension des risques d'instabilité thermique des bétons* Ph.D. thesis UPPA

IOP Conf. Series: Materials Science and Engineering **596** (2019) 012020 doi:10.1088/1757-899X/596/1/012020

Influence of internal composition of fiber reinforced cement composites on resistance to near field blast

J Zíma and M Foglar

Czech Technical University in Prague, Faculty of Civil Engineering, Prague, Czech Republic

E-mail: jakub.zima@fsv.cvut.cz

Abstract. *This paper describes the influence of internal composition of fiber reinforced cement composites on resistance to near field blast. The influence of multiple basalt meshes and dispersed fibers on a damage induced by a near-field blast is studied and numerically evaluated. Experimental measurements performed in the Boletice military area in 2014, 2015 and 2016 are evaluated by numerical simulations. The evaluation of the results is mainly focused on the stress in the cement composite, propagation of the overpressure caused by the blast and velocity of the ejected parts from the specimen. The influence of velocity of ejected parts coming out of the specimen, originally caused by the impact of a steel plate which bears TNT charges, is examined. And last, but not least, the influence of presence and position of basalt meshes in the specimen on its damage induced by delamination is also examined.*

1. Introduction to the experimental program

The main objective of this article is to describe the experimental program and to present the numerical simulations that evaluate it. However, numerical simulations are closely related to the experimental program (or experimental measurements). Composition of the concrete mixtures and heterogeneity level included in the numerical material model is given by experimental measurements. And therefore it is necessary to provide a reader with some basic and important information.

Experimental program focused on resistance of concrete structures subjected to blast loading (blast resistance of concrete structure) has been taking place continuously for many years. In the last three years (2014, 2015 and 2016) the basalt mesh has been put into the concrete specimens. The mesh increases heterogeneity of the cross section. High heterogeneity of the concrete elements (or concrete cross section) is important for improving the blast resistance of these elements.

The experiments were performed in the Boletice military training area, in cooperation with the Armed Forces of the Czech Republic. The location and layout of the experiment was very similar to the authors' previous experiments [1, 2, 3]. The shape and size of the specimens were the same as the specimens used in the previous (and next) tests. Specimens were concrete slabs 6 m long, 1.5 m wide and 0.3 m thick. Each specimen had the same amount of steel reinforcement, i.e. 11 pcs ∅ 16 mm reinforcing bars every 140 mm on both sides, ∅ 10 mm every 150 mm as an outer transverse reinforcement, while the shear reinforcement was provided by ∅ 8 mm links (9 pcs/m^2). Concrete cover of 50 mm was taken to the surface of the transverse reinforcement. The scheme of the experiment is shown in Fig. 1. The ground beneath the slabs was excavated to the depth of 2 m in order to avoid the results being influenced by a shock wave bouncing off the ground. The 25 kg TNT charges were placed on the steel chairs (steel plate which is used for placing the charges is shown in Fig. 1 too) in the middle of each slab. The steel chairs provided 450 mm standoff from the slab. The other research has found that the steel chairs (steel plate) which bear the TNT charges behave like a projectile. It will be introduced as a projectile in next parts of the article. Information on the instrumentation used during the experiment can be found in [4,5,6].

Figure 1. Scheme of the experiment – specimen P19 with basalt mesh in five layers

Used specimens were made of ultra-high performance fiber-reinforced concrete (UHPFRC). Two types of high-strength steel (HSS) fibers were used: 35 mm and 13 mm in length. The yield strength of the fiber material to be higher than 2200 MPa is guaranteed by the manufacturer.

The specimens contained basalt meshes in multiple layers along the depth of the specimen in distances about 50 mm (Fig. 2). Mesh size of basalt mesh was 30 x 30 mm. The melting point of the basalt mesh is 1350 °C and the tensile strength is about 4200 MPa with tensile modulus of elasticity of about 85 GPa. The unit weight of the mesh is 250 g/m^2 with weight density of 2.67 g/cm^3.

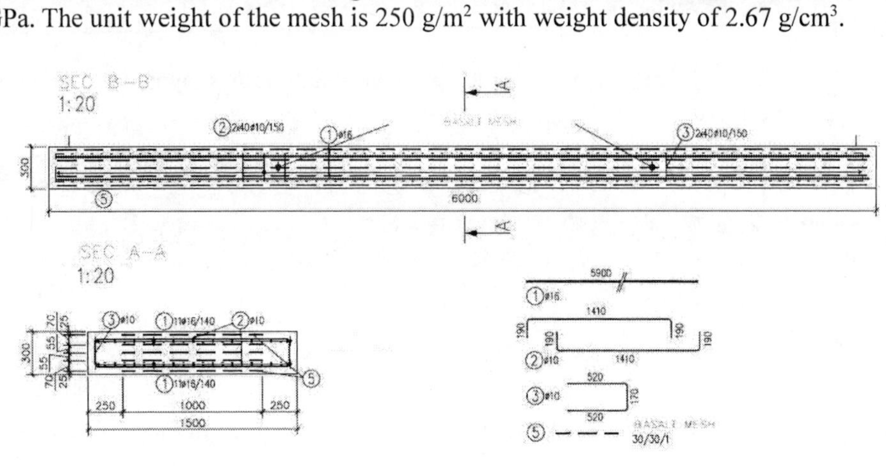

Figure 2. Concrete details drawing of a Specimen No. 18 – specimen with basalt mesh in five layers

2. Numerical evaluation

Numerical simulations were performed in the LS-DYNA software – the explicit finite elements method was used. Numerical models were created for the evaluation of a shock wave propagation, velocity of flying elements (ejected parts) and influence of heterogeneity on delamination of the specimen. During the evaluation of these main effects, other values were checked in order to verify the correctness of the numerical models. Numerical models are used to understand and describe the phenomenon inside the specimen which includes different cross-sectional heterogeneities [7,8].

2.1. Description of the model

Exact specification of the material models was taken from [1, 2, 7]. Heterogeneity of specimen is included in each material model.

Concrete was defined by the material model *MAT_CSCM. Reinforcement and basalt mesh were defined by the material model *MAT_PLASTIC_KINEMATIC which is a standard material model of the LS-DYNA [7, 8]. All material models were assigned with the appropriate parameters based on real material properties.

Using axis symmetry of the experiment, one quarter of a specimen is considered for a shorter calculation time. This simplified model is also easier and more friendly to interpret the results.

Three types of numerical models were created (with different concrete mixtures):

- Model with reinforced concrete and dispersed fibers (Fig. 3)
- Model with reinforced concrete, dispersed fibers and one layer of basalt mesh at the soffit of the specimen in the half of the reinforcement cover (Fig. 4)
- Model with reinforced concrete, dispersed fibers and five layers of basalt meshes (Fig. 5)

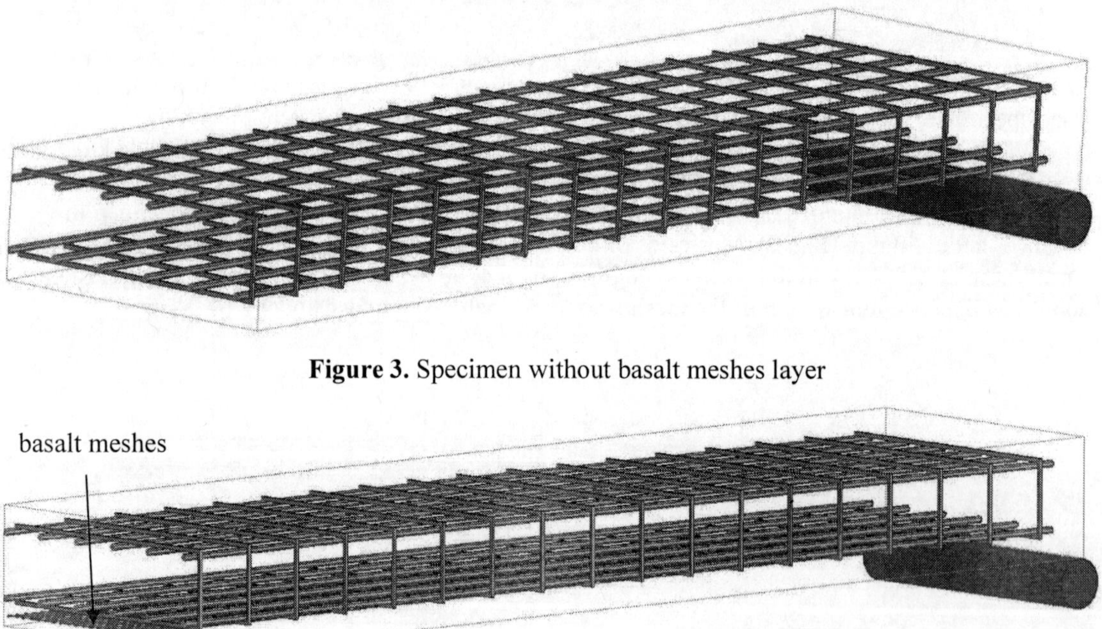

Figure 3. Specimen without basalt meshes layer

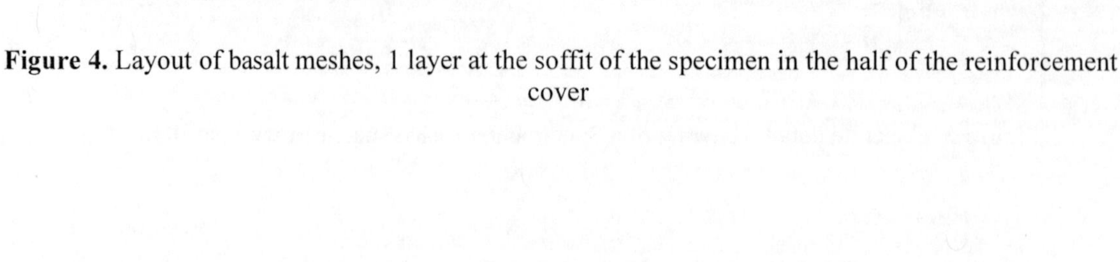

Figure 4. Layout of basalt meshes, 1 layer at the soffit of the specimen in the half of the reinforcement cover

IOP Conf. Series: Materials Science and Engineering **596** (2019) 012020 doi:10.1088/1757-899X/596/1/012020

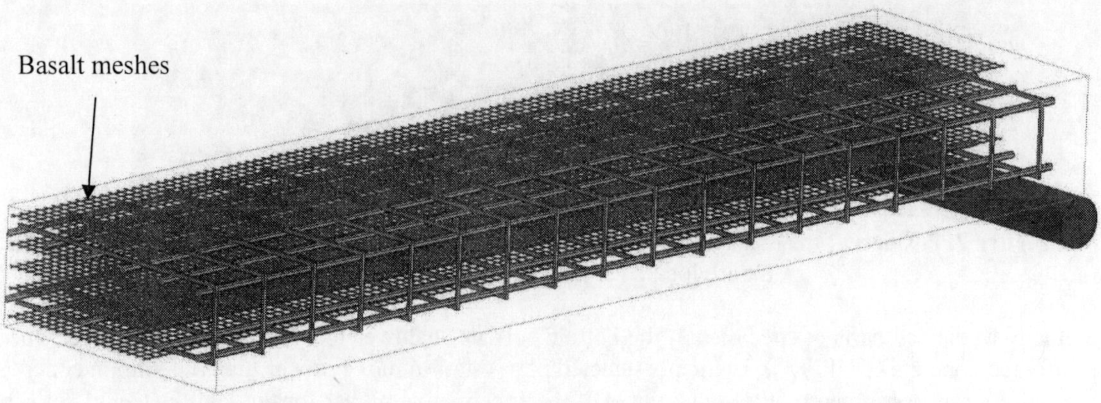

Figure 5. Layout of basalt meshes, 5 layers, as presented in Figure 2

Near-field blast was simulated by the function *LOAD_BLAST_ENHANCED (LBE) which converts the blast loading into pressure impulse. Input data of this function are the distance between charges and specimen, weight and type of charges. This method was selected on the basis of [7] where compliance with other load method is verified.

2.2. *Velocity of ejected parts*

For the velocity evaluation of ejected parts of the specimen, there were points determined on the lower surface of the specimen. During the experimental measurement, the velocity was measured in the middle of the specimen span, 150, 300 and 450 millimetres from the center. Coordinates consider the coordinate system of a quarter of a specimen. The velocity of the ejected parts in the middle of the span only was measured during the experimental measurement (Fig. 6 and 7).

- Channel position CH1, Node 332, Coordinate [3000; 0; 0]
- Channel position CH2, Node 317, Coordinate [2850; 0; 0]
- Channel position CH3, Node 302, Coordinate [2700; 0; 0]
- Channel position CH4, Node 289, Coordinate [2550; 0; 0]

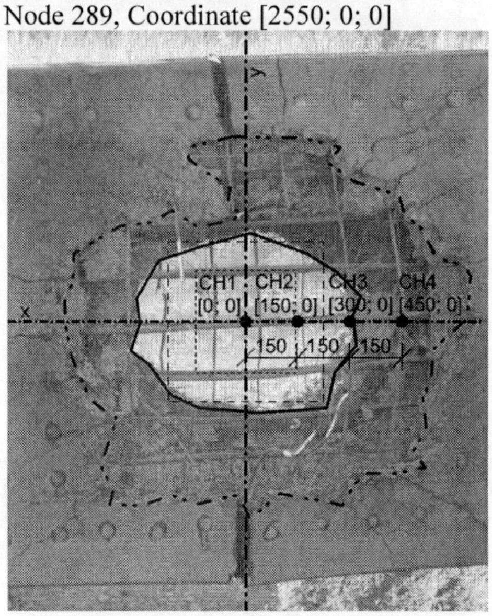

Figure 6. Experimental results for the specimen no. 18 – Damaged specimen; channel position measurement scheme

Figure 7. Position of the measurement points, view from the soffit of the specimen; measurement point are located at the centre-line of symmetry

Velocity of ejected parts of specimen at the bottom surface is slower for ultra-high performance fiber-reinforced concrete (UHPFRC). Elements (ejected parts) start to move earlier than if common concrete is used. Energy consumed by the shock wave of the specimen is higher for the UHPFRC and velocity of the elements is lower. It is obvious from the experiment that the velocity of the ejected parts decreases away from the centre of the specimen for each concrete mixture used. Also, in the figures below there can be seen that using the basalt mesh in one or five layers can reduce the velocity. By using the basalt mesh in the specimen (along the depth of the specimen) appropriately, the velocity can be reduced up to 15%.

In the following figure 8, the influence of the basalt mesh layers in the specimen can be observed with respect to the velocity of the ejected parts.

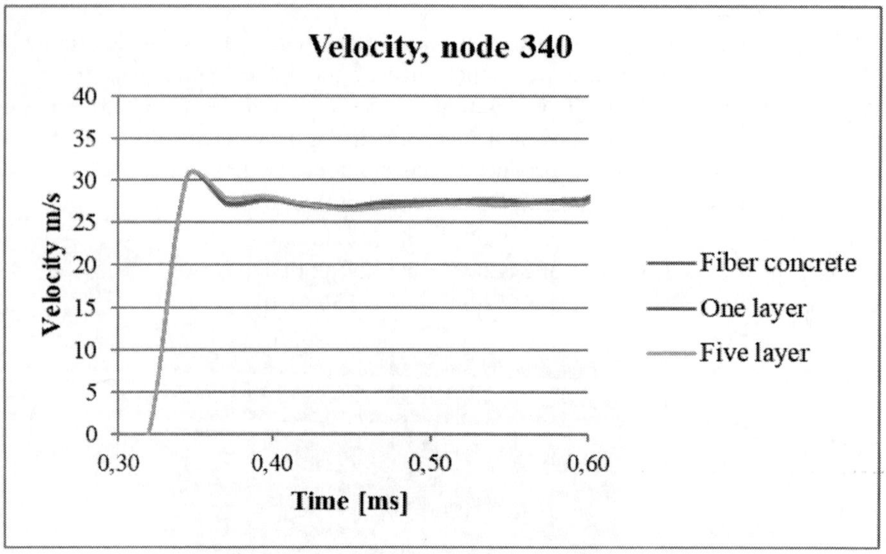

Figure 8. Velocity of the ejected parts, in the middle of the span

2.3. Shock wave propagation

According to the numerical model, there was found that the shock wave propagation is faster than commonly used concrete if the UHPFRC concrete is used. The high strength concrete has higher bulk density. Also, it is more homogeneous than common concrete. Fine aggregate fraction, more cement and steel fibers make faster passage of the shock wave. The time difference of the shock wave propagation depends on used material is about 0.01ms.

During the passage of the shock wave, there is a concentration of stress around the reinforcement and reflections of the shock wave at the interface between the concrete and reinforcement. The stress in the reinforcement is redistributed from the centre (away from the epicentre of explosion) to the edge of specimen. By the concentrating the stress the cohesion between the concrete and the reinforcement is reduced.

For the specimen with basalt mesh, it is possible to observe stress concentration near this reinforcement. At the same time, it is possible to observe the redistribution of generated stress after reflection of the shock wave from the lower surface of the specimen. The stress is then redistributed in the concrete cover from the centre to the edge of the specimen.

2.4. Shock wave propagation

In Figures 9 to 12, delamination of the specimen due to use of basalt meshes can be observed. When passing a shock wave of a specimen piece, the energy is absorbed by basalt mesh and reflects a part of the shock wave. In the specimen without basalt mesh, the section is delaminated only at the steel reinforcement where stress is concentrated. In the specimen with one layer basalt mesh, a delamination occurs in the area of the steel reinforcement at the top surface and at the basalt mesh. At the specimen with five layers of basalt meshes, there is a clear delamination near all the basalt meshes.

Figure 9. Specimen without basalt meshes – delamination at steel reinforcement

Figure 10. Specimen with basalt mesh in the middle of the concrete cover, delamination of the specimen due to the placement of basalt meshed and at the steel reinforcement

Figure 11. Specimen with five basalt meshes, delamination of the specimen due to the basalt meshes and steel reinforcement

Figure 12. Delamination of the specimen due to the presence of basalt meshes, experimental finding

3. Conclusion

Numerical models were made by the LS DYNA software. Blast loading was made by the Load Blast Enhanced function. Models served primary to evaluate shock wave propagation and determination of velocity of ejected parts of specimen taking into the material properties of the specimens.

Appropriate increase of heterogeneity of the element - using some basalt meshes at along the depth of the specimen, can reduce the velocity of ejected parts up to 15% against the specimen without these non-steel reinforcements in this case. From the models it is also evident that passage of the shock wave is faster for ultra-high performance concrete (with steel fibers in this case) than concrete which is commonly used. However, the velocity of ejected parts is lower for the UHPFRC. Using these findings, the protection of civilian population against the effects of ejected parts of structures exposed to extreme loads due to near-field blast can be successfully increased.

From numerical models it is also clear that there is a concentration of stress around the steel reinforcement. Stress concentrated in the reinforcement is redistributed from the center to the edge of specimen (away from the centre of explosion). In place of the stress concentration, the cohesion between concrete and steel reinforcement is disrupted. This phenomenon is also verified by the numerical model with basalt meshes. Around the basalt meshes, there is a concentration of stress, reflection of the shock wave and thereby delamination of the specimen.

Commonly available literature does not mention concrete elements with basalt meshes and their influence on its resistance under near-field blast. The same problem is with describing the behaviour of a projectile (steel chairs which bears TNT charges). This topic is very specific and unique. Therefore, this article and whole work is not easy to compare with another measurement or results of other experimental programs.

Acknowledgements
Financial support from Grant Agency of the Czech Republic project no. 17-23067S is gratefully acknowledged.

IOP Conf. Series: Materials Science and Engineering **596** (2019) 012020 doi:10.1088/1757-899X/596/1/012020

References

[1] Foglar M, Kovar M 2013 Conclusions from experimental testing of blast resistance of FRC and RC bridge decks *International Journal of Impact Engineering*, **59** 18-28.
[2] Foglar M, Hajek R, Kovar M, Štoller J 2015 Blast performance of RC panels with waste steel fibers *Construction and Building Materials* **94** 536-546.
[3] Foglar M, Hajek R, Fladr J, Pachman J, Stoller J 2017 *Full-scale experimental testing of the blast resistance of HPFRC and UHPFRC bridge decks* Construction and Building Materials.
[4] Kovar M, Foglar M, Hajek R 2014 Blast performance of real-scale reinforced concrete specimens with varying fiber type and content. In *WIT Transactions on The Built Environment* **141** 159-169.
[5] Foglar M, Pachman J, Pelikan V, Kunzel M, Kovar M 2014 The structural response of a reinforced concrete specimen subjected to adjacent blast loading. In *WIT Transactions on the Built Environment* **141** 171-179.
[6] Hajek R, Kovar M, Foglar M, Pachman J, Stoller J 2015 Field testing of concrete members subjected to contact and adjacent blast. In *Advanced Materials Research* **1106** 164-167.
[7] LS-DYNA 2006 *Theory Manual. Livermore Software Technology Corp* (LSTC).
[8] LS-DYNA 2007 *Keyword User's Manual, Version 971* (LSTC).

IOP Conf. Series: Materials Science and Engineering **596** (2019) 012021 doi:10.1088/1757-899X/596/1/012021

Modelling the damping effect for structures in reinforced concrete frames

F Tepes-Onea and M Popa

University Ovidius of Constantza, Faculty of Construction

E-mail: tflorin@univ-ovidius.ro

Abstract. In this paper we study the influence of the damping on the stress and displacements response for a reinforced concrete structure P + 4. In this sense, it was used a mesh consisting of two overlapping layers with common junctions. A concrete layer and a reinforcement layer Therefore, a numerical study was carried out by the help of an especially made program (a rewriting of the NONSAP PC program where a graphic postprocessing software was created). Finally, we obtained the structural response from the effect of the concrete degradation in the areas of the plastic joints. The paper presents our recommendations regarding the use of damping for reinforced concrete structures.

1. Viscous damping

Resistant damping forces, currently used in calculating the dynamic response of moving mechanical systems, correspond to the following models for the damping phenomenon:

-viscous damping (resistant force proportional with velocity) $F_d = -c \cdot \dot{x}$

-hysteretic structural damping (resistant force proportionalwith displacement but out of phase from itwith $\pi / 2$, $F_d = igkx$

-colombian damping (dry friction-resistant force proportional to the normal surface reaction on which the displacement of a point occurs $F_d - fN \operatorname{sgn} \dot{x}$

Each of these models, for damping forces, has a specific scope of application, so that substitution of one type of strong force through another can only be made in full knowledge of the phenomena that take place.

For frame structures we are interested in viscous damping, which, as it is well known, can be modeled by the Rayleigh model to model energy dissipation inside the structure. It is possible to estimate the rate of effective viscous damping from structural tests.

One of the methods is to apply a static displacement by applying a cable to the structure and then suddenly removing the cable load. Obviously we can measure the values of two subsequent successive amplitudes, through the known relationships, while obtaining the fraction of critical damping.

In elastic dynamic analysis it is very common to use a 5% damping. Always, this value is in many cases very little experienced or justified theoretically. In the Rayleigh viscous damping model [1], the damping matrix [C] is calculated using multiplication constants for the mass matrix [M] and for the stiffness matrix [K]:

$$[C] = \alpha[M] + \beta[K] \tag{1}$$

The constants α and β are calculated according to the fractions of the critical damping ξ_i corresponding to their own vibration mode "i", and then α , β satisfy the relation (see in figure 1):

$$\xi_i = \frac{\alpha}{2\omega_i} + \frac{\beta\omega_i}{2} \tag{2}$$

We will admit that the sum of constants α and β is almost constant for a certain range of circular frequencies.

Using the fractions of critical damping and corresponding circular frequencies of two vibration modes, we determine the constants α and β and solve the system of equations (2).

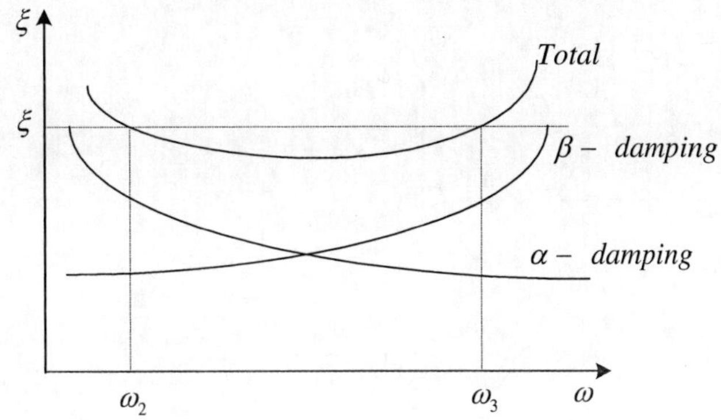

Figure 1. The damping-circular frequency relationship.

The problem of damping, and in particular the problem of structural damping, should be more broadly defined, both for materials with linear and non-linear behavior. In the case of structures in reinforced concrete frames, nonlinear behavior is well-known: the yielding of the reinforcement and the crushing of the concrete.

2. The level of damping in structures

The appreciation of the level of damping of structures in their different modes of oscillation still has a subjective character.

Although there are numerous papers on the values of the critical damping fraction in the different modes of oscillation, one cannot speak of a theoretical substantiation of the phenomenon. That is why the evaluation of the damping of a structure can only be done by considering as many of the parameters defining the phenomenon as possible (the shape of the structure, the nature of the materials used, the foundation ground, the excitation, etc.)

Taking into account the important contribution of the damping in determining the dynamic response of a structure, it is admitted that the damping effects are considered to come from a viscous damping (possibly a viscous damping equivalent to a structural damping) for which the critical damping fraction has the same values in all modes of oscillation.

3. Case study

The case study consists of a linear elastic time-history analysis of a P + 4. For the modeling of reinforced concrete (beams and pillars), two layers were used with joint displacements at the nodes. A concrete layer made of 8-nodes rectangular elements and a reinforced layer made of "Truss" elements. The reinforcement areas obtained from the sizing were used, using static seismic forces equivalent to seismic grouping [3], [6].

The accelerogram used is one recorded and scaled to 0.16g and the duration of the calculation step is 0.02s and the accelerogram duration is 16s (figure 2).

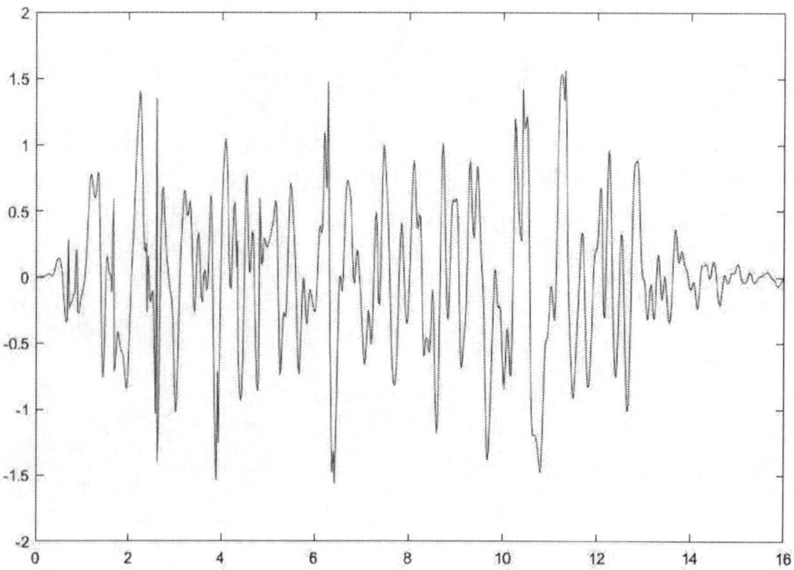

Figure 2. Recorded accelerogram (1977) in m/s^2, duration 16s.

For use, the variation of nodalforce applied over time (figure 2), the graph in the figure was resolved to 0.2g and amplified with the nodal masses.

One of the major advantages of the Direct Numerical Integration Method consists in its application without difficulty to the dynamic analysis of the structures allowing the introduction of the damping in analysis.

Another advantage of integration over time is that the mathematical equation of motion is the approach method to approximate the real phenomenon.

It is to be noted that most seismic design rules do not recognize the variation in damping depending on the type of material and the level of effort in the structure, specifying in all cases a fraction of the critical damping of 5%

In order to obtain the coefficients α and β a modal analysis was performed, and the values of the own vibration modes obtained are presented in Table 1.

Tabel 1. Vibration modes.

Vibration mode	ω_i [rad/s]	T_i [s]
1	15.97	0.39
2	51.94	0.12
3	89.91	0.06

IOP Conf. Series: Materials Science and Engineering **596** (2019) 012021 doi:10.1088/1757-899X/596/1/012021

Using two successive frequencies and the relationship $\alpha + \beta \cdot \omega_i^2 = 2 \cdot \omega_i \cdot \xi_i$ and solving the system [2], the proportional constants of the damping matrix are obtained, according to Table 2.

Tabel 2. Rayleigh damping coefficients.

Damping	α	β
3%	0.72	8.86E-4
5%	1.21	1.45E-3
7%	1.7	9.06E-3

Response comparisons between the three critical damping fractions are shown in Table 3.

Tabel 3. Comparisons between critical damping factions.

Damping	Dymax (at the top)	Maximum compression effort at the ends of the beams daN / cm^2	Maximum compression effort at the base of the pillars daN / cm^2
without damping	0.602E-1	382.2	355.4
3%	0.259E-1	172.9	162.6
5%	0.21E-1	146.2	138.8
7%	0.188E-1	131.3	125.8

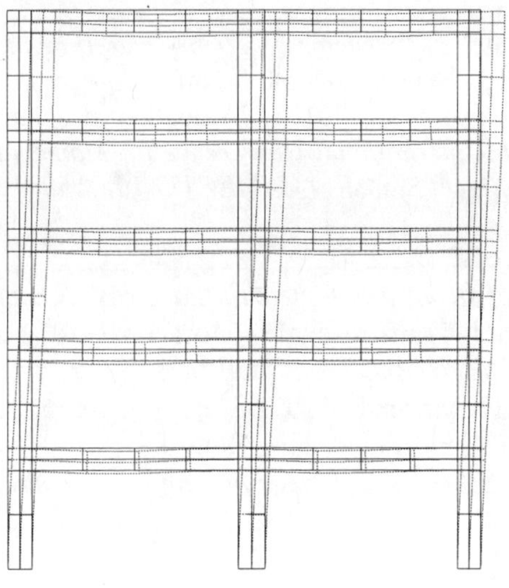

Figure 3. Deformed shape, calculation step 544, Dymax = 0.21E-1, 5% damping.

Whenever a material, element, or structure is required by cyclic loads, the cyclic loading-deformation curve (strenght-strain) is not a uniquely defined function but forms a hysteresis curve [5].

The structural-hysteretical damping, in the present case, can occur due to phenomena such as: the non-linear behavior of the concrete, the yielding of the reinforcement.It is well known that plastic entry of the reinforcement and cracking of the concrete increase the structural damping.

The second part of the study introduces the effect of degradation of concrete in the areas of plastic joint formation. The maximum compressed stresses were used. A two-stage calculation was performed: in the first stage efforts were made in elements, concrete and reinforcement (5% damping efforts); in the second stage the stiffness at the ends of the beams and at the base of the pillars was reduced.

The elastic modulus has been operated in areas where tensions have reached the limit, obviously in the sense of reducing it [4].

In this situation, the maximum displacement, shown in figure 3, increased from 0.021m to 0.037m.

4. Conclusion

Using a 5% or 7% damping in terms of the percentage of solutions means from the point of view of the displacement a difference around 10%. It is obvious that there are a number of factors that can falsify the response much more. One of them is the solution without damping. Other factors of influence of the response can be structural degradation during the seismic action and the influence of the foundation ground.

From the point of view of the structural degradation, on the chosen model, the displacements increase compared to the chosen reference value with 76%. A reference value was obtained in the case of 5% damping where maximum displacement is Dy = 0.021m.

In conclusion, it is obvious that in case of integration in time even if numerically we have to deal with a more elaborate solution, the impossibility of controlling all the terms involved can generate solutions errors.

Thus, the use of cover values for all the involved terms represents one of the most viable solutions.

References

[1] Chopra A K 2012 *Dynamics of Structures –Theory and Aplications to Earthquake Engineering* (Upper Saddle River, New.Jersey.07458: Prentice Hall)

[2] Ţepeş Onea F , Gelmambet S 2010 Evaluation of damping in dynamic analysis of Structures*12th WSEAS International Conference on Mathematical Methods, Computational Techniques and Intelligent Systems (MAMECTIS'10), Matematical Methods, Computational Techniques*, ed WSEAS Press pp 67-70

[3] P100-1/2013 *Design of structures for earthquake resistence, part 1*

[4] Postelnicu T, Damian I, Zamfirescu D, Morariu E 2012 *Design of reinforced concrete structures in seismic areas* (Bucharest: MarLink)

[5] Paulay T, Bachmann H, Moser K 1997*Design of reinforced concrete structures for seismic actions* (Bucharest: Technical Publishing House)

[6] SR EN 1992-1-1/2006 *Eurocod 2 Design of reinforced concrete structures. Part 1*

IOP Conf. Series: Materials Science and Engineering **596** (2019) 012022 doi:10.1088/1757-899X/596/1/012022

A testing method for studying the serviceability behavior of reinforced UHPFRC tensile ties

M Khorami[1,2], J Navarro-Gregori[1], P Serna[1] and M A Navarro-Laguarda[1]

[1]Institute of Science and Concrete Technology, ICITECH, Universitat Politècnica de València, València, 46022, Spain.

[2]Universidad UTE, Facultad de Arquitectura y Urbanismo, Calle Rumipamba s/n y Bourgeois, Quito, Ecuador

E-mail: juanagre@cst.upv.es

Abstract. Structural control under the serviceability limit state is a requirement of design codes to ensure the durability of structural elements. As it is possible to consider fibers to be reinforcement in concrete, UHPFRC can be used to guarantee properly distributing cracks and limiting crack width in the serviceability limit state. This research presents an experimental testing method for direct tensile tests on UHPFRC specimens. The results obtained from the proposed method, such as the specimen's average tensile stress-strain curve, tensile stress in concrete, number and width of cracks, can be used to consider the behavior and design requirements of UHPFRC under serviceability conditions.

1. Introduction

Ultra high-performance fiber-reinforced concrete (UHPFRC) is a material with high compression stress, and great ductility and toughness due to the inclusion of steel fibers. The addition of fibers to the concrete matrix leads to significant energy being absorbed and helps control crack opening [1,2]. The presence of fibers in cracks limits the width of cracks, and results in their proper distribution and increases their serviceability [3,4]. In all concrete structure design codes, controls related to the serviceability state and cracking control are important. Many research works have been conducted to study only the tensile behavior of UHPFRC [5–9]. They have focused on the mechanical properties of UHPFRC without reinforcements. However in real structural elements, concrete is usually accompanied by reinforcement. One method for recognizing this behavior is to test the direct tensile tie (a prismatic concrete member with a high length-to-width ratio and a reinforcing steel bar in the center of the cross-section). In order to understand the tensile behavior of concrete as a simple material, and of concrete combined with conventional reinforcement, parameters like crack width, distance between cracks and tension stiffening have been studied for many years [10–13]. As adding fibers to concrete limits the cracking phenomenon by narrowing cracks with short distances among them, it is important to analyze how the aforementioned parameters could be affected when UHPFRC is combined with a steel bar. In recent years, very few studies have analyzed the tensile behavior of UHPFRC by including steel bar reinforcement [14–16].

The present paper focuses on providing an appropriate testing method to investigate the UHPFRC cracking behavior under the serviceability limit state (SLS). For this purpose, tensile tie experiments were proposed and developed. A new method is provided to measure the ability to transfer tensile force to a concrete member. Some parameters were evaluated by this testing method, such as the element's average tensile stress-strain diagram, the element's strain behavior in small segments and the development of cracks and their frequency.

2. Research Significance

Performing a direct tensile test on reinforced concrete specimens and applying tensile force to concrete specimens are not easy tasks. This present work provides a simple testing method for studying the behavior of reinforced UHPFRC under uniaxial tension load. This method is able to analyze the multicracking behavior of a UHPFRC tensile tie using different types of measurements equipment. It also helps us to analyze the interaction between concrete and steel reinforcement and the possible synergy between them under SLS.

3. Experimental Program

The experimental program was considered for one type mixture to evaluate the behavior of the UHPFRC tensile tie. Due to the capacity limitation of the laboratory mixer and the number of molds, five batches with the same dose were prepared (C1 to C5). The mixture design of UHPFRC is reported in Table 1. The achieved average 28-day compressive strength test of concrete cubs of 10 cm x 10 cm was 158.41 MPa and the average Young's Modulus was 48.88 GPa. The length and diameter of applied steel fibers in this study were 13mm and 0.2mm respectively and their tensile strength was more than 2000MPa.

Table 1. The UHPFRC Mixture Design.

Component	Content (kg/m^3)
Cement I 42.5 R/RS	800
Silica Fume 940 D Elkem UD	175
Silica Flour U-S500	225
Fine Sand 0.5 mm	302
Medium Sand 0.6-1.2 mm	565
Water	160
Superplasticizer, Viscocrete 20 HE	30
Fiber	160

All the specimens were made as follows: the considered specimen length was 1000 mm and a central bar was located over the entire element with a length of 1450 mm. Two complementary rebars were located at both ends with a length of 450 mm. These reinforcement bars were welded to the main bar. This experimental research was conducted for one cross-section type (i.e. 80 × 80 mm) and three tie series with rebars ratios of 1.23 %, 1.77% and 3.14% (rebars were 10 mm, 12 mm and 16 mm). The nominal yield stress of the rebars was 500 MPa. To identify specimens, the first number represents the reinforcement diameter, whereas the second number refers to the number of series per specimen; e.g., for the tensile tie with a 10-mm reinforcing bar, the ID specimens were named (T-10-1), (T-10-2) and (T-10-3). Figure 1. illustrates specimen details.

During the manufacturing procedure, an attempt was made to locate the rebar exactly on the center section to avoid eccentricity and bending effects. Specimens were stored in a curing chamber at 95% relative humidity and a temperature of T = 20±2°C until 2 or 3 days before testing. UHPFRC was cast horizontally so that the casting process would start from one extreme to the other extreme by applying a uniform velocity to ensure a good alignment of the steel fibers in the concrete mixture.

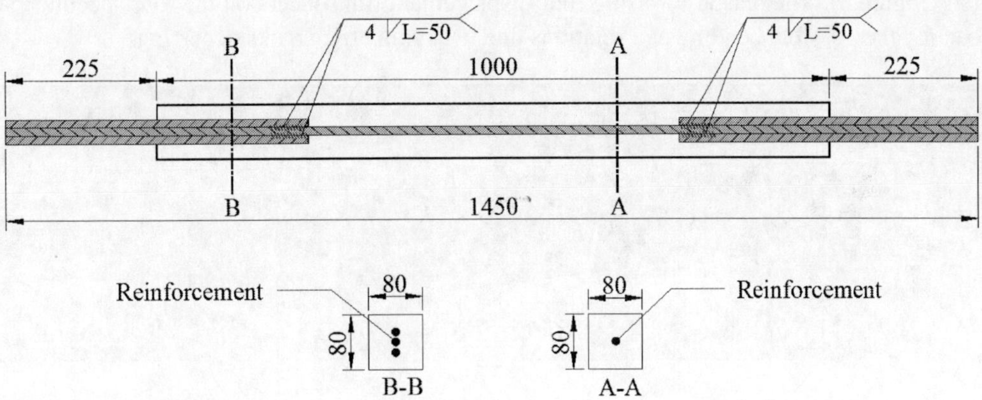

Figure 1. Specimen reinforcement details (units in mm).

4. Experimental setup and method

4.1. Description of the test system

In order to perform this test, a 2-meter steel structural frame was used, as outlined in [17] 'Figure 2.a'. This structure was made using two plates whose thicknesses were 5 cm on both extremes, along with four 60 × 60-mm rectangular tube sections of 8 mm thickness. As tensile force had to be transferred to the specimen, two two-piece steel jaws (2 mm high) with indented corrugations were used as shown in Figure 2.b.

In order to prevent the main bar from yielding and failing on both ends without concrete, two 45-cm long rebars were used in the external region of the specimen, and penetrated 22.5 cm into the concrete specimen (Figure 1). The test zone lay in the center part of the member where only one rebar existed. As previously mentioned, two st52 steel jaws were used to transfer tensile force from the system to both the concrete specimen and the ending reinforcing bars. Each jaw consisted of two segments with six high-strength bolts whose diameters were 13 mm. By tightening these bolts, tensile force was transferred by the hydraulic jack to the bar embedded in the concrete specimen because of the frictional force between the rebar and the interior side of the jaw. In order to distribute the force caused by tightening the bolts uniformly, this process was done in a zigzag manner. Figure 2. b. provides details of the jaw and the assembled connection.

Hydraulic jacks are designed as cylinders and a bar can be passed to their centers. A Ø20-mm high-strength steel bar (fy=1000MPa) was used to transfer the tensile force caused by the movement of the hydraulic jack. At the end of these bars, a circular steel rod end bearing was embedded (Figure 2.c). At both ends of the jaws, there was a Ø25-mm bolt that allowed the steel jaw to be connected to the bars through the steel rod end bearing. This connecting system displayed a hinge behavior and allowed bending or twisting at both specimen ends (Figure 2.d).

4.2. Instrumentation and experimental procedure

As previously mentioned, this experiment aimed to study the tensile behavior of the UHPFRC reinforced tie. To conduct this study, the behavior of the member was studied at both the general and local levels. To investigate the general behavior, displacement transducers were employed, while local behavior was evaluated using demountable mechanical gauges (DEMEC points). The position of the measurement equipment is illustrated in Figure 3.

For this purpose, specimens were tested under the displacement control. On both sides of the concrete specimen and on four surfaces, four 35-cm long displacement transducers were installed (Figure 3). In this method, it was assumed that the strain of the rebar located in the center of the section would equal the average value of the strains recorded for four external surfaces. For the local level measurement, #16 DEMEC steel discs were installed at a 1-cm distance from the upper and lower edges on each specimen

IOP Conf. Series: Materials Science and Engineering **596** (2019) 012022 doi:10.1088/1757-899X/596/1/012022

edge (see Figure 3). The reason for using four displacement transducers on the four specimen surfaces was to study the possible bending and rotations due to asymmetric cracks occurring.

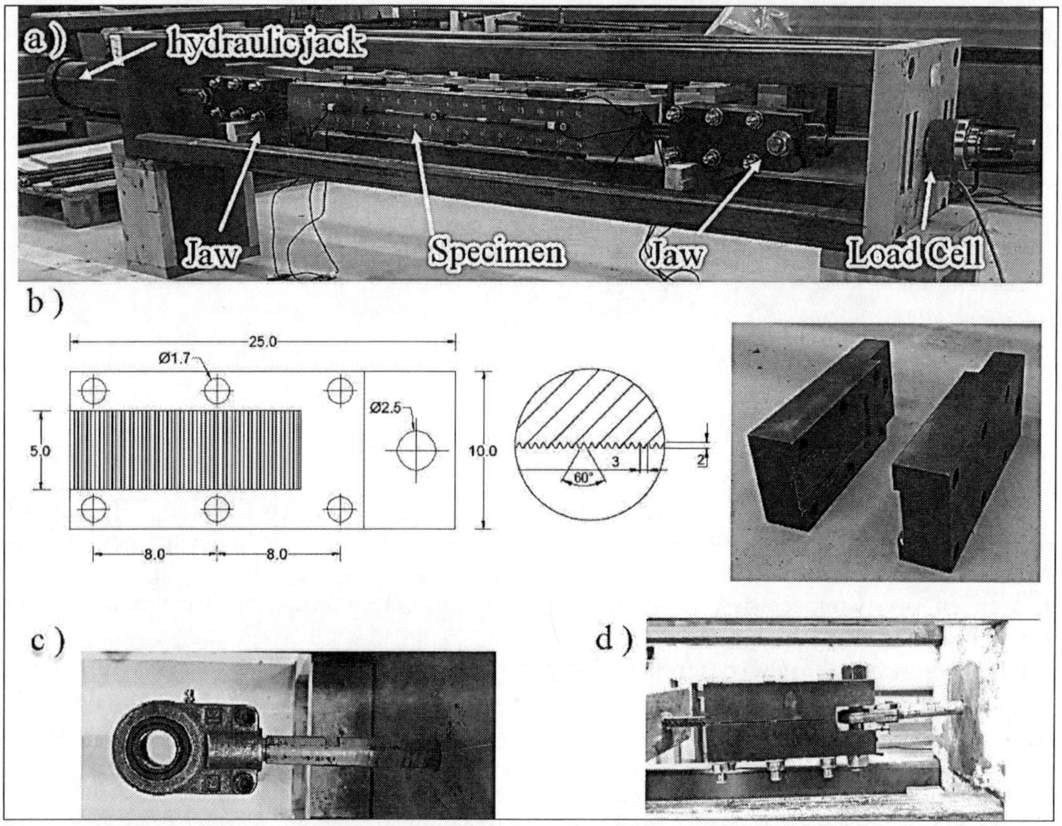

Figure 2. a) Tensile test equipment. **b)** Jaw details. **c)** rod end bearing. **d)** assembled Jaw connection.

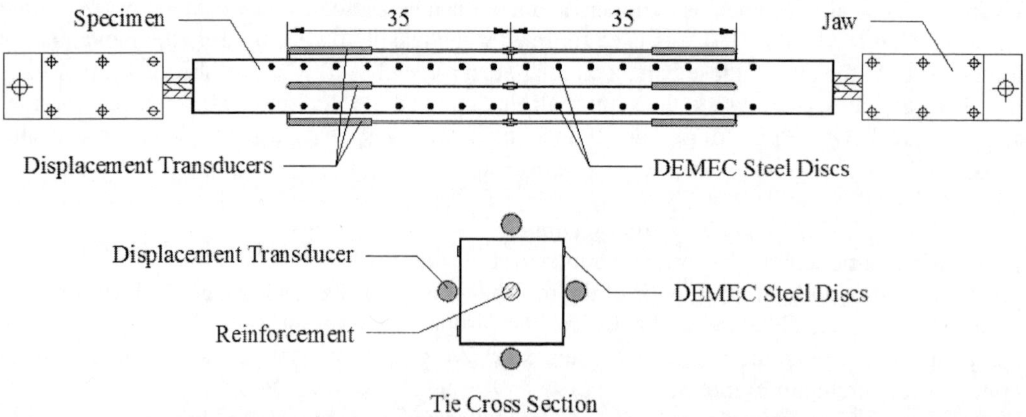

Figure 3. Measurement equipment: positions of the displacement transducers and the DEMEC points along the element.

After installing the concrete specimen within the test equipment, force was applied by the hydraulic jack and the increase in force was stopped in strains of 0.03‰, 0.05‰, 0.10‰, 0.15‰, 0.5‰, 1‰, 1.5‰ and 2‰. Then the changes in length between the DEMEC disks were measured. While testing, an attempt was made to maintain force at constant levels.

As the average displacement recorded by the displacement transducers reached an equivalent strain of 2‰, the experiment was stopped, and the number of cracks on the upper and lower edges between the DEMEC points was recorded by plotting the crack pattern on the specimen surfaces. For the specimens containing the Ø16 bars, it was difficult to reach the tension strain of 2‰ because the bars slipped at a high load value. Therefore, the experiment continued for these specimens until a displacement corresponding to a strain of 1.5‰ took place.

5. Results and Discussion

A tensile stress-strain diagram of specimens was obtained from the average of the four displacement transducers located on the left and right specimen sides. Figure 4. illustrates the displacement-force curves due to the tensile force on both the left and right sides of specimen T-12-2, respectively.

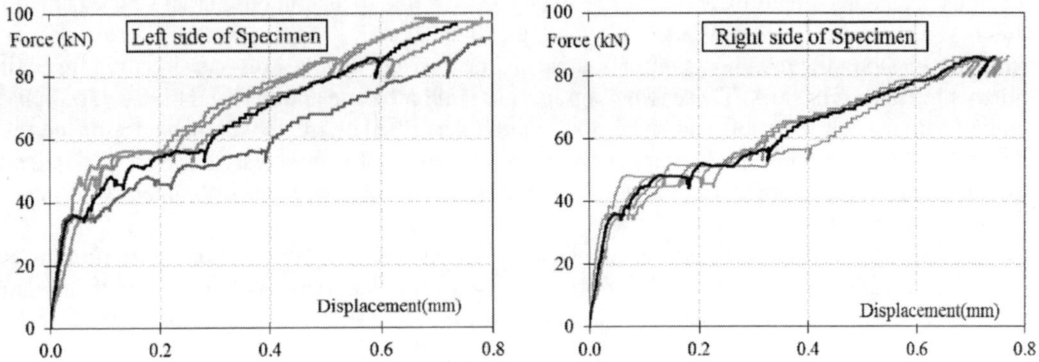

Figure 4. Displacement-force relationship for specimen #T-12-2.

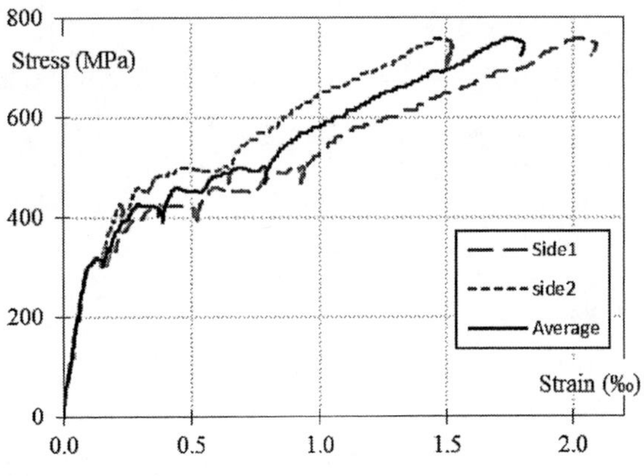

Figure 5. Tensile stress-strain diagram for specimen #T-12-2 and the average values.

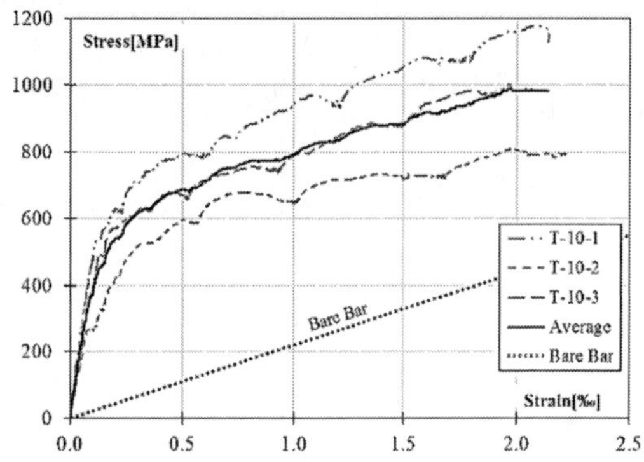

Figure 6. Stress-strain diagram based on the rebar area for tie 8x8cm with the Ø10-mm bar.

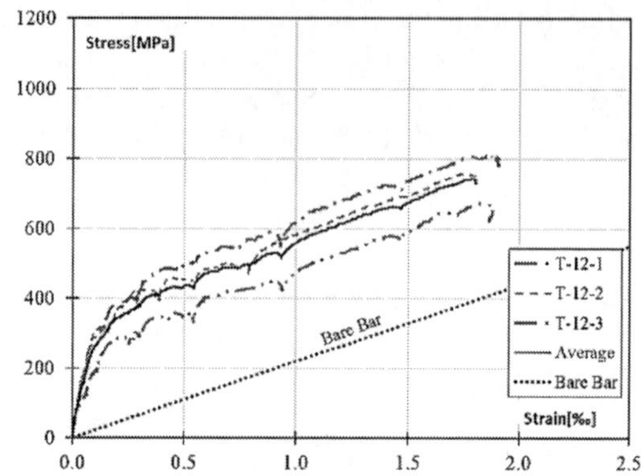

Figure 7. Stress-strain diagram based on the rebar area for tie 8x8cm with the Ø12-mm bar.

As previously stated, a slight difference in the diagrams was due to asymmetric cracks appearing on the specimen's cross-section, as well as very small localized bending because of the difference in the cracking inertia along the member. Figure 5. presents the average tensile stress-strain curve for both the specimen's left and right sides. These two diagrams are similar to one another, and the average behavior diagrams for the specimen's left and right sides can be described as the tensile behavior of the tensile tie. The stress expressed in this diagram is shown as the equitant steel stress, which was obtained by dividing the total tensile force by the area of the reinforcing steel. The results obtained for each series of specimens are shown in Figures 6, 7 and 8.

As the equivalent stresses shown in the above diagrams are based on the area of the reinforcing steel, the numerical value of this stress is higher for the bars with smaller areas. As observed in the diagrams, the tie's stress-strain behavior is relatively in parallel to that of the steel bar after the member cracked under tensile force, and reducing of the stiffness and change in slope of the stress-strain diagram.

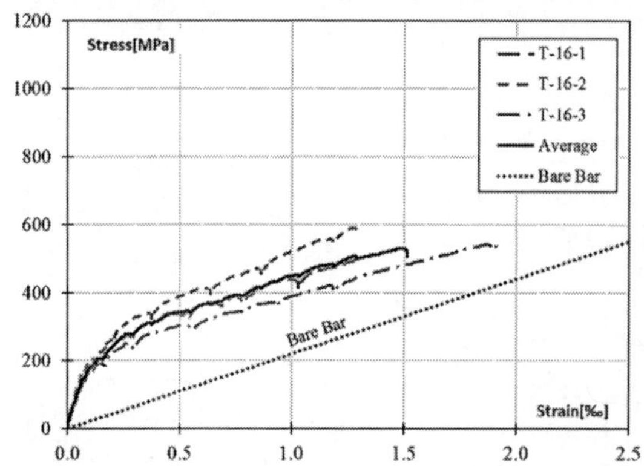

Figure 8. Stress-strain diagram based on the stress on rebar area for tie 8x8cm with the Ø16-mm bar.

Regarding the above diagrams, the elastic stiffness in the linear region varied from 39 to 42 GPa for the ties with the Ø10-mm bar, from 19 to 25 GPa for the ties with the Ø12-mm bar, and from 15 to 18 GPa for the ties with the Ø16-mm bar. By assuming the simultaneous participation of concrete and steel in tension and with the same strain, it was possible to obtain tensile stresses in concrete by subtracting the strain-stress curve of the bare rebar from the total stress-strain curve; e.g., the tensile stress of concrete is calculated in Figure 9. The average tensile stresses for all three tie series are shown in Figure 10. It should be noted that no significant strain-hardening behavior was observed for UHPFRC. With this diagram, it is possible to compute the tensile stresses that correspond to the desired strains and to compare them with the allowable SLS values.

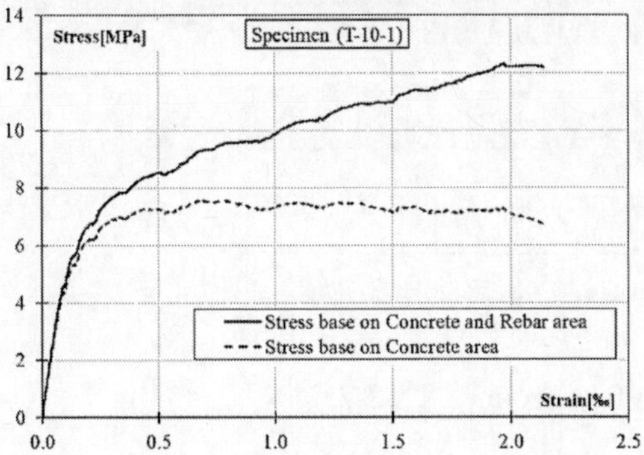

Figure 9. Tensile stress-strain curve for specimen T-10-1 (stress based on concrete area).

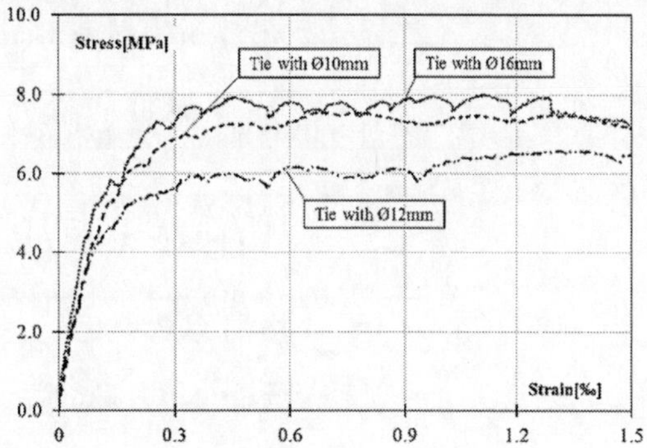

Figure 10. Average tensile stress-strain curve calculated according to the concrete area for all three specimens per series.

It should be noted that due to the asymmetry and uniformity of cracking in concrete, strains would not be identical on the element's four edges.

By way of example, for specimen T-12-2 at the force level of 2.31 kN, the average value of the strains of all the elements obtained by DEMEC was $\varepsilon = 0.066$ ‰ and the average value obtained by the displacement transducer was $\varepsilon = 0.059$ ‰, with a difference of 10.5%.

It is worth noting that in some specimens, it was not possible to record the increased length using DEMEC within the required strain intervals because cracks and a sudden increase in the longitudinal strain occurred.

At the end of the test and after accessing the tensile strain of 2 ‰ (1.5 ‰ for the specimen with the Ø16-mm bars), the number of cracks between the disks installed on the specimen's six edges was recorded by wetting the surface with water. Due to the rough surface created on the specimen surface that did not come into contact with the steel mold, no cracks were seen (see Table 2).

Table 2. The average number of cracks recorded between DEMEC disks and the average width of cracks for specimen T-12-2 (force value 2.31kN, strain obtained= 2 ‰).

Point Number		Total Number of Cracks	Achieved Elongation (mm)	Mean Crack Width (μm)
Up side	Edge1	70	0.632	9
	Edge2	91	0.632	7
Front Side	Edge1	69	0.632	9
	Edge2	86	0.632	7
Back Side	Edge1	92	0.632	7
	Edge2	56	0.632	11
			Mean	8

The number of cracks and the average crack widths are shown for specimen T-12-2. A large number of cracks and the short distance among them are some of the important characteristics of UHPFRC, as evidenced by the crack pattern shown in Figure 11.

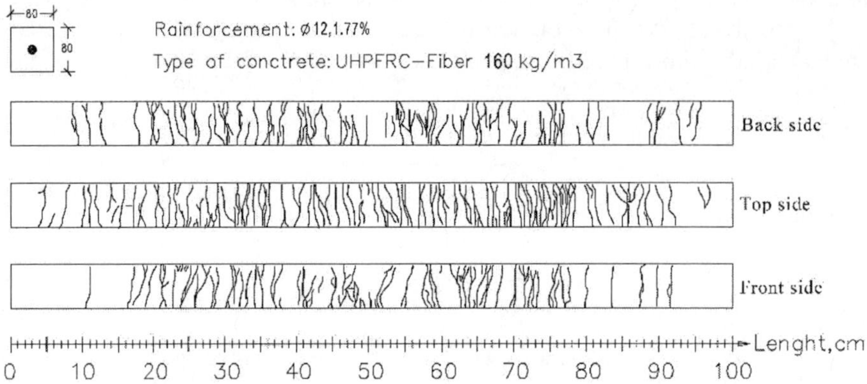

Figure 11. Cracks pattern for specimen T-12-2.

6. Conclusion

In order to study the necessary UHPFRC design requirements under SLS, a new testing method was developed to perform direct tensile tests. This method is able to examine the behavior of these elements under serviceability conditions and can determine tensile strains from the multicracking phase to the localized crack development stage.

In this method, the average tensile stress-strain curve of tensile ties was obtained by installing four displacement transducers on the specimen's four surfaces on both sides. Some parameters were obtained, such as the stiffness of elastic region, the tensile stress-strain diagram of concrete, and the width and number of cracks. The tensile strain on the 5-cm elements was acquired by installing DEMEC equipment on the four edges of the cross-section. Using the data obtained by this method, the tensile strain variations along the member were studied. In the final loading stage, after reaching the strain of 2 ‰ recorded by the displacement transducers, the cracking pattern along the specimen, the number of cracks on each 5-cm element on four edges and the average crack widths of the specimens were obtained. The proposed testing method and the obtained results will enable the study of UHPFRC behavior under direct tension under serviceability conditions.

Acknowledgement. This work forms part of Project "BIA2016-78460-C3-1-R" supported by the State Research Agency of Spain

References

[1] Wille K, El-Tawil S and Naaman A E 2014 Properties of strain hardening ultra high performance fiber reinforced concrete (UHP-FRC) under direct tensile loading *Cem. Concr. Compos.* **48** 53–66

[2] Japan Society of Civil Engineers 2008 Recommendations for Design and Construction of High Performance Fiber Reinforced Cement Composites with Multiple Fine Cracks (HPFRCC) *Concr. Eng. Ser.* **82** Testing Method 6-10

[3] di Prisco M, Plizzari G and Vandewalle L 2009 Fibre reinforced concrete: new design perspectives *Mater. Struct.* **42** 1261–81

[4] R. Yu et al. 2015 Development of Ultra-High Performance Fibre Reinforced Concrete (UHPFRC) : towards an efficient utilization of binders and fibres *Constr. Build. Mater.* 273–82

[5] Naaman A E 2006 Proposed classification of HPFRC composites based on their tensile response *Mater. Struct.* **39** 547–55

[6] Kusumawardaningsih Y 2015 Tensile strength behavior of UHPC and UHPFRC *Procedia Eng.* **125** 1081–6
[7] Hun S, Joo D, Sung G and Taek K 2012 Tensile behavior of Ultra High Performance Hybrid Fiber Reinforced Concrete *Cem. Concr. Compos.* **34** 172–84
[8] Eduardo J. Mezquida-Alcaraz, Juan Navarro-Gregori, Juan Ángel López and Pedro Serna-Ros 2019 Validation of a non-linear hinge model for tensile behavior of UHPFRC using a Finite Element Model *Comput. Concr.* **23** 11–23
[9] López J Á, Serna P, Navarro-Gregori J and Coll H 2016 A simplified five-point inverse analysis method to determine the tensile properties of UHPFRC from unnotched four-point bending tests *Compos. Part B Eng.* **91** 189–204
[10] Bernardi P, Michelini E, Minelli F and Tiberti G 2016 Experimental and numerical study on cracking process in RC and R/FRC ties *Mater. Struct. Constr.* **49** 261–77
[11] Rimkus A, Jakstaite R and Kupliauskas R 2017 Experimental identification of cracking parameters of concrete ties with different reinforcement and testing layouts *Procedia Eng.* **172** 930–6
[12] Bernardi P, Cerioni R, Ferretti D and Michelini E 2014 Role of multiaxial state of stress on cracking of RC ties *Eng. Fract. Mech.* **123** 21–33
[13] Tan R, Eileraas K, Opkvitne O, Giedrius Ž, Hendriks M A N, Geiker M and Terje D B 2018 Experimental and theoretical investigation of crack width calculation methods for RC ties 1–12
[14] Redaelli D and Muttoni A 2007 Tensile Behaviour of Reinforced Ultra-High Performance Fiber Reinforced Concrete Elements Symp. Dubrovnik. *Concr. Struct.* 267–74
[15] Sigrist V and Rauch M 2010 Deformation behavior of reinforced UHPFRC elements in tension *Tailor Made Concr. Struct.* 106–106
[16] Redaelli D 2006 Testing of reinforced high performance fibre concrete members in tension Proc. 6th Int. Ph.D. Symp. Civ. Eng. 9
[17] Caro L A, Serna P and Marti J R 2013 Experimental Technique for Measuring the Long-term Transfer Length in Prestressed Concrete 125–34

IOP Conf. Series: Materials Science and Engineering **596** (2019) 012023 doi:10.1088/1757-899X/596/1/012023

Experimental analysis of crack development of an UHPC wall element under shear loading

V Příbramský[1], M Kopálová[1] and L Dlouhý[2]

[1] Department of concrete and masonry structures, Faculty of Civil Engineering, Czech Technical University in Prague, Thákurova 7/2077, 166 29 Praha 6, Czech Republic
[2] SCIA CZ s.r.o., Evropská 2591/33d, 160 00 Praha 6, Czech Republic

E-mail: vladimir.pribramsky@fsv.cvut.cz

Abstract. UHPC (Ultra High-Performance Concrete) is an innovative material that enables design of lightweight and structurally optimized, long-lasting structures. In this paper is presented experimental analysis of precast webs of "butterfly web" box-girder bridge on scaled-down specimens. Pretensioned beam specimens were analysed in 2 variants – with continuous web and with lightened web. Based on the experimental results are both variants compared and are presented numerical and material models suitable for UHPC modelling in software SCIA Engineer. In SCIA Engineer is implemented modified Mazarz material damage model which is applicable for material with residual strength typical for FRC and UHPFRC.

1. Introduction

This paper describes experimental analysis of initiation and propagation of cracks in webs of bridge with box cross-section. The web is composed of thin UHPC precast walls rigidly connected to bottom and top slab of the bridge cross-section. The shape of these UHPC wall element was optimized in order to achieve the most favourable stress distribution. The optimal shape was determined from optimisation based on principal tensile and compressive stress distribution with distinctive shape of tensile and compressive diagonal, which are caused by shear force transfer between top and bottom slab. The web of the box cross-section is lightened and viable for prefabrication so these elements may be used on bridges over 100 m span (concept shown on Figure 1). The design of lightened web elements is based on concepts of so called "butterfly web" bridges [1].

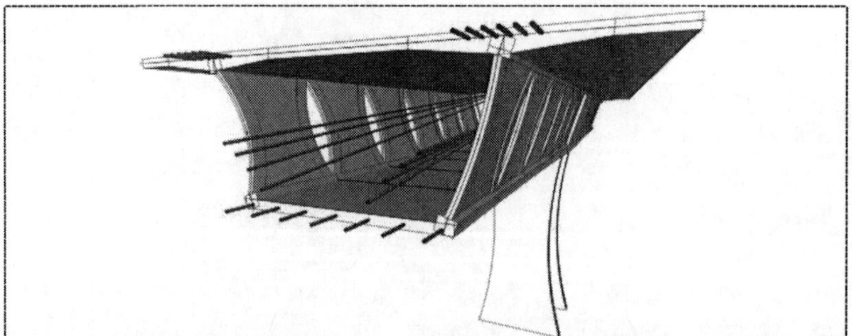

Figure 1. "Butterfly web bridge" concept visualisation.

Precast panels connected with top and bottom slab by composite action do not transfer shear force as continuous web would and their behaviour is well differentiated with tensile and compressive diagonals and is closer to Warren truss system [2]. Based on research in this paper, the viable description of the behaviour is the analogy of Vierendeel beam, where discrete web elements act as vertical members with rigid connection to top and bottom slab (flange) in the longitudinal direction. Vierendeel frame are more fitting to describe the behaviour with UHPC webs, due to different UHPC properties in plain tensile strength and bending tensile strength.

2. Analysis methods implemented in SCIA Engineer for UHPFRC analysis

Physical non-linear analysis represents a very powerful tool for analysing any kind of structure in civil engineering created not only from UHPFRC but also from other materials. Generally, there could be significant differences of results compared to a linear analysis, especially in case of hyper-static structures. In case of linear analysis, only the E-modulus of the material is considered for the preparation of the stiffness matrix. There is no stress distribution based on increasing the strain in the component. On the other side the non-linear analysis provides a stress distribution and increase of the bearing capacity after reaching the ultimate strain until the collapse mechanism. This mechanism is based on different values of stress and strain, dependent on a predefined nonlinear stress-strain relationship in the material diagram.

Generally, two approaches are usually used for considering fibres in nonlinear analysis. The first one uses the fibre and concrete matrix independently. The second and more common one considers the steel fibre directly in the behaviour of the concrete material. The second, more exact option was selected in this paper.

A typical shape of the stress-strain diagram is best described with a parabolic behaviour in compression in the same way as for standard concrete. The tensile behaviour displays a narrow peak expressed by the mean tensile strength. From the point of strain, this peak works very well as a crack localizator. After the crack formation, the toughness of steel reinforced fibre concrete allows to keep a certain level of tensile stress with increasing of the strain up to failure.

2.1. Modified Mazarz material damage model

The nonlinear calculation uses a very efficient damage material model called the Mazars model [3]. This material model is very well applicable for a material diagram with peaks and descending stress-strain diagrams typical for steel fibre concrete. The combination of elasticity and damage behaviour is combined in this material model. Moreover, the damage description was initially considered as isotropic and directly affecting the stiffness matrix. The modified Mazars model [4] is modified to very simples anisotropic damage model to better respect different behaviour of the fibre concrete in tension and compression. This leads to decomposition of stress tensor ($\boldsymbol{\sigma}$) to tension ($\boldsymbol{\sigma}^{\boldsymbol{t}}$) and compression ($\boldsymbol{\sigma}^{\boldsymbol{c}}$) part.

$$\boldsymbol{\sigma} = \boldsymbol{\sigma}^{\boldsymbol{t}} + \boldsymbol{\sigma}^{\boldsymbol{c}} \tag{1}$$

The resultant stress tensor is based on so called "effective damage parameter" in tension (d^t) and in compression (d^c) which determines changes of the stiffness depending on the elastic estimation of stress ($\boldsymbol{\sigma}^{\boldsymbol{trial}}$) from the loading.

$$\boldsymbol{\sigma} = (1 - d^t) \cdot \boldsymbol{\sigma}^{\boldsymbol{t}} + (1 - d^c) \cdot \boldsymbol{\sigma}^{\boldsymbol{c}} = [(1 - d^t) \cdot \boldsymbol{P}^{\boldsymbol{t}} + (1 - d^c) \cdot \boldsymbol{P}^{\boldsymbol{c}}] : \boldsymbol{\sigma}^{\boldsymbol{trial}} \tag{2}$$

This elastic estimation of stress can be expressed using constitutive tensor ($\boldsymbol{C}$) as follows:

$$\boldsymbol{\sigma}^{\boldsymbol{trial}} = \boldsymbol{C} : \boldsymbol{\varepsilon} \tag{3}$$

The damage parameters are calculated based on equivalent Mazars strains (ε_t; ε_c) which help for determining of actual values of stress from stress-strain diagram of material.

$$d^t = 1 - \frac{\sigma(\varepsilon_t)}{\sigma^{trial}(\varepsilon_t)};\ d^c = 1 - \frac{\sigma(\varepsilon_c)}{\sigma^{trial}(\varepsilon_c)} \tag{4}$$

As the stress-strain diagrams of UHPC has typically descending branch of stress-strain in tension, it is not possible to use tangential constitutive tensor ($\boldsymbol{C}$) but it is recommended to use secant one ($\boldsymbol{C^s}$) to fulfil its positive definition which is finally calculated as below.

$$\boldsymbol{C^s} = [(1 - d^t) \cdot \boldsymbol{P^t} + (1 - d^c) \cdot \boldsymbol{P^c}] : \boldsymbol{C} \tag{5}$$

Additionally, the effect of cracks must be considered during application. Here an analogy with the thermodynamic variable is applied for the two main damage states which are cracking of concrete in tension and crushing in compression. In case of plotting the surface failure of the steel fibre reinforced concrete the very well-known curve described by Kupfer [5] for biaxial loading is obtained which is also typical for regular reinforced concrete. The standard Newton-Raphson method is used for solving of this physical nonlinear problem.

3. Experimental analysis of UHPC web elements

3.1. Description of tested UHPC specimens

For experimental analysis of slender structural members of UHPC under shear loading were designed 2 types of beam specimens – one with full, continuous web and one with longitudinally lightened web. These beams have I cross-section with sufficiently designed flanges with longitudinal pretensioned tendons in order to mitigate effects of bending. The topology of these beams is apparent on figure 2 below. The aim of this experimental setup is to verify behaviour of lightened specimens and viability of application of similar larger scale precast and pretensioned web elements in greater magnitude and on bridges of span over 100 m.

The dimensions of the beams were chosen with respect to prefabrication, manipulation and transportation possibilities. Beams were 2.3 m long and 0.39 m high. Bottom flange was prestressed with 2 straight tendons with initial prestressing stress 900 MPa.

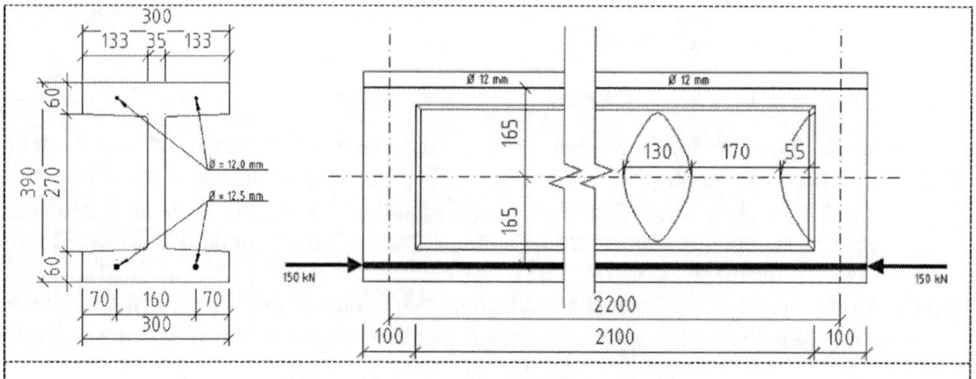

Figure 2. Shape of UHPC specimens.

In the structural analysis software SCIA Engineer was the specimen with loading mechanism modelled as half (from support to midspan) with corresponding boundary conditions for reduction of analysis time. Furthermore, FEM mesh was refined in areas where cracks were expected to develop.

3.2. UHPC beam specimens and UHPC mixture characteristics

In total, 6 beam specimens were casted and loaded in 4-point bending test, 3 specimens with continuous web labelled P1, P2, P3 and 3 beams with lightened web labelled V1, V2, V3. First beam from each set (P1, V1) was loaded till failure. On other 2 specimens in each set was first applied cyclical loading and after that the specimens were loaded till failure. Compressive strength of the UHPC mixture was measured on cubes 100x100x100 mm and average compressive strength was 157.1 MPa. Tensile strength in bending was measured on prismatic specimens of 40x40x160 mm and was in average

28.0 MPa. Modulus of elasticity was measured on cylindrical specimens 300 mm high with diameter 150 mm and the average modulus of elasticity was 51.6 GPa.

4. Results and experiment evaluation

In this section are presented results of the experiment for bothy types of beam specimens. Results are presented in form of force – deflection diagrams and are approximated by numerical model in SCIA Engineer 18 where was used nonlinear material model with damage. Basic characteristics of the material model were set according to experimental results on cubes (strength in compression), prisms (tensile strength in bending) and cylinders (modulus of elasticity). These specimens had the same age in the time of testing as the bigger beam specimens. For both beam specimens was in the numerical model counted with the effects of longitudinal prestressing with estimated short-term and long-term losses of 15%. The age of the specimens at the time of testing was in average 90 days.

4.1. Results on beam with full web

Beams with continuous web were tested in 4-point bending in 2 separate scenarios. In the first scenario was the first beam (P1) loaded by continuous increase of displacement till failure. In the second scenario were 2nd and 3rd beams (P2 and P3) loaded with cyclic loading pattern. Five loading cycles of approximately 0 – 100 kN were applied and the beams were loaded till failure. Force – displacement diagram of this experiment is presented on figure 3 below.

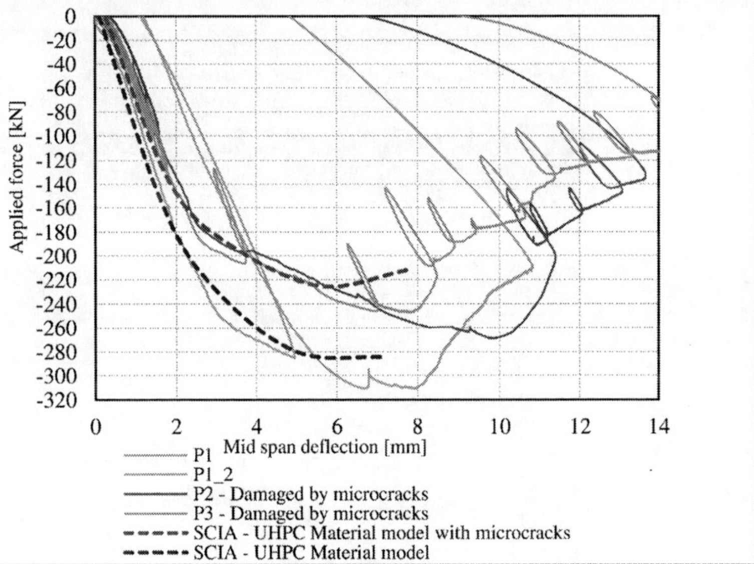

Figure 3. Force – displacement diagram: beams with continuous webs.

From the force – displacement diagram it is apparent softening of the beams with applied cyclical loading to magnitude of applied load of approximately 150 kN when compared to the beam which was loaded without the cyclic scenario. Furthermore, specimens with applied cyclical loading appeared fist visible cracks at much lower magnitude of acting force. Cause of this behaviour is initiation of microcracks when cyclical loading was applied. From the figure 3 is apparent, that this effect has a significant effect on both mean and residual tensile strength of the UHPC.

For numerical model of beams with continuous webs were used material models of UHPC with characteristics shown below in figure 4. Magnitudes of mean and residual stresses were obtained iteratively in order to achieve behaviour consistent with measured force – displacement diagram in figure 3. Magnitudes of residual strain in the material model are dependant on FEM mesh size. In this case the size of mesh in areas of crack initiation was set to 10 mm and crack width was limited to 5 mm.

This corresponds to upper limit of residual strain for the mesh elements to be 50% (for better readability of values in figure 4 is displayed only section below 3% strain).

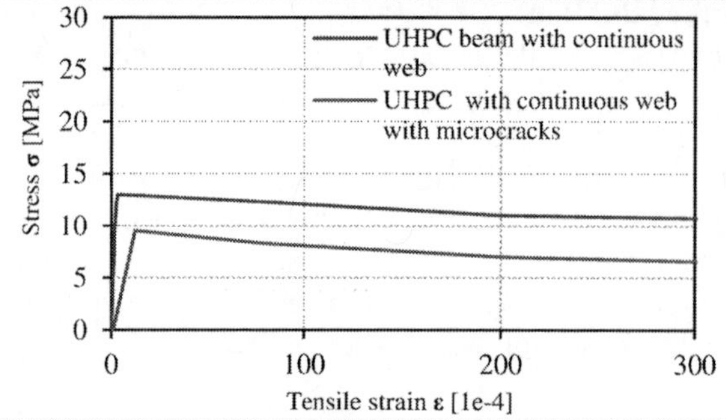

Figure 4. Stress – strain diagrams of UHPC – tension.

Type of failure and shape of the developed cracks is apparent on figure 5 below. Photography taken during the experiment is compared to shape of developed crack in program SCIA Engineer 18, where on figure 5 (b) and (c) are shown possible modes of shear failure obtained by numerical analysis. These modes are consistent to shape on tested specimen.

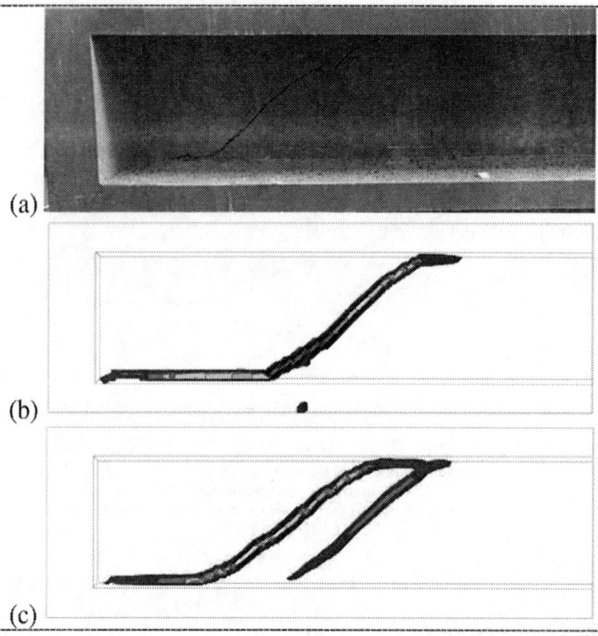

Figure 5. Analysis of crack shape: photo (a); development of one (b), or 2 (c) macrocracks in the numerical model.

4.2. Results on beam with lightened web

Beams with lightened web were tested in 4-point bending in 2 separate scenarios the same way as beams with continuous web. In the second scenario was applied cyclic loading pattern on specimens V2 and V3. Five loading cycles of approximately 0 – 40 kN were applied and the beams were loaded till failure. Force – displacement diagram of this experiment is presented on figure 6 below.

From the force – displacement diagram is apparent softening of the beams with applied cyclical loading to magnitude of applied load of approximately 50 kN when compared to the beam which was

IOP Conf. Series: Materials Science and Engineering **596** (2019) 012023 doi:10.1088/1757-899X/596/1/012023

loaded without the cyclic scenario. Furthermore, specimens with applied cyclical loading appeared fist visible cracks at significantly lower magnitude of acting force.

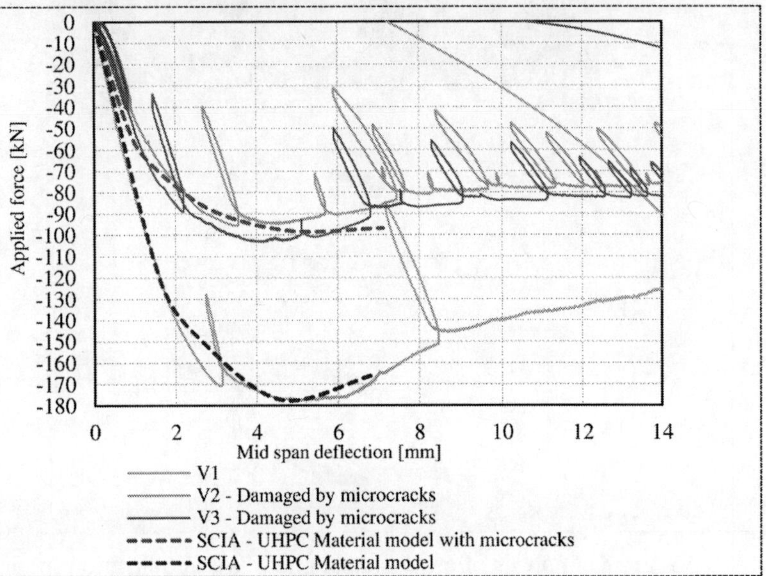

Figure 6. Force – displacement diagram: beams with lightened webs.

For the specimen which was not cyclically loaded is apparent much better performance in comparison with the beam that was subjected to cyclical loading. The increase of the load bearing capacity between the scenarios is much greater than in the case of specimens with continuous web. This effect is very important and is influenced by the fact, that tensile strength of UHPC in tension is few times greater when subjected to bending rather than plain tension. Tension strength in bending was experimentally verified on prismatic specimens 40x40x160 mm, which represents cross-section with comparable dimensions as thickness of the web of beam specimens to mitigate the size effect. For numerical model of beams with lightened webs were used material models of UHPC with characteristics shown below in figure 7.

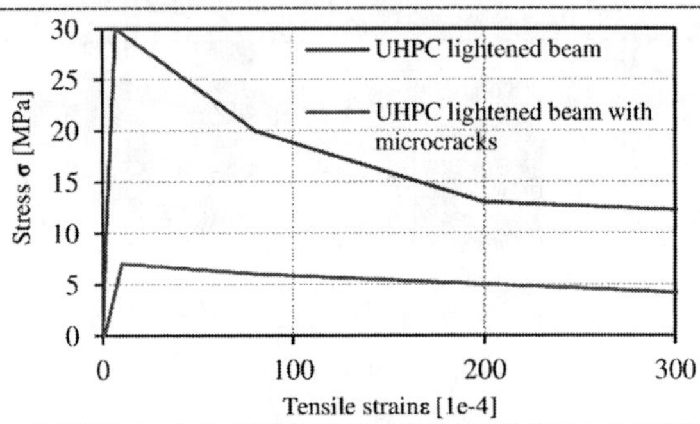

Figure 7. Stress – strain diagrams of UHPC – tension.

Type of failure and shape of the developed cracks is apparent on figure 8 below and again shows good correlation between observed crack distribution and numerical model.

IOP Conf. Series: Materials Science and Engineering **596** (2019) 012023 doi:10.1088/1757-899X/596/1/012023

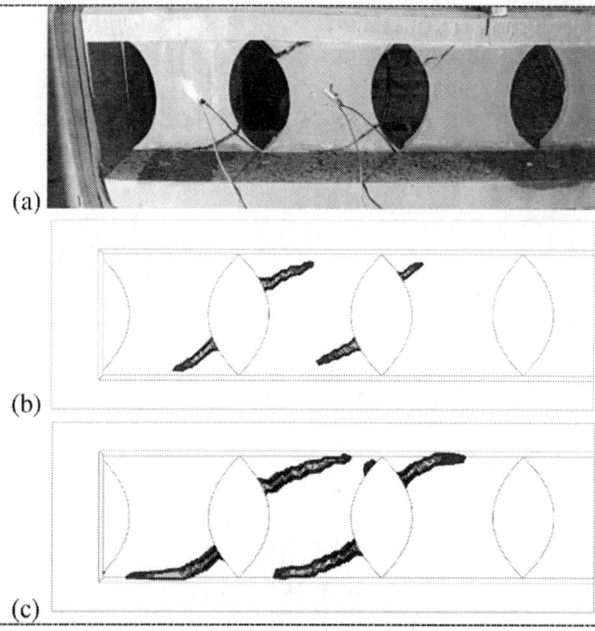

Figure 8. Analysis of crack shape: photo (a); initialization (b) and development (c) of macrocracks in the numerical model.

4.3. Comparison of solid and lightened web beams

For an objective comparison of beams with continuous and lightened webs which were loaded till failure (loading scenario 1) it is prudent to determine values of maximal applied loading on such lightened specimens, that would require identical amount of UHPC as the specimen with continuous web. Such modified specimen would have lightened web with thickness 48.4 mm increased by 38%. On figure 9 below is shown the comparison of magnitude of applied forces on level of macrocrack initiation and with maximal applied load.

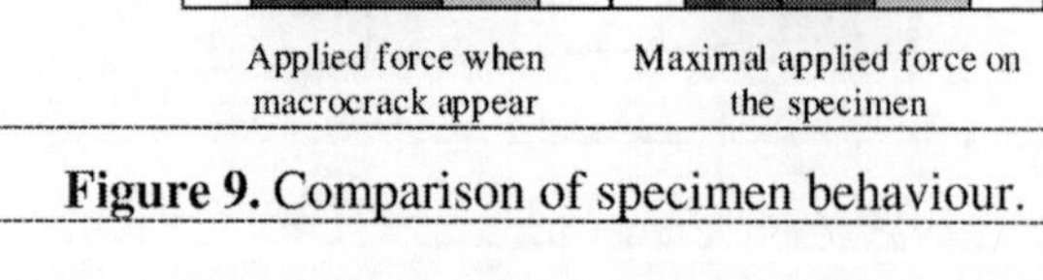

Figure 9. Comparison of specimen behaviour.

The recorded effect of reduction of both mean and residual strength of UHPC beam specimens loaded with cyclical loading when compared to beam loaded straight till failure without cyclical loading is more severe in the case of the lightened specimen. This effect is caused by higher magnitude of tensile stresses in the specimens when the cyclical loading is applied. In localised areas of the lightened web specimen rise the magnitudes of tensile stresses to 16.5 MPa when compared to tensile stresses in specimen with continuous beam 9.0 MPa. Localised tensile areas in lightened beam with greater magnitude lead to higher initiation of microcracks and more severe damage of the specimens, before any visible macrocracks are visible. On figure 10 below are shown distributions of principal tensile stresses from combination of self-weight, prestressing and amplitude of applied cyclical load.

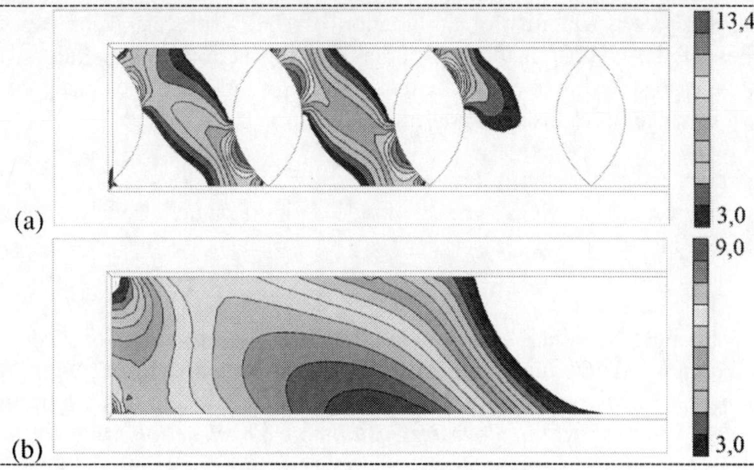

Figure 10. Maximal tensile stresses in web of (a) lightened beam and (b) beam with continuous web during cycled loading.

Furthermore, it can be concluded, that due to initiation of microcracks in the UHPC structure and consecutive development of macrocrack, the UHPC loses its high tensile strength caused by the bending characteristic of lightened web element failure. Tensile strength reduction of UHPC damaged by fatigue cyclical loading have significant effect on the performance of the beam specimens. For web segments used on structures with realistic proportions, it is however possible to add pretensioned tendons in the web segments in the direction of the tensile diagonal in order to eliminate tensile stresses in serviceability combinations.

5. Discussion

The comparison of beam with lightened and continuous web is made on specimens with the same amount of material required. Beam with continuous web with web thickness 35mm (as real specimen) and lightened beam with recalculated web thickness 48,4 mm. First macrocracks appear in the case of beam with lightened web on load level 15% lower and maximal loading 22% lower than the beam with continuous web. Despite the magnitudes are lower, there are following important aspects:

- Lightened beam shape may be yet further optimized (increase thickness on edges of the wall segments, where macrocracks first appear).
- Possibility of prefabrication of separate web segments to ensure superior quality.
- Due to clear and consistent shear force transfer by the web segments, these segments may be provided with prestressing tendons in the direction of the tensile diagonal to mitigate tensile stresses.
- Contradictory to analysis performed prior the experiments and based on available studies of butterfly-web bridges [1],[2] where web segment behaviour was described as an approximation with tensile and compressive diagonal, more suitable seems analogy with Vierendeel beam. Wall segments are thus approximated as frame members and their action is bending in longitudinal direction. Given the excellent UHPC properties in tensile strength in bending, this behaviour is most convenient.

6. Conclusions

In this paper were presented results of an experimental analysis of UHPC beams with lightened webs and numerical verification in program SCIA Engineer. These specimens demonstrate in smaller scale behaviour of precast web segments of a bridge with box cross-section, which are connected to top and bottom monolithic slabs of the cross-section by composite action. Discrete behaviour of the web segments was compared to continuous behaviour of beam with solid web with constant thickness. Conclusion of the analysis is viability of application of UHPC precast webs especially due to the excellent properties of the UHPC in tension under bending action. When these web segments are provided with efficient prestressing to eliminate tensile stresses on their edges, the precast web segments prove superior performance to beams with continuous web.

Acknowledgements
This work was supported by SGS grant of CTU in Prague, Czech Republic, grant No. SGS19/036/OHK1/1T/11.

References

[1] Kasuga A, Nagamoto N, Kata K and Asai H 2010 *Study of a bridge with a new structural system using ultra high strength fiber reinforced concrete, Proc. of 3rd fib Cong.* (Washington)
[2] Kata K, Ashizuka K, Miyamoto K and Nakatsumi K 2013 *Design and Construction of Butterfly Web Bridge, Proc. of 3rd Int. Conf. on Sustainable Construction Materials and Technologies* (Kyoto)
[3] Mazars J 1986 *A description of micro and macroscale damage of concrete structure. Eng. Fract. Mech.* **25** pp 729–37
[4] Němec I, Trcala M and Rek V 2018 *Nelineární mechanika*, (Brno: Vutium)
[5] Kupfer H, Hilsdorf H K and Rusch H 1969 *Behaviour of concrete under biaxial stresses, ACI Journal* **66-52**

IOP Conf. Series: Materials Science and Engineering **596** (2019) 012024 doi:10.1088/1757-899X/596/1/012024

Analysis of interfacial interaction in UHPFRC-strengthened reinforced concrete beams

C Zanuy, G S D Ulzurrun and I M Díaz

Dept. Continuum Mechanics & Structures, Universidad Politécnica de Madrid, ETS Ingenieros de Caminos, 28040 Madrid, Spain

E-mail: carlos.zanuy@upm.es, https://orcid.org/0000-0001-9044-8666

Abstract. The behaviour and failure mode of reinforced concrete elements strengthened with a thin tensile layer of ultra-high performance fiber-reinforced concrete (UHPFRC) is governed by the interface interaction between the UHPFRC and conventional concrete. In order to analyze the relative displacements at the interface and its influence on the crack pattern evolution and monolithic response, digital image correlation (DIC) has been used in this contribution. With the help of DIC, the distinct stages of the behaviour of UHPFRC-strengthened beams have been correlated with the crack formation and development. Special attention has been paid to the progressive debonding due to the strain incompatibility between UHPFRC and conventional concrete when cracks develop. It is shown that the utilization of the capacity of the UHPFRC can be achieved from a certain thickness of the strengthening layer for shear-critical elements, thereby providing a gain of shear strength and ductility. In contrast, a thinner UHPFRC layer has been more beneficial for flexure-critical elements.

1. Introduction

The use of ultra-high performance fiber reinforced concrete (UHPFRC) to strengthen concrete structures with a thin layer at the tensile part has shown to be an effective structural solution [1]. In such composite members, the interaction between the UHPFRC and reinforced concrete components is responsible for the monolithic behaviour and the failure mode [2]. UHPFRC and conventional concrete have a distinct fundamental behaviour, especially in tension: while conventional concrete cracks without appreciable post-peak softening capacity, UHPFRC is able to develop a strain-hardening stage before the formation of macro-cracks and a significant softening stage thereafter [3]. Progressive relative displacements (tangential slip and normal debonding) along the UHPFRC-conventional concrete interface develop as each layer enters a distinct stage, which characterizes the failure mode of the composite member [4]. The efficiency of composite structures (including steel-concrete, FRP-concrete, FRP-steel, etc) typically relies on the interface behaviour.

The present contribution aims at analyzing the influence of the global response of UHPFRC-strengthened reinforced concrete beams paying attention at the evolution of the interface between both layers. Results of quasi-static tests are presented and digital image correlation (DIC) technique is employed to follow the relative interface displacements and crack pattern development.

2. Experimental approach

2.1. Description of test configuration

IOP Conf. Series: Materials Science and Engineering **596** (2019) 012024 doi:10.1088/1757-899X/596/1/012024

Six beams have beam tested under quasi-static loading to failure in a three-point bending configuration as represented in Figure 1. The external dimensions (2000 x 125 x 250 mm) and longitudinal reinforcement (2Φ16 at bottom and top) were the same for all specimens. Three beams were designed without stirrups (referred to as RC-series) and three other beams with Φ8 stirrups at 15 cm spacing (referred to as RCS-series), thereby covering shear- (RC-series) and flexure-critical members (RCS-series). Reference specimens RC and RCS contained no UHPFRC strengthening layer, while specimens xxx-U35 and xxx-U55 included a UHPFRC layer at the tensile face of 35 and 55 mm thickness, respectively. The longitudinal reinforcement and the stirrups were partially embedded by the UHPFRC layer when its thickness was 55 mm, but not when the thickness was 35 mm. It is noted that some authors [1] have recommended the introduction of additional longitudinal bars within the UHPFRC layer and an eventually higher member height in order to increase the member's resistance instead of just repairing it. Nevertheless, additional reinforcement and increase of section height have not been considered in the present research in order to allow direct comparison between strengthened and conventional beams.

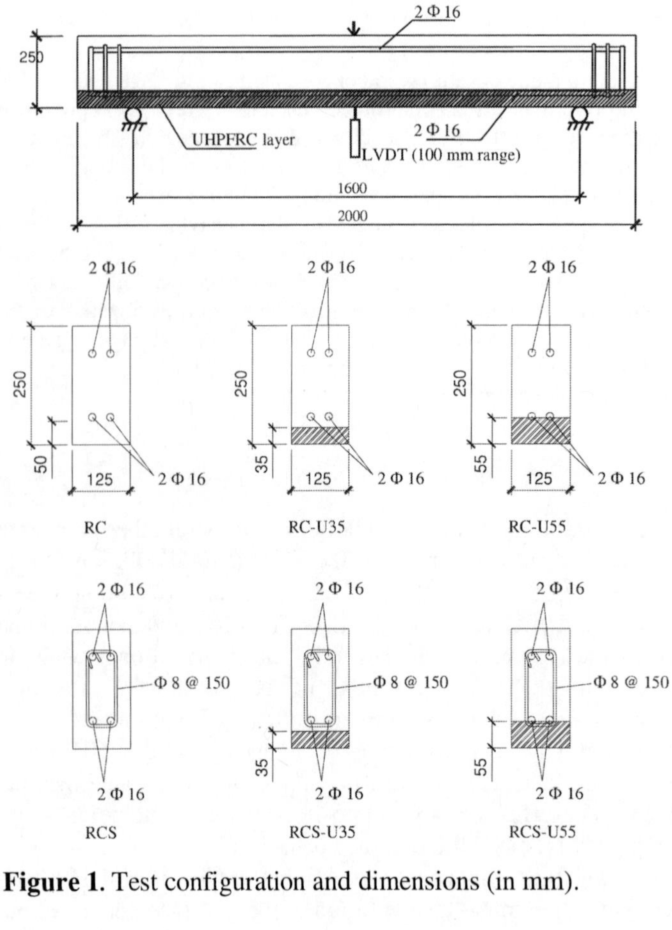

Figure 1. Test configuration and dimensions (in mm).

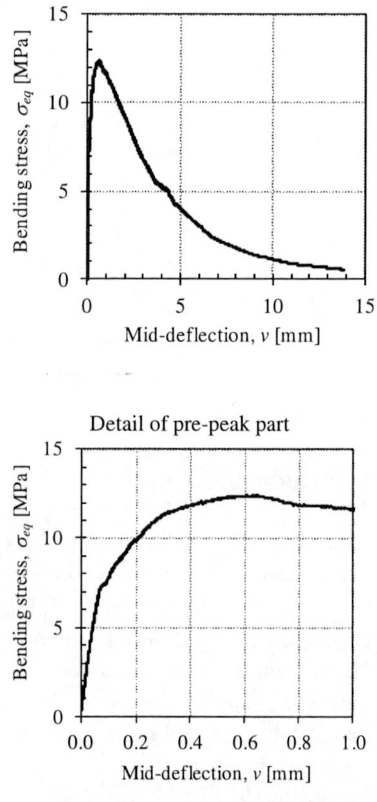

Figure 2. Example of material test (4-point bending on 500 x 100 x 100 mm prisms).

The specimens were manufactured and tested at the Laboratory of Structures of the Technical University of Madrid (UPM), Spain. The conventional concrete was supplied by a local contractor. The average cylinder strength at testing age was 41 MPa and the maximum aggregate size was 12 mm. For the specimens to be strengthened with UHPFRC, the total height was not concreted, i.e. the height of 35 or 55 mm was left empty instead of concreting and demolishing. In the later case, the beams were manufactured upside down. The upper surface of the RC layer was not levelled nor conditioned, so that the roughness of the interface was the natural surface of concrete. Before pouring the self-compacting UHPFRC, the surface was washed, slightly wire brushed to remove loose material, and kept moist. The age of the conventional concrete was 15 months when the UHPFRC layer was manufactured. The UHPFRC was mixed in the laboratory from CRC premix kindly supplied by Hi-Con A/S (Denmark). The dry-mortar premix was mixed with the water and the fibers following the instructions of the supplier. A 2% volumetric amount of short straight steel fibers (12 mm length, 0.4 mm diameter) was used. The average cubic compressive strength at testing age (2 months after manufacturing) was 135 MPa. Flexural tests on prismatic specimens were also carried out to check the tensile behaviour of the UHPFRC: strain-hardening could be observed by formation of microcracks and a noticeable pre-peak stage (e.g. see Figure 2).

The tests were carried out by applying a monotonic displacement on the top side of the midspan section with a hydraulic actuator at a rate of 0.02 mm/s. The applied load was measured with a built-in load cell. An LVDT of 100 mm range was used to measure the midspan deflection at the bottom side (refer to Figure 1). In addition, a speckle pattern was painted on the front side of specimens in order to take pictures for DIC post-processing of test results. The photographs were taken by a Nikon D90 camera with lens Nikon 18-200 mm f/3.5-5.6, taking pictures of 4288 x 2848 px resolution every 5 seconds (approx.).

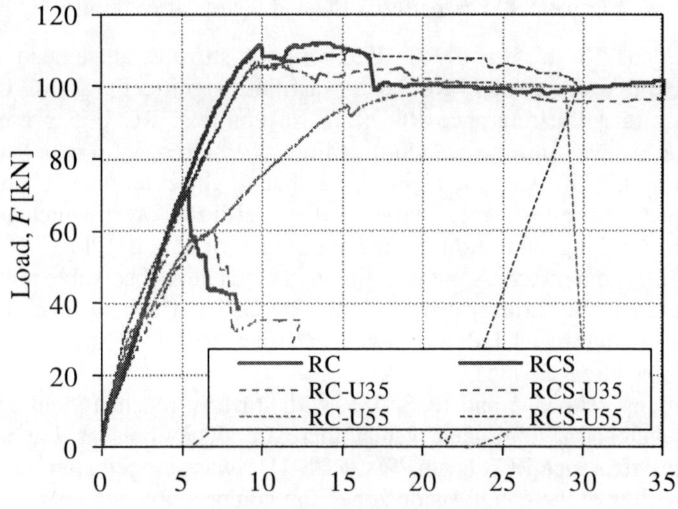

Figure 3. Load-midspan deflection of tested beams.

2.2. Test results

The behaviour of the specimens can be classified into two groups according to the failure mode: the beams without stirrups failed by shear and the beams with stirrups presented a flexural failure mode. The load-deflection curves of all tested specimens are represented in Figure 3. The crack patterns after failure are plotted in Figure 4. It has to be noted that cracks painted in red colour were developed before the tests started, while cracks formed during the tests are remarked in blue colour. Pre-test cracks were probably due to the shrinkage of the UHPFRC during the 2 months from manufacturing to testing. During that time, the beams had been supported horizontally on the laboratory floor in upside down position. As observed from red cracks in Figure 4, pre-test cracks included a longitudinal crack along the interface near the ends. For the beams with a 55 mm UHPFRC layer, the length of the interface crack was smaller than 10 cm, while for beams with a 35 mm thick layer, the length was around 25 cm, thereby entering in the span between supports. Moreover, some very thin transverse cracks were observed in both RC and UHPFRC layers close to the interface. For beams with a 35 mm UHPFRC thickness, those cracks (1 at midspan for RC-U35 and 2 away from the midspan for RCS-U35) crossed the whole height of the UHPFRC. For breams with a 55 mm UHPFRC thickness, only 1 transverse crack was observed at the UHPFRC, away from the midspan and not crossing fully its thickness. In all cases, the largest crack opening of pre-test longitudinal and transverse cracks was 0.1 mm. This description of pre-test cracks can help in interpretation of test results.

The reference reinforced concrete beams RC and RCS presented shear and flexural failure, respectively. Test of beam RCS was stopped after a midspan deflection of 35 mm (span length/45) had been reached and some concrete spalling at compression zone started to occur.

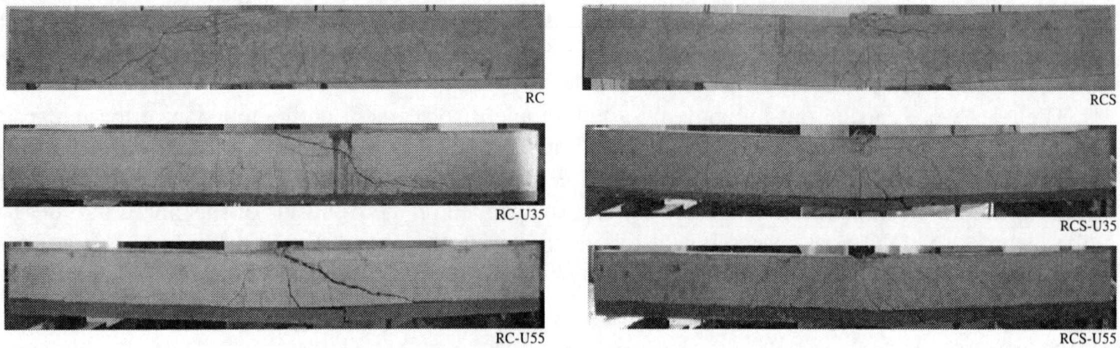

Figure 4. Crack pattern of tested beams after failure.

Composite specimens RC-U35 and RC-U55, without stirrups, developed shear failures. The influence of the thickness of the UHPFRC layer was quite significant, as RC-U35 showed a shear strength 15% smaller than the reference RC beam. In contrast, RC-U55 presented a 43% higher strength and was able to develop some pseudo-ductility. It is interesting to note that the initial stiffness of both RC-U35 and RC-U55 was not higher than that of the reference RC beam, which can be attributed to the effect of pre-test cracks crossing the UHPFRC layer which might avoid the full utilization of the pre-cracking and strain-hardening stages of the UHPFRC. This effect is further analyzed in section 3.3. From the crack pattern (Figure 4), it is also observable that the tensile capacity of the UHPFRC could be fully utilized by specimen RC-U55 but not by specimen RC-U35 (see the wide macro-crack crossing the UHPFRC layer in beam RC-U55 and the associated interface debonding, not developed by RC-U35).

Composite specimens RCS-U35 and RCS-U55, with stirrups, reached yielding of the longitudinal reinforcement and developed a significant plateau thereafter. They presented approximately the same loading capacity as the reference RCS beam. Test RCS-U35 was stopped after a midspan deflection of 35 mm with some spalling at the compression zone. The compression zone of test RCS-U55 exploited at a midspan deflection of 30 mm. This higher brittleness is due to the higher equivalent reinforcement ratio provided by both the tensile reinforcement and the 55 mm thick UHPFRC layer. Other authors also showed higher brittleness of flexure-critical composite members [5]. The initial stiffness of RCS-U35 and RCS-U55 was slightly higher than that of the reference RCS beam. In addition, a wide macro-crack accompanied by interface debonding developed and crossed the UHPFRC layer in both RCS-U35 and RCS-U55 specimens, showing high utilization of the UHPFRC layer.

Because interface debonding and crack development were responsible for the response of strengthened specimens, it is interesting to complete a deeper study of the interaction between the UHPFRC and conventional concrete layers. In this paper, the analysis is carried out with the help of DIC technique in section 3.

3. Analysis of results

3.1. DIC technique

Digital Image Correlation (DIC) is a robust and adaptable tool to analyze the deformed pattern of 2-D or 3-D problems from the photogrammetric follow-up of a speckle pattern painted on a solid. In the present contribution, 2-D DIC has been employed to study the in-plane deformational response of tested beams from a speckle pattern painted on the front side (2000 x 250 mm^2) and a camera placed perpendicularly in front of the midspan. The focal length of the lens was set to 24 mm. According to the camera resolution (section 2.1), 1 px corresponded to 0.5 mm approx. The scale was provided with a line gauge on the beams' top side. The speckle pattern has consisted on randomly-painted black points of 1-3 mm size over a white background. GOM Correlate software [6] has been used for the DIC analysis, with a facet size of 30 px and a point distance of 20 px.

DIC can be more efficient than the installation of several transducers at different locations of the specimens in order to follow the evolution of e.g. displacements or strains. In the present contribution, the study has focused on the evolution of the UHPFRC-concrete interface and the crack pattern in correlation with the members' response and the stages of the load-displacement diagrams of Figure 3.

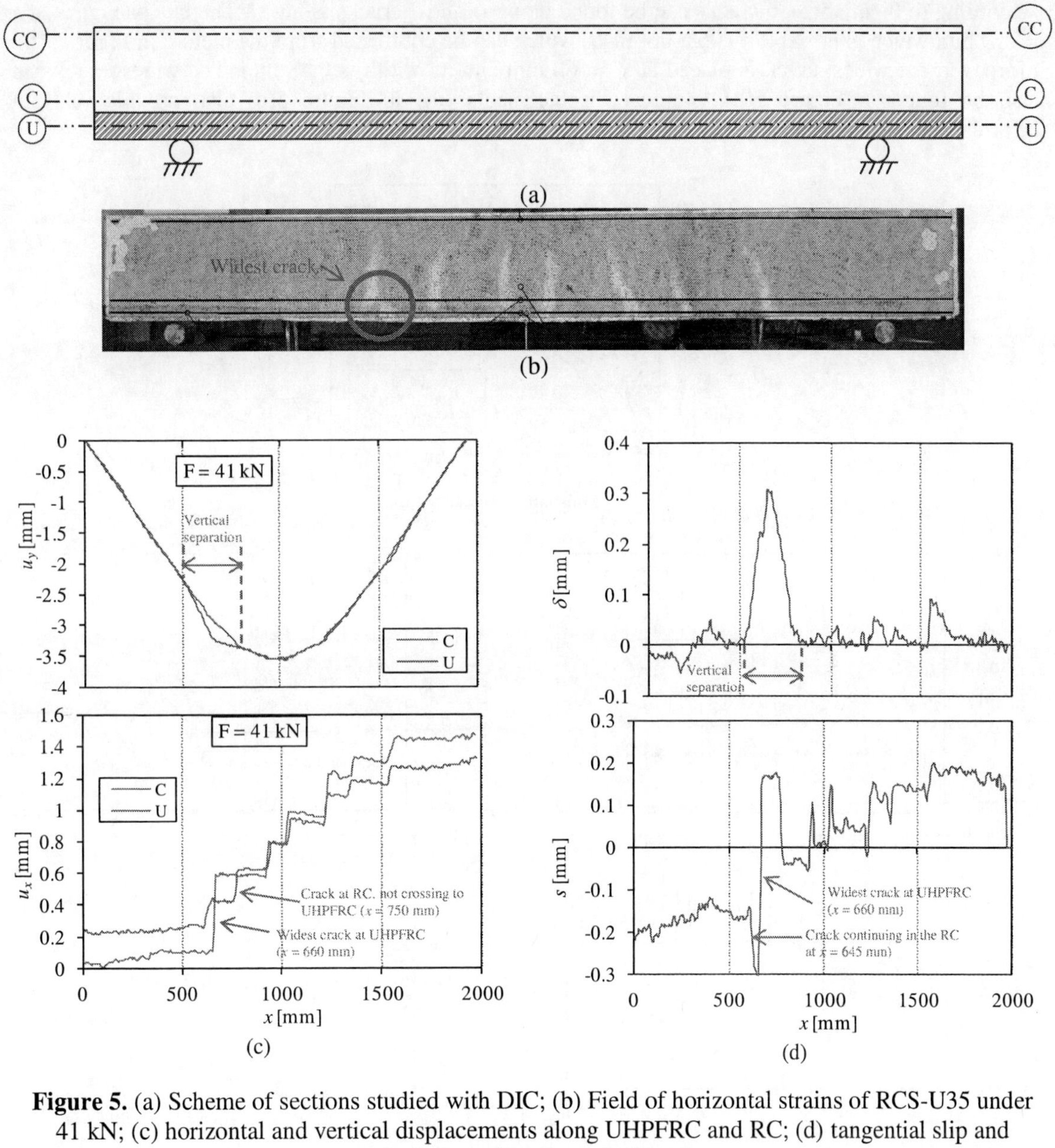

Figure 5. (a) Scheme of sections studied with DIC; (b) Field of horizontal strains of RCS-U35 under 41 kN; (c) horizontal and vertical displacements along UHPFRC and RC; (d) tangential slip and vertical separation along the interface.

The interface state can be characterized by the relative tangential slip and normal displacement between the UHPFRC and conventional concrete at the interface. Such relative displacements can be calculated by extracting the horizontal and vertical displacements along two horizontal sections at both

IOP Conf. Series: Materials Science and Engineering **596** (2019) 012024 doi:10.1088/1757-899X/596/1/012024

sides of the interface. Here, the sections at the centroid of the UHPFRC layer and 12 mm above the interface have been taken, which are denoted as *U* and *C* in Figure 5(a).

As an example to explain the methodology followed in this paper, the state of beam RCS-U35 under a load of 41 kN (service load level) is studied. The horizontal strain field is represented in Figure 5(b). Highly concentrated strains typically indicate cracks. The horizontal and vertical displacements along both sides of the interface are plotted in Figure 5(c). From the horizontal displacements (u_x), each vertical jump indicates the presence of a crack and the corresponding crack width can be calculated as the difference between the horizontal displacements at both sides. According to Figure 5(c), 6 cracks can be found along *U* and 7 cracks along *C*. There is one crack at x = 750 mm which is present in *C* but not in *U*, which can be confirmed from the picture in Figure 5(b). Moreover, the widest crack is placed at x = 660 mm and its width is 0.48 mm. The widest crack can also be found in Figure 5(b). Vertical displacements (u_y) in Figure 5(c) also provide relevant information, as the difference between u_y at *U* and *C* shows vertical debonding.

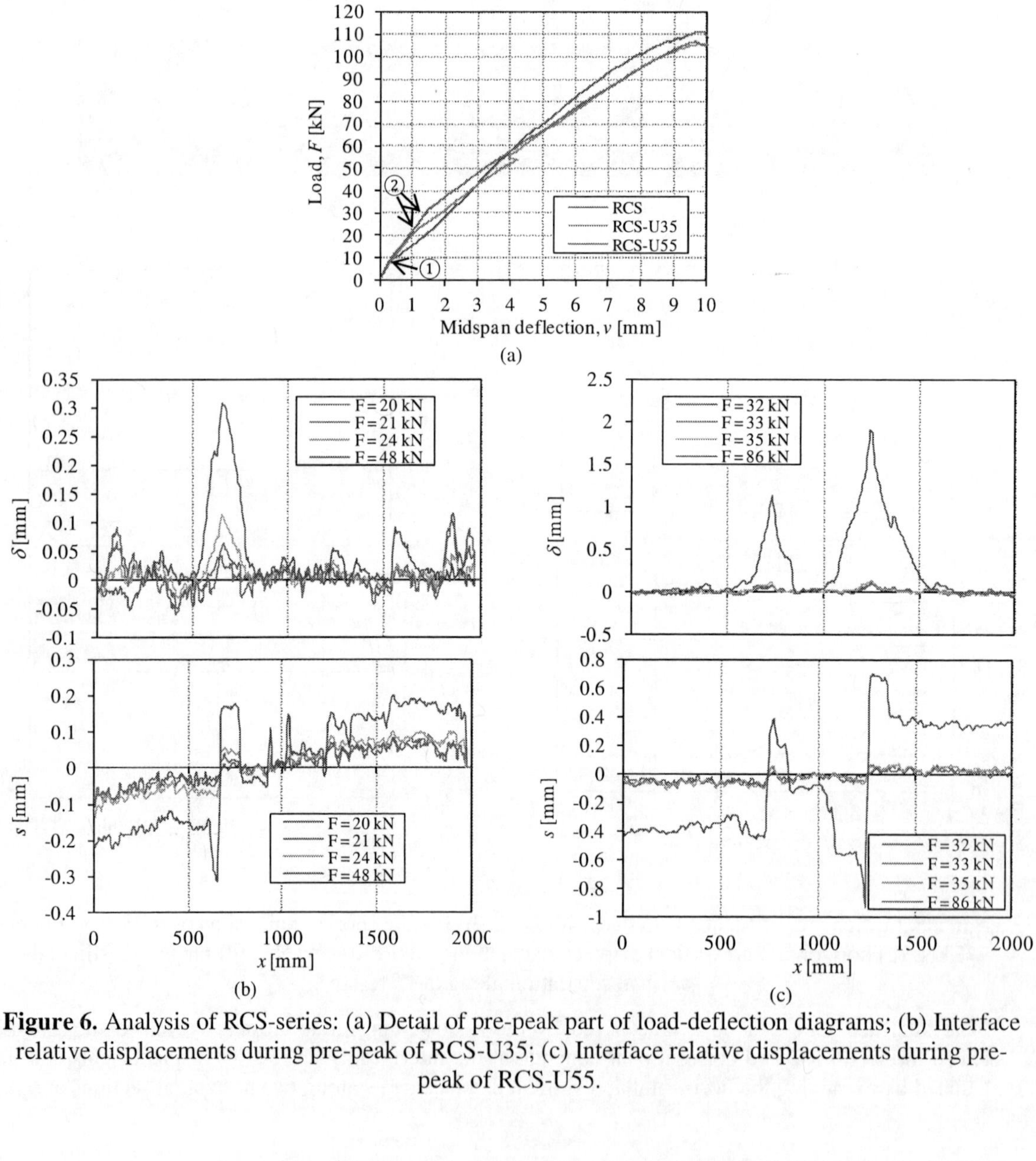

Figure 6. Analysis of RCS-series: (a) Detail of pre-peak part of load-deflection diagrams; (b) Interface relative displacements during pre-peak of RCS-U35; (c) Interface relative displacements during pre-peak of RCS-U55.

Relative displacements at the interface are represented in Figure 5(d). The tangential slip (s) is obtained by the difference between the horizontal displacements at U and C. Jumps of s represent cracks at either UHPFRC or conventional concrete and the value of the jump is the corresponding crack width. The detail of the analysis provides information like the fact that the widest crack at U (x = 660 mm, width = 0.48 mm) does not continue to C at the same position, but 15 mm to the left (x = 645 mm, width = 0.15 mm), which can be also observed by carefully looking at u_x in Figure 5(c) and the strain field in Figure 5(a). Analogously, the relative normal displacements at the interface (δ) are calculated by the difference of the vertical displacements at U and C and they can be used to study the vertical debonding. According to Figure 5(d), vertical separation produced by the widest crack at UHPFRC extends over a length of 300 mm and has a maximum value of 0.3 mm. Smaller separation also exists around each crack. The curvature of the deflected shape (u_y) along U indicates negative bending of the UHPFRC layer in the zone where vertical debonding develops, which has been considered as other authors to explain the contribution of the strengthening layer [4].

From the variables explained above, a detailed analysis of the interface state can be carried out. In the following sections, the evolution of the interface is studied in correlation with the global response of tested composite members.

3.2. Analysis of flexure-critical beams

The composite specimens that contained stirrups (RCS-U35 and RCS-U55) failed in a similar way as the reference reinforced concrete beam RCS, but with a smaller ductility due to compression zone destruction. The DIC technique is employed to analyze the crack pattern evolution and the behaviour of the interface in correlation with the load-deflection diagram.

The pre-peak part of the load-deflection diagram of RCS-series is represented in Figure 6(a). Relative displacements at the interface between UHPFRC and conventional concrete are displayed in Figure 6(b) and (c) for different load levels of RCS-U35 and RCS-U55, respectively. The first load level represented in Figure 6(b) and (c) is the load below which cracks are not detected with the DIC.

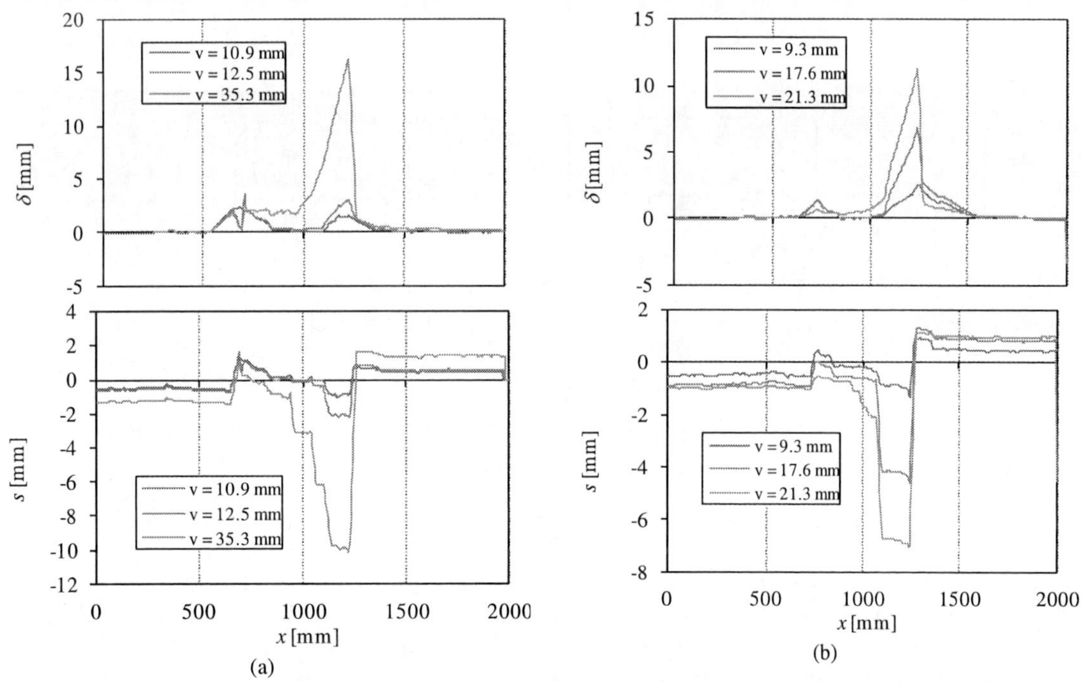

Figure 7. Interface relative displacements during post-peak: (a) RCS-U35; (b) RCS-U55.

In the initial stage of the load deflection diagram, two key points can be distinguished. Point ① in the RCS beam represents the cracking load, i.e. the tensile strength of conventional concrete is reached at the bottom of the beam. The load is around 10 kN. At that load level, a slight change of stiffness can be also observed in the load deflection diagram of RCS-U35 and RCS-U55, but the bottom of the beam is of UHPFRC. Because Figure 6(b) and (c) shows that no cracks open until 21 kN and 33 kN for RCS-U35 and RCS-U55, respectively, point ① indicates the matrix microcracking of the UHPFRC and the beginning of the strain hardening stage. Indeed, from point ①, the response of UHPFRC-strengthened beams is stiffer than that of RCS beam.

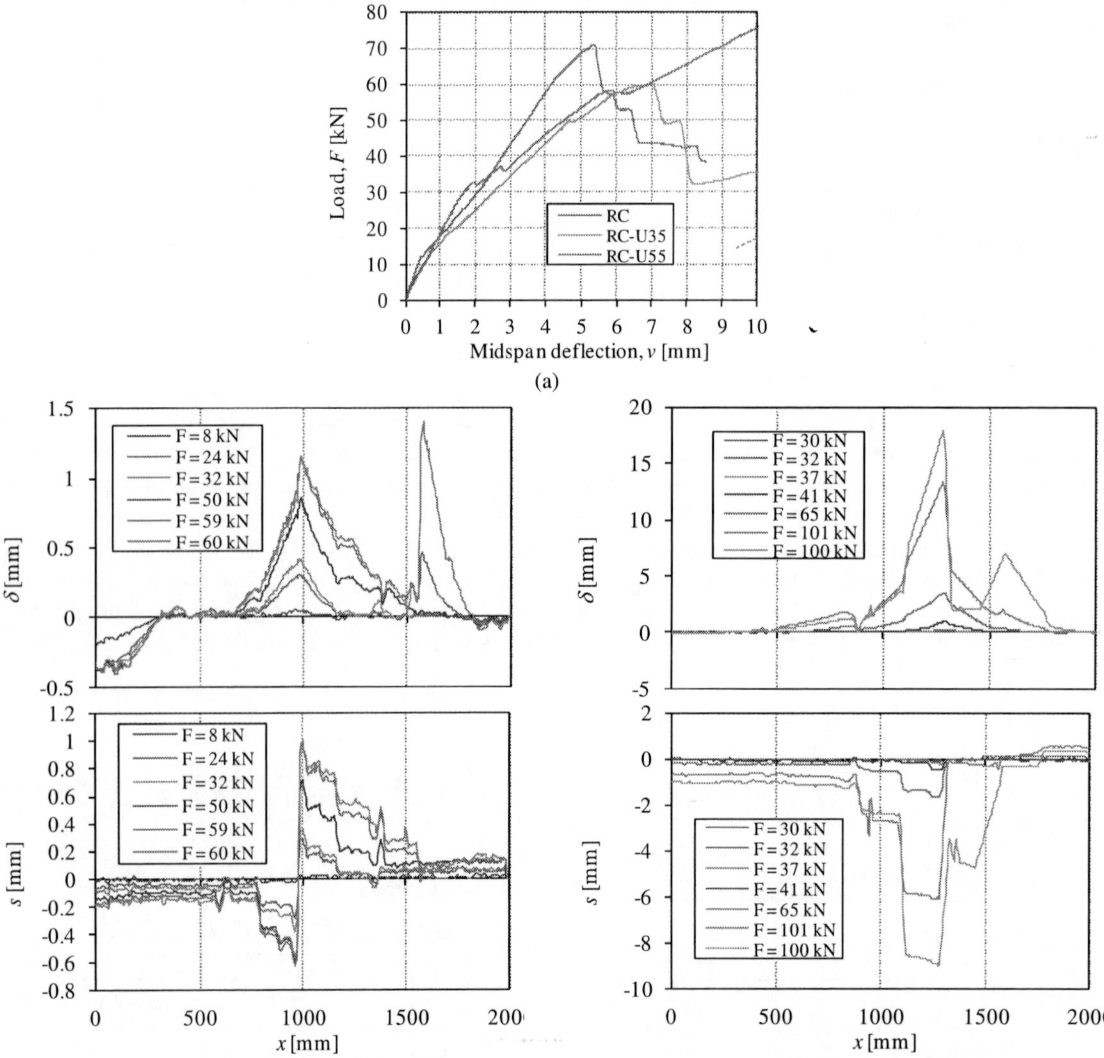

Figure 8. Analysis of RC-series: (a) Detail of load-deflection diagrams; (b) Interface relative displacements of RC-U35; (c) Interface relative displacements of RC-U55.

A second characteristic point ② in the response of UHPFRC-strengthened specimens shows decrease of stiffness. The load level of point ② (21 kN and 33 kN for RCS-U35 and RCS-U55, respectively) corresponds with the detection of the first macro-crack in the UHPFRC, as noted from the jumps detected in the relative displacements at the interface. For increasing load levels, more macro-cracks develop in the UHPFRC until a number of 6 macro-cracks for RCS-U35 and 2 macro-cracks for RCS-U55. Additional cracks form progressively in the conventional concrete layer (e.g. refer to Figure 5). Accompanying the formation of cracks, the progression of vertical debonding can be followed by analyzing the relative vertical displacement at the interface. It can be observed that a longer debonding zone is obtained for larger crack widths, which seems logical as vertical separation is a result of the incompatibility of strains produced by cracks. In addition, crack widths have been significantly smaller in the specimen with 35 mm thick UHPFRC layer than that with 55 mm thickness, due to the higher number of cracks formed. The 2 cracks formed in specimen RCS-U55 had a considerable width in the pre-peak part of the test

The evolution of the relative displacements at the interface during post-peak stage is studied in Figure 7. Each curve is denoted by the midspan deflection of the specimen (*v*). At the yielding point, a shear plug was visible in the midspan zone of RCS-U35. The shear plug is characterized by two cracks with an inclination of 70-80° in the concrete zone, continuing across the UHPFRC with 2 macro-cracks. The largest crack width at the UHPFRC was around 2 mm at the peak load. Form that value, it then increased until one of the cracks reached 2.5-3.0 mm, when the UHPFRC layer divided into two parts. In spite of that, the beam was still able to carry loads without release. The test was stopped at a midspan deflection of 35 mm. The response of beam RCS-U55 was similar, but the lesser number of cracks resulted in a larger width of them, as their width was already 2.5 mm under the peak load and the UHPFRC layer divided into two parts at that load level. The specimen was able to continue carrying the load despite the breakage of the UHPFRC layer with the formation of a shear plug in the central region, until final failure of the compression zone.

From the above result, it has been observed how the UHPFRC layer has been able to provide carrying capacity even after its breakage during the post-peak stage, when a tied-arch mechanism seems to bridge the load. It is noted that the widest macro-cracks in the UHPFRC were not the pre-cracks formed by shrinkage prior to testing, i.e. in these tests pre-test cracks have not influenced the response of the beams.

3.3. Analysis of shear-critical beams

The composite specimens without stirrups (RC-U35 and RC-U55) failed by shear, as well as the reference reinforced concrete specimen RC. The evolution of those specimens is studied with the help of load-deflection curve and relative displacements at the interface, represented in Figure 8.

The first cracking load of the reference specimen RC was at a load of 10-11 kN. The behaviour of RC-U35 was dominated by the presence of a pre-crack crossing the UHPFRC layer at the midspan section (x = 1000 mm). From the relative displacements at the interface it is clear that the pre-crack has governed the global behaviour. On the one hand, vertical separation around that crack developed from the beginning of the test (see F = 8 kN in Figure 8(b)). On the other hand, the initial stiffness of specimen RC-U35 was smaller than that of the reference RC beam. The difference between measured midspan deflection of beams RC-U35 and RC (difference of green and blue curves in Figure 8(a)) has resulted equal to the relative vertical separation at the midspan obtained with the DIC (Figure 8(b)). It becomes apparent that the pre-crack has reduced the initial uncracked and strain-hardening stages. At a load of 24 kN, 2 additional macro-cracks opened at the UHPFRC, one of them at x = 1350 mm. From the later crack, significant vertical debonding extended to the midspan at a load of 50 kN. From that load, significant principal tensile strains in the web from the UHPFRC at x = 1350 mm to the compression zone at midspan were detected with the DIC. Finally, a diagonal crack opened at a load of 59 kN, which can be observed by the deformed shape obtained with the DIC along *CC* (refer to Figure 5(a)). The deformed shape along *U*, *C* and *CC* at a load of 59 kN are plotted in Figure 9, where it becomes clear not only the distortion produced by the diagonal crack, but also the significant debonding between the UHPFRC and the conventional concrete between x = 1000 mm and 1350 mm.

Regarding specimen RC-U55, its pre-cracks did not cross the UHPFRC layer. The initial stiffness was somewhat smaller than the reference RC beam, but higher than the one of RC-U35. The first change of stiffness of the load-deflection diagram was observed at a load of 32 kN. At that instant, the first macro-crack at UHPFRC was detected by the relative displacements at the interface at x = 1300 mm. That crack extended upwards in the conventional concrete at a different location (x = 1230 mm), thereby leading to beginning of vertical separation along the distance in-between. A second macro-crack at the UHPFRC opened at 38 kN at x = 880 mm. The peak of the load-deflection diagram at 41 kN was due to the formation of a shear plug in the centre of the reinforced concrete layer, with two cracks inclined 70-80° between x = 900 and 1100 mm. From 41 kN, vertical debonding was observable around the crack at x = 1300 mm and progressed until its crack width in the UHPFRC layer reached 2-2.5 mm when the load was 58 kN. That point can be observed by a change of stiffness in the load-deflection curve and was characterized by the division of the UHPFRC into two parts from that section. Nevertheless, the load carrying capacity was not affected and the test load could be significantly increased as the interface progressively debonded to the right and the left of x = 1300 mm. The beam seemed to work as a tied arch with the longitudinal reinforcement bridging the stresses at the point of the macro-crack at UHPFRC. Simultaneously the cracks of the shear plug progressively opened further. The peak load was reached when the compression zone at midspan started to spall but a plateau could still be developed from a deflection of 20 mm to 29 mm, when a diagonal crack extended fully from the compression zone at midspan to the debonded UHPFRC-conventional concrete interface. The UHPFRC with 55 mm thickness has provided an additional load carrying mechanism and pseudo-ductility with respect to the reference RC beam that could not be achieved with the 35 mm thickness.

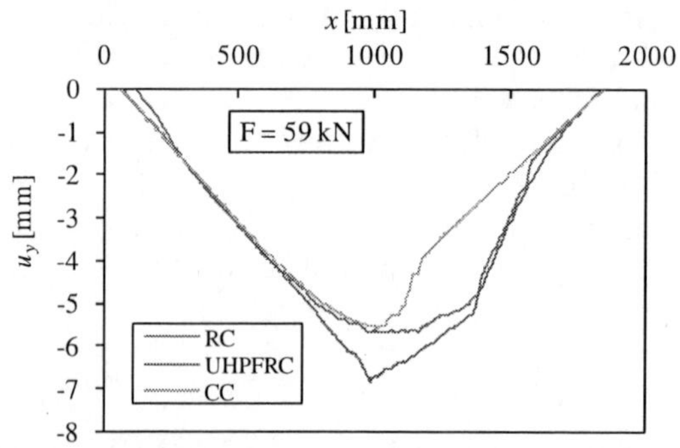

Figure 9. Deformed shape of beam RC-U35 under a load of 59 kN.

4. Conclusions

Digital Image Correlation (DIC) has been used in this research to study the interfacial behavior of UHPFRC-strengthened reinforced concrete beams. The study has focused on the identification of the different stages of the load-deflection diagrams by analyzing the crack pattern evolution and relative displacements between the UHPFRC and conventional concrete.

It can be concluded that DIC is a reliable technique to follow the horizontal slip and vertical separation at the interface, as well as the formation and widening of cracks. It has allowed following the evolution of the crack pattern including formation of shear plug at centre of beams, development of diagonal cracking, tangential slip and vertical debonding at the interface. It has been possible to detect the progressive vertical separation produced by the strain incompatibility due to different cracking at the UHPFRC and conventional concrete layers. For shear-critical beams, the properties of the UHPFRC have been utilized for the thicker UHPFRC layer, which has provided significant improvement of shear strength and ductility with respect to reference reinforced concrete beam. For flexure-critical beams, a better behaviour has been obtained with the thinner UHPFRC layer, as the compression zone is less sensitive to spalling.

Acknowledgement

The financial support provided by the Spanish ministry for economy and competitiveness (Project ID BIA2016-74960-R AEI/FEDER, UE) is gratefully acknowledged. The CRC pre-mix and fibers have been kindly provided by Hi-Con A/S (Denmark). Also thanks to "Fundación José Entrecanales Ibarra" for funding the PhD fellowship of the second author.

References

[1] Habel K, Denairé E and Brühwiler E 2006 *J Struct Eng* Structural response of elements combining ultrahigh-preformance fiber-reinforced concretes and reinforced concrete **132** 1793-1800.
[2] Noshiravani T and Brühwiler E 2013 *ACI Struct J* Experimental investigation on ultra-high-preformance fiber-reinforced concrete composite beams subjected to combined bending and shear **110** 251-61.

[3] Wille K, El-Tawil S and Naaman AE 2014 *Cem Concrete Comp* Properties of strain hardening ultra high performance fiber reinforced concrete under direct tensile loading **48** 53-66.
[4] Noshiravani T and Brühwiler E 2014 *J Struct Eng* Analytical model for predicting response and flexure-shear resistance of composite beams combining reinforced ultrahigh performance fiber-reinforced concrete and reinforced concrete **140** 1-10.
[5] Pimentel M and Nunes S 2016 *Proc. 4th Int. Symp. Ultra-High Performance Concrete and High Performance Materials (Kassel)* Experimental tests on RC beams reinforced with a UHPFRC layer failing in bending and shear (Kassel University Press) p 101-2.
[6] GOM. GOM Correlate 2018. Free version. 2018.

IOP Conf. Series: Materials Science and Engineering **596** (2019) 012025 doi:10.1088/1757-899X/596/1/012025

Macro synthetic fibre reinforced concrete: Influence of the matrix mix design on interfacial bond behavior

C Del Prete, N Buratti, S Manzi and C Mazzotti

University of Bologna, ITALY

E-mail: clementina.delprete2@unibo.it

Abstract. Since the composite nature of Fibre Reinforced Concrete materials, their performance is strictly dependent on the mechanical properties of their components, matrix and fibres, but, above all, on their interaction. The collaboration of the two materials is directly responsible of the force transferring mechanism from concrete to fibres that reflects their contribute in the residual flexural strength performance. This paper aims at understanding how the concrete admixture components might affect the bond of polypropylene crimped fibres, and presents the preliminary results of an experimental campaign consisting of pull-out and compression tests. Four water cement ratios and three cement-sand ratios were considered. The experimental analysis is also supported by the calibration of a numerical model that simulates the pull-out behaviour of a single fibre.

1. Introduction

The design of concrete structures requires particular attention to the cracking problem. The cracked state of plain concrete might be controlled adding fibres to the matrix [1]. The advantage in using fibres is the possibility of gaining a residual tensile strength: the progressive crack opening activates the action of fibres which connect the sides of cracks. The mechanical performance of FRC during the cracked state is controlled by the crack bridging phenomenon which depends on the bond between fibres and matrix. On one hand, a weak bond can cause the slippage of fibres at low stress levels, and, on the other hand, a very strong bold might lead to the failure of fibres in tension. Intermediate bond values, for which most fibres pull out without breaking represent the most desirable behaviour because that will correspond to the highest apparent ductility of FRC elements [2]. There are many factors that might influence the bond of fibres to concrete: chemical adhesion, surface roughness, material properties and shape of fibres, strength and stiffness of the matrix [3].

Zaho et all studied the bond slip mechanism comparing hooked-end and straight shape of steel fibres by means of pull-out tests using acoustic emission techniques to investigate deeply the interfacial bond [4]. This research has revealed a better adhesion for hooked-end fibres due to his shape. Moreover, the strength of the matrix might play an important role in the pull-out mechanism, a higher concrete strength involves a greater pull-out force for steel fibres.

Besides, in case of micro PVA fibres, the interfacial properties of mortar designed with a lower water/cement ratio and higher strength are better than those with higher water/cement. In addition, the shape of the pull-out curve changes with decreasing of water cement ratio because the bond strength increases and so the peak of the curve is higher, while the pull-out displacement decreases [5].

Macro Synthetic (MS) fibres adoption is becoming widespread because of their benefits in terms of corrosion problems and ease in handling, despite their potential problems connected with temperature and creep [6]. These fibres are mostly produced using various blends of polypropylene that does not feature a hydrophilic nature. Therefore, chemical adhesion is weaker compared to PVA fibres.

Aside from understanding which factors influence the fibre bond in FRCs, it could be important also to analyse the behaviour of the single fibre in terms of constitutive law. In particular, in order to develop models to simulate the bending capacity of structural elements [7]. The behaviour of a single fibre subjected to an increasing pull-out load can be separated into two phases: there is a debonding phase after which the fibre starts to slip from the matrix following a softening or hardening branch. The load may tend to decrease during the slippage phase because of the Poisson effect that reduces the section of the fibre with its elongation, this effect is remarkable especially for polymeric materials [8]. The different trend of the second phase could be dependent on the fibre material and shape, and in general it is related to the different strength of the fibre.

The present work concerns the study of the bond of a MS fibre, with the main objective of analysing the effect of different concrete mix design parameters on the interfacial bond properties. In particular the effect of W/C ratio, Cement Sand ratio (C/S) and type of cement. The effect of this fibre on the flexural behaviour of FRC beams has been already analysed by Del Prete et al.[9]. The research is carried on by means of experimental direct pull-out tests, the experimental results are used for the calibration of a constitutive debonding law through a numerical one-dimensional model. The purpose of the analysis developed is also linked to an appropriate design of the FRC material, since the strictly connection of the fibre failure mechanism with the crack-bridging phenomenon and, consequently, the material performance in the cracking phase. In addition, the calibration of the numerical model makes possible a correct estimation of the mechanical parameters characterizing the bond behaviour of this fibre type and the definition a numerical constitutive law for the single fibre pull-out behaviour.

2. Materials and methods

2.1. Materials

The present work analyses the interfacial bond properties of crimped polypropylene fibres. This shape was adopted because it is associated to better bond performances than straight fibres [10]. The geometrical and mechanical properties of fibres are listed in Table 1; the aspect ratio is a parameter relating the diameter and the length and it is specific for geometrical characterization of short discontinuous fibres.

The fibre diameter was estimated weighing many groups of fibres, with a different length for each group, in order to obtain a more accurate value. The tensile strength and, consequently, the elastic secant modulus, were calculated from the results of direct tensile tests on single fibres.

Table 1. Mechanical and geometrical properties of the fibres used.

Fibre property	Value
Fibre length [mm]	39
Fibre diameter [mm]	0.92
Fibre Aspect ratio [-]	43
Fibre material	PP
Fibre tensile strength [MPa]	391
Elastic secant modulus [GPa]	2.25

In order to study the effect of the cementitious matrix on fibres bond, different mix designs were considered. Sand (0-4 mm) was the only aggregate used because coarse aggregates can compromise the failure mechanism when placed close to the fibre. The effect of the coarse aggregates in FRCs mix design is reflected more on the random fibres placement in the concrete matrix [11]. Two different cements were adopted, i.e. CEM I 42.5 R and CEM I 52.5 R. Different mixes were defined by considering various values of Cement/Sand (C/S) ratio and of Water/Cement (W/C) ratio. In particular, for each cement type, four W/C ratios ranging from 0.4 to 0.6 and three C/S ratios, from 0.25 to 0.5 were used. Table 2 shows an outline the various mixes considered. A superplasticizer was used in order to obtain a similar workability for all the mixes. For each combination of cement type, W/C and C/S, three different batches of two specimens were produced, casting a total of six specimens.

Table 2. Experimental campaign outline.

Mix code	Cement	W/C	C/S	Mix code	Cement	W/C	C/S
G1	42.5R	0.4	0.5	H1	52.5R	0.4	0.5
G2	42.5R	0.4	0.3	H2	52.5R	0.4	0.3
G3	42.5R	0.45	0.5	H3	52.5R	0.45	0.5
G4	42.5R	0.45	0.3	H4	52.5R	0.45	0.3
G5	42.5R	0.45	0.25	H5	52.5R	0.45	0.25
G6	42.5R	0.5	0.5	H6	52.5R	0.5	0.5
G7	42.5R	0.5	0.3	H7	52.5R	0.5	0.3
G8	42.5R	0.5	0.25	H8	52.5R	0.5	0.25
G9	42.5R	0.6	0.5	H9	52.5R	0.6	0.5
G10	42.5R	0.6	0.3	H10	52.5R	0.6	0.3
G11	42.5R	0.6	0.25	H11	52.5R	0.6	0.25

2.2. Specimens casting

The experimental campaign performed consists of pull-out and compression tests; the specimens used for the tests are 100 mm x 100 mm x 100 mm cubes with a single fibre inserted at the middle of one of their faces for a depth 25 mm. All the specimens were produced using a planetary mixer (Figure 1a), according to UNI EN 197-1 that indicates also the mixing time. Workability was controlled measuring the slump with a 60 mm high cone. The dosage of superplasticizer was adjusted in order to obtain values ranging from 20 to 50 mm (Figure 1b).

(a) *(b)* *(c)*

Figure 1. Casting procedure:(a) Specimens preparation; (b) Slump test; (c) Fibres tested.

IOP Conf. Series: Materials Science and Engineering **596** (2019) 012025 doi:10.1088/1757-899X/596/1/012025

Figure 2. Specimens cast and then covered avoiding superficial shrinkage.

After mixing, concrete was cast in cubic polyurethane moulds and then vibrated (Figure 2). Fibres were positioned at the centre of the upper face of the moulds using aluminium supports screwed to their walls (Figure 2). Specimens were vibrated in order to remove entrapped air, in particular in the surroundings of the fibre. Moulds were then wrapped with plastic sheet. After two days, the specimens were carefully demoulded and aged in water, to limit shrinkage, for 26 days at a constant temperature of 20 °C (Figure 3). Each specimen was then used for pull out tests and then tested in compression. The choice of using the same specimen for both tests is justified by the low values of pull-out forces that do not influence the compressive strength of concrete.

(a) *(b)*

Figure 3. Specimens after 28 days curing in water (a) and samples for porosity tests (b).

2.3. Pull-out tests

Pull-out tests were performed using a servo-hydraulic MTS Landmark loading frame, controlled by a MTS Flextest 40 controller. Specimens were placed between two square steel plates and blocked by four threaded bars at their corners (Figure 4). These bars were tensioned at the beginning to each test. A threaded bar was attached to the centre of the bottom plated and was connected to the testing machine using its hydraulic grips. The upper plate had a 60 mm diameter circular opening at the top in order to grip fibres. Two aluminium plates with sand paper glued to their surface were used to this aim. These plates were tightened by the hydraulic grips of the testing machine. The test was carried in displacement control at a rate of 0.06 mm/sec. During the tests the force applied and the actuator displacement were measured. In some preliminary tests the fibre pull-out was measured using two LVDTs but their measurements were very consistent with the displacement of the actuator and therefore they were removed in order to simplify the test setup.

(a) (b)

Figure 4. Test set-up used to perform the (a) pull-out tests and (b) compression tests.

2.4. *Porosity tests*

A Mercury Intrusion Porosimetry (MIP) test was used on samples obtained by some of the specimens used for the pull-out tests. Porosity tests are a good way to investigate the different interfacial state due to the different mix design, being a good predictor of the concrete state [12]. These tests use the properties of mercury (Hg), i.e. it penetrates inside the voids of the matrix samples with pressure that increases as soon as the pores size decreases. Since the failure type is expected to be connected to the microstructure of concrete, porosity tests were carried out in order to investigate if concrete porosity can influence the pull-out behaviour of fibres. MIP tests need samples of about 1 cm^3 representative of the microstructure (Figure 3b). They were extracted from the undamaged core of cubes after compression tests.

3. Experimental results

3.1. *Pull-out tests*

In general, two failure modes could be observed in the tests; either fibre pull-out or fibre tensile failure. The typical pull-out curve observed is reported in Figure 5a. It has a quasi-linear first branch up to the peak strength, corresponding to the deboning phase. After the peak strength there is a sudden strength decrease followed by a decreasing undulating branch until complete pull-out. This last part of the curves is dominated by the friction and the mechanical interlock between the fibre and concrete. In particular the undulating behaviour is related to the shape of the fibres tested. When the pull-out strength is higher than the tensile strength of the fibre, this latter will fail in tension, as in Figure 5b. When a fibre breaks in tension it will fibrillate, the steps in the descending branch are associated to the failure of the various fibrils (Figure 5b). Table 3 lists the mean experimental results in terms of peak forces for pulled-out and ruptured fibres and number of fibres for each failure mode.

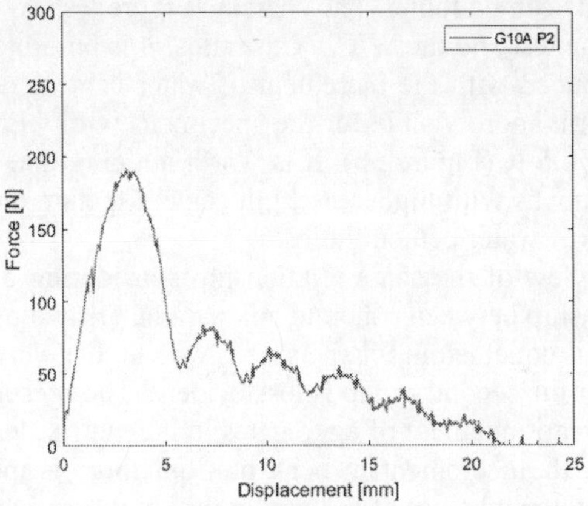

(a)

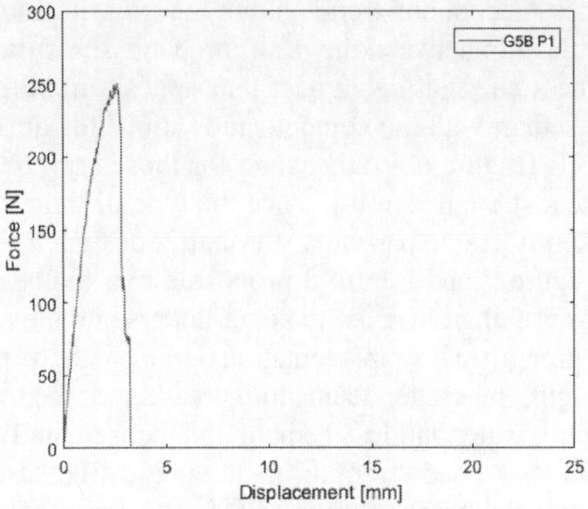

(b)

Figure 5. Mechanical behaviour of (a) a pulled-out and (b) a broken fibre.

Table 3. Mean values and coefficient of variation (COV) of the peak strength ($F_{p,m}$) and number of pulled out and broken fibres for the different concrete mixes tested.

Mix Code	Pulled out fibres			Ruptured fibres		
	$F_{p,m}$ [N]	COV [%]	Number [-]	$F_{p,m}$ [N]	COV [%]	Number [-]
G1	232	/	1	260	12	5
G2	/	/	0	261	12	6
G3	228	8	3	243	16	3
G4	236	11	2	246	14	4
G5	250	/	1	277	17	5
G6	221	12	5	281	/	1
G7	231	18	4	302	/	1
G8	273	/	1	246	15	5
G9	159	18	6	/	/	0
G10	198	20	6	/	/	0
G11	190	10	5	290	/	1
H1	/	/	0	298	5	6
H2	234	/	1	290	7	5
H3	/	/	0	293	6.5	6
H4	/	/	0	280	3	6
H5	/	/	0	298	8.6	5
H6	236	/	1	290	3	5
H7	207	/	1	283	9.8	5
H8	/	/	0	294	7.4	6
H9	256	10	3	262	3.7	3
H10	262	14.7	6	/		0
H11	260	21.5	3	233	4.2	3

The results reported in Table 3 point out that the peak forces measured on pulled out and broken fibres are very similar. Therefore for this embedment length (25 mm) it is not easy to clearly understand the effects of W/C, C/S and cement type on the bond properties. Nevertheless, analysing the possible relations between the bond strength in terms of peak forces and the variables involved in the debonding process, a common trend among the experimental results can be found. The charts in Figure 6 show the relationship between the peak force measured during the test and the W/C – C/S ratios. It is possible to notice a strength decrement that appears more pronounced with the increment of water-cement ratio rather than with the cement-sand ratio. This decrement is more visible for the specimens with CEM I 42.5 R (Figure 6a) rather than for those with CEM I 52.5 R (Figure 6b). It is worth noticing that the cement strength can influence the type of failure, specimens with higher strength cement feature more broken fibres, in particular for matrix designed with a low water cement ratio.

Figure 7 and Figure 8 propose a two dimensional view of the same relationship considering only pull-out failure in order to better understand the relationship between concrete mix parameters and pull-out strength. This representation agrees with the previous conclusion: for specimens with higher cement strength, the scatter seems to increase and, the trend for the second group is less evident. The presence of more water and less cement and, consequently, the rearrangement of aggregates in the matrix, leads to a lower bond strength of the single fibre. Besides, the decrement in peak pull-out force is more pronounced with C/S ratio of 0.51, it seems that a greater sand amount in the mix design determines a different internal organization and, consequently, an higher scatter of results in terms of peak forces.

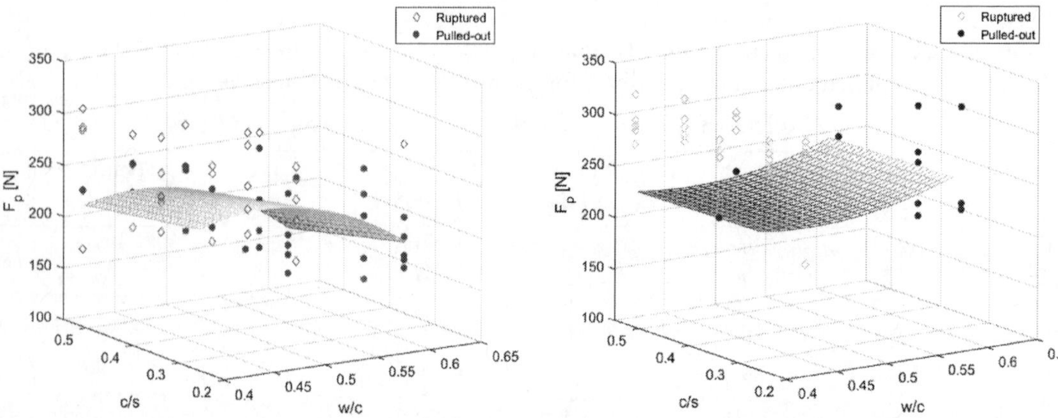

Figure 6. Correlation of peak forces for pulled-out and ruptured fibres with C/S and W/C rations for specimens with (a) CEM I 42.5R and (b) CEM I 52.5R.

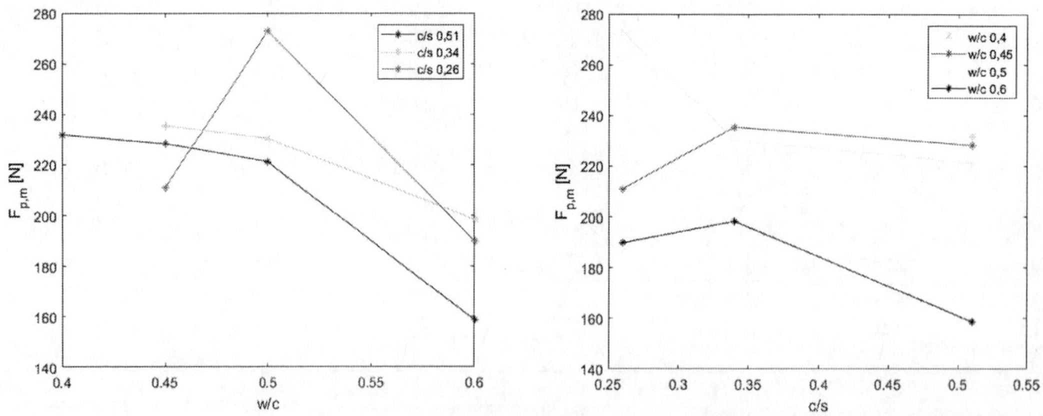

Figure 7. 2D representation of peak forces mean value relation with (a) w/c and (b) c/s ratio for specimens with CemI 42.5R

IOP Conf. Series: Materials Science and Engineering **596** (2019) 012025 doi:10.1088/1757-899X/596/1/012025

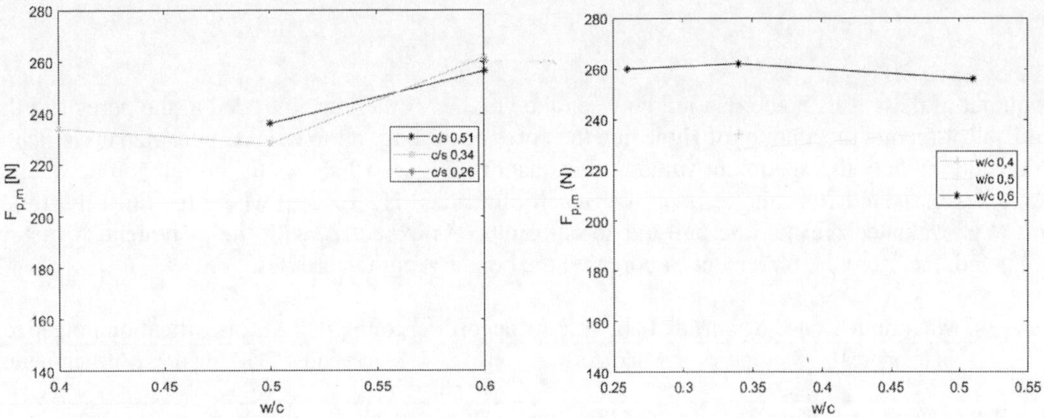

(a) *(b)*

Figure 8. 2D relations between mean value of forces and (a) w/c and (b) c/s ratio for specimens with CemI 52.5R.

3.2. *Compressive tests*

The matrix mix design, W/C, C/S ratios and cement strength, defines certainly its strength. For this reason, compressive tests results are considered for a more detailed analysis of the cement strength effect. The plot in Figure 9a, relating the peak pull-out force to the mean value of mortar compressive strength for each batch, makes possible to conclude that a higher matrix strength, associated mostly with a lower W/C ratio in the mix design, implies a higher pull-out force. On the other hand, for the specimens with CEM 52.5R (Figure 9b) the trend is less clear but in this case the number of pulled-out fibres is much lower and the regression line is not very reliable.

Taking into consideration the batches exhibiting more pulled out fibres in both groups, G9, G10, G11 and H9, H10, H11, designed with W/C ratio 0.6 and progressive C/S ratio, and respectively CEM I 42.5 R and 52.5 R, have different compressive strength. With equal mix design, the first group is placed in a range of variability between 27 – 35 MPa while the second covers a range of 40 – 45 MPa. Thus, it is directly perceptible that, a higher cement strength assures a high bond strength indeed the same compressive strength (i.e. 40 MPa) is achieved with a W/C ratio of 0.5 by the first group while the second gains the same compressive strength with a W/C ratio of 0.6.

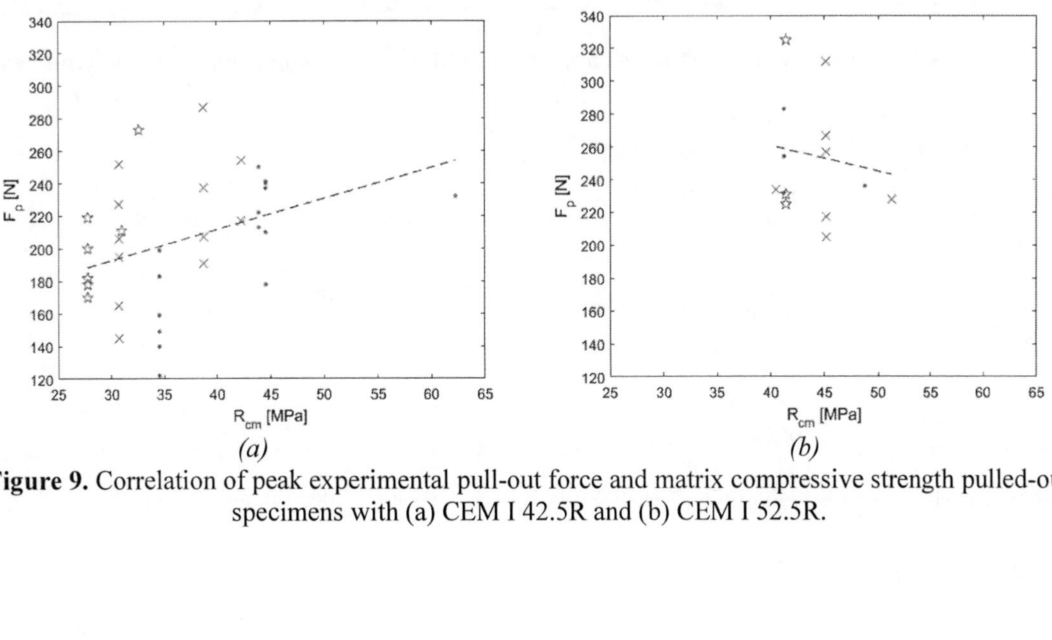

(a) *(b)*

Figure 9. Correlation of peak experimental pull-out force and matrix compressive strength pulled-out specimens with (a) CEM I 42.5R and (b) CEM I 52.5R.

3.3. Porosity tests

The output of these tests is the relation between the volume of mercury intruded in the pores and their size. Table 4 reports the amount of Hg inside the pores according to the IUPAC pore size classification but considering only the maximum value of this quantity. Figure 10 shows the correlation between the peak forces measured from the tests on the Specific Intruded Hg volume where the linear regression reported is evaluated only on the pull-out force results. As expected, with the increment of mercury amount and, therefore the increment of porosity, the bond strength decreases.

Table 4. Maximum values of intruded Hg volume according to the IUPAC classification pore size.

ID sample	Macropores	Mesopores	Micropores	ID sample	Macropores	Mesopores	Micropores
G1A_P1	0.82	26.06	102.48	G7C	0.99	17.61	79.01
G1B_P1	0.69	5.29	67.16	G7C	0.69	6.73	68.91
G2B_P1	0.71		75.92	G8C	0.65	35.14	108.79
G2B_P2	0.59	29.16	82.44	G9C	0.84	9.16	85.90
G3C_P1	0.66	6.23	84.81	G10C	0.82	3.65	84.47
G4C_P2	0.67	46.38	108.76	G11C	0.66	25.94	96.98
G5C	0.61	7.74	74.70	G11C	0.64	26.06	102.48
G6C	0.8	9.55	85.80				

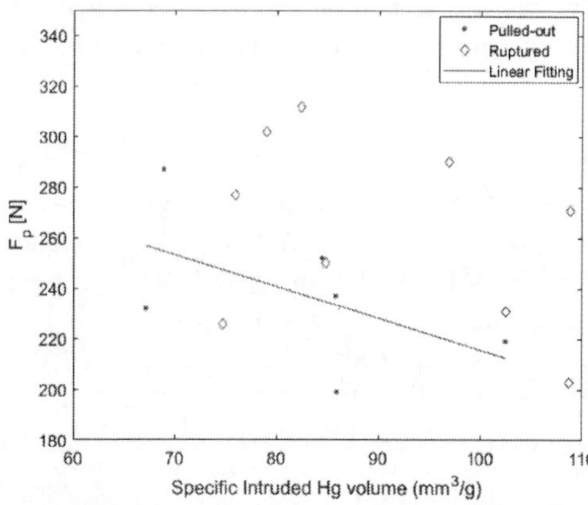

Figure 10. Relation between peak forces of pulled out fibres with volume intruded in micropores.

4. NUMERICAL BOND SLIP MODEL

A 1D FEM model was used in order to simulate the experimental results. The embedded portion of a general fibre was modelled by means of a series of truss elements (each of them 1 mm long) with nonlinear springs at their ends in order to simulate the bond-slip behaviour. The constitutive law of these springs (Figure 11a) is based on a modified version of the Cosenza et al.[13] model, first proposed for FRP bars in concrete. This model describes the bond-slip behaviour with two different equations corresponding to the debonding and pull-out phases, respectively. The relationship between bond strength (τ) and slip (s) in the debonding stage is described by the following power-law:

$$\tau = \tau_m \left(\frac{s}{s_m}\right)^{\alpha} \quad (1)$$

where τ_m represents the maximum bond strength, s_m the corresponding slip and α is a shape parameter.

After the peak bond-strength τ_m the Cosenza et al. model was modified in order to consider a sudden stress drop followed by a friction-dominated behaviour. This second part of the constitutive relationship is bilinear and is defined by the parameters τ_f, s_f and s_0, as shown in Figure 11a. Therefore the adopted bond-slip model has a total of six unknown parameters. Their values were identified by means of an inverse analysis procedure, applied on all the experimental results corresponding to pulled-out fibres.

This procedure made use of the Matlab™ Pattern Search algorithm that was used to minimize a cost function defined as a linear combination of the differences between experimental and numerical results at a set of reference points. Figure 11b shows a comparison of the results given by the numerical model against the experimental curve used as input of the inverse analysis. It is possible notice that there is a very good agreement between the two curves. Of course the numerical model is not able to reproduce the undulating shape of the last branch of the experimental curves being based on a purely frictional behaviour. Figure 12 shows the relationship between the τ_m values obtained from the inverse analyses based on the experimental results for specimens with CEM I 42.5 R against the W/C ratio. The trend that can be observed follows that of the experimental results, a decrement of the peak force with the increment of water cement ratio.

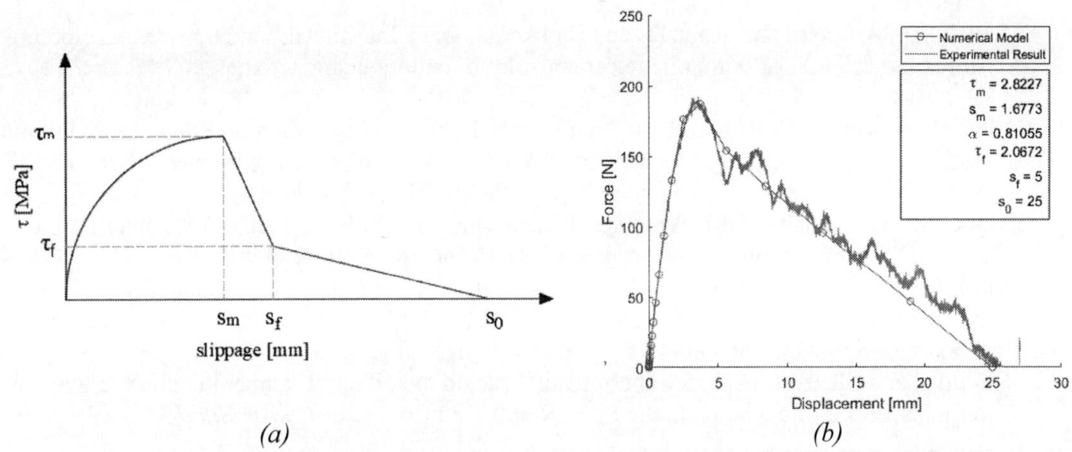

Figure 11. (a) Local bond behaviour characterizing the single spring and (b) global behaviour of the pull-out failure mechanism for the entire fibre.

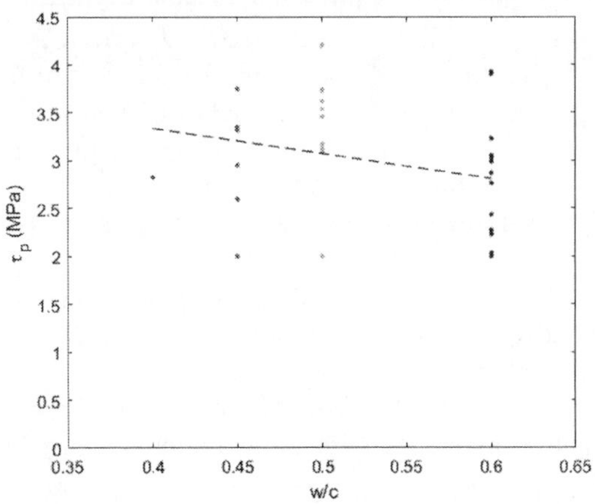

Figure 12. Relation between the maximum bond strength obtained from inverse analysis and W/C ratio, for specimens with CEM I 42.5R.

IOP Conf. Series: Materials Science and Engineering **596** (2019) 012025 doi:10.1088/1757-899X/596/1/012025

5. CONCLUSIONS

This paper presented the preliminary results of an experimental campaign aimed at investigating the effect of mix design parameters, namely Water/Cement and Water/Sand ratios and cement strength, on the pull-out behaviour of a macro-synthetic fibre. Results for different concrete mixes and a single fibre embedment length were discussed. It was found that increasing the W/C the fibre bond strength decreases while higher C/S ratio values have a positive effect on the bond properties. Also the cement type, affects the interfacial behaviour enhancing the bond strength.

It is important to notice that in the present work only one fibre embedment length was considered and for this length a significant number of fibres failed in tension without pulling-out. Therefore a better understanding of the bond behaviour will be possible after analysing different embedment lengths.

Acknowledgments

The support of the staff of the Interdepartmental Center for Industrial Research (CIRI) Building and Construction, University of Bologna, is gratefully acknowledged.

References

[1] ACI 544.5R 2010 Report on the Physical Properties and Durability of Fibre-Reinforced Concrete

[2] Jewell RB, Mahboub KC, Robl TL and Bathke AC 2015 Interfacial bond between reinforcing fibres and calcium sulfoaluminate cements: Fibre pullout characteristics *ACI Mater J* **112** pp 39-48

[3] Zhao G, Verstrynge E, Prisco M and Vandewalle L 2012 Investigation on Single Fibre Pullout and Interfacial Debonding Mechanisms With Acoustic Emission Techniques *8th RILEM Int Symp Fibre Reinf Concr challenges Oppor (BEFIB 2012)* pp 1-11

[4] Markovich I, Van Mier JGM, Walraven JC and Microlab W, Structures C 2001 Single fibre pullout from hybrid fibre reinforced concrete *Heron* **46** pp 191-200

[5] Liu J, Li C, Liu J, Yang Z and Cui G 2012 Pullout behavior of polyvinyl alcohol fibre from cementitious matrix during plastic state *8th RILEM Int Symp Fibre Reinf Concr challenges Oppor (BEFIB 2012)* pp 480-488

[6] Buratti N and Mazzotti C 2015 Experimental tests on the effect of temperature on the long-term behaviour of macrosynthetic Fibre Reinforced Concretes *Constr Build Mater* **95** pp 133-142

[7] Manca M, Ciancio D and Dight P 2017. Fibre reinforced concrete in flexure and single fibre pull-out test: A correlation *IOP Conf Ser Mater Sci Eng.***246**

[8] Redon C, Li VC, Wu C, Hoshiro H, Saito T and Ogawa A 2001 Measuring and Modifying Interface Properties of PVA Fibres in ECC Matrix *J Mater Civ Eng.* **13** pp 399-406

[9] Del Prete C, Tilocca A, Buratti N and Mazzotti C 2018 Effect of fibre dosage and matrix compressive strenght on msfrc performance *3rd FRC International Workshop Fibre Reinforced Concrete: from Design to Structural Applications* pp 174-175

[10] Tang CS, Li J, Wang DY and Shi B 2016 Investigation on the interfacial mechanical behavior of wave-shaped fibre reinforced soil by pullout test *Geotext Geomembranes* **44** pp 872-883

[11] Ostrowski K et al. 2018 The Effect of the Morphology of Coarse Aggregate on the Properties of Self-Compacting High-Performance Fibre-Reinforced Concrete *Materials (Basel)* **11** p 1372

[12] Claisse PA, Cabrera JG and Hunt DN 2001. Measurement of porosity as a predictor of the durability performance of concrete with and without condensed silica fume *Adv Cem Res. 001* **13** pp 165-174

[13] Cosenza E, Manfredi G and Realfonzo R 2002. Development length of FRP straight rebars *Compos Part B Eng* **33** pp 493-504

IOP Conf. Series: Materials Science and Engineering **596** (2019) 012026 doi:10.1088/1757-899X/596/1/012026

Numerical modelling of failure on brick masonry strengthened with FRCM overlays

João A P P Almeida, Eduardo B Pereira and Joaquim A O Barros

ISISE, University of Minho, Department of Civil Engineering, School of Engineering; 4810-058 Guimarães, Portugal

E-mail: j.almeida@civil.uminho.pt

Abstract. The behaviour of masonry elements under in-plane loads can be improved by using strengthening overlays. However, the changes in stiffness of the elements after strengthening may have various effects at the level of the behaviour of the entire structure. Advanced numerical simulations can be a helpful and cost effective strategy to guide the efficient design of strengthening systems. These models should simultaneously show the ability to deal with the complexities associated with the composite behaviour of panels at a smaller scale, and allow the simulation of models that can include the entire structure. A constitutive model for a 3D interface finite element, implemented in a FEM-based computer program, was applied for the simulation of the interfaces between substrate and the Fabric Reinforced Cementitious Matrix (FRCM) overlay. The mechanical behaviour of the interfaces between the different layers of the strengthened masonry elements were obtained from direct shear tests on couplet specimens. Subsequently, these results were used to define the constitutive laws of the interface elements. Finally, the numerical model developed was used to simulate the in-plane structural response of the strengthened masonry specimens.

1. Introduction

The use of strengthening techniques based on the application of reinforcing overlays to masonry substrates is widespread. To enhance the tensile behaviour and ductility, different types of materials can be used as reinforcement, in the form of meshes made of steel, polymers, carbon or glass [1,2].

The in-plane behaviour of masonry infill panels gives an important contribution to the strength of structures subjected to horizontal actions. The in-plane characterization of masonry panels can be performed by means of the diagonal tensile test (DTT), using specimens built in the laboratory, [3], or in-situ [4]. Different types of strengthening systems have been evaluated using this type of test [5,6]

Fabric reinforced cementitious matrix systems (FRCM) are considered of great potential, because the increase of ductility can be achieved with simple and economic strengthening procedures. In order to assess the effectiveness of these strengthening systems the in-plane behaviour must be studied [7]. However, these systems often show vulnerability at the level of the interface due to the sharp gradient of mechanical properties between the substrate and the strengthening system [8].

This work discusses the numerical modelling strategy adopted for simulating the in-plane tests carried in unreinforced and strengthened masonry. Further details about the experimental programme, as well as the discussion of the main results obtained, may be found in Almeida et al.[9].

Considering the importance of testing and developing numerical tools for supporting the design process of masonry structures strengthened by FRCM overlays, the numerical strategy adopted was devised trying to achieve the optimum balance between modelling accuracy and computational cost.

IOP Conf. Series: Materials Science and Engineering **596** (2019) 012026 doi:10.1088/1757-899X/596/1/012026

1.1. Experimental characterization of the in-plane behaviour

As discussed by Almeida et al. in [9], the cyclic effect on the shear behaviour of masonry panels can be characterized by means of DTT's. The contribution of the strengthening effect provided by FRCM overlays to the in-plane structural behaviour under cyclic loading was accessed in specimens with geometry similar to figure 1(a). Cyclic DTT´s were carried out by Brignola et al. [4], under force control, considering a load increment of 200 N/s, and at least four cycles for each specimen. Santa-Maria et al. [10] used this type of test but considering a different procedure, which started by the loading and unloading of the first diagonal of the specimen, followed by the loading and unloading of the second diagonal.

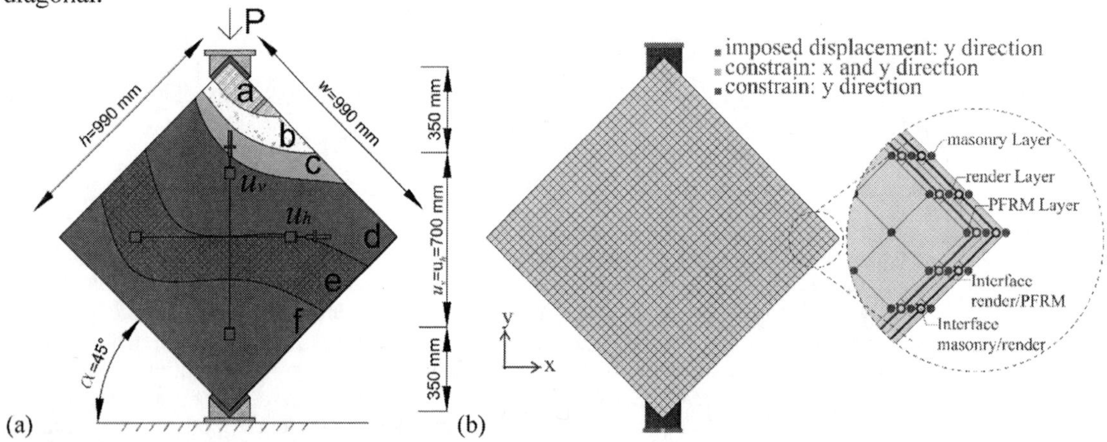

Figure 1. (a) General set-up of a DTT. (a – masonry; b – roughcast mortar; c – render layer of 15 mm thickness; d – Polypropylene fibre reinforced mortar (PFRM) of 12.5 mm thickness; e – carbon fibre mesh; f – PFRM of 12.5 mm thickness.); (b) Finite element model, imposed displacements and constrains, and detail of the layers.

1.2. Numerical modelling of in-plane tests

Masonry walls exhibit a complex structural behaviour since masonry is a composite material formed by constituents showing brittle behaviour. Two main numerical approaches have been adopted by researchers to describe the mechanical behaviour of masonry, macro-modelling and micro-modelling [11]. In macro-modelling, masonry is considered as a homogeneous material, while in micro-modelling, masonry is considered as an assembly of units connected by joints. When considering masonry strengthened with overlay systems, two types of approaches can be considered: those that model the behaviour of the strengthening layer and interface in a coupled manner by means of homogenization; approaches that model the masonry, the reinforcement and the interface separately [12]. In the second case interface finite elements with nonlinear constitutive laws are used. The nonlinear behaviour of masonry was reproduced in various studies using a smeared crack approach [3,12]. Typically, the masonry is modelled as an isotropic continuum material before any nonlinear effect, and orthotropic or even anisotropic continuum material after cracking, by adopting constitutive laws for modelling the nonlinear behaviour in tension and compression, and considering also the degradation of shear stiffness and strength with the smeared cracking propagation.

2. Materials, test procedures and relevant results

The complete discussion of the experimental program may be found elsewhere ([8,9]).

The specimens presented a square geometry with approximately 990 mm. The thickness was 140 mm for reference and 190 mm for the strengthened specimens. The specimens were assembled using hollow ceramic bricks (Length x Height x Thickness: 28.5x19.5x11.0 cm^3) and cement mortar for the joints. Subsequently a render layer of a cementitious mortar with a thickness of 15 mm was applied on both faces. In the case of the strengthened specimens, after the stages previously described a first layer of polypropylene fibre reinforced mortar (PFRM) with 12.5 mm thickness was applied, then a carbon

IOP Conf. Series: Materials Science and Engineering **596** (2019) 012026 doi:10.1088/1757-899X/596/1/012026

fibre mesh was placed over that layer and finally a second layer of PFRM with the same thickness was applied, see figure1. All materials were characterized experimentally. In the case of the carbon mesh, the mechanical properties were obtained from the supplier, see table 1.

Table 1. Properties of the masonry components, render and FRCM.

Masonry components	
Compressive strength of units, parallel to hollows *	6.55 (8%) N/mm^2
Compressive strength of units, normal to hollows *	2.21 (9%) N/mm^2
Compressive strength, mortar joints and render *	8.58 (7%) N/mm^2
Flexural strength, mortar joints and render *	2.11 (3%) N/mm^2
Adhesion strength to render layer to brick *	0.51 (22%) N/mm^2
Polypropylene fibre reinforced mortar	
Density	2050 kg/m^3
Elasticity modulus, [13]	38240 N/mm^2
Compressive strength*	36.46 (6%) N/mm^2
Flexural strength, [13]*	6.25 (5%) N/mm^2
Adhesion strength to render layer*	1.48 (20%) N/mm^2
Carbon Fibre Mesh [14]	
Carbon fibres in both directions	50 threads/m
Elasticity modulus	≥240 kN/mm^2
Tensile strength	≥4300 N/mm^2
Elongation at rupture	1.75 %
Ultimate tensile force	185 kN/m

*The values in parenthesis represent the coefficient of variation.

The elasticity modulus of the render, E_{ci}, was estimated using equation (1), as suggested by the Model Code 2010 [15], considering the compressive strength obtained from compression tests on cylindrical mortar specimens, which resulted in E_{co}=20431 MPa.

$$E_{ci} = E_{c0} \times \alpha_E \times \left(\frac{f_{cm}}{10}\right)^{1/3} \tag{1}$$

where: E_{co} is a constant value (21.5x10^3 MPa); α_E represents the effect of the type of aggregates (1.0 for quartzite aggregates) and f_{cm} is the average compressive strength at 28 days (8.58 MPa).

The compressive strength and the elasticity modulus of the masonry weakest direction were deduced considering the provisions of Eurocode 6, [16]. The characteristic compressive strength value, f_k, may be estimated by equation (2).

$$f_K = K \times f_b^{0.7} \times f_m^{0.3} \tag{2}$$

where: k is a constant (k=0.35); f_b is the normalized mean compressive strength of the brick units in the direction of the applied action effect (f_b=2.21MPa); f_m is the compressive strength of the mortar joint, which should not be considered as greater than 20 MPa or greater than $2f_b$ when units are assembled using general purpose mortar; f_m=4.42 MPa.
Following these assumptions, f_k=0.95 MPa. For estimating the elasticity modulus of masonry, E_m, Eurocode 6 [16] suggests $700f_k > E_m > 1000f_k$, ($667 > E_m > 952$ MPa). In this case, E_m=952 MPa was adopted, once more isotropic behaviour was assumed.

The load vs displacement responses obtained for the reference and the strengthened specimens are presented in figure 4. The vertical and horizontal displacements represent the average of the measurements obtained from a pair of LVDT's placed in opposite sides of the specimens. The strengthened specimens, FRCM_EXP, have reached considerably higher strength than the reference specimens, REF_EXP. The load vs displacement responses obtained for the strengthened specimens presented, in general, a low scatter of results. Besides the increase of the pre-peak load carrying capacity observed, the peak loads were reached for higher displacements than the ones registered for the reference specimens. After reaching the peak load the strengthened specimens presented a relatively smooth load decay with the increase of the horizontal deformation.

3. Characteristics of the Finite Elements models

3.1. Finite element mesh and boundary conditions

The finite element analysis was carried out using the computer code FEMIX 4.0 [17]. All the nonlinear static analyses were performed using the Newton–Raphson iteration method. The numerical simulations used a macro-modelling approach, where the masonry and the FRCM layers were considered as continuum materials. Due to symmetry conditions, the model was simplified, only one half of the specimen was discretized along the direction of the thickness. Consecutive material layers were connected by interface elements. The simulation of each material layer was done through the Reissner-Mindlin theory adapted to layered shells [18]. Finite elements of 4 nodes and 2x2 Gauss-Legendre integration points were used in the numerical model, figure 1(b). Surface interface Lagrangian finite elements of 8-node with 2 integration points in each direction where adopted to simulate the interfaces between different layers (masonry/render and render/FRCM). Finite elements of 25x25 mm^2 were used for all the layers and interfaces. All rotations were kept free.

3.2. Masonry, render and FRCM overlay constitutive models

The contribution of microfibres present in the mortar matrix and the strengthening carbon mesh for the post-cracking behaviour of the FRCM strengthening overlay was simulated in the smeared cracking model by using a softening law exclusively for fracture mode I ,figure 2(a), since their effect in fracture mode II was assumed of marginal impact [19]. The tension-softening or tension-stiffening behaviour of materials can be simulated by changing the normalized coordinates (ξ_i; α_i) that define the transition points of the diagram, the tensile strength, f_{ct}, and the mode I fracture energy, G^I_f. The ultimate crack strain, $\varepsilon^{cr}_{n,u}$, is defined as a function of the parameters α_i and ζ_i, the mode I fracture energy, the tensile strength, ($\sigma^{cr}_n = f_{ct}$), and the crack bandwidth, l_b [20]. The crack bandwidth aims to guarantee that the results are mesh objective [21].

In the case of the masonry and of the render, a smeared cracking approach was also used to simulate their post-cracking behaviour. However, the quasi-brittle behaviour of both the masonry and the render in tension, is better simulated by adopting a bilinear law, or the Cornelissen softening law, having this last one been adopted. The Cornelissen´s softening law simulates the tension-softening behaviour by considering an exponential function for the crack opening vs tensile strength [22]. The parameters used in the constitutive model were the tensile strength, f_{ct}, the mode I fracture energy, G^I_f, and the crack bandwidth, l_b, as shown in figure 2(b).

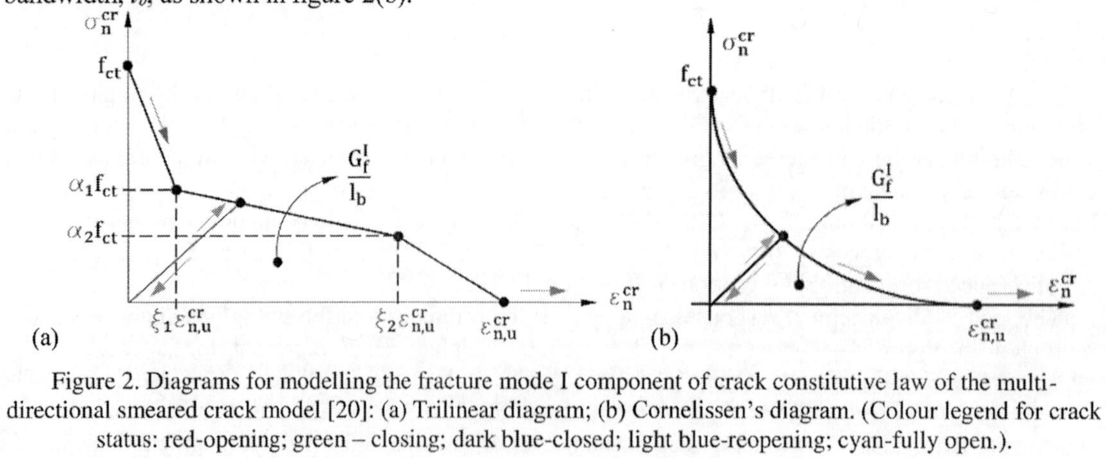

Figure 2. Diagrams for modelling the fracture mode I component of crack constitutive law of the multi-directional smeared crack model [20]: (a) Trilinear diagram; (b) Cornelissen's diagram. (Colour legend for crack status: red-opening; green – closing; dark blue-closed; light blue-reopening; cyan-fully open.).

3.3. Constitutive model for the surface interface finite elements

Both the interfaces between the masonry and the render, and between the render and the FRCM layers have been simulated by following a discontinuous strategy based on the adoption of the surface interface finite elements represented in figure 3(b). The constitutive model (CM) is governed by the yield function (f) and the yield surface gradient (n), the plastic potential (g) and the plastic potential variables (ψ), plastic flow direction (m), the hardening law (Δk) and the hardening variables (Φ). Work hardening and non-associated flow were assumed. The yield surface is governed by three hardening variables, which are the tensile strength (χ), the shear strength (c) and the friction angle ($\tan\phi$), as shown in equation (3) and figure 3(a).

$$f = f\left[\sigma_N,\sigma_T\right] = \sigma_T^2 - \left(c - \sigma_N \times \tan\phi\right)^2 + \left(c - \chi \times \tan\phi\right)^2 \tag{3}$$

Where σ_N and σ_T represent the normal and tangential stress respectively.

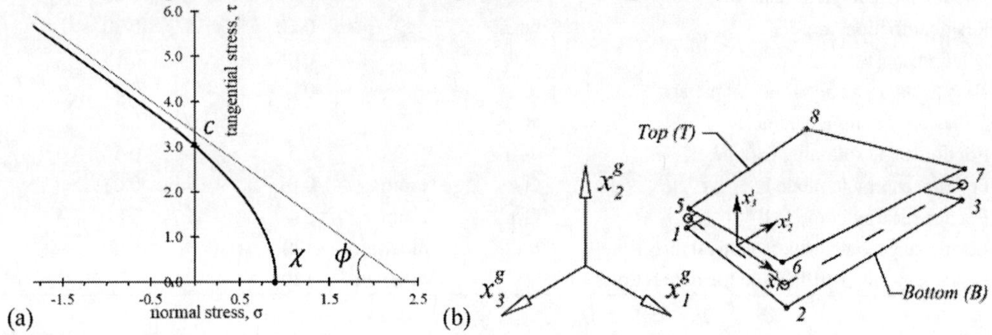

Figure 3. (a) Representation of the Mohr failure criterion when used to characterize the interaction between the normal stress and the peak values for positive tangential stresses. (b) Lagrangian 8-node surface interface element used in the numerical models.

A dilation stress (σ_{dil}) which corresponds to the normal stress at which the dilatancy vanishes when compression and shear stresses occur at the same time, was also considered in the constitutive model.

The evolution of the yield surface and the evolution of the hardening parameters depend on the evolution of the plastic work (W). Additionally, the increase of the plastic work is reflected by means of a dimensionless parameter that is related to the portion of the fracture energy (G_f) dissipated. Two dimensionless parameters are associated to the fracture modes I and II. The constitutive model used is designated as CM I/II_2D by Coelho et al. [23].

4. Numerical simulation of the reference prototype

4.1. Analysis of masonry and render properties

The tensile strength (f_{ct}) of the masonry and of the render were estimated based on the ratio (f_{ct}/f_c) equal to 0.10, according to Angelillo et al. [24]. Conversely, the mode I fracture energy of the masonry was considered equal to 0.01 N/mm, taking the results obtained by Basili et al. [12]. The mode I fracture energy of the render layer was calculated following the experimental procedure of RILEM 50-FMC, [25]. A value of 0.12 N/mm with a coefficient of variation of 16% was obtained.

The Cornelissen's softening law was used for both materials. A parametric analysis was carried out to ascertain the sensitivity of the numerical responses obtained with respect to the mode I fracture energy (G_f), as well as their approximation to the experimental results. Within the range of values considered, 0.005 to 0.03 for the masonry, and 0.05 to 0.30 N.mm for the render, the parametric analysis shown that the values that better matched the experimental response where 0.02 N.mm for the masonry and 0.20 N.mm for the render.

The mechanical characterization of the interfaces responses was carried out by performing direct shear tests in couplet specimens according to the procedure and set-up presented in Almeida *et al.* [8].

The cohesion and the friction angle values, obtained from the experimental results, were used to define the interface constitutive model, see table 2. Additionally, mode II fracture energy was obtained from the shear stress vs slip experimental responses [8].

The resulting load vs displacement numerical responses are plotted in figure 4. The simulation that showed a better agreement to the experimental results was the REF_NUM_03, presenting similar values of the peak-load and a similar post-peak behaviour.

Table 2. Proprieties defining interface behaviour.

Description	Symbol	Units	Interface masonry/render	Interface render/PFRM
Tensile strength	χ	MPa	0.51	1.48
Cohesion in yield function	c	MPa	4.8	6.0
Normal stress at which the dilatancy vanishes	σ_{dil}	MPa	0.1	0.3
Friction angle in yield function	$\tan\phi$	–	0.43	0.80
Residual friction angle	$\tan\phi_r$	–	0.10	0.30
Dilation angle	$\tan\beta$	–	0.05	0.15
Tensile strength softening parameter	α_χ	–	0.5	0.5
Cohesion softening parameter	α_c	–	0.5	0.5
Friction angle softening parameter	$\alpha_{\tan\phi}$	–	0.5	0.5
Fracture energy in mode I	G^I_f	N/mm	0.01	0.03
Fracture energy in mode II	G^{II}_f	N/mm	0.4	0.8
Elastic tangential stiffness in local direction l_1	k^e_1	MPa/mm	120	150
Elastic tangential stiffness in local direction l_2	k^e_2	MPa/mm	120	150

4.2. Load vs displacement response

The numerical load vs displacement response obtained for the reference specimens is presented in terms of vertical and horizontal displacement between two points of the mesh in the render layer, equivalent to the LVDT's position during testing, see figure 1. The response obtained for the reference model, REF_NUM_03, shows a first branch characterized by a linear evolution considering both the horizontal and the vertical displacements. After this phase, two different behaviours can be observed; considering the horizontal displacements, the load vs displacement response is characterized by an increase of the load until the peak load is reached, and the subsequent mild softening, following closely the response obtained experimentally; when considering the vertical displacements.

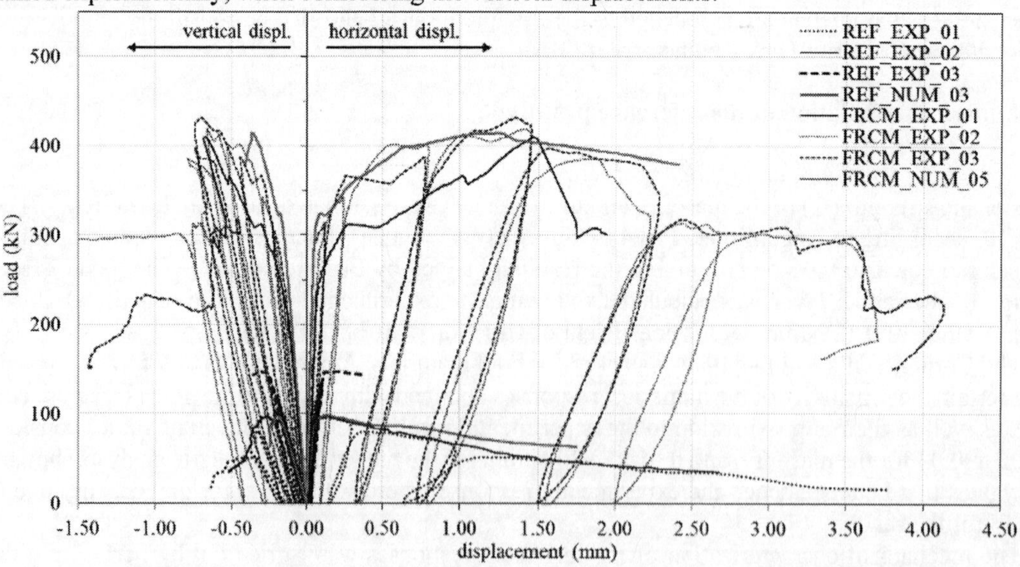

Figure 4. The load vs displacement response of the numerical simulation and experimental results.

5. Numerical simulation of the strengthened prototype

5.1. Analysis of FRCM properties

The values of the f_{ct} and of the mode I fracture energy (G^I_f) for the FRCM system were estimated considering an effective composite behaviour resulting from the combination of the carbon fibre mesh and of the PFRM. According to the Model Code 2010 [15], the flexural tensile strength ($f_{ctm,flex}$) can be obtained as a function of the axial tensile strength (f_{ctm}), resulting in f_{ctm}=1.75 MPa. For mode I fracture energy a sensitivity analysis was carried out considering values between 1.00 N/mm and 2.00 N/mm. Also the shape of the softening diagrams for modelling the crack opening in mode I were analysed. Two different softening diagrams were considered, the (ξ_i, α_i) values for each point of the diagrams were assumed taking into account the composite nature of the strengthening system. Diagram 1 was defined by (0.05,0.7;0.6,0.5) and diagram 2 by (0.02,0.6;0.6,0.2) see figure 2(a). For both diagrams three different values of fracture energy were considered, 1.00, 1.25 and 1.50 N/mm. The response considering diagram 2 and fracture energy of 1.25 N/mm, FRCM_NUM_05, presents the best agreement with the experimental, see figure 4.

5.2. Load vs displacement response

The response registered for the strengthened specimens, FRCM_NUM_05, is shown in figure 4. The load-displacement response in terms of the horizontal displacement shows a first linear branch with a slope similar to the experimental, followed by a non-linear phase that also presents a good agreement with the experimental. At the initial part of this phase a local peak occurs, and this behaviour is justified by the steep shape of the first branch of the tensile softening diagram adopted for modelling the crack opening in mode I. When the peak load is reached, both the load and displacement are in agreement with the average experimental values. After the peak, a softening behaviour is observed, where the load response presents a relatively smooth load decay with the increase of the horizontal deformation. For higher deformations the load decays observed in the experimental responses were slightly steeper than the one obtained in the numerical simulation. The load-displacement numerical response considering the vertical displacements also presents a first branch with a slope matching the experimental. After this stage, when the non-linear behaviour begins, an increasing lag between numerical and experimental responses can be observed. This lag reaches approximately 0.3 mm at the peak stage. Probably this difference may be justified by some degree of crushing existing at the interfaces between masonry units, joints and overlays, which is not conveniently simulated by the numerical model since elastic behaviour in compression is considered for the sake of computational efficiency. After the peak, an abrupt decay occurs, and a similar response was observed in the experimental tests.

6. Discussion of results

The average shear stress, S_s, in accordance with the ASTM E519 [26], is obtained by S_s = 0.707 P/An, where P is the applied load and An is the net area of the section of the panels (96740 mm^2 for reference, and 147778 mm^2 for FRCM model). The average vertical and horizontal strains, ε_v and ε_h have been computed by dividing the average displacement along the compressive and tensile diagonals, respectively, by the gauge lengths in each case. The shear strain, γ, was finally computed as $|\varepsilon_v| + \varepsilon_h$. The obtained shear stress vs shear strain responses are presented in figure 5.

The shear stress, shear strain and modulus of shear stiffness that were obtained for the elastic branch and at peak are presented in table 3. The average peak shear stress deduced from the experimental tests and the one obtained from the numerical simulation differ less than 17% for the reference and less than 1% for the strengthened models. In the case of the modulus of shear stiffness the difference is of 14% and 20% when considering the elastic branch until 1/3 of the peak stress is reached. When considering the peak shear stress values, as recommended by ASTM E519 [26], the differences are respectively 12% and 32%.

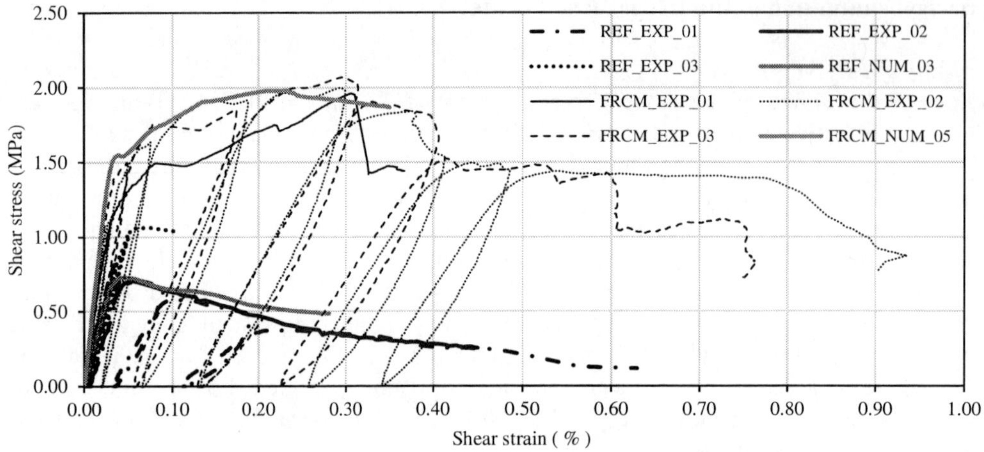

Figure 5. Shear stress vs shear strain response of experimental tests and numerical simulations.

Table 3. Average of the limit values obtained for the experimental and numerical responses.

Response range	Type of specimen	Shear Stress, s_s (MPa)		Shearing Strain, γ (%)		Modulus of shear stiffness, G (MPa)	
		Exp.	Num.	Exp.	Num.	Exp.	Num.
Elastic branch	Reference	0.185	0.186	0.007	0.008	2716	2325
	Strengthened	0.532	0.536	0.012	0.010	4458	5360
Peak	Reference	0.871	0.725	0.054	0.049	1696	1485
	Strengthened	2.002	1.978	0.299	0.231	651	856

The shear stress field at the interface elements between masonry/render and render/FRCM layers for peak load are presented in figure 6. The interface shear stresses are presented for local direction l1. For direction l2 an anti-symmetric shear stress field was obtained. The normal stresses for direction l3, perpendicular to the plane XY, where null in all cases. The shear stresses obtained in the elements for both interfaces are lower than the value of the cohesion adopted for the interface model, see table 2, which demonstrates that the interfacial behaviour remained elastic throughout the entire loading sequence. Nonetheless, for the interface masonry/render it is possible to identify values of shear stress higher than for render/FRCM interface, which indicate that delamination is more likely between the masonry and the render layers. The interfacial stresses do not reach sufficiently high values to produce the interfacial delamination. In contrast, during the experimental testing the delamination was identified. The delamination observed experimentally may have result from possible initial stress states at the interfaces due to shrinkage, which were not were not considered in the numerical model.

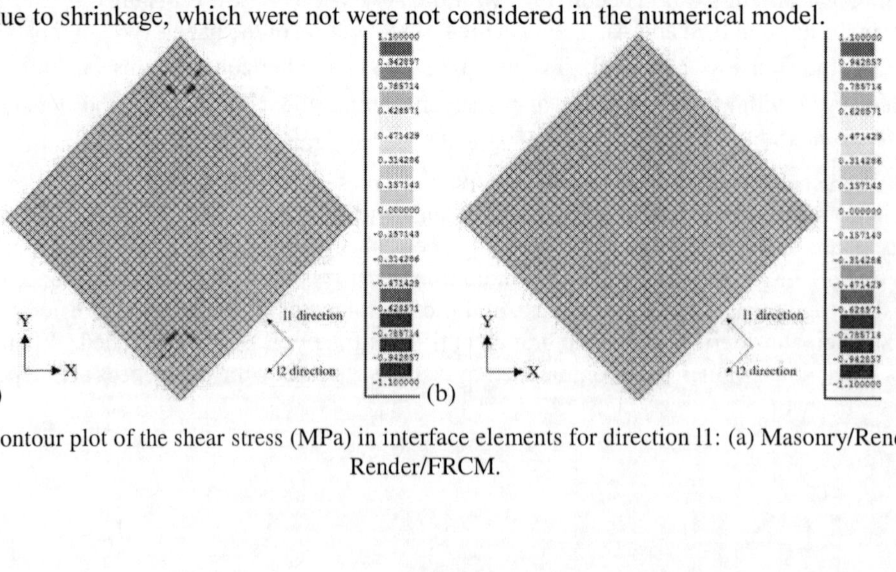

Figure 6. Contour plot of the shear stress (MPa) in interface elements for direction l1: (a) Masonry/Render; (b) Render/FRCM.

7. Conclusions

In summary, the numerical simulations of the reference and the strengthened masonry models have produced reasonable and accurate results when compared to the experimental responses, without the need to resource to excessively refined or computationally costly models.

The numerical load-displacement responses have reproduced quite well the experimental ones when the horizontal displacement is regarded. In the case of the experimental load-displacement responses regarding the vertical displacement, the numerical results have shown a general tendency to underestimate the displacements observed experimentally. This is most likely related to the fact that all materials have been assumed to behave linear elastically in compression. However, this effect was not significant and the computational cost was kept lower than if nonlinear behaviour in compression was additionally considered.

The overall shear stress vs shear strain responses obtained from the numerical models have accurately simulated the experimental responses observed, which is particularly interesting for supporting simplified design procedures of the masonry strengthening systems based on the application of FRCM overlays, such as the procedures presented in ACI 549.4R-13 [7].

Acknowledgements
The first author wishes to acknowledge the grant SFRH/BD/115090/2016, provided by FCT.

References

[1] Papanicolaou C G, Triantafillou T C, Karlos K and Papathanasiou M 2007 Textile-reinforced mortar (TRM) versus FRP as strengthening material of URM walls : in-plane cyclic loading *Mater. Struct.* **40** 1081–97

[2] Corradi M, Borri A, Castori G and Sisti R 2014 Shear strengthening of wall panels through jacketing with cement mortar reinforced by GFRP grids *Compos. Part B Eng.* **64** 33–42

[3] Milosevic J, Gago A S, Lopes M and Bento R 2013 Experimental assessment of shear strength parameters on rubble stone masonry specimens *Constr. Build. Mater.* **47** 1372–80

[4] Brignola A, Frumento S, Lagomarsino S and Podestà S 2008 Identification of Shear Parameters of Masonry Panels Through the In-Situ Diagonal Compression Test *Int. J. Archit. Herit.* **3** 52–73

[5] Dehghani A, Fischer G and Nateghi Alahi F 2013 Strengthening masonry infill panels using engineered cementitious composites *Mater. Struct.* 1–20

[6] Parisi F, Iovinella I, Balsamo A, Augenti N and Prota A 2013 In-plane behaviour of tuff masonry strengthened with inorganic matrix – grid composites *Compos. Part B* **45** 1657–66

[7] ACI Committee 549 2013 ACI 549.4R-13: Guide to Design and Construction of Externally Bonded Fabric- Reinforced Cementitious Matrix (FRCM) Systems for Repair and Strengthening Concrete and Masonry Structures (Farmington Hills, Michigan)

[8] Almeida J A P P, Bordigoni D, Pereira E N B, Barros J A O and Aprile A 2016 Assessment of the properties to characterize the interface between clay brick substrate and strengthening mortar *Constr. Build. Mater.* **103** 47–66

[9] Almeida J A P P, Pereira E B and Barros J A O 2015 Assessment of overlay masonry strengthening system under in-plane monotonic and cyclic loading using the diagonal tensile test *Constr. Build. Mater.* **94** 851–65

[10] Santa-Maria H, Duarte G and Garib A 2004 Experimental Investigation of masonry panels externally strengthened with CFRPlaminates and fabric subjected to in-plane shear load *13th World Conference on Earthquake Engineering* (Vancouver, B.C., Canada)

[11] Lourenço P B 1996 Computational strategies for masonry structures (Delft University)

[12] Basili M, Marcari G and Vestroni F 2016 Nonlinear analysis of masonry panels strengthened with textile reinforced mortar *Eng. Struct.* **113** 245–58

[13] Frazão C, Barros J and Gonçalves D Durability of cementitious mortars reinforced with synthetic fibers (in Portuguese) *IIº Encontro Luso-Brasileiro de Degradaçãode Estruturas de Betão, Lisboa, LNEC, 27-29 de Setembro de 2016*

[14] S&P Clever Reinforcement Company AG 2013 S&P ARMO-System FRCM Design Guidelines - Fiber Reinforced Cementitious Matrix (VER12.03.2013/IND) (Seween)

[15] fib - federation internationale du Beton 2013 *fib Model Code for Concrete Structures 2010* (Wiley-VCH Verlag GmbH & Co. KGaA)

[16] Comité Européen de Normalisation 2005 EN 1996-1-1:1995 Eurocode 6 - Design of masonry structures - Part 1-1: General rules for reinforced and unreinforced masonry structures

[17] Azevedo A F M, Barros J A O, Sena Cruz J M and Ventura Gouveia 2003 Software in teaching and structures design (in Portuguese) *Proceedings of III Congresso Luso`_Moçambicano de Engenharia* (Mozambique) pp 81–92

[18] Ventura-Gouveia A, Barros J A O and Azevedo A F M 2011 Crack constitutive model for the prediction of punching failure modes of fiber reinforced concrete laminar structures *Comput. Concr.* **8** 735–55

[19] Barros J A O, Baghi H, Dias S J E and Ventura-Gouveia A 2013 A FEM-based model to predict the behaviour of RC beams shear strengthened according to the NSM technique *Eng. Struct.* **56** 1192–206

[20] Sena-Cruz J M 2004 Strengthening of concrete structures with near-surface mounted CFRP laminate strips (University of Minho)

[21] Edalat-Behbahani A, Barros J A O and Ventura-Gouveia A 2016 Application of plastic-damage multidirectional fixed smeared crack model in analysis of RC structures *Eng. Struct.* **125** 374–91

[22] Cornelissen H A W, Hordijk D A and Reinhardt H W 1986 Experimental determination of crack softening characteristics of normalweight and lightweight concrete *HERON, Fract. Mech. Struct. Asp. Concr.* **31** 45–56

[23] Coelho M R F 2017 *Bond behaviour of NSM FRP systems in concrete* (University of Minho)

[24] Angelillo M, Lourenço P B and Milani G 2014 Masonry behaviour and modelling *Mechanics of Masonry Structures* ed M Angelillo (Springer Vienna)

[25] RILEM TCS 1985 Determination of the fracture energy of mortar and concrete by means of three-point bend tests on notched beams *Mater. Struct.* **18** 287–90

[26] American Society for Testing and Materials 2002 E 519-02 Standard test method for diagonal tension (shear) in masonry assemblages

IOP Conf. Series: Materials Science and Engineering **596** (2019) 012027 doi:10.1088/1757-899X/596/1/012027

Behaviour of recycled aggregates RC columns strengthened with CFRP under uniaxial compressive loadings

A R Khan and S Fareed

Department of Civil Engineering, NED University of Engineering & Technology

E-mail: sfareed@neduet.edu.pk

Abstract. The recent development in the construction industry has significantly influenced the socio-economic growth and the built environment of the mega cities around the globe. However, these developments have also resulted in millions of tonnes of construction demolition waste, which is mainly dumped, in the already populated landfills. As a result, the construction waste is not only causing a threat to the natural resources but also to the already deteriorating environment. Over recent past, studies have been conducted which investigated the use of concrete based construction waste as recycled coarse aggregates in preparing concrete for structural purposes. Based on these studies, in general it was found that the concrete compressive strength decreases with the increase in the percentage of recycled aggregates. Furthermore, studies have also been carried out in which use of Carbon Fiber Polymers wrapped along the outer surfaces of the structural elements resulted in significant increase of load carrying capacity. Therefore, in this study, investigation aims at the behaviour of reinforced concrete columns made with recycled aggregates and wrapped with Carbon Fiber Reinforced Polymers under uniaxial static compressive loading. For this purpose, 12 rectangular specimens having cross section dimensions of 150mm x 300mm and height of 600mm reinforced both longitudinally and transversely were used. The specimens were made with three different mixes of concrete having 0%, 30% and 50% of recycled aggregates. It was found that the recycled aggregates reinforced concrete specimens wrapped with CFRP exhibited higher compressive loads as compared to their corresponding specimens made with natural aggregates.

1. Introduction

The development of the infrastructure according to the modern era is of prime importance to ensure smooth economic growth of any developing nation. However, these developments may outcome problems which not only threat the environment as well as natural resources. Construction demolition waste which is one of the negative outcomes of the ongoing urban developing activities around the globe has become the major concern. The major part of the construction waste goes to solid waste landfills which in turn are requiring more sites and ultimately influencing the already polluted ambience.

Over past recent decades studies have been conducted which focuses on the use of the concrete demolition waste as recycled coarse aggregates in preparing concrete for structural purposes. Zhou and Chen [1] studied the mechanical properties of concrete made with different types of recycled coarse aggregates under static loading. For that purpose, Natural Aggregate Concrete (NAC) was made with coarse aggregates extracted from crushed rock and pebbles. For comparison purposes, the Recycled Aggregates Concrete (RAC) was also made with parent concrete having same type of crushed rock and pebbles aggregates as used for the NAC. The specimens were made with concrete having different percentages of recycled crushed rock aggregates (RCRA) and recycled pebbles aggregates (RPA). The percentage of both types of recycled aggregates (RCRA and RPA) used for making concrete varied between of 0% to 100% with water to cement ratio of 0.49 for concrete made with natural and recycled crushed rock aggregates and 0.47 for concrete made with natural and recycled pebble aggregates. In general, it was observed that the relationship of concrete compressive strength with respect to replacement percentage (RP) is scattered and no definite relationship can be formed between concrete compressive strength and percentage of recycled aggregate used irrespective of type of recycled aggregates used for making concrete. Furthermore, it was observed that when using 100% RP of recycled pebbles concrete (RPC), the compressive strength shows slight improvement when compared to its counterpart concrete having 0% recycled aggregates.

Etxeberria and Mari *et al.* [2] carried out an investigation to study the influence of amount of RCA and its production process on the properties of concrete made with recycled aggregates (RAC). For this purpose, four different concrete mixes were used to study the behavior of concrete made with RCA under uniaxial static loading. In general, it was observed that with the increase in the curing period the concrete compressive strength also increases irrespective of the production type. It was also observed that for the case of specimens made with 25% (RC25) and 50% (RC50) of RCA, the concrete compressive strength decreases as compared to natural aggregate concrete irrespective of curing period.

Based on the experimental studies, it has been generally observed that the concrete having replacement percentage of recycled aggregates above 30% showed decline in its compressive strength. Therefore, this study focuses on investigating the use of the concrete made with recycled aggregates in the reinforced concrete columns. Furthermore, study also aims at rectifying the decline compressive strength trend of the recycled aggregate concrete using Carbon Fiber Polymers (CFRP) wraps as studies [3-6] have also been carried out in which use of Carbon Fiber Polymers wrapped along the outer surfaces of the structural elements resulted in significant increase of load carrying capacity. For this purpose, 12 rectangular specimens having cross section dimensions of 150mm x 300mm and height of 600mm reinforced both longitudinally and transversely were tested under static uniaxial compressive loads. The specimens with and without wraps were made with three different mixes of concrete having 0%, 30% and 50% of recycled aggregates.

2. Experimental Programme

In order to investigate the behavior of reinforced concrete columns made with recycled aggregates wrapped with and without CFRP, 12 specimens were tested under static uniaxial compressive loads. The specimens were made with concrete having 0%, 30% and 50% of recycled aggregates. Four specimens (two each with and without wraps) were casted for each percentage of recycled aggregates and the results were averaged out to investigate its behaviour under uniaxial compressive loading.

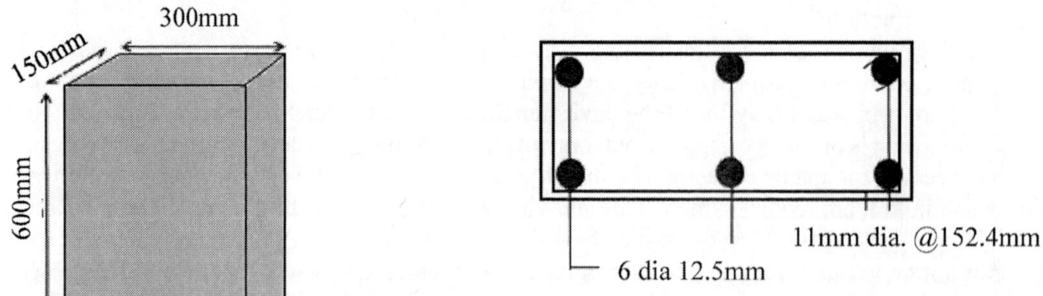

Figure 1. (a) Specimen cross sectional dimensions and (b) reinforcement details.

2.1. Specimen details

The rectangular shape specimens having cross section dimensions of 150mm x 300mm and height of 600mm as shown in Figure 1(a) were used. The specimens were reinforced with six longitudinal bars of 12.5mm diameter with longitudinal steel ratio of 1%. The shear reinforcement was provided in the form of stirrups having diameter of 11mm spaced at 152.4 mm centre to centre therefore having a concrete cover of 25.4mm as shown in Figure 1(b).

2.2. Material Properties

The specimens were prepared with three different mixes of concrete having a constant mix and water cement ratios of 1:1.24:2.60 and 0.43 respectively. Four out of twelve specimens used in this investigation were prepared for each mix having 0% (NAC), 30% (RAC-30) and 50% (RAC-50) of recycled aggregates. For preparing concrete, the total weight of coarse aggregates (both natural and recycled) consist of 60% of aggregates retained on 9.5mm and 40% retained on 12.7mm sieves. The details of constituent used for each mix for preparing specimens are provided in Table 1.

Table 1. Details of mix proportions used for preparing specimens.

	OPC cement (kg/m^3)	**Fine Aggregates (kg/m^3)**	**Natural Coarse Aggregates (kg/m^3)**	**Recycled Coarse Aggregates (kg/m^3)**	**water (kg/m^3)**
NAC	454.01	567.51	1183.26	0	195.22
RAC-30	454.01	536.34	782.79	335.48	195.22
RAC-50	454.01	517.40	539.39	539.39	195.22

3. Results and discussions

3.1. Peak load exhibited

Figure 2 shows the peak loads exhibited by specimens made with different percentages of recycled aggregate concrete (RAC) wrapped with and without CFRP under uniaxial static compressive loading. In general, it was observed that for the case of specimens without CFRP wrapping the peak load exhibited by specimens decreases with the increase in the percentage of the recycled aggregates. However, for the case of specimens with CFRP wrapping no definite relationship can be formed as initially specimens having up to 30% of recycled aggregates the peak load increases, which then decreases, for the specimens made with 50% of recycled aggregates. It was also observed that for the case of specimens made with 0% recycled aggregates, the CFRP wrapping does not influence the peak load response as similar peak load was observed for both wrapped and unwrapped specimens (see Figure 2).

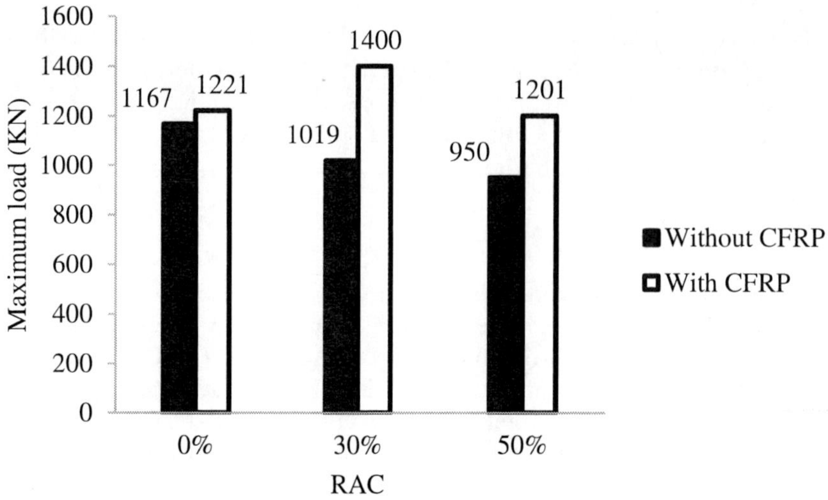

Figure 2. Peak load exhibited by specimens made with different percentages of recycled aggregate concrete (RAC) wrapped with and without CFRP under uniaxial static compressive loading.

When comparing the peak loads exhibited by specimens with and without CFRP wrapping, it was observed that the specimens with CFRP exhibited higher compressive loads as compare to its counterpart for the case of specimens without CFRP wrapping irrespective of the percentage of aggregates used. Furthermore, it was also observed that CFRP significantly increased the peak load capacity of the specimens made with recycled aggregates as higher load was recorded for the case of specimen made with 30% recycled aggregates and similar load was also recorded for the case of 50% specimens when compared with specimens made with 0% recycled aggregates.

Table 2 gives the increase in load carrying capacity of recycled aggregate reinforced concrete specimens with and without CFRP wrapping. It can be seen from Table 2 that use of CFRP with recycled aggregate reinforced concrete columns proved to be beneficial as the decline strength effect of compressive strength with increase in the percentage of the recycled aggregates was overcome as significant increase in the load carrying capacity was observed.

Table 2. Increase in load carrying capacity of recycled aggregate reinforced concrete specimens with and without CFRP wrapping.

Specimens	**Maximum load (kN)**		**Increase in load carrying capacity (%)**
	Without CFRP	**With CFRP**	
NAC	1167	1221	4.42
RAC-30	1019	1400	27.21
RAC-50	950	1201	20.89

3.2. Failure pattern

Figure 3 shows the damage caused to the specimens made with different percentages of recycled aggregate concrete (RAC) wrapped with and without CFRP under uniaxial compressive loading. For the case of specimens without CFRP wrapping and made with 0%, 30% and 50% of recycled aggregates, it was observed that the failure initiated due to the formations of crack at the outer edges of the cross section which extends along the whole length. For the case of unwrapped specimen with 50% recycled aggregates, the cracks initiated at the mid-height of the specimens then extending along the top and bottom, similar to the behaviour of cylindrical specimen under uniaxial compression test. It was also observed that the damage caused to the specimens without CFRP wrapping only caused damage to the concrete in the cover region as reinforcement was not exposed.

For the case of specimens having CFRP wrapping, in general, it was observed that the failure occur due to crushing of the concrete cover and CFRP rupture irrespective of the percentage of recycled aggregate used for preparing the specimen. It was also observed that except for the specimen made with 0% recycled aggregates, the damage was only limited to the concrete cover as reinforcement were not exposed. Furthermore, the use of CFRP, also limited the failure to very localized area as compared to the case of specimens without CFRP wrapping.

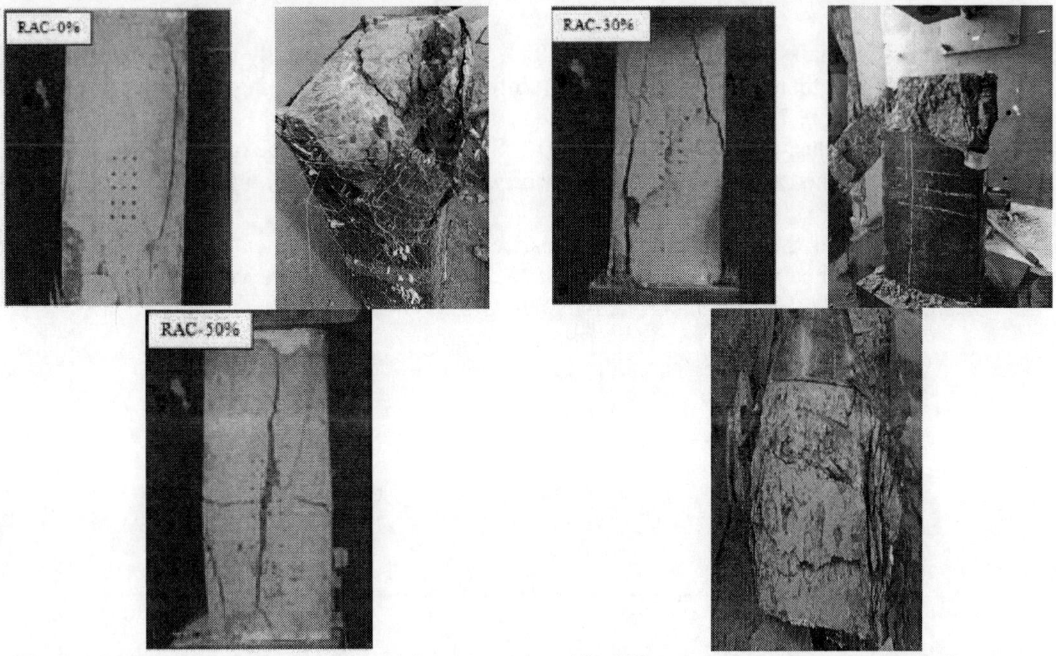

Figure 3. Damage caused to the specimens made with different percentages of recycled aggregate concrete (RAC) wrapped with and without CFRP under uniaxial compressive loading.

4. Conclusions

Based on the experimental investigation carried out herein which investigates the behaviour of recycled aggregate reinforced concrete columns under uniaxial static compressive loadings following conclusions can be deduced:

- The peak load exhibited by specimens without CFRP wrapping decreases with the increase in the percentage of the recycled aggregates.
- The specimens with CFRP exhibited higher compressive loads as compared to its counterpart for the case of specimens without CFRP wrapping irrespective of the percentage of aggregates used.
- The use of CFRP, also limited the failure to very localized area as compared to the case of specimens without CFRP wrapping.

5. Acknowledgement
The authors are indebted to the undergraduate students Tabish Jamil, Maha Amber, Ayesha Karim, Zarak Khan, Mohammad Junaid Saleem, Irfan Rustami Asrabadi and Department of Civil Engineering, NED University of Engineering & Technology, Karachi, Pakistan and Pakistan Science Foundation (PSF) in the pursuit of this work.

References
[1] Zhou, C. & Z.J.C. Chen, 2017, Mechanical properties of recycled concrete made with different types of coarse aggregate. *B. Materials*. **134**: p. 497-506.

[2] Etxeberria, M., Vázquez, E., Marí, A., Barra M., 2007, Influence of amount of recycled coarse aggregates and production process on properties of recycled aggregate concrete. *Cement and Concrete Research* **37**(5): p. 735-742

[3] Abdel-Hay, A. S. , 2014, Partial strengthening of RC square columns using CFRP. *HBRC Journal,* **10,** 279-286.

[4] Chaallal, O., Shahawy, M. & Hassan, M., 2003, Performance of axially loaded short rectangular columns strengthened with carbon fiber-reinforced polymer wrapping. *Journal of Composites for Construction,* **7,** 200-208.

[5] Chikh, N., Gahmous, M. & Benzaid, R., 2012, Structural performance of high strength concrete columns confined with CFRP sheets. *Proceedings of the World Congress on Engineering*, 4-6.

[6] Khan, A. & Fareed, S., 2013, Behaviour of concentrically and eccentrically loaded short concrete columns confined with CFRP wraps. *SMAR 2013 Second Conference on Smart Monitoring, Assessment and Rehabilitation of Civil Structures*, 9-11.

IOP Conf. Series: Materials Science and Engineering **596** (2019) 012028 doi:10.1088/1757-899X/596/1/012028

Utilisation of date palm fibres in cement-based composites: A feasibility study

Wafa Abdelmajeed Labib
College of Engineering, Prince Sultan University, P.O.Box No. 66833 Rafha Street, Riyadh 11586 Saudi Arabia

E-mail: wlabib@psu.edu.sa

Abstract. The renewed interest in the preservation of non-renewable resources has made the use of plant-based natural fibres as an environmentally friendly replacement to their synthetic counterparts in construction materials a necessity. This is due to the ability of plant-based natural fibres to enhance the mechanical properties of building components, their relatively low cost, and their derivation from a renewable resource. In this paper, the feasibility of using date palm fibres in cement composites as a replacement for synthetic fibres is reported. This study has reported on the production, types, and comparison between the chemical and mechanical properties of date palm fibres and synthetic fibres.

1. Introduction

Three main types of natural fibre are derived and based on different materials, such as animal, mineral, and plant fibres that can be used for cement-based composites [1]. Animal-based fibres constitute proteins, such as wool, silk, and hair fibres. Similarly, fibres derived from minerals comprise wollastonite, palygorskite, and asbestos. Lastly, fibres based on plants include flax, cotton, jute, hemp, sisal, palm, bagasse, ramie, and processed speciality fibres derived from wood.

Date palm fibres are a natural alternative reinforcement that is considered a principal agricultural product in the arid and semi-arid regions of the world, especially the Middle Eastern and North African regions. More than 120 million date palm trees worldwide yield several million tons of dates per year, apart from secondary products including palm midribs, leaves, stems, fronds, and coir. The majority are in the Kingdom of Saudi Arabia (KSA), Egypt, Iraq, Iran, Algeria, Morocco, Tunisia, the United Arab Emirates, Pakistan, India, Australia, and the United States. The KSA is a land famous for date production, and its palm tree is regarded as one of the oldest trees in the world. A recent estimate shows that the kingdom has more than 23 million date palm trees, which produce about 1 million tons of dates per year. Date palm trees produce a huge amount of agricultural waste in the form of dry leaves, stems, pits, seeds, etc. A typical date tree can generate as much as 20 kilograms of dry leaves per annum, as the date pits account for almost 10% of the date fruits [2]. As indicated in [3], the KSA alone generates more than 200,000 tons of date palm biomass each year. Biomass has generally been described as all organic matters or compounds either produced from crops, forestry, or marine life. In addition to the availability of the date palms in the KSA in huge quantities, date palms have the advantage of being considered a renewable natural resource because they can be replaced in a relatively short period [4]. Usually, date palm waste is burned on farms or disposed of in landfills, which causes environmental pollution in date-producing nations, and can cause environmental

IOP Conf. Series: Materials Science and Engineering **596** (2019) 012028 doi:10.1088/1757-899X/596/1/012028

hazards, such as fire, bait for insects, and diseases [2,3,5-7]. Normally, the date palm waste is left outdoors, which could provide a good environment for certain pests, such as rodents and insects.

Therefore, from environmental and economic points of view, this research aims to study the feasibility of using date palm waste for producing a fibre cement-based composite as an alternative to synthetic fibres. The paper compares the chemical and mechanical properties, eco-friendliness, and cost-effectiveness characteristics of date palm natural fibres and polypropylene synthetic fibres.

2. Types of Date Palm Fibres

Most common types of natural-based plant fibres are stem, leave, and seed that is further categorized like seed-fibres, bast-fibres, and leave-fibres [1]. The extraction of the blast-fibre takes place to form the plant stems outer bark. Figure 1 has shown that the tree stem in the date palm is surrounded by the mesh made from single fibre. Generally, the fibre is used for the creation of different diameters from the natural crossed fibres woven mat. Based on the traditional practices, the annual removal of the mat from the trees takes place for the removal of the ropes. The general properties of bast-fibres are represented as the high tensile strength in the bundles made of long fibre.

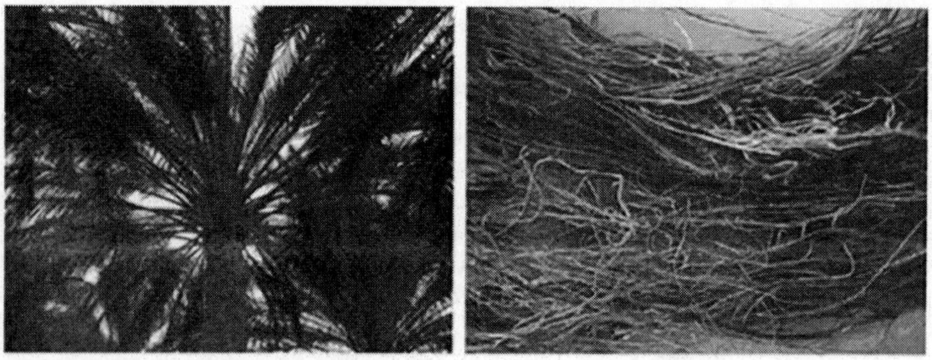

(a) (b)

Figure 1. (a) tree of date palm (b) Fibres of date palm encircling stems.

3. Fibre Production

The estimated production volumes of the commonly used plant-based on natural fibres have been shown in Figure 2 [8, 9]. Figure 1 has shown that date palm possesses the largest yield of more than 4 million tons per year, followed by Bamboo and Jute fibres that have increased yield every year like the Herrera-Franco and Valadez-González [10] study results. Considering the high production yield of date palm fibres, these serve as enough sources to be utilized in the construction industry.

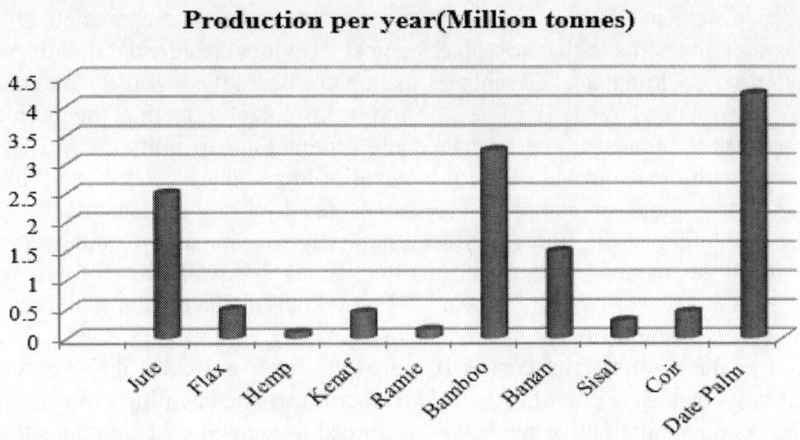

Figure 2. Average volume of the globally produced natural fibers (million metric tons/year).

Date palm is reflected upon as a renewable natural resource, which production can take place in a short-run [11]. The time required for the date palm for bearing fruit varies from four to eight years when it is replaced, whereas the time for the commercial harvesting is said to range from seven to ten years. The presence of 120 million trees of the date palm is expected to increase its fibre production, where the input from the Arab countries is expected to be significant i.e. more than two-thirds (Figure 3) [8]. An average of 20 kilograms of waste can be produced by a date palm tree every year. Generally, the waste of the date palm is either disposed of in the landfills and sometimes by burning into farms, which can lead to detrimental consequences on the bait for insects, fire, and other diseases [12-15]. Therefore, it is recommended to utilize such wastes in producing fibre composite concrete to save the environment from pollution.

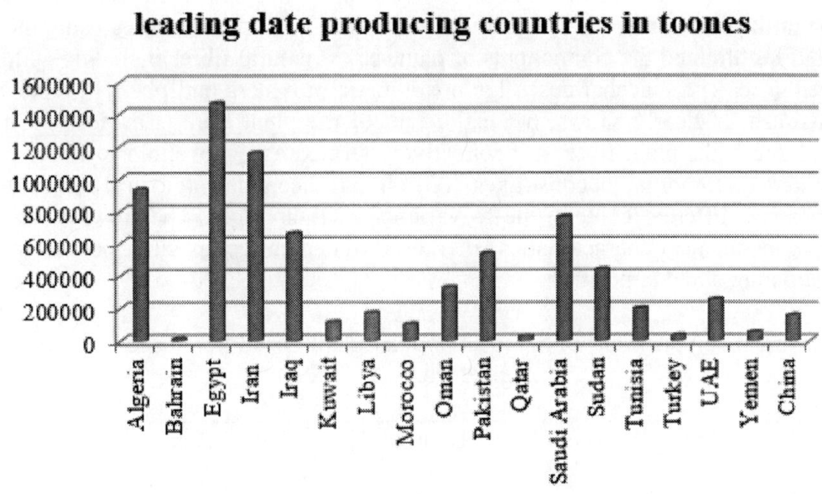

Figure 3. Countries with significant date-production.

It has been reported by a UN organization (i.e. FAO) that majority of these trees exist in Egypt, Iran, Algeria, KSA, Iraq, Morocco, Tunisia, United Arab Emirates, Pakistan, India, Australia, and United States. The reason behind such huge production is the increased positive acceptance of the tree socially, along with the support provisioned at governmental level. This extent of government support is provided to the date palm trees because they produce valuable food for human beings. These trees are also used as an ornament for the garden and are recognized as well as appreciated for the increased health as well as social benefits to the individual in most individual residential buildings in KSA.

In addition to their environmental advantages, data palm fibres possess the economic advantage, as the cost of raw materials is recognized as the major stimulant affecting the performance of the industry. It is because the industries performance and production capability are heavily dependent on the cost associated with the acquiring of the raw material impacting its cost-benefit plans. The factor that adds value to the goods or company property is the low cost which directs the construction industry for the adaptation of the fibres produced naturally as compared to other synthetic material. The production of huge amounts of date palm waste ensures the availability of date palm fibres at a relatively low cost. Ghosh et al [16] and Lewin [17] stated that the date palm fibres are considered the cheapest among several natural fibre's types.

The variation in the raw material cost is based on its type; however, the cheapest of the fibres among the various types is date palm fibres. This further draws the attention towards the utilization of the date palm fibres as an alternative for being reinforced in cement based composites. A significant energy quantum is required for the synthetic fibre (also known as polypropylene) production in the manufacturing process to be produced from non-renewable recourses, unlike the production of date palm fibres.

Moreover, it generates pollution and health risk when inhaled as it requires the use of compounds of phenol such as ultraviolet stabilizer (i.e. amines and antioxidants). It acts like flame retardant, which characteristics vary from the sustainable materials [18]. Consequently, the integration of the activities pertaining to the synthetic material recycles and production causes pollution which causes a negative influence on the environment [19]. One of the natural alternative reinforcements to the synthetic complements is date palm waste which can be used for producing fibre cement composites. These composites could provide a good mechanical performance of cement composites.

4. Fibre Chemical Structure and Mechanical Characteristics

Fibre based on natural plant induces in such properties both mechanical as well as a chemical which allows it to be utilized in numerous applications within the building industries using these fibres [20-22]. Table 1 has highlighted the components of plant-based natural fibres including cellulose (glucose units), hemicelluloses (polysaccharides), lignin (aromatic polymers and phenyl propane monomers), extractives, and ash. It clearly shows that major part of the plant fibre is made from hemicellulose, cellulose, and lignin. The plant fibres are collectively termed as lignocellulosic fibres because of the presence of three major organic constitutes [23]. It has been depicted in Table 1 that the three constituents possess different content due to variance in their types. The fibres derived from plant exhibit different mechanical characteristics when used in cement composites, because of their variety in molecular structure and composition.

Table 1. Chemical compositions of the most popular plant-based fibre used in cement composites (adopted from [8, 24-29]).

Fibre	Cellulose	Hemicellulose	Lignin	Extractives	Ash	Reference
Abaca	62.5	21	12	-	-	Jawaid & Abdul Khalil [24]
Flax	72.5	14.5	2.5	-	-	Jawaid & Abdul Khalil [24]

Jute	33.4	22.7	28	-	--	Ramakrishna & Sundararajan, [25]
Hemp	81	20	4	-	-	Jawaid & Abdul Khalil, [24]
Kenaf	53.5	21	17	-	-	Jawaid & Abdul Khalil, [24]
Ramie	72	14	0.8	-	-	Jawaid & Abdul Khalil [24]
Bamboo	34.5	20.6	26	-	-	Jawaid & Abdul Khalil [24]
Hibiscus	28	25	22.7	-	-	Ramakrishna & Sundararajan, [25]
Banana- trunk	31.48	14.98	15.07	4.46	8.65	Arsene et al [26]
Banana	60-65	6-8	5-10	-	4.7	Reddy & Yang, [27]
Sisal	38.2	26	26	-	-	Ramakrishna & Sundararajan, [25]
Sisal	73.11	13.33	11	1.33	.33	Arsene et al, [26]
Sisal	43-88	10-11	3/8-9.9	-	-	[Li, 2000; Agopyan & John,[28]
Banana	25.65	17.04	24.84	9,84	7.02	[29]
Pineapple	70-82	18	5-12	-	0.7-0.9	Reddy & Yang, [27]
Corn	38-40	28	7-21	-	3.67	Reddy & Yang, [27]
Date Palm	46	-	20	-	-	AL-Oqla & Sapuan [8]

Plant-based natural fibre mechanical characteristics are significantly affected due to their chemical composition. The induced cellulose is recognized as the most strengthening and strongest organic compound supplementing the stiffness, strength, and stability in the fibre [30]. The study has demonstrated positive association between the cellulose and the tensile strength, Young's modulus and specific strength [30]. Whereas, the sturdy shows that there is a correlation between the hemicellulose with the moisture gain, specific strength, and Young's modulus. The negative correlation is observed between the microfibril angle, tensile strength, density, and natural fibre strain failure. The moisture gain is found to have a sturdy positive correlation with the Lignin. Subsequently, specific strength, the strength of the tensile, density, and specific Young's modulus has a negative correlation with lignin [30].

The variation in mechanical properties of lignocellulosic fibres is a great challenge for designing of components in a specific industry as the engineers are adapted to the reproducible and precise properties of synthetic fibres. In the hydrophobic polymer matrix, low moisture resistance and incompatibility results from the lignocellulosic fibres hydrophilic nature. As a result of the incompatibility, the deficiency occurs in the fibre interface which mitigates the mechanical characteristics present in the bio composites. It clearly depicts the significance of matrix bonding with fibres essential for improving mechanical strength. The existence of the hydroxyl group induced in hydrophilicity in the fibres which form poor interfacial bonding of hydrophobic polymer matrices with hydrophilic fibres derived naturally.

The content of the three polymers is highly variable between the plant fibres. Figure 3 has shown the content of the three main components in the majority of the common natural fibres. Figure 4 exhibits that cellulose content of date palm fibres is greater than the majority of the common plant-based fibre naturally, including bamboo, hibiscus, sisal, and jute. Therefore, their higher strength of the tensile and Young's modulus is observed, which is considered as an added value for date palm fibres. The natural fibres tensile strength and Young's modulus increase with the increase in cellulose content. The content increase in the cellulose impacts the alignment of fibrils, resulting in a decline of angles of micro-fibril as reported by Li and Ye [31].

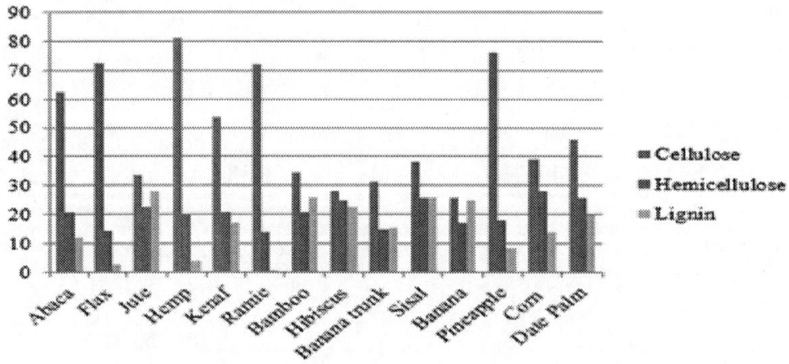

Figure 4. Natural fibre three main components content.

The key mechanical properties for selecting suitable reinforcing fibres in cement composites are two named as tensile strength and Young modulus. Figures 5 and 6 have shown that date palm fibres have an acceptable degree of tensile strength (300 MPa) and Young's modulus (9 GPa). The date palm derived fibre holds low position based on the Young modulus and strength of the tensile occupy a low position, as compared to other fibre obtained naturally. The date palm fibre strength of tensile (32 MPa) and Young modulus are extremely high as a synthetic counterpart, as compared to polypropylene. This increases its utilization prospects as reinforcement in the cement composites.

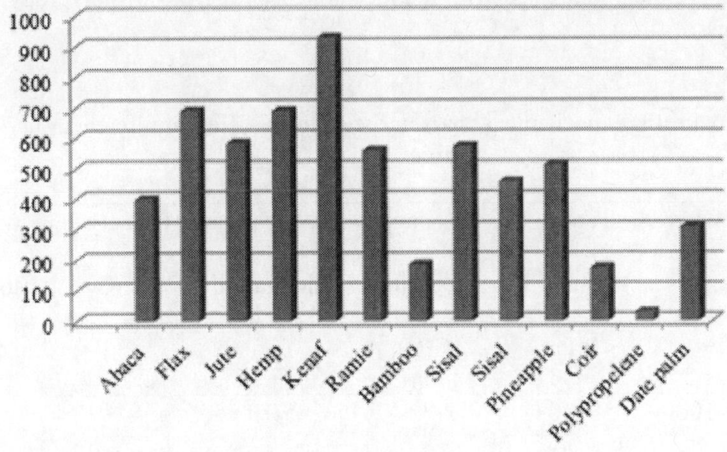

Figure 5. Tensile Strength comparison with other naturally derived fibres in construction industry.

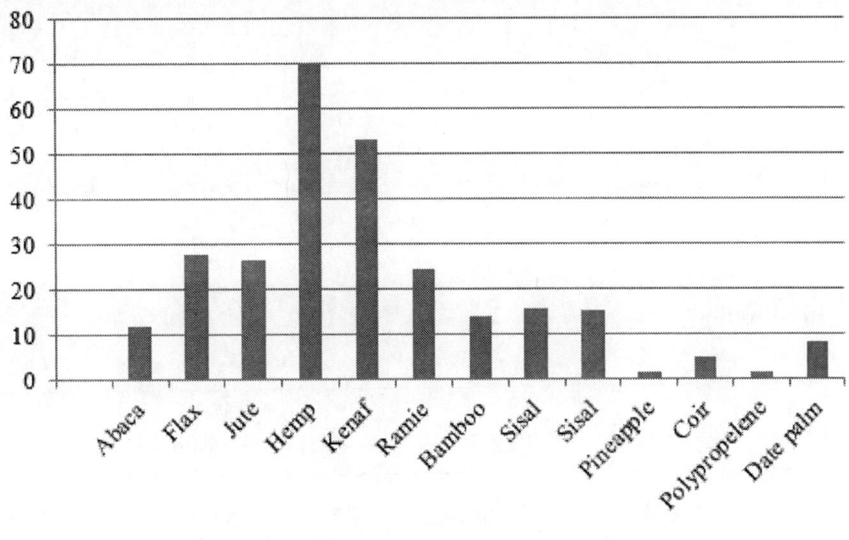

Figure 6. Comparison between the date palm's Young modulus with other types of natural fibres used in concrete industry.

The fibre mechanical characteristics are regarded as crucial parameters which improve the capability for selection of the appropriate fibres as reinforcement in cement composite. Ramakrishna and Sundararajan [25] and Reddy and Yang [27] reported that the determination of the concentration and its components is substantially impacted by the fibre type, its growth, position on the plant, age, dimensions, extraction, and ways adopted for its processing. Therefore, there is a difference in the mechanical characteristics of the same type of fibre obtained through various sources. It has been illustrated in Table 2 that the fibres of the date palm comprising of the mechanical properties are acquired through different sources, which lead to different mechanical properties when used in cement-based composites. Similarly, Table 3 has highlighted the most observed mechanical characterizes of the natural fibres derived from the plant which are incorporated in the composites of the cement, which are then contrasted with the mechanical properties of polypropylene fibres as their synthetic counterpart.

Table 2. Date palm fibre mechanical characteristics from different references and polypropylene fibres [14, 32-36].

Length mm	Diameter (μm)	Tensile strength (MPa)	Young modulus (GPa)	Break Elongation (%)	Density g/cm3	Water absorption %	Origin
-	100–1000	58–203	2-7.5	5-10	-		Saudi Arabia
100	-	170+40	4.74+2	16+3	0.5-1.08	97-203	Algeria
60	-	240+30	5.00+2	12+2	0.5-1.08	97-203	Algeria
20	-	290+20	5.25+3	11+2	0.5-1.08	97-203	Algeria
20-250	100-1000	-	7	2-19	0.9-1.2	-	
-	-	107.36	2.57	4.5	1.08	75	Algeria
-	-	125-200	70	-	0.46	60-65	Iraq
		70-170				60-84	
Poly-propylene		21-40	0.5-2.4	50-800	0.85-0.95	0.01-0.035	

IOP Conf. Series: Materials Science and Engineering **596** (2019) 012028 doi:10.1088/1757-899X/596/1/012028

Table 3. Mechanical properties of most common plant-based natural fibres used in cement composites and polypropylene fibres.

Fibre source	Tensile strength MPa	Young modulus GPa	Elongation at break %	Density g/cm3
Abaca	400	12	3-10	1.5
Flax	345-1035	27.6	2.7-3.2	1.5
Jute	393-773	26.5	1.5-1.8	1.3
Hemp	690	70	1.6	1.48
Kenaf	930	53	1.6	-
Ramie	560	24.5	2.5	1.5
Bamboo	140-230	11-17	-	0.6-1.1
Sisal	511-635	9.4-22	2-2.5	1.5
Pineapple	400-627	1.44	14.5	0.8-1.6
Coir	175	4-6	30	1.2
Polypropelene	21-40	0.5-2.4	50-800	0.85-0.95

The determination of the adequate fibre for reinforcing in terms of cement composite is significantly dependent on the induced density in the natural fibre. The fibre low level of density emerges because of low composite weight. Elongation at break plays a significant role in selecting suitable reinforcing fibres in cement composites. Figure 7 has shown the highest value (13%) of the elongation is observed at the break in the date palm fibre and it occupies the third highest position after coir (29%), and pineapple (14%), as compared to other plant-based natural fibres. The comparison of it with polypropylene concerning break elongation, it is observed to be low showing value of 410% (Figure 8). However, the moderate value of elongation at break of date palm fibre is considered suitable for application in different industries from the mechanical perspective incorporating the concrete industry.

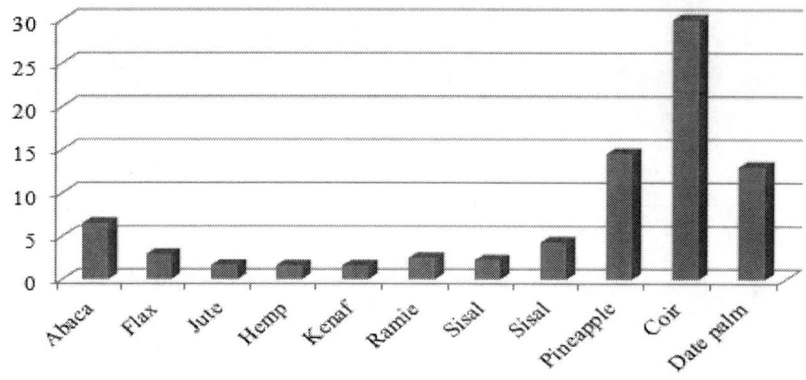

Figure 7. Concrete industry elongation at break comparison of date palm and other natural fibres.

Figure 8 has compared densities of various fibers derived from nature and polypropylene fibres using their average values as depicted in Table 3. The density value of the date palm fibre is found to be lowest amongst palm fibres that make it competitive to produce lightweight composites as compared with polypropylene fibres and other types of natural based fibres (Figure 9). This present study has recommended the utilization of the fibre obtained from the date palm to be used as reinforcement within the concrete industry.

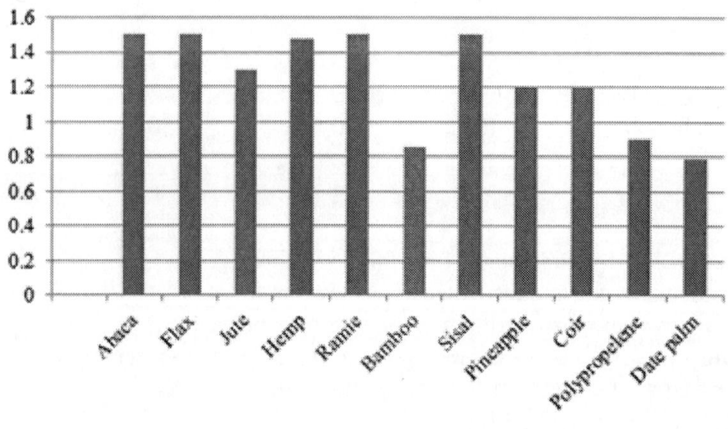

Figure 8. Comparison between densities of natural fibres and polypropylene fibres.

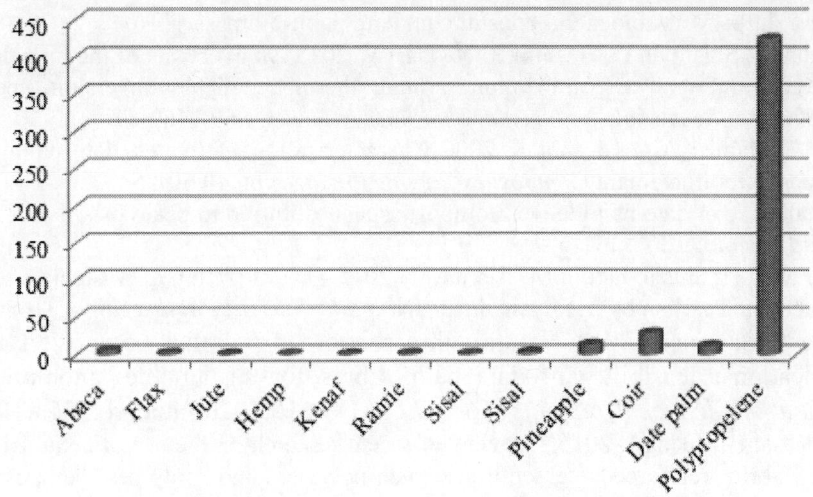

Figure 9. Comparison between elongation at break of different types of natural fibres used in concrete industry and polypropylene synthetic fibres.

5. Conclusions

The focus of the modern world is on the development of materials that are environmentally friendly and renewable. Plant-based fibres have become a common topic for discussion regarding the replacement of their synthetic counterparts. This is due to the ability of these fibres to reduce the weight of a building along with sustaining composite effective mechanical properties and feasible economic advantages. Cement-based composites still extensively use synthetic fibres obtained from non-renewable sources. However, several advantages are achieved with the utilisation of natural fibres.

This study discussed the potential of using date palm fibres to replace polypropylene synthetic fibres. Thus, the study reports different types, production, and chemical and mechanical properties of date palm fibres and polypropylene synthetic fibres. The increased inclination towards using the date palm fibre is due to the economic and environmental advantages providing usable structural properties, low manufacturing energy requirements, derivation from a renewable resource, and ease of disposal upon completion of their lifecycle. The chemical structure of the fibre and its associated characteristics assist in determining its resistance, degradability, and recyclability, which determine how its integration assists in the reinforcement of satisfactory cement composites. In addition, mechanical characteristics allow the assessment of its applications within the building industry. These properties make it a potential competitor for synthetic fibres, particularly polypropylene.

The comparison of the date palm fibre as a natural fibre to polypropylene as a synthetic fibre indicates that strength, eco-friendliness, and cost-effectiveness characteristics are greater in natural fibres. They can be obtained from renewable resources and help to achieve sustainable construction.

Acknowledgement
This research was supported by Prince Sultan University.

References

[1] Onuaguluchi O, Banthia N 2016 Plant-based natural fibre reinforced cement composites: A review *Cement and Concrete Composites* **68** 96-108

[2] Hussain A, Farooq A, Bassyouni M, Sait H, El-Wafa M, Hasan S, Ani F 2014 Pyrolysis Of Saudi Arabian Date Palm Waste: A Viable Option For Converting Waste Into Wealth *Life Science Journal* **11** 667-71

[3] Zafar S 2018 Date Palm as Biomass Resource *BioEnergy Consult - Powering Clean Energy Future (*http://www.bioenergyconsult.com/date-palm-biomass/)
[4] Awalludin M, Sulaiman O, Hashim R, Nadhari W 2015 An overview of the oil palm industry in Malaysia and its waste utilization through thermochemical conversion, specifically via liquefaction, *Renewable and Sustainable Energy Reviews* **50** 1469-84
[5] Alawar A, Hamed A, Al-Kaabi K 2009 Characterization of treated date palm tree fiber as composite reinforcement *Composites Part B: Engineering* **40** 601-6
[6] Ali Y 2008 Use of date palm leaves compost as a substitution to peatmoss *American Journal of Plant Physiology* **3**(4) 131-6.
[7] Joardder M U H, Shazib Uddin M, Islam M N 2015 The utilization of waste date seed as bio-oil and activated carbon by pyrolysis *Advanced in Machanical Engineering* **4** 316806
[8] AL-Oqla F, Sapuan S 2014 Natural fibre reinforced polymer composites in industrial applications: feasibility of date palm fibres for sustainable automotive industry. *Journal of Cleaner Production* **66** 347-54. Doi: 10.1016/j.jclepro.2013.10.050
[9] Yan L, Kasal B, Huang L 2016 A review of recent research on the use of cellulosic fibres, their fibre fabric reinforced cementitious, geo-polymer and polymer composites in civil engineering *Composites Part B: Engineering* **92** 94-132 Doi: 10.1016/j.compositesb.2016.02.002
[10] Herrera-Franco P J, Valadez-Gonzalez A 2004 Mechanical properties of continuous natural fibre-reinforced polymer composites *Composites Part A: applied science and manufacturing* **35** 339-45 Doi: 10.1016/j.compositesa.2003.09.012
[11] Awalludin M F, Sulaiman O, Hashim R, Nadhari W N 2015 An overview of the oil palm industry in Malaysia and its waste utilization through thermochemical conversion, specifically via liquefaction *Renewable and Sustainable Energy Reviews* **50** 1469–84 Doi: 10.1016/j.rser.2015.05.085
[12] Ali Y 2008 Use of date palm leaves compost as asubstitution to Peatmoss.American *Journal of Plant Physiology* **3** 131–6 Doi: 10.3923/ajpp.2008.131.136
[13] Joardder M U H, Uddin M S, Islam M N 2012 The utilization of waste date seed as bio-oil and activated carbon by pyrolysis process *Advances in Mechanical Engineering* **4** 316806
[14] Alawar A, Hamed A M, Al-Kaabi K. 2009 Characterization of treated date palm tree fibre as composite reinforcement *Composites Part B: Engineering* **40** 601–6
[15] Hussain A, Farooq A, Bassyouni M I, Sait H H, El-Wafa M A, Hasan S W, Ani F N 2014 Pyrolysisof Saudi Arabian date palmwaste: A viable option for converting waste into wealth *Life Science Journal* **11** 667-71
[16] FAOSTAT 2014 *Food and Agriculture Organization of the United Nations*. Accessible at: http://faostat.fao.org/. (Retrieved on 25th May 2017.)
[17] Ghosh S K, Nayak L K, Day A, Bhattacharyya S K 2007 Manufacture of particle board from date-palm leaves-A new technology product *Indian Journal of Agricultural Research* **41** 132
[18] Lewin M 2007 *Handbook of Fibre Chemistry* (Boca Raton: CRC/Taylor & Francis)
[19] Berge B 2007 The ecology of building materials. 2nd ed: Architectural Press, Elsevier
[20] Mahjoub R, Yatim JM, Sam ARM, Hashemi SH 2014 Tensile properties of kenaf fibre due to various conditions of chemical fibre surface modifications *Construction and Building Materials* **55** 103-13 Doi: 10.1016/j.conbuildmat.2014.01.036
[21] Kalia S, Kaith B S and Kaur I 2011 *Cellulose Fibers: Bio- and Nano-Polymer Composites Green Chemistry and Technology* (Berlin: Springer)
[22] Ochi S 2008 Mechanical properties of kenaf fibres and kenaf/PLA composites *Mechanics of materials* **40** 446-52 Doi: 10.1016/j.mechmat.2007.10.006
[23] Sapuan S M, Leenie A, Harimi M, Beng Y K 2006 Mechanical properties of woven banana fibre reinforced epoxy composites *Materials & design* **27** 689-93
[24] Jawaid M H P S, Khalil H A 2011 Cellulosic/synthetic fibre reinforced polymer hybrid composites: A review *Carbohydrate polymers* **86** 1-18

[25] Ramakrishna G, Sundararajan T 2005 Studies on the durability of natural fibres and the effect of corroded fibres on the strength of mortar *Cement and Concrete Composites* **27** 575-82

[26] Arsène M A, Okwo A, Bilba K, Soboyejo A B, Soboyejo W O 2003 Cementitious composites reinforced with vegetable fibres *First Inter-American Conference on Non-conventional Materials and Technologies in the Eco-construction and Infrastructure*, Joao-Pessoa, Brazil 13-16

[27] Reddy N, Yang Y 2005 Biofibres from agricultural by-products for industrial applications *TRENDS in Biotechnology* **23** 22-7

[28] Li Y, Mai Y W, Ye L 2000 Sisal fibre and its composites: a review of recent developments *Composites science and technology* **60** 2037-55 Doi: 10.1016/s0266-353800.00101-9

[29] Agopyan V, John V 1992 Durability evaluation of vegetable fibre reinforced materials *Building Research and Information* **20** 233–5 Doi: 10.1080/09613219208727212

[30] Komuraiah A, Kumar N S, Prasad B D 2014 Chemical composition of natural fibres and its influence on their mechanical properties *Mechanics of composite materials* **50** 359-76

[31] Li Y, Mai Y W, Ye L 2000 Sisal fibre and its composites: a review of recent developments *Composites science and technology* **60** 2037-55 Doi: 10.1016/s0266-353800.00101-9

[32] Abani S A, Kriker A, Bali A 2008 Effect of Curing and Mix Design Types on Performance of Date Palm Fibers Reinforcement Concrete Under Hot Dry Environment *11TH International Inorganic-bonded fiber composites Conference* 5-7

[33] Djoudi A 2012 Performance of date palm fibres reinforced plaster concrete *International Journal of the Physical Sciences* **7** 2845-53

[34] Al-Khanbashi A, Al-Kaabi K, Hammami A 2005 Date palm fibres as polymeric matrix reinforcement: Fibercharacterization *Polymer Composites* **26** 486–97 Doi: 10.1002/pc.20118

[35] Li Z, Wang X, Wang L 2006 Properties of hemp fibre reinforced concrete composites *Composites Part A: Applied science and manufacturing* **37** 497-505

[36] Cheung H Y, Ho M P, Lau K T, Cardona F, Hui D 2009 Natural fibre-reinforced composites for bioengineering and environmental engineering applications *Composites: Part B* **40** 655-63. Doi: 10.1016/j.compositesb.2009.04.014

IOP Conf. Series: Materials Science and Engineering **596** (2019) 012029 doi:10.1088/1757-899X/596/1/012029

Experimental verification of functionality of fibre-reinforced concrete submersible piers

J Buchlák[1], J Matějka[1], P Ryjáček[1], P Bílý[1], J Procházka[1], J Pollert[1] and J Fabel[2]

[1]Faculty of Civil Engineering, Czech Technical University in Prague, Thákurova 7, 166 29 Praha 6, Czech Republic
[2]C&COM Advertising, Veslařský ostrov 62, 147 00 Praha 4 – Podolí, Czech Republic

E-mail: petr.bily@fsv.cvut.cz

Abstract. Low capacity of river banks is a problem of many world cities. Extension can be realised in many ways. One of the ways is to use a system of floating piers. Usual types of piers are filled with floating material, which supports the pier for the whole lifetime. The system of piers described in this article is innovative, because it is supported by an air bag, which can be deflated and then the whole system sinks down to the bottom of the river. This can be helpful in case of danger of floods, because there will be no need to transport the piers to a secure dock. Piers are designed for easy modular connection in various groups. The main types of groups are linear and areal. This article briefly describes the design of fibre reinforced concrete pier and other support constructions which are necessary for the right function of the system. The design of the pier was verified by hydraulic experiments on models in scale 1:10 to real pier. The article contains the description and results of the experiments that have proven the system to be feasible.

1. Introduction

The river bank is a very frequently visited place in Prague. It is searched by tourists and citizens of the city because of its significance and surroundings. Lots of cultural and sport events are organized here. During these events there is not enough space on the river bank for people. The operator of Prague river banks is therefore looking for a way to increase its capacity. Logical improvement of the river bank would be its extension. Permanent river bank extension would be very expensive and difficult. Extension by floating piers could be a solution, but it is connected with another problem. The usual type of concrete pier is filled with floating material (for example polystyrene) and it floats all the time. In case of floods, all floating objects from the river must be moved to secured harbors to prevent any damage they would cause if carried away by the stream. The transport of objects to these harbors is technically demanding and the capacity of the harbors is limited. If the usual type of piers would be used for the extension, new secured harbors would have to be build, but there is a lack of both space and finance for this solution.

Because of these problems, the authors are trying to develop a new system of concrete piers with no need of transportation in case of floods. The key idea of that system is replacing the polystyrene with an air bag, which could be deflated if needed and the pier would sink down to the bottom of the river. A group of piers would be connected by a system of air pipes that would allow to inflate and deflate the whole group together using one blower. The submerged group would remain connected with the river bank by a pipe for inflating the air bags after the flood. The submerged pier would lie down on the

IOP Conf. Series: Materials Science and Engineering **596** (2019) 012029 doi:10.1088/1757-899X/596/1/012029

bottom for the whole period of a flood. After the flood, the air bag would be re-inflated with air and the pier would float again.

Usual types of piers are made from three mostly used materials or their combination. These materials are wood, steel and concrete. Concrete piers are usually used for making big floating platforms. In our case, concrete was chosen also for the architectural reasons.

There are two existing types of submersible concrete piers (patents no. KR101550453 [1] and US4938629 [2]). In both these systems, the piers are submerged by flooding a cavity in a concrete shell. This is connected with many problems related to securing the leaktightness of the shell, pumping the water inside and outside the cavity, cleaning the cavity from dirt etc. These problems should be solved by using the system with air bags.

The following Figures 1, 2 and 3 show the main idea of the developed system of submersible piers and the visualization of piers on Prague river bank.

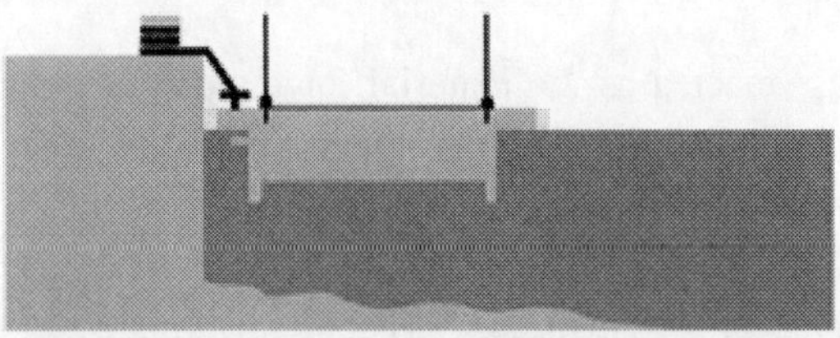

Figure 1. Floating concrete pier with inflated air bag.

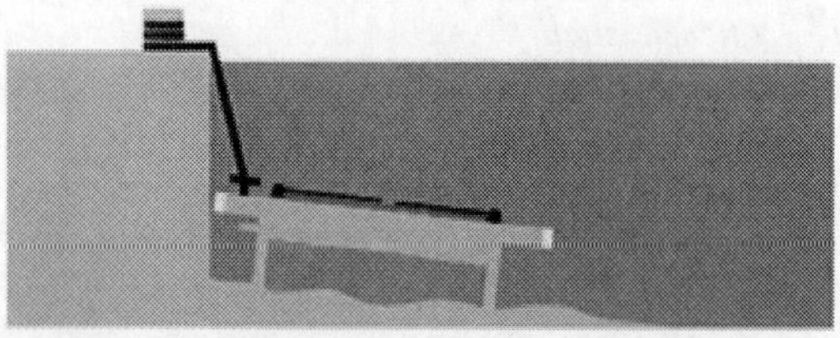

Figure 2. Submerged concrete pier with deflated air bag.

Figure 3. Visualisation of linearly connected piers for cyclists on Prague river bank.

Figure 4. Example of numerical analysis of deflections of a group of piers.

2. Intended use

In the original architectural study, the system of piers was used to create a bicycle path. Further considerations have shown that the use of the piers for this purpose would not be ideal. There would be small gaps between two neighbouring piers, the depth of immersion of differently loaded piers would be different, leading to uneven surface of a group of piers unsuitable for cycling. Another problem is that the piers will be attractive for pedestrians and collisions with cyclists could occur. After a discussion with the operator of Prague river banks, the main purpose of the system was changed to a path that will be used only by pedestrians.

The system should be modular, so the owner will be able to connect the piers to different linear or areal units. The connection of piers needs to be designed for fast disconnection, so the whole system would be as variable as possible. Figures 5 and 6 show two examples of arrangement of piers.

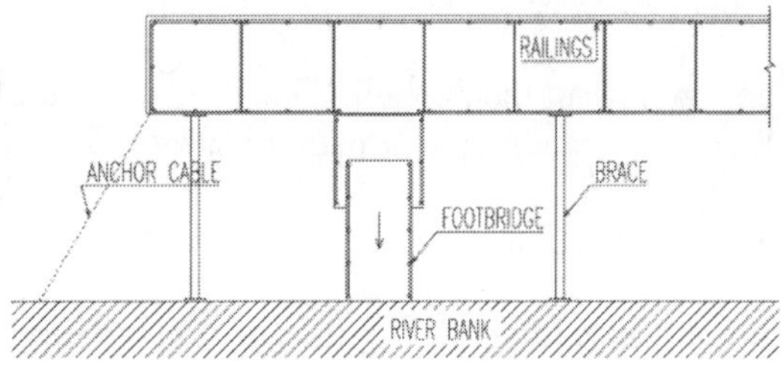

Figure 5. Linear group of piers, used for example as a pathway.

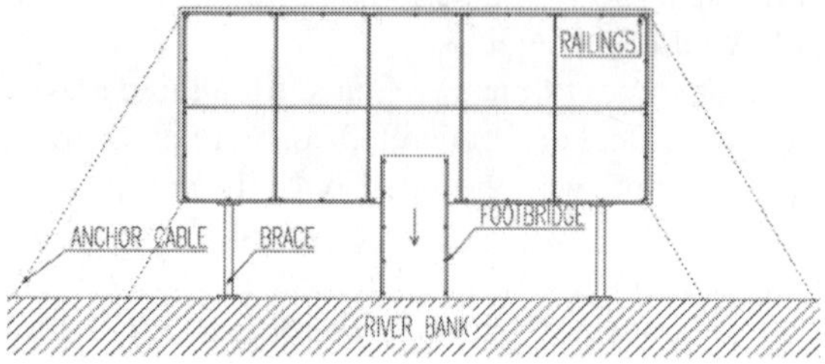

Figure 6. Areal group of piers, used for example as a restaurant seating.

3. Materials

3.1. Concrete shell

Fibre reinforced concrete (FRC) of C30/37 class was selected as the material for the piers (mixture composition is shown in Table 1). Polymer fibres will increase the resistance of the material to shrinkage, freeze-thaw cycles, abrasion and other environmental effects. The shell will be reinforced by steel wire mesh (10 mm bars per 150 mm in both directions). The bar reinforcement needs to be covered by a layer of concrete. The thickness of the cover layer (30 mm) defines the thickness of all pier components. Piers will be made as precast elements. This gives an advantage of precise production with better quality.

3.2. Air bag

After discussion with producer of air bags, PVC/Nitril with nylon insertion was chosen as the bag material. Thickness of the air bag is 2.5 mm. This should provide sufficient toughness and tightness of the air bag. In Figure 7, there is a similar air bag from the same producer.

Table 1. FRC composition.

Compound	Specification	Amount [kg/m^3]
cement	CEM I 42.5 R	400
water	-	180
w/c	0.45	-
fine aggregates	0/4 mm	950
coarse aggregates	4/8 mm	600
metakaolin	Mefisto L05	40
superplasticizer	Stachement	3
polymer fibres	Forta Ferro 54 mm	4.8

Figure 7. Example of an air bag of a similar type.

4. Design of geometry and size of the pier

The design of the piers was optimized by numerical analysis (Figure 4) based on hydraulic laws and defined operational conditions of the piers. For the details, please refer to [3]. The maximum allowed tilt of the pier was defined as 4° and the minimum clear distance between the water surface and pier deck as 150 mm based on the information obtained from certifying authority.

The aim was to design the piers with maximum stability and loading capacity. These two conditions go against each other. The heavier is the pier, the more stable it is, but the lower is the loading capacity as the air bag has to support the self-weight of the pier. The size of the deck was designed to be 3.3 x 3.3 m; the height of the pier element was set to 0.95 m. The thickness of all components is 80 mm, which is defined by reinforcement diameters and cover layers (2x10 + 2x30 = 80 mm). The geometry is shown in Figures 8 and 9.

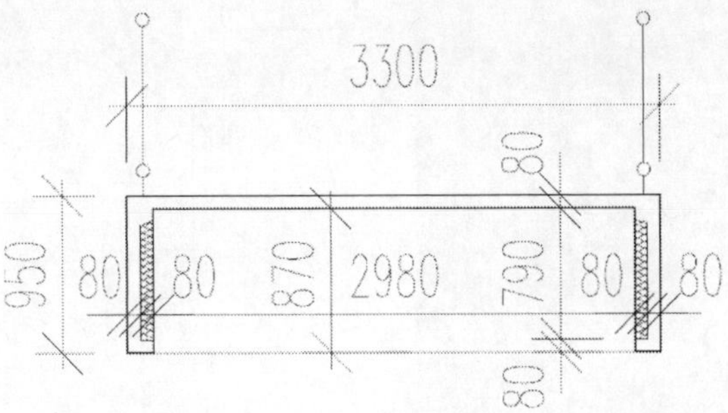

Figure 8. A section of the pier without the air bag showing the final dimensions.

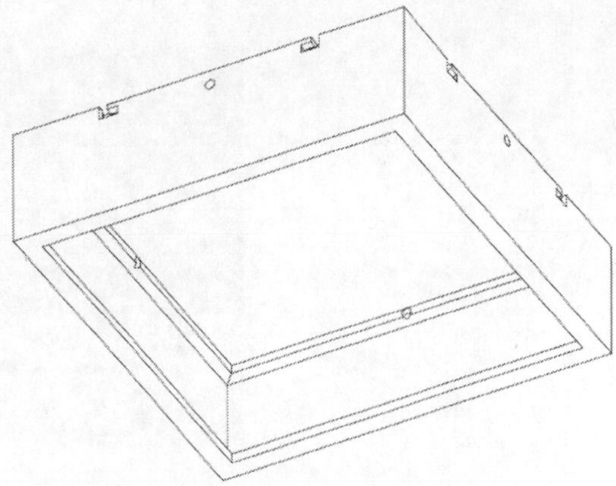

Figure 9. Lower axonometry of the pier without the air bag.

A capacity test was carried out in order to find out how many people can theoretically fit in 3.3 x 3.3 m area, resulting in 51 people (approximately 4000 kg when 80 kg per person is considered, Figure 10). According to the calculations, the pier of the abovementioned geometry is able to carry 55 people when the load is evenly distributed over the whole area. However, the situations when a lower number of people is unevenly distributed is more critical. Therefore, the maximum allowed capacity of one pier element was set to 15 people (1200 kg, Figure 11). If this capacity is exceeded, the pier will not sink or overturn, but the tilt will increase and make the use of the pier uncomfortable.

Figure 10. Maximum theoretical load of the pier – 51 people (4000 kg) on 3.3 x 3.3 m area.

Figure 11. Maximum allowable load of the pier – 15 people (1200 kg) on 3.3 x 3.3 m area.

5. Connection system

The connection system must allow fast connection and disconnection of the piers. Steel hinged joints were designed and verified by loading tests (Figures 12, 13, 14). The joints consist of three parts: a case embedded in the concrete shell, a connection bar with cylindrical ends and a cap preventing the connection bar from vertical movements.

A mock-up of the connection and the surrounding part of the pier was created in 1:1 scale and tested in the laboratory to verify the bearing capacity. For the details, please refer to [3]. The tests have shown that the capacity of the joints is sufficient for the given purpose. The capacities reached during the tests were 3 – 4 times bigger than the maximum expected forces in the connections. The FRC mixture specified in Table 1 was used and bar reinforcement was placed in the mock-up in order to test also the concreting of the structure. No problems were detected.

IOP Conf. Series: Materials Science and Engineering 596 (2019) 012029 doi:10.1088/1757-899X/596/1/012029

Figure 12. Connection system.

Figure 13. Load-bearing capacity test on 1:1 mock-up.

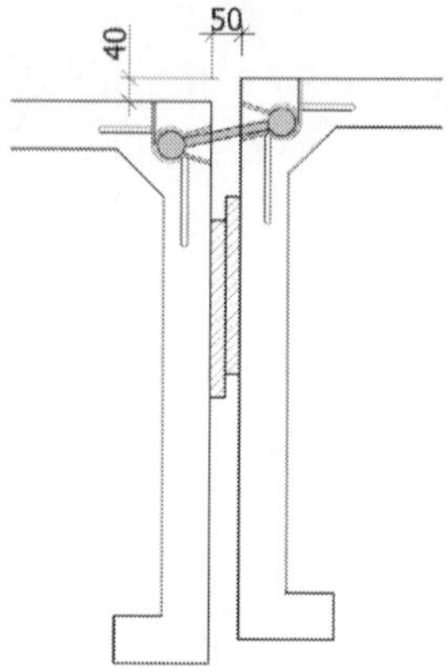

Figure 14. A section of two connected piers.

6. Design and production of models for hydraulic experiments

Hydraulic experiments are very important in design of floating structures. Firstly, not all situations can be properly simulated by numerical analysis. Secondly, the numerical models should always be verified by experiments.

The scale of hydraulic experiments was selected as 1:10. Pier model size was 330 x 330 x 95 mm with thickness of components 8 mm (Figure 16). This thickness was extremely low, so it was necessary to design a special concrete mixture (Table 2) and mould (Figure 15). The models needed to be created as precisely as possible, because every weight inaccuracy leads to unintended initial tilt of the pier.

Connection of piers was simulated by soft steel wires fastened in electrical chocolate block connectors embedded in concrete. The behavior of the connection was very similar to the real proposed connection system. The air bag was simulated by PVC foil glued to the bottom of the concrete model (Figure 17).

Table 2. Material used for 1:10 models.

Compound	Specification	Amount [kg/m^3]
cement	CEM I 42.5 R	400
water	-	260
w/c	0.65	-
fine aggregates	0/4 mm	1600
metakaolin	Mefisto L05	80
superplasticizer	Sika 0.5	2

Figure 15. A mould prepared for concreting.

Figure 16. The model for hydraulic tests in scale 1:10 without the air bag.

Figure 17. The model with an "air bag" simulated by PVC foil.

7. Hydraulic experiments with models of piers

The hydraulic experiments that were performed can be divided in two groups. First group contains static hydraulic experiments. These experiments were carried out on calm water surface with no movements in order to verify and extend the numerically obtained values of tilt and draught of the piers. The models were loaded with different values of load in different loading patterns. The load from pedestrians was simulated by steel plates placed on the whole area or on one half of the pier. Around every tested group steel bars representing the railings were placed. All loads were in 1:1000 scale to be consistent with scaling of dimensions (1:10 in all three directions).

Dynamic hydraulic experiments were carried out on moving water surface. The surface was either waving, flowing or waving and flowing at the same time. Tilts and forces in anchorage cables and braces were measured. Detailed results can be found in [4].

7.1. Static hydraulic experiments

During static hydraulic experiments, 26 different loading states were tested. The differences between the states were:

- Arrangement of elements: lonely pier, pair of piers, linear group (5x1), areal group (3x3)
- Value of load on one element: 4000 g (max. theoretical) or 1200 g (max. allowable)
- Arrangement of load: load over the whole area, load on one half of the area
- Failure state – failure of an air bag in one of the elements simulated

Some of the states were numerically analysed during the design of the pier before the hydraulic experiments. The experiments confirmed the results of numerical analyses.

The biggest tilt of 12.5° was observed on a lonely pier with 2000 g load placed on one half of the area (Figure 19). Even in this extreme case, the pier did not sink or overturn. When the piers were connected into linear or areal groups, the tilt decreased significantly (Figure 18). For example, in case of linear group loaded by maximum allowable load on one half of the area, the tilt was only 1.7°. As the piers are generally intended to be used in groups, the static tests proved that the design is functional.

Figure 18. Areal group of piers loaded by 4000 g on the middle pier of the group.

Figure 19. Lonely pier loaded by 2000 g on one half of the pier.

7.2. Dynamic hydraulic experiments

During dynamic hydraulic experiments (Figures 20 and 21), 27 different loading states were tested. The differences between states were:

- Arrangement of elements: lonely pier, pair of piers, linear group (3x1), areal group (3x2)
- Value of load on one element: 4000 g (max. theoretical) or 1200 g (max. allowable)
- Arrangement of load: load over the whole area, load on one half of the area
- Movement of water surface: just waves, just flow, flow and waves together

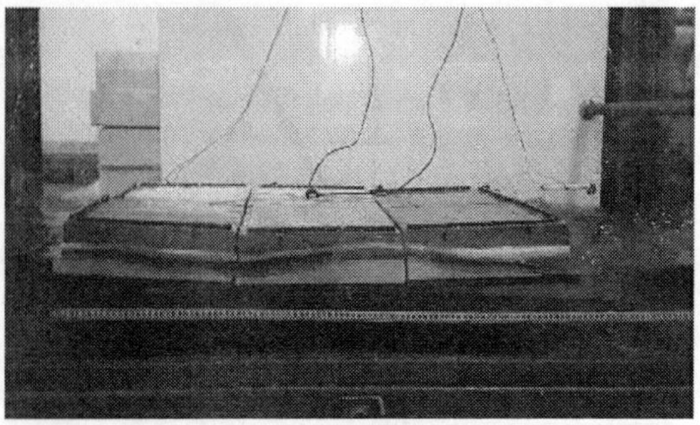

Figure 20. Dynamic hydraulic experiments with just waving water surface.

Figure 21. Dynamic hydraulic experiments with just flowing water surface.

The biggest variance of tilt was measured during experiments with just waving water surface. Height of the waves was set to 50 mm and length of the waves was 600 mm. This corresponds to the size of waves caused by the largest tugboats cruising on Vltava river. In loading state with lonely pier without any load, extreme tilt variance of 20.4° was measured. When piers were connected to linear group (3x1) with same conditions, the measured tilt variance decreased to 11°. After connecting to areal group (3x2), further decrease to just 4° was observed. These results proved that connecting the piers to bigger groups is very beneficial for the stability of the whole group. Another encouraging result was that the higher was the load of the pier, the more stable it was (i.e. the lower was the tilt variance).

Figure 22 shows static scheme during dynamic hydraulic experiments (Figure 23). Group of piers was anchored by elements preventing it from flowing away. Forces in these elements were measured by strain gauges.

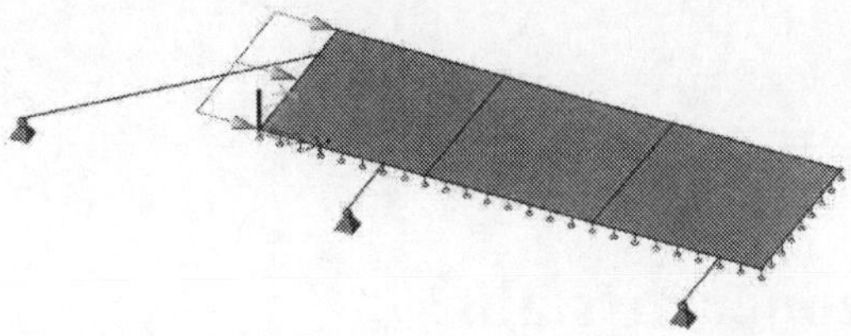

Figure 22. Static scheme during dynamic hydraulic experiments.

Figure 23. Real dynamic hydraulic experiments.

The biggest forces in anchorage elements were measured in loading states where waving and flowing of the surface were combined. The speed of water flow corresponded to the initial phase of the flood when the piers would have been submerged under water (1.0 m/s in reality, i.e. 0.3 m/s in the test after scaling). The biggest measured force in tensioned cable was 46 N, the biggest force in compressed brace was 36.9 N. These forces couldn´t be counted numerically, the experiments were the only way to obtain them. Anchorage elements will be designed according to these measurement.

8. Conclusions

The development, design and experiments have shown that the idea of submersible concrete piers is feasible. Hydraulic experiments have proven that the piers are in general sufficiently stable when used in groups. Dynamic hydraulic experiments have provided data for design of anchorage elements. Based on the results of the research, patent application for the system of submersible concrete piers was submitted to the Czech Patent Office.

Development of concrete piers will continue in 2019. After partial amendments of the design in accordance with the needs of the production company, two prototypes of the piers will be made in real scale. These prototypes should be tested in the river, with the focus on their stability and process of inflating and deflating the air bags. If the experiments confirm that the realization of the system is possible, it could be used on Prague river banks from the year 2020.

Acknowledgement

The paper was prepared thanks to the support of the City of Prague, Operational Programme Prague – Growth Pole of the Czech Republic, project no. CZ.07.1.02/0.0/0.0/16_040/0000377 „Concepts of the Faculty of Civil Engineering CTU for Prague 2017“, concept DK-05 „Submersible fibre-reinforced concrete pier with anti-flood pneumatic protection“.

References

[1] Submerging type floating pier having bridge. Patent no. KR101550453. Available online at: https://worldwide.espacenet.com/publicationDetails/biblio?II=0&ND=3&adjacent=true&locale=en_EP&FT=D&date=20150908&CC=KR&NR=101550453B1&KC=B1#. Cited on 4th January 2019.

[2] Floatable and sinkable wharf structure. Patent no. US4938629. Available online at: https://worldwide.espacenet.com/publicationDetails/biblio?II=0&ND=3&adjacent=true&locale=en_EP&FT=D&date=19900703&CC=US&NR=4938629A&KC=A#. Cited on 4th January 2019.

[3] Matějka J 2019 *The fibre-concrete floating pontoon* (Faculty of Civil Engineering, Czech Technical University in Prague) Diploma thesis

[4] Buchlák J 2019 *Experimental verification of functionality of submersible concrete pier* (Faculty of Civil Engineering, Czech Technical University in Prague) Diploma thesis

IOP Conf. Series: Materials Science and Engineering **596** (2019) 012030 doi:10.1088/1757-899X/596/1/012030

Blast resistance of multi-layered concrete slabs

J Fládr[1], R Chylík[1], K Šeps [1], J Štoller[2] and T Trtík[1]

[1] Faculty of Civil Engineering CTU Prague, Thákurova 2077/7, 166 29 Praha 6
[2] Faculty of Military Technology University of Defence, Kounicova 65, 662 10 Brno

E-mail: roman.chylik@fsv.cvut.cz

Abstract. This paper compares blast resistance of multi-layered and single-layered concrete panels having different thickness and reinforcement and subjected to various explosive charges. The panels were analysed by ultrasound measurements before the application of blast loading; the velocity of transfer of ultrasound waves through the panels was determined. The same measurement was conducted after exposing the panels to contact explosion and the damage was evaluated based on the determined changes. Nine panels were tested, four of them single-layered and five of them multi-layered. The weights of Pentrite explosive charges were 150 g, 300 g and 500 g. The results show that multi-layered panels are suitable for the protection of soft targets or as cladding of load-bearing structures.

1. Introduction

With increasing geopolitical stress in the world in recent years, blast loading becomes a very important type of accidental load of structures that should not be neglected. It is of particular importance mainly for critical infrastructures such as dams, bridges or power plants [4, 5] where it is normally taken into account in the design phase. However, the terrorists do not aim just on these key engineering structures. Very often, they attack public buildings or open spaces where there is a great accumulation of people. Therefore, it is important to focus on the development of materials suitable for increasing blast resistance of existing structures or for protective barriers.

The presented research compared blast resistance of ultra-high performance concrete (UHPC) panels with and without internal damping layer and with different types of dispersed reinforcement. The panels were exposed to contact blast of Pentrite explosive charges weighing 150 g, 300 g or 500 g. The main observed parameter was the velocity of transfer of ultrasound waves through the panel in specified points (orange dots in figure 1) and its change after the explosion.

2. Experiments

2.1. Specimens

The size of tested panels was 1x1 m. Nine different panels were tested. Further characteristics of particular specimens are summarized in table 1. An example of one panel before and after loading is given in figures 1 and 2.

Table 1. List of tested panels.

Type	Mixture	Damping layer	Thickness [mm]	Explosive charge [g]
A1	Z1	-	50	300
A2	Z1	-	50	150
A3	Z2	-	50	300
A4	Z2	-	50	150
A5	Z3	Rubber	150	300
A6	Z3	Rubber	150	500
A7	Z3	Rubber	150	300
A8	Z3	Rubber	150	500
A9	Z2	Gang Nail plate	100	500

Figure 1. A panel before the explosion.

Figure 2. A panel after the explosion.

2.2. *Concrete*

Three different UHPC mixtures were used. The composition of the mixtures is given in table 2.

Table 2. UHPC mixtures.

Compound	Z1 [kg/m^3]	Z2 [kg/m^3]	Z3 [kg/m^3]
Cement	600.0	600.0	600.0
Water	140.0	140.0	140.0
Basaltic aggregate	1710.0	1710.0	1710.0
Superplasticizer	22.0	22.0	22.0
Microsilica	80.0	80.0	80.0
Dramix OL13 steel fibres	80.0	80.0	80.0
Dramix 80/30 steel fibres	-	30.0	60.0
Krampe Harex PM6/18 steel fibres	2.0	2.0	2.0
Recycled steel fibres (cord wires)	60.0	30.0	-
Fibrex A1 steel fibres	5.0	5.0	5.0

Compressive strength of UHPC at the age of 28 days f_c was measured on 150 mm cubes according to [1]. Flexural tensile strength at the age of 28 days $f_{t,fl}$ was tested on 100x100x400 mm beams by four-point bending test according to [2]. The results are summarized in table 3.

Table 3. Mechanical properties of UHPC mixtures.

Mixture	f_c [MPa]	$f_{t,fl}$ [MPa]
Z1	124.2	7.9
Z2	123.4	8.9
Z3	126.0	9.8

2.3. Explosive

Pentrite explosive was used. The charges were attached to the surface in the central part of the panels by industrial adhesive tape.

Pentrite is preferred for this type of tests due to its high explosion strength, explosiveness and chemical stability. The explosion energy of Pentrite is 5810 kJ/kg and detonation velocity is 8266 m/s. Pentrite is widely used mainly for the production of fragmentation and blasting-fragmentation charges.

3. Results

3.1. Panels without damping layer

In the first stage, panels with different types of dispersed reinforcement and without damping layer (A1 to A4) were tested. The charge of 300 g of Pentrite was used in case of panels A1 and A3. Smaller charge of 150 g was applied to panels A2 and A4 to compare the effects of different charges. The evaluation of damage was carried out by analysing measured ultrasound pulse velocities with the use of SVD_{ult} software developed by Štefan and Foglar [3].

Figures 3 and 4 show just the resulting damage extent for panels A3 and A4. The higher is the difference between pulse velocities before and after the explosion, the higher is the damage of the panel in the given point. In figures 5 and 6, detailed analysis of panels A1 and A2 is presented. Each column shows (from top to bottom): the ultrasound pulse velocity before the explosion; after the explosion; damage extent; and composed image of the real panel and measured damage extent.

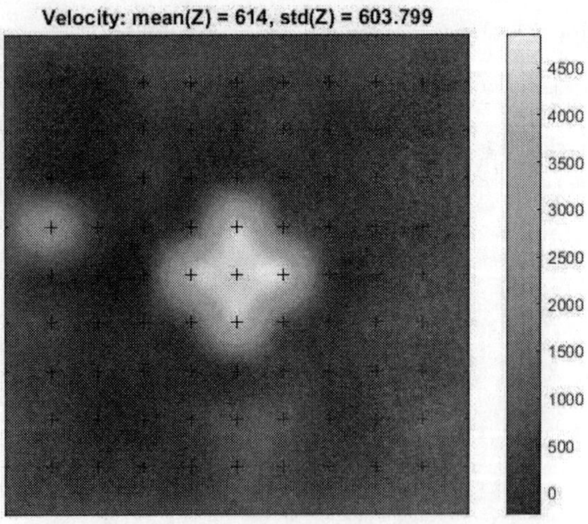

Figure 3. Panel A3.

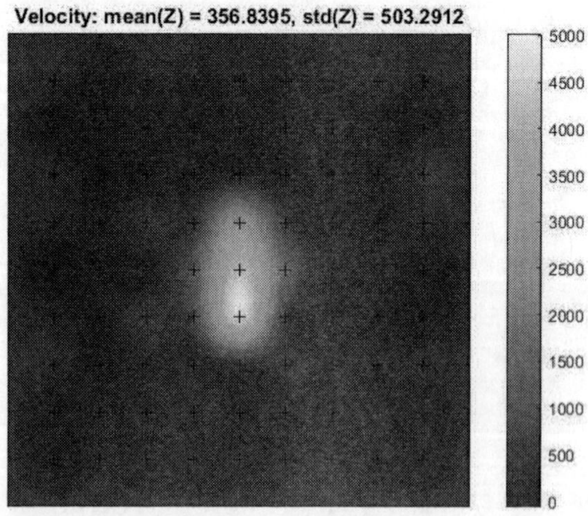

Figure 4. Panel A4.

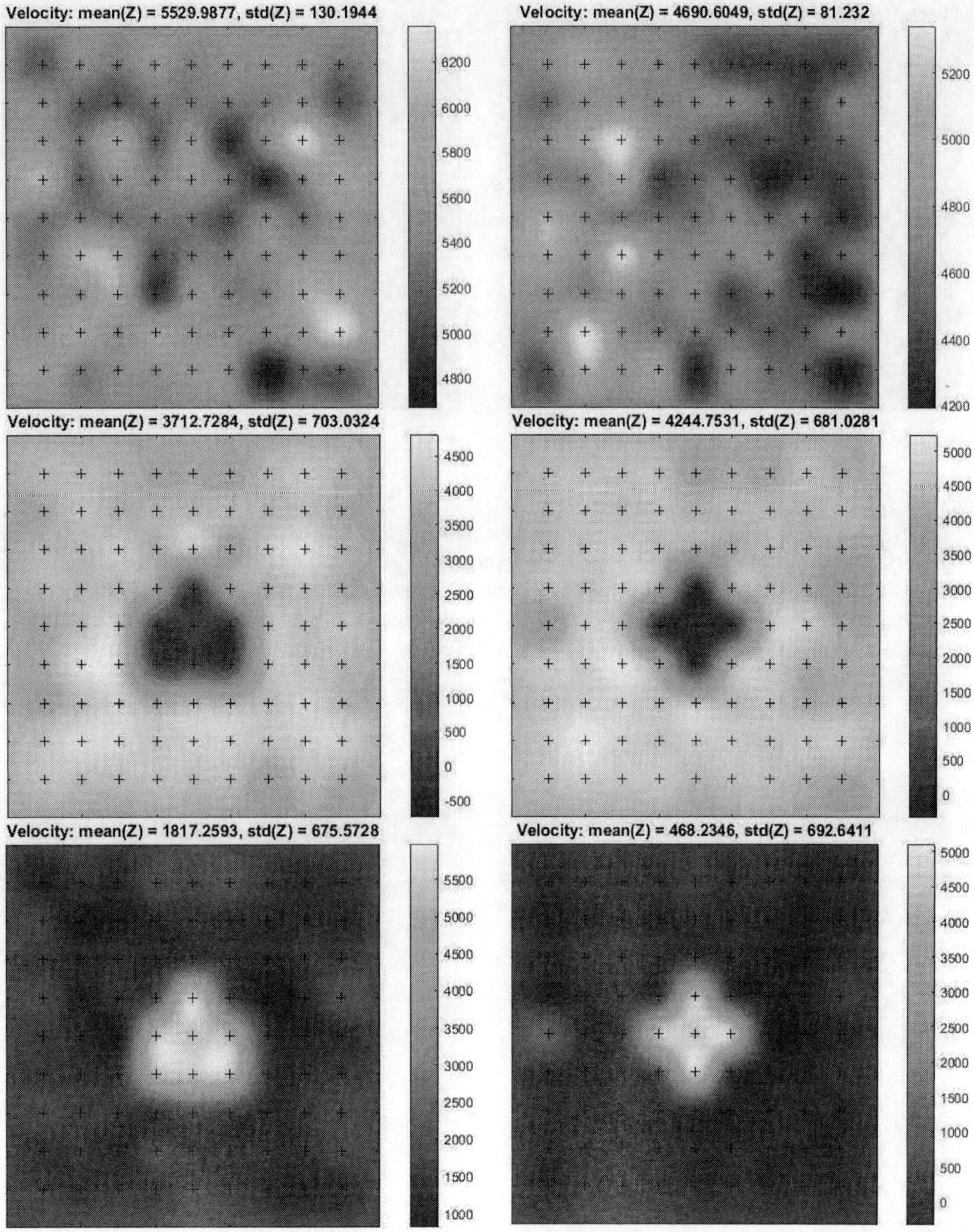
Velocity: mean(Z) = 5529.9877, std(Z) = 130.1944
6200
6000
5800
5600
5400
5200
5000
4800
Velocity: mean(Z) = 4690.6049, std(Z) = 81.232
5200
5000
4800
4600
4400
4200
Velocity: mean(Z) = 3712.7284, std(Z) = 703.0324
4500
4000
3500
3000
2500
2000
1500
1000
500
0
-500
Velocity: mean(Z) = 4244.7531, std(Z) = 681.0281
5000
4500
4000
3500
3000
2500
2000
1500
1000
500
0
Velocity: mean(Z) = 1817.2593, std(Z) = 675.5728
5500
5000
4500
4000
3500
3000
2500
2000
1500
1000
Velocity: mean(Z) = 468.2346, std(Z) = 692.6411
5000
4500
4000
3500
3000
2500
2000
1500
1000
500
0

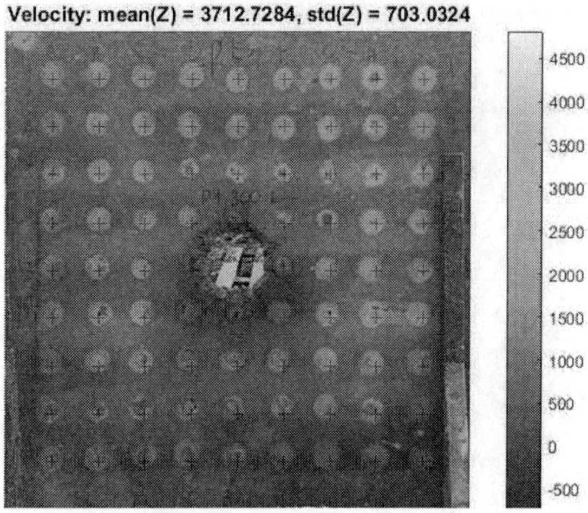

Figure 5. Panel A1.

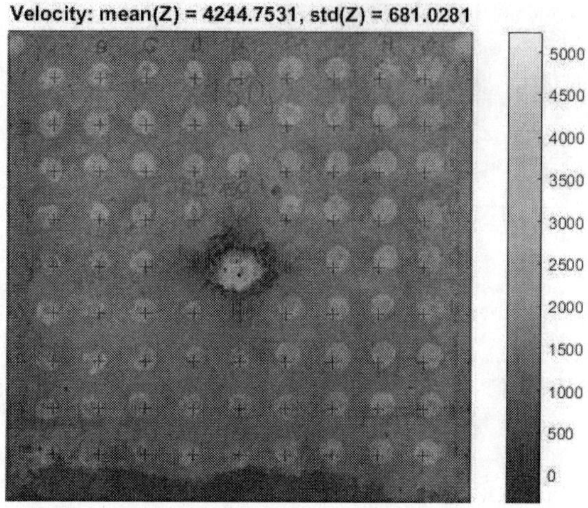

Figure 6. Panel A2.

3.2. Panels with damping layer

In the second stage, tests of multi-layered panels were carried out. Panels A5 (figure 7) and A7 (figure 9) that contained rubber damping layer were exposed to 300 g charge. The sequence of particular images is the same as in figures 5 and 6. Panels A6 (figure 8) and A8 (figure 10) with the same damping layer were loaded by 500 g Pentrite. Just the damage extents are shown in figures 9 and 10.

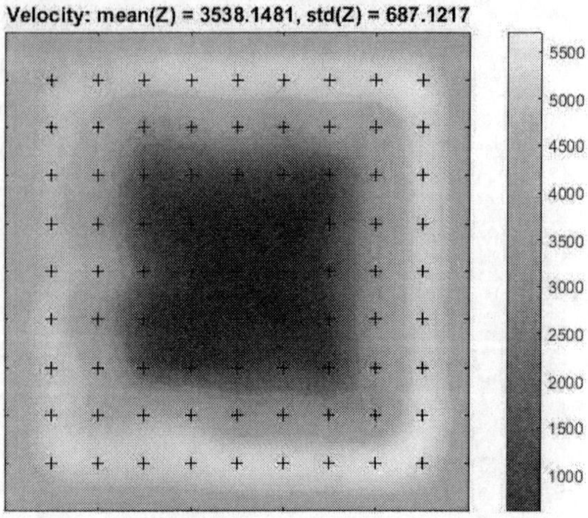

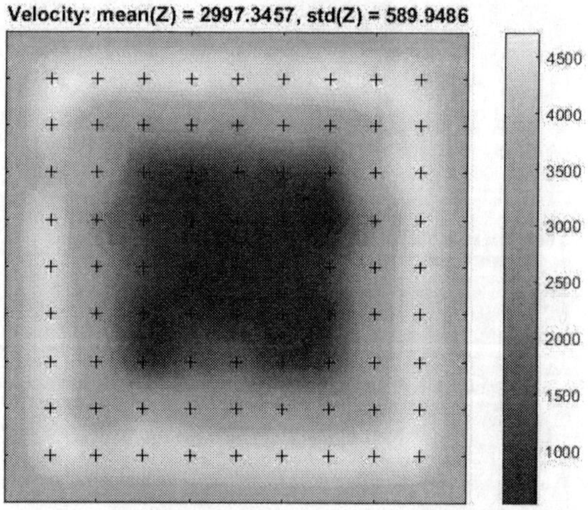

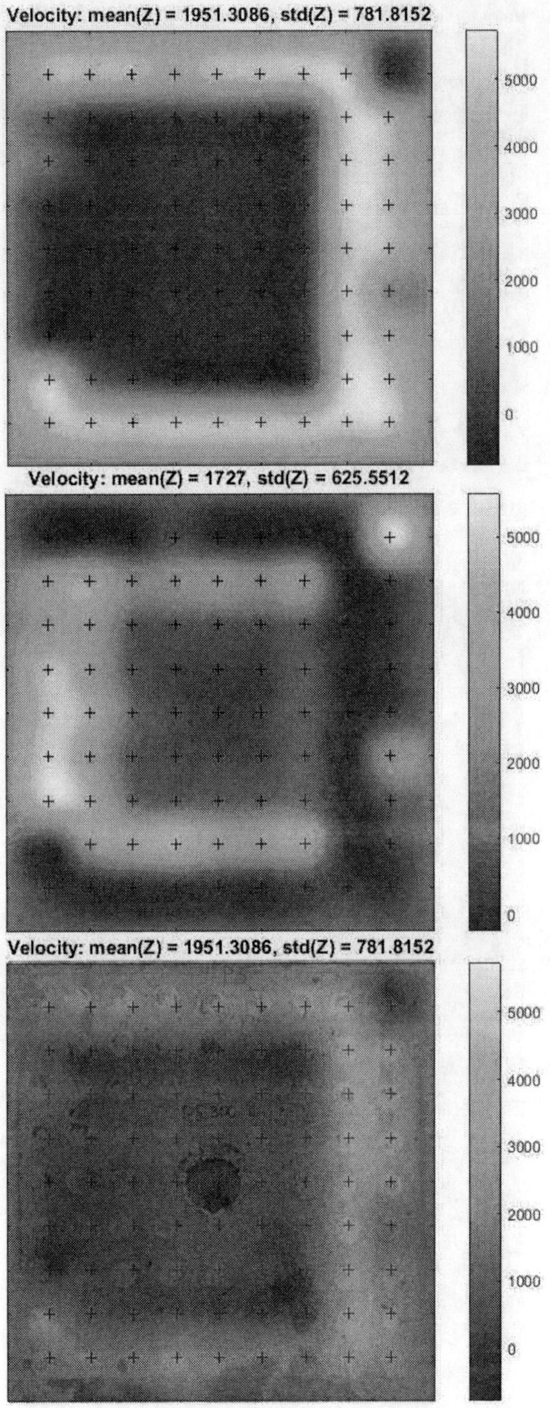

Figure 7. Panel A5.

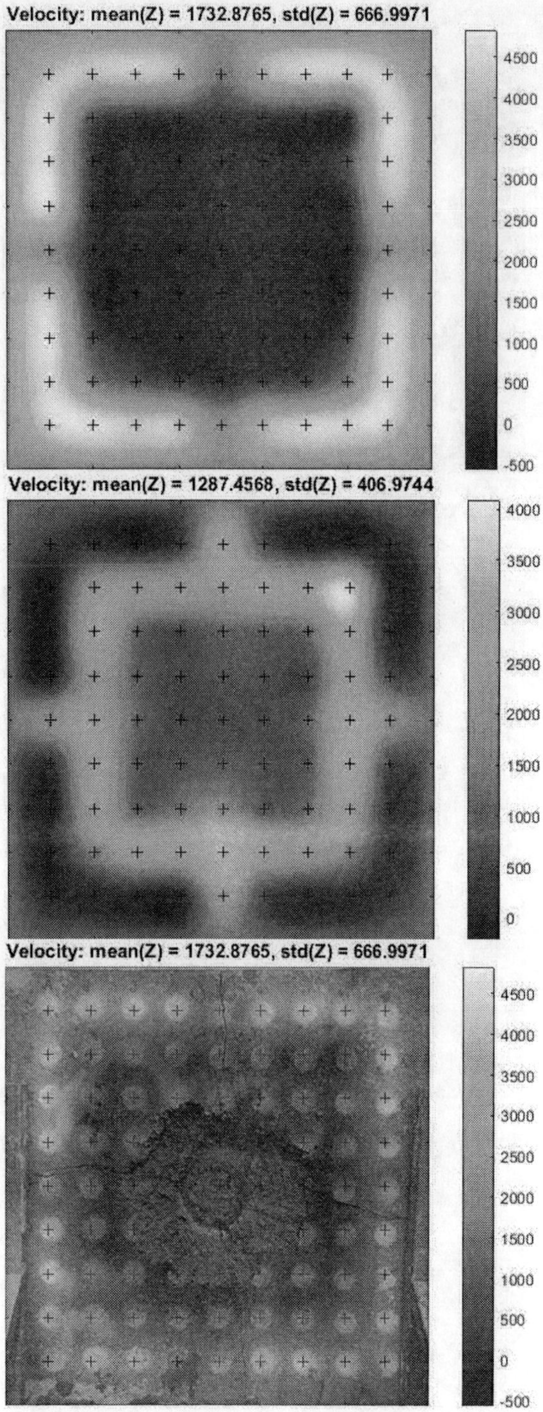

Figure 8. Panel A6.

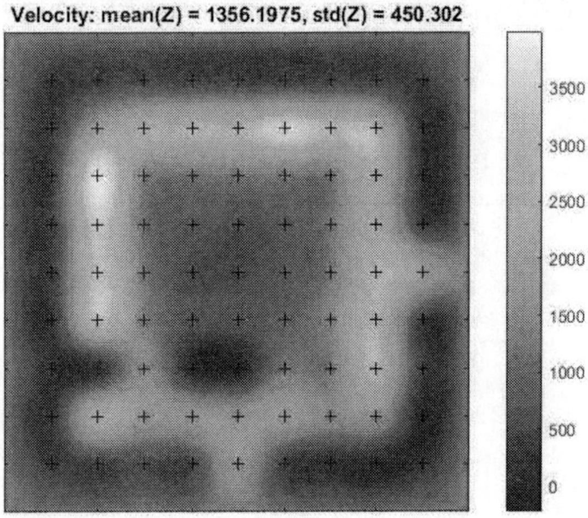

Figure 9. Panel A7.

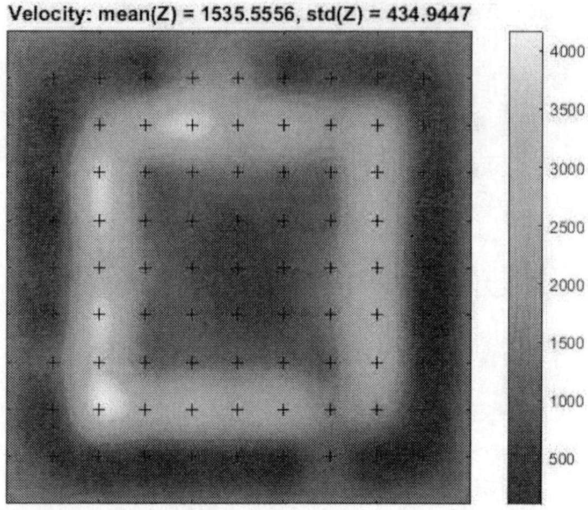

Figure 10. Panel A8.

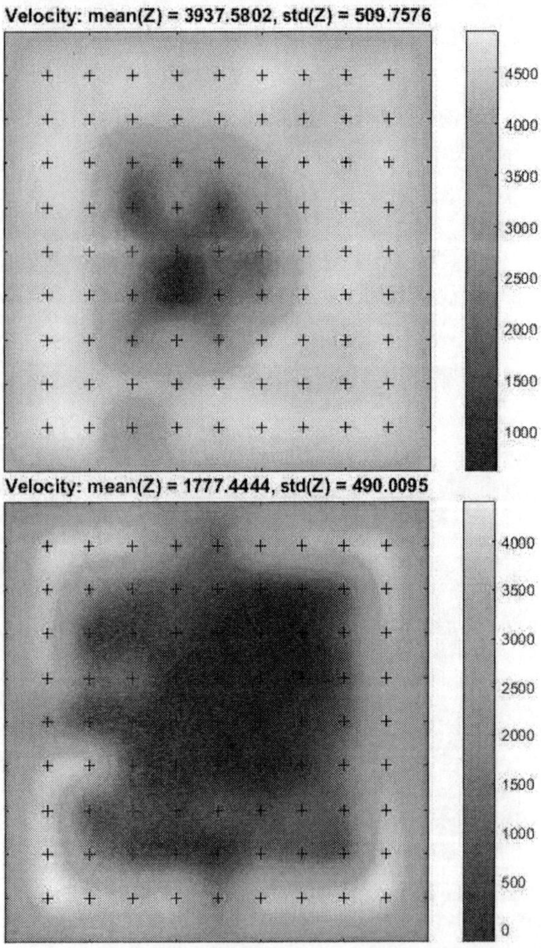

Figure 11. Panel A9.

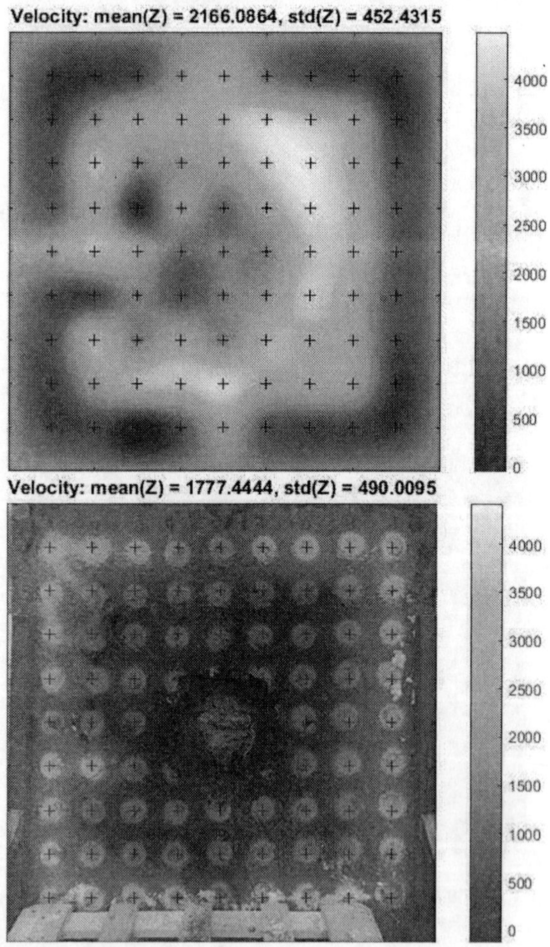

Figure 12. Panel A9.

Figures 11 and 12 show multi-layered panel A9 with damping layer made of gang-nail steel plates. The order of the images is the following: ultrasound pulse velocity before the explosion (figure 11 top); after the explosion (figure 11 bottom); damage extent (figure 12 top); composed image of the real panel and measured damage extent (figure 12 bottom).

4. Conclusions

In case of panels without the damping layer, the damage extent was directly proportional to the weight of explosive used. The damage was concentrated in the blast area.

For single-layered panels the influence of different type of dispersed reinforcement was also observed. The damage of panels A1 and A2 (made of Z1 concrete containing more recycled steel fibres from cord wires from discarded tyres) was slightly higher when compared to panels A3 and A4 (made of Z2 concrete containing half the amount of recycled steel fibres).

The damage of multi-layered panels seems to be smaller, but distributed over larger area around the point of explosion. Moreover, horizontal and vertical cracks occurred in these panels. The total absorbed explosion energy is higher and the amount of released debris is lower in case of multi-layered panels. Therefore, the multi-layered panels are more suitable as protective barriers or claddings of load-bearing structures.

Acknowledgments

This paper was prepared thanks to the support of the Technology Agency of the Czech Republic (TAČR), project (no TH02010888), project (no. SGS18/115/OHK1/2T/11)

References

[1] European Committee for Standartization (CEN) 2009 Testing hardened concrete; Compressive strength of test specimens Part 3 (EN 12390)e

[2] ČSN EN 14651:2005+A1:2007 Test method for metallic fibre concrete. Measuring the flexural tensile strength (limit of proportionality (LOP), residual)

[3] Štefan R, Foglar M 2015 SVD_{ult} – Software for visualisation of data from ultrasound measurements of concrete samples (Prague: Department of Concrete and Masonry Structures, Faculty of Civil Engineering, Czech Technical University in Prague)

[4] Bílý P, Kohoutková A 2014 Numerical analysis of anchorage between steel liner and prestressed nuclear containment wall *Proceedings of the 10th fib International PhD Symposium in Civil Engineering* (Quebec: Université Laval) pp 529-534

[5] Bílý P, Kohoutková A 2018 An estimation of the effect of steel liner on the ultimate bearing capacity of prestressed concrete containment *Nuclear Engineering and Design* **328** pp 197-208

Study of boundary conditions for design of new types of fibre concrete energy dissipators in hydraulic structures

P Fošumpaur, T Kašpar, M Králík and M Zukal

Department of Hydraulic Structures, Faculty of Civil Engineering, CTU in Prague, Thákurova 7, 166 29 Prague 6, Czech Republic

E-mail: fosumpaur@fsv.cvut.cz

Abstract. The paper describes experiments aimed at setting parameters for an efficient design of new types and shapes of energy dissipators on chutes of dam spillways. During major rainfall events, large water inflows into the reservoir induce a rise in the reservoir level, and a spillway system must be installed to spill safely the flood waters. Two key challenges during the spillway design are conveyance and energy dissipation. Energy dissipation on dam spillways can be achieved by a range of dissipator designs. One type of spillway is a block ramp. Block ramps are hydraulic structures which are often used in practical applications to assure a correct balance between hydraulic functioning and the environmental impact. One of the main peculiarities of this approach is the capacity to dissipate a larger energy amount than other traditional structures. Thus, significant efforts were spent by the scientific community around the world in order to optimize their energy dissipation efficiency. The use of fibre concrete for the design of reinforced block ramps can improve their resistance significantly. The presented experiments describe the boundary condition for the design of new fibre concrete dissipators based on the measurements performed on reinforced block ramps.

1. Introduction

Reinforced block ramps are structures designed to cross drops in elevation along the longitudinal profile grade of streams. Compared to drop structures, they offer a number of advantages, e.g. better integration in the landscape, preservation of the sediment control regime on the stream and the water habitat migration permeability. Block ramps are therefore often used as fish ladder [1]. The basic requirements for their design are related to sufficient capacity, kinetic energy dissipation and stability of their lining [2]. Block ramps are either designed on streams instead of drop structures or they can also be purpose-designed for spillway tailrace constructions of water reservoirs.

Conveying water across a drop in elevation is always associated with the need to efficiently dissipate the kinetic energy of the water flow on the chute. The boulder chutes provide very good energy dissipation efficiency, which is advantageous in terms of the stability of the reinforcements of the downstream section of the watercourse, which may already be stabilized by another suitable stabilization measures, for example by means of geonets [3]. The first structures designed specifically for water energy dissipation and installed directly on chutes, known as "baffled aprons", are described in [4, 5, 6]. The study [4], in particular, provided instructions and procedures that have been followed in applications commonly occurring on minor hydraulic structures to-date. The dissipation efficiency of water energy in the torrent control regime was studied e.g. [5, 7]. The spillways of large dams are mostly designed as smooth chutes ended by a ski jump [8, 9, 10]. The reason is easier maintenance

IOP Conf. Series: Materials Science and Engineering **596** (2019) 012031 doi:10.1088/1757-899X/596/1/012031

and reconstruction of simple surfaces. Other options for energy dissipation are a smooth chute ended by a stilling basin or a cascade chute or a smaller stepped chute whose hydraulics are generally discussed e.g. in [11]. As classic energy dissipation by a water jump in the stilling basin of dams may, in some cases, pose problems, the water flow energy is often dissipated by its aeration e.g. in a ski jump. Therefore, in structural terms, chutes are relatively demanding structures as they involve dealing with high velocity water flows accompanied by pressure pulsations. This is associated with greater demands for the structural materials used and their higher resistance. For example, fiber concrete which is suitable for constructions with higher requirements can be used advantageously on the one side for small structures (e.g. furniture [12]) and on the other side for huge constructions, e.g. dams. Fiber reinforced concrete meets these increased demands better than plain concrete and, therefore, it has been used in hydraulic engineering for several decades [13, 14]. The main domain where fiber concrete is applied are sections exposed to fast flowing water, or water with sediment transport where abrasion risks arise [15, 16]. It is also used in parts of constructions where the limitation of shrinkage cracks appearance and concrete deformations are desirable [17].

In the chapters below, we will concentrate on the quantification of the efficiency of kinetic energy dissipation and, particularly, on the description of hydrodynamic loading of reinforced block ramp constructions caused by pressure pulsations due to the flowing water.

2. Energy dissipation on block ramps

The quantification of the energy dissipation efficiency on the block ramp can be analyzed using the procedure designed by Pagliara and Chiavaccini [18] and presented in the diagram in figure 1. Let us assume a block ramp with the bed slope *S*, which overcomes the difference in elevation *H* along the length *L*. The critical depth h_k occurs at the ramp entrance in profile No. 0 and, according to the critical flow equation, the total energy of the flowing water above the reference plane $E_0 = H + 1.5h_k$. After profile No. 0 on the chute, a drawdown curve with a super-critical flow regime occurs. Provided the chute is long enough, the uniform flow depth h_u can be reached at the chute end in profile No. 1 and the total energy in profile No. 1 above the reference plane $E_1 = h_u + \alpha v_1^2/(2g)$. Hydraulic losses on the chute are defined by the variable $\Delta E_1 = E_0 - E_1$ and, thus, the relative energy dissipation $\Delta E_r = \Delta E_1/E_0$. Based on the research [18] came to the conclusion that the relative energy dissipation on the chute ΔE_r primarily depends on the following variables:

$$\Delta E_r = f\left(\frac{h_k}{d_{50}}, \frac{h_k}{H}, S\right) \quad (1)$$

where d_{50} is the average grain size of the chute material.

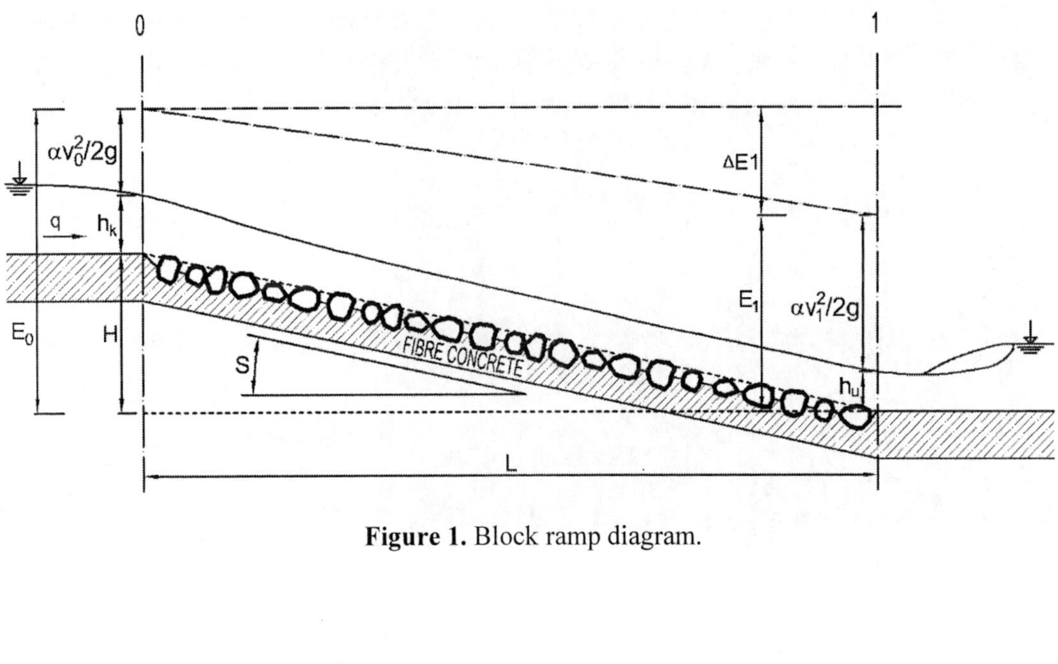

Figure 1. Block ramp diagram.

Hydraulic losses on a block ramp depend on the chute bed roughness (expressed by the Manning coefficient), which can be calculated using the formula devised by [19]. The respective formula is valid for slopes S ranging from 0.1 to 0.4 and for quarried stone:

$$n = 0.029(d_{50}S)^{0.147} \tag{2}$$

where d_{50} is the average grain size of the block ramp in millimetres.

The experiments carried out during our study validated energy dissipation in a block ramp with boulders stabilized in concrete. This application is desirable for large longitudinal bed slopes and in dam spillway tailraces. A characteristic ramp bed cross section is presented in figure 2. The dimensions of boulders are 0.64 to 1.28 m and they are embedded in concrete to half of their depth. Thus, the average boulder size d_{50} = 0.96 m, and the mean height of protruding boulders h_{50} = 0.48 m. Part of the experiments also dealt with the verification of an alternative with the mean height of protruding boulders h_{50} = 0.24 m and a smooth concrete chute. The alternative with the protruding boulder height h_{50} = 0.24 m was prepared after the gaps in the preceding alternative of h_{50} = 0.48 m had been additionally filled with concrete.

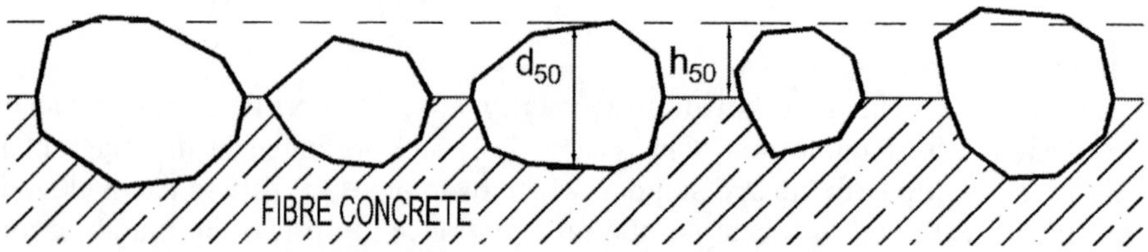

Figure 2. Block ramp cross section.

Experience gained from many practical applications shows that physical modelling is a useful tool to increase the efficiency of hydraulic research [20]. The experiments were performed on a hydraulic model made in a scale of 1:20. The other block ramp parameters in reality are as follows: the bed slope S = 0.21, the chute length L = 56 m, the chute height H = 11.7 m, the chute width b = 15 m. The tested chute will serve as a side channel tailrace of a reservoir and, therefore, a series of discharge rates from Q_{20} to $Q_{10\,000}$ was verified. These discharges are related to specific discharge rates in the chute in the interval from q = 1.7 to 7.1 m^2.s^{-1}. The $Q_{10\,000}$ discharge is the maximum discharge serving, at the same time, to verify the dam safety in floods. Figure 3 below presents a photo of the laboratory hydraulic model in a 1:20 scale.

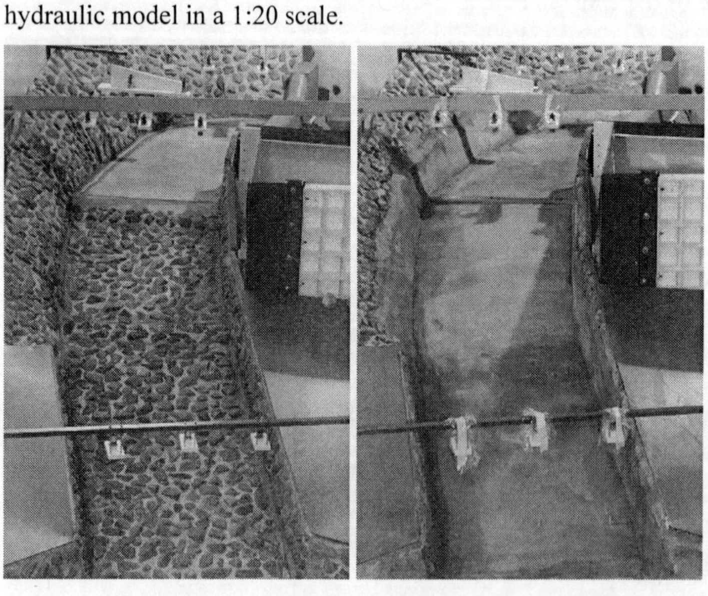

Figure 3. Block ramp scale model (1:20): average boulder size d_{50} = 0.96 m (left) and smooth chute (right).

The experiments included the assessment of the depth and velocity values at the end of the chute in profile No. 1 as shown in figure 1. The results obtained in this way were compared with the mathematical model, which employed the water level pattern on the chute calculated by the standard section-by-section method in the super-critical flow regime e.g. [21]. This calculation must be performed as it cannot be guaranteed that the very depth of uniform flow will be reached at the end of the chute in profile No. 1 given the respective chute length. Figure 4 displays the comparison of energy dissipation on the chute for the studied roughness alternatives. The plotted charts present the relative energy dissipation on the chute ΔE_r in relation to the relative height h_k/H. The chart also compares the calculated energy dissipation pattern on the chute against the values measured on the scale model.

The chart in figure 4 implies that the relative energy dissipation is the lowest for a smooth surface of the concrete chute and it quickly approaches zero with the growing relative depth h_k/H. The growing relative depth h_k/H, in this case, corresponds to the situation of a growing critical depth h_k and thus discharge through the chute, or a falling chute height H. The falling chute height at the constant slope S corresponds to a shorter chute length. The chart in figure 4 further shows excellent agreement of the relative energy dissipation pattern for a smooth concrete chute and for a chute made with concreted boulders with the protruding height $h_{50} = 0.48$ m. For the protruding boulder depth $h_{50} = 0.24$ m, the energy dissipation is by ca 10 % lower than that obtained by the calculation. This circumstance, most likely, results from the fact that the bed roughness n according to the used formula (2) relies on the value d_{50}, which was substituted by the depth of the gap between the boulders. For the case of the alternative with boulders half-embedded in concrete, this assumption still well applies according to the measurement, but for the boulders embedded in concrete to three quarters of their height, the actual chute roughness is lower than the calculated one. This results in lower energy dissipation on the chute as it has a smoother surface.

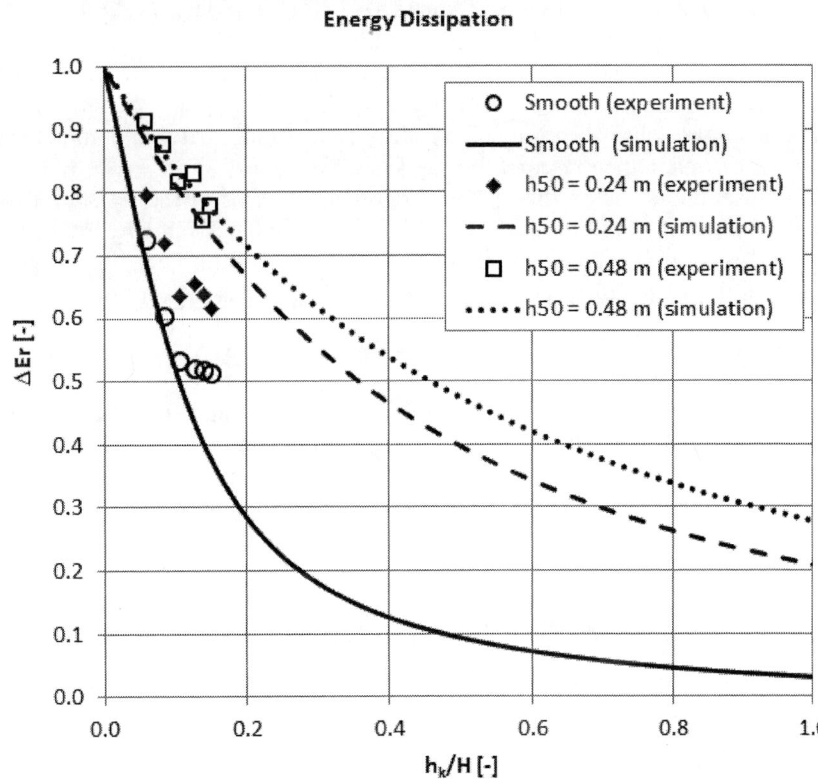

Figure 4. Relative energy dissipation ΔE_r related to the relative depth h_k/H.

3. Hydrodynamic loading of reinforced block ramps

The primary purpose of using reinforced block ramps is efficient water conveyance across drops in elevation along the longitudinal profile grade of a stream, which is always associated with the need for efficient dissipation of the flowing water kinetic energy. In Chapter 2, the basic approach to the quantification of water energy dissipation on the chute was described in relation to its bed roughness and other parameters. The presented facts imply that the most efficient dissipation can be reached on chutes with macro-roughness elements. The chute roughness can be achieved by the application of man-made dissipators, which are usually made of concrete and possess optimized geometric characteristics. This paper addresses natural block ramps stabilized by mounting individual boulders in concrete. It is evident that dissipation grows with the increasing grains size of the used boulders, but, at the same time, the hydrodynamic loading of the whole chute construction grows, too. This loading is represented mainly by pressure pulsations which can gradually cause the degradation of the chute material reducing its life cycle and serviceability. For this reason, we must try to quantify this oscillating loading and modify the concrete chute construction to resist this load. By integrating pressure pulsations over a defined chute area, the forces acting on the chute as a slab can be successively quantified. The immediate force is defined by the sum of the mean force $\overline{F}$ and the fluctuating force component F'. Hence, the pulsating component of the loading force is:

$$F' = \int_A p' \mathrm{d}A \tag{3}$$

where p' is the fluctuating pressure component and A is the defined area for which the resultant compressive force must be specified. The quantification of the resultant compressive force per defined area is important in the stability assessment of slab constructions of hydraulic engineering structures in terms of resistance to lifting due to upward hydrostatic pressure. An example is the quantification of the hydrodynamic loading of the stilling basin bed slab below the weir. In [22] is states that this force loading can be quantified by two procedures. The first approach is direct compressive force measurement (e.g. by means of a transducer) in a part of the construction separated from the rest. In their research, they measured this effect with an experimental device under laboratory conditions. The second approach is based on a statistical assessment of pressure pulsations measured over the loaded area and the quantification of spatial correlations between immediate pulsations. This approach was adopted in their research by e.g. [23, 24].

The pressure pulsation values on the bed are important in terms of the resistance of the concrete parts of the chute. Twelve pressure sensors were installed on the block ramp hydraulic model allowing the monitoring of pressure pulsations for the time T = 1 hour with a reading frequency of 1 Hz. The pressure pulsations in individual sensors were standardized using the formula:

$$y = \frac{p - \bar{p}}{\sigma_\mathrm{p}} \tag{4}$$

where p is the immediate pressure value, $\bar{p}$ is the mean pressure and σ_p is the standard deviation of pressure. The standardized pressure values show the probability characteristics of the Gaussian probability distribution with a very low deviation of skewness from zero and excess kurtosis from 3. The analysis was processed for all the three chute bed roughness values tested and for a set of values of tested discharges. In this respective case, these were discharges corresponding to Q_{100}, $Q_{1\,000}$ and $Q_{10\,000}$, which served for testing the chute behaviour in conveying major flood events. The specific discharges corresponding to the above flood discharges are q = 2.9; 5.4 and 7.1 $\mathrm{m^2.s^{-1}}$. Figure 5 displays the Q-Q charts (normal probability plot) of standardized pressure pulsations y.

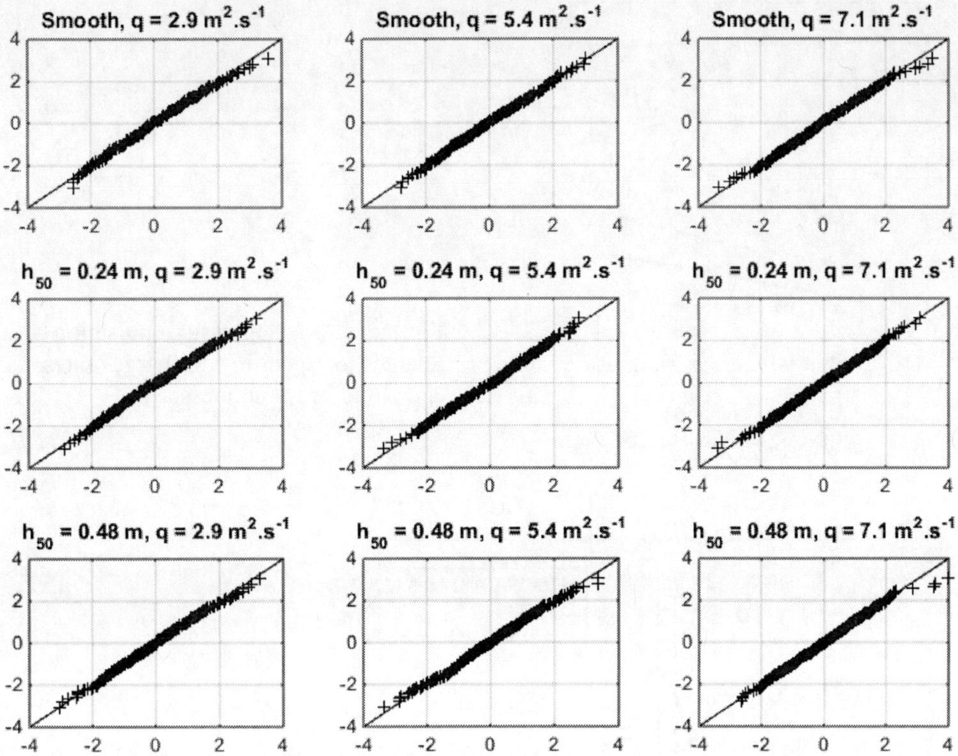

Figure 5. Q-Q charts of standardized pressure pulsations for studied chute bed roughness values and three discharge situations. Standardized pressure pulsations are plotted on the x axis, and the corresponding z-scores of the standard random variables on the y axis.

The charts in figure 5 manifest that pressure pulsations show the Gaussian probability distribution, and the maximum pressure pulsation values can reach 4 times the RMS (root mean square) value defined by the formula:

$$RMS = \sqrt{\frac{1}{T}\int_0^T [(p'(t)]^2 \, dt} \quad (5)$$

The RMS value depends on the chute bed roughness and the conveyed discharge rate. This relationship was investigated in a case study of a spillway chute. Pressure pulsations on the chute bed were read on the hydraulic model for all the three bed roughness values tested in conveying a series of specific discharges q [$m^2.s^{-1}$]. The results of this analysis are documented in an illustrative way in the chart in figure 6. It is evident from the chart that RMS in the case of a smooth bed takes the lowest value increasing only a little with growing discharge rates. In the alternative with the roughest chute bed composed of boulders d_{50} = 0.96 m embedded in concrete to half of their depth (gap depth h_{50} = 0.48 m), on the contrary, RMS takes the highest value growing significantly with growing discharge rates. In the case of this alternative, the pressure pulsations may amount up to 4 times the RMS value, i.e. ca 1.6 m of the water head (16 kPa), as is documented in figure 7. In the case of low mean pressures or water depths on the chute, total pressure values lower than the atmospheric pressure can be reached on the surface of concrete constructions with some probability. This phenomenon can successively trigger cavitation risks on the concrete surface reducing its life cycle. An efficient preventive measure consists in the reinforcement of concrete parts of chutes with fibre concrete.

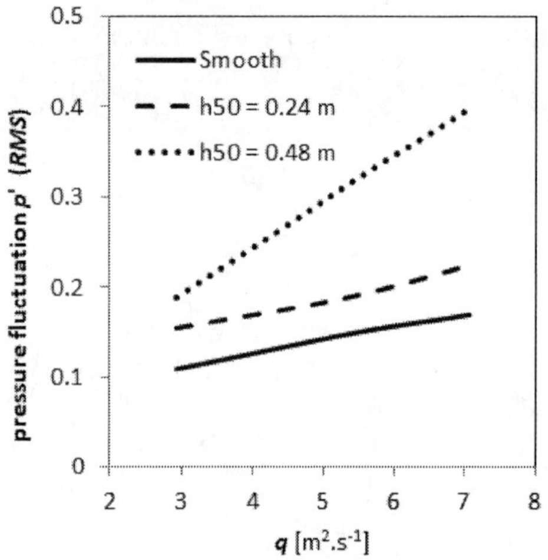

Figure 6. *RMS* pressure pulsations related to specific discharge rates and chute bed roughness.

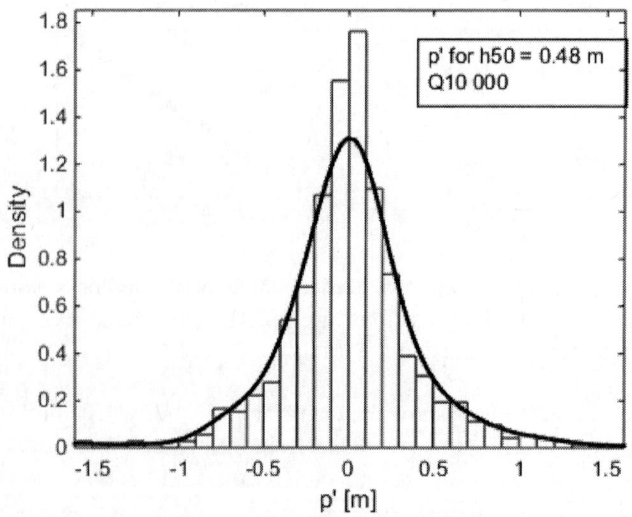

Figure 7. Pressure pulsations on the block ramp with a mean gap depth between boulders $h_{50} = 0.48$ m for the $Q_{10\,000}$ discharge ($q = 7$ m^2.s^{-1}).

4. Conclusion

The presented study brought the following main conclusions: (1) The application of block ramps represents a very beneficial technical solution in conveying water across high drops in elevation along the profile grade of streams. Its benefits include, above all, the preservation of the landscape natural pattern, water habitat migration permeability and the preservation of the sediment control regime. (2) Block ramps provide very good energy dissipation efficiency, which is favourable for the lining stability of the adjacent stream section. (3) Higher energy dissipation is achieved with greater roughness values of the chute. This is accompanied by higher hydrodynamic loading of the chute bed in the form of pressure pulsations. (4) It has been manifested that pressure pulsations have the Gaussian probability distribution and the maximum pressure pulsation values reach up to 4 times the *RMS* (root mean square) value. (5) The RMS relationship to the chute roughness and the specific discharge has been derived, which may serve for dimensioning chute surfaces made of fibre concrete.

Acknowledgement

In the paper some results of CTU projects (No. SGS18/121/OHK1/2T/11 and No. SGS19/046/OHK1/1T/11A) were used and this fact is gratefully acknowledged.

References

[1] Možiešik Ľ, Šulek P and Orfánus M 2014 *Int. Multidisciplinary Scientific GeoConference Surveying Geology and Mining Ecology Management, SGEM* vol **1** pp 447–454
[2] Chanson H 2015 *Energy Dissipation in Hydraulic Structures* (London: CRC Press) p 178
[3] Slezingr M and Zelenakova M 2011 *Colloquium on Landscape Management*, ed M Kravka (Brno: Mendel Univ) pp 9–11
[4] Peterka A J 1984 *Hydraulic design of stilling basins and energy dissipators* (Denver: Bureau of Reclamation) p 240
[5] Rhone T J 1977 *A baffled apron as spillway energy dissipator* (Denver: Bureau of Reclamation) p 32
[6] George R L 1979 *T or C Baffled Apron Spillway* (Denver: Bureau of Reclamation) p 33
[7] Pařílková J, Zachoval Z and Tureček J 2008 *Proc. of 22nd Symp. on Anemometry* (Praha: Institute of Hydrodynamics ASCR, v.v.i.)
[8] 1987 *Spillways for dams* (Paris: Commission Internationale des Grands Barrages) p 172
[9] 1992 *Spillways. Shockwaves and air entrainment* (Paris: Commission Internationale des Grands Barrages) p 117
[10] 2015 *Proc. 25th International Congress on Large Dams* (Paris: Commission Internationale des Grands Barrages) p 117
[11] Khatsuria R M 2005 *Hydraulics of spillways and energy dissipators* (New York: Marcel Dekker)
[12] Šonka Š, Frantová M, Štemberk P, Havrda J and Janouch P 2017 *IOP Conf. Ser.: Materials Science and Engineering* vol. **246** p 6
[13] Schrader E K 1988 *Fiber reinforced concrete* (Paris: Commission Internationale des Grands Barrages) p 51
[14] Zollo R F 1997 *Cement and Concrete Composite* **19** pp 107–122
[15] Horszczaruk E 2005 *Wear* **259** pp 62–69
[16] Ji T, Ji G, Wang S and Liu Y 2010 *J. of Southeast University (Natural Sicience Edition)* **40** pp 192–196
[17] Soutsos M N, Le T T and Lampropoulos A P 2012 *Construction and Building Materials* **36** pp 704–710
[18] Pagliara S and Chiavaccini P 2006 *J. of Hydraulic Engineering* **132** pp 41–48
[19] Rice C E, Kadavy K C and Robinson K M 1998 *J. of Hydraulic Engineering* **124** pp 179–185
[20] Fošumpaur P, Králík M and Zukal M 2010 *Proc. Int. Conf. on Modelling and Simulation (MS'10)* ed P Stemberk (Prague: CTU in Prague) pp 109–113
[21] Chow V T 1959 *Open-channel hydraulics* (New York: McGraw-Hill) p 680
[22] Farhoudi J and Narayanan R 1991 *J. of Hydraulic Engineering* **117** pp 64–82
[23] Bribiesca J L S and Mariles O A F 1979 *Proc. 13th Congress on Large Dams* (Paris: Commission Internationale des Grands Barrages) **3** pp 85–103.
[24] Spoljaric A, Maksimovic L and Hajdin G 1982 *Proc. Int. Conf. on Hydraulic Modeling* (Coventry: BHRA) pp 97–107

IOP Conf. Series: Materials Science and Engineering **596** (2019) 012032 doi:10.1088/1757-899X/596/1/012032

Air-entrainment as an alternative to polypropylene fibers and its effect on the compressive strength of concrete at high temperatures

J Holan, J Novák and R Štefan

Czech Technical University in Prague, Faculty of Civil Engineering, Department of Concrete and Masonry Structures, Thákurova 7, 166 29 Prague, Czech Republic

E-mail: jakub.holan@fsv.cvut.cz

Abstract. Although much research has been done in the area of the fire response of concrete in the past decades, only few studies are focused on the mechanical properties of air-entrained concrete at high temperatures. This paper presents preliminary results of an experimental investigation focused on the effect of air-entrainment on the compressive strength of concrete at various high temperatures. Within the scope of this work, heat treatments and compression tests have been performed on reference and air-entrained concrete specimens. The results obtained from the experiments have been analysed and show that the air entrainment seems to have an adverse effect on the compressive strength of the concrete at high temperatures when exposed for a prolonged period of time.

1. Introduction

Although concrete does outperform other commonly used building materials in terms of high temperature response, severe irrecoverable changes still happen in its internal structure when it is subjected to high temperature. We, therefore, look for means of improving concrete so that it is less affected when subjected to high temperatures. One of the responses of concrete to high temperatures is spalling, which most probably happens due to evaporation of water which causes the pore vapour pressure growth and the spalling of concrete [1]. One of the possible and common ways of decreasing the pore vapour pressure at high temperatures is incorporating polypropylene fibres into the concrete mix [2]; however, polypropylene fibres greatly reduce the slump flow, requiring more effort in placement and compaction [2, 3]. It would, therefore, be beneficial to find a suitable alternative to polypropylene fibres.

An alternative to polypropylene fibres in terms of reducing the spalling of concrete could be the air-entrainment of concrete which intrinsically provides the much-needed porosity. Existing studies investigating the effect of air entrainment [4–7] were conducted only on specimens at an ambient temperature after a high-temperature exposure and a cool down phase, and thus, provide insight only into the residual strength of air-entrained concrete. These results cannot be related to the mechanical properties of air-entrained concrete at high temperatures as it has not been fully understood whether the results correspond with the properties of air-entrained concrete during the high temperature exposure. Only a single study focused on the mechanical properties of air-entrained concrete at high temperatures [3] was found. Moreover, only high-strength concrete (HSC) was investigated within this study, and thus, the results and conclusions may differ from those applicable to normal-strength concrete. It is also worth to mention that the air-entrainment of concrete has an adverse effect of reducting the strength of concrete at ambient temperature due to the increase in porosity [8].

The state-of-the-art review of the fire response of air-entrained concrete shows that most relevant experimental investigations have been focused on either the residual properties of air-entrained concrete or on the high-temperature properties of HSC. The results presented in these studies may not accurately describe the material properties of normal-strength concrete at high temperatures. The main objective of this experimental investigation is to determine the effect of air-entrainment on the compressive strength of normal-strength concrete at high temperatures.

In this paper, a preliminary results of an experimental investigation focused on air-entrained normal-strength concrete at high temperatures are presented. The work presented in this paper is part of an ongoing extensive research on the Czech Technical University in Prague, Faculty of Civil Engineering – see our previous work [2, 9–15].

2. Methods

2.1. Specimens

To investigate the effect of air-entrainment on the compressive strength of concrete at high temperatures, experiments were carried out on 15 reference specimens and 15 air-entrained specimens. The experiments were carried out at an ambient temperature of 20 °C and at elevated temperatures of 200 °C, 400 °C, 600 °C, and 800 °C.

Cylindrical specimens with the diameter of 100 mm and height of 200 mm were used in this experiment in order to obtain as uniform temperature distribution throughout the specimen as possible. The concrete mixture used for the specimens was composed of commonly used components, see table 1.

Table 1. Concrete mixtures.

Mixture component [kg/m^3]	Reference specimens	Air-entrained specimens
Cement CEM I 42,5 R	370	370
Water	135	135
w/c ratio	0.36	0.36
Silica sand 0/2	863	863
Fine silica aggregate 4/8	195	195
Coarse silica aggregate 8/16	755	755
Plasticizer	2.6	2.6
Air-entraining agent	0	0.4

2.2. Heat treatment

Before the compresion test could be executed, it was first necessary to subject the specimens to the high temperature and achieve a uniform heating of the specimens. The heat treatment of the specimens was performed using a control machine, ceramic pads, and K type thermocouples, see figure 1.

Before the heat treatment, each specimen was covered with two ceramic pads, which provided the heat, and wrapped by high-temperature resistant glass wool insulation, which worked as thermo-box ensuring heat accumulation. Subsequently, a thermocouple was placed between the ceramic pads and a specimen in order to monitor the specimen surface temperature. The thermocouple and the ceramic pads were connected to the control machine.

During the heat treatment, the control machine both provided and regulated the electric current for the pads and displayed the temperature measured by the thermocouple. The temperature was always steadily gradually increased until the target temperature was reached. Then, the target temperature was held for 3 hours to ensure uniform temperature distribution throughout the specimen as recommended by [16] and stated in [2]. During the heat treatment, the temperature measured by thermocouple was regularly checked by the observing researcher, and an infrared thermometer was used to monitor the surface temperature of insulated specimens and to check for possible heat leeks.

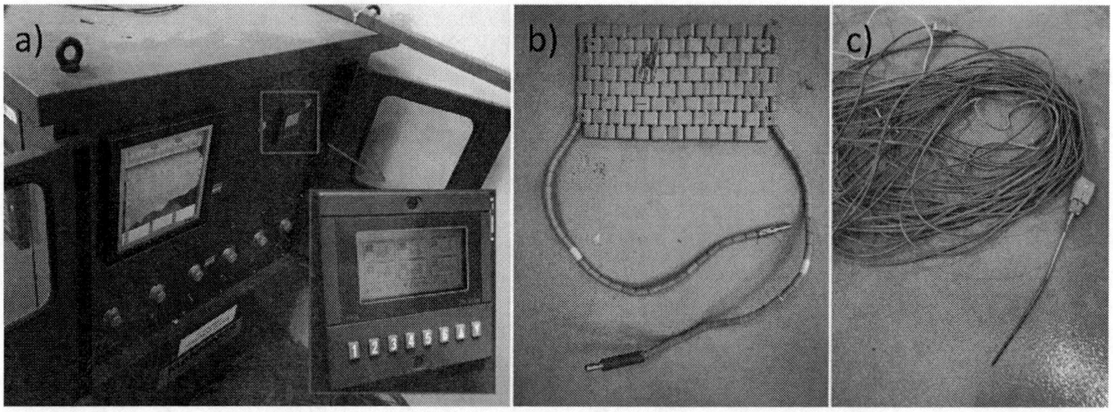

Figure 1. Setup used for the heat treatment: (a) control machine, (b) ceramic pads, (c) K-type thermocouples.

Four types of heating programs were executed, each for one of the target temperatures of 200 °C, 400 °C, 600 °C, and 800 °C. The rate of temperature increase was 600 °C per hour for all target temperatures except the target temperature of 800 °C where a rate of 1000 °C per hour was used. Three reference specimens and three air-entrained specimens were subjected to each target temperature. Overall 24 specimens were subjected to high temperatures – 12 reference specimens and 12 air-entrained specimens.

2.3. Compression test

Right after the heat treatment was finished, a compression test of each specimen was conducted in order to determine the compressive strength of the concrete specimen at the target temperature. The test was executed in accordance with Czech standard ČSN EN 12390-3 [17]. A computer-controlled compression machine was used to execute the compression test, and the test was driven by a constant rate of deformation 0.02 mm/s until the failure of the specimen. The vertical deformation of the specimen was measured by a control computer, and the relative strain was later calculated from the deformation.

3. Results and discussion

The data obtained by compression tests show that the effect of temperature on the compressive strength is similar for reference and air-entrained concrete at temperatures above 400 °C. However, the effect differs for the temperatures up to 400 °C; in this temperature range, the strength of reference concrete first increases before it starts decreasing, whereas the strength of air-entrained concrete decreases right from the beginning, see figure 2. The two processes causing this phenomenon, i.e. the different behaviour of reference and air-entrained concrete,

IOP Conf. Series: Materials Science and Engineering **596** (2019) 012032 doi:10.1088/1757-899X/596/1/012032

are further hydration of cement grains and water evaporation. Those processes both happen in both types of concrete up to the temperature of 300 °C [18] but affect each concrete to different extents. The hydration of cement grains, which results in an increase of compressive strength, is more dominant in low permeability concrete. The water loss, which adversely affects the compressive strength of concrete, is more dominant and faster in more porous concrete [19]. As the air-entrained concrete is more permeable and porous, the water loss is more dominant than the hydration of cement grains, and thus, the sharp decline of the strength is justified. In the case of reference concrete, which is less permeable and porous, the hydration of cement grains is more dominant and the water loss is delayed into higher temperatures, and thus, the initial increase of the strength followed by a sharp decrease is justified. No further hydration nor any water loss happen above the temperature of 400 °C, and thus, neither affects the strength of concrete.

In the temperature range of 400 °C to 800 °C, more degradation processes, such as micro-cracks in the cement matrix and calcium carbonate decarbonation, happen in the structure of concrete. These processes affect the reference and air-entrained concrete in the same manner – i.e., all decrease the strength of the concrete.

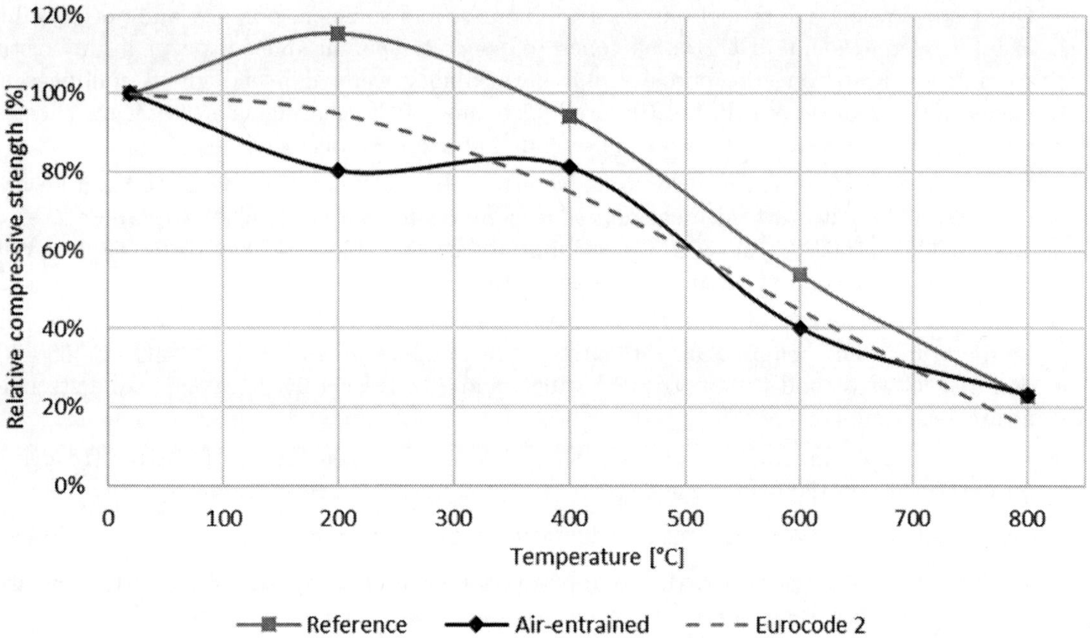

Figure 2. Relative mean strength of reference concrete and air-entrained concrete at various temperatures compared to Eurocode 2 recommended values.

The presented values of the strength of the reference concrete differ from the values recommended by [20] for siliceous aggregate concrete. The recommended values are lower and, therefore, more conservative than the presented values, see table 2.

Table 2. Relative mean strength of reference concrete and air-entrained concrete at various temperatures compared to Eurocode 2 recommended values.

	20 °C	200 °C	400 °C	600 °C	800 °C
Eurocode 2	100%	95%	75%	45%	15%
Reference	100%	115%	94%	54%	23%
Air-entrained	100%	80%	81%	40%	23%

4. Conclusions

In this paper, a preliminary results of an experimental investigation focused on the compressive strength of reference and air-entrained concrete at high temperatures has been presented. Heat treatments of specimens and compressive tests on those specimens were conducted. From the obtained results, the following observations are highlighted.

- Due to the low enough rate of temperature increase and sufficient porosity and permeability of both reference and air-entrained concrete, concrete spalling did not occur in any of the specimens. Moreover, the porosity and permeability were high enough to conduct and release almost all vapour during the prolonged time of the heating. No significant effect of the remaining vapour on the strength of concrete has been observed.
- With increasing temperature, compressive strength of the non-air-entrained concrete first increases before it starts decreasing. The compression strength at the temperatures of 200 °C, 400 °C, 600 °C, and 800 °C was found to be 115%, 95%, 54%, and 23%, respectively, of the initial strength at an ambient temperature.
- Compressive strength of the air-entrained concrete generally decreases with increasing temperature. The compression strength at the temperatures of 200 °C, 400 °C, 600 °C, and 800 °C was found to be 80%, 81%, 40%, and 23%, respectively, of the initial strength at ambient temperature.
- At a very high temperature (800 °C), the difference between the compressive strength of air-entrained and the compressive strength of non-air-entrained concrete is negligible.

From the highlighted observations, the following conclusions are made.

- With increasing temperature, the compressive strength of air-entrained concrete decreases more rapidly than the compressive strength of non-air-entrained concrete.
- Air-entrainment has no effect on the compressive strength of concrete at a very high temperature (800 °C).
- Generally, air entrainment has an adverse effect on the compressive strength of normal-strength concrete when the concrete is exposed to high temperature for a prolonged period of time, which was required for the uniform heating of specimens.

Based on the preliminary finding, it is currently concluded that air-entrainment has an adverse effect on the compressive strength of concrete when the concrete is subjected to high temperatures for an extended period of time. As no vapour is present in the pores after the prolonged exposure to high temperatures, the increased porosity of air-entrained concrete, which would otherwise have a beneficial effect of decreasing the pore vapour pressure, only has an adverse effect on the compressive strength of the concrete. A more detailed study aimed at the air-entrainment and it's effect on the compressive strength of normal-strength concrete at high temperatures will be done in our future work.

Acknowledgments
This paper was supported by the Czech Science Foundation, project no. GA17-23067S, and by the Grant Agency of the Czech Technical University in Prague, project no. SGS19/034/OHK1/1T/11. The support is gratefully acknowledged.

References
[1] Dwaikat M B and Kodur V K R 2009 *Fire Safety Journal* **44** 425–434
[2] Novák J and Kohoutková A 2018 *Fire Safety Journal* **95** 66–76
[3] Khaliq W and Waheed F 2017 *Construction and Building Materials* **150** 747–757
[4] Khaliq W, Waheed F and Khushnood R A 2018 *Construction and Building Materials* **115** 425–435
[5] Drzymala T, Jackiewicz-Rek W, Galaj J and Šukys R 2018 *Journal of Civil Engineering and Management* **24** 138–144
[6] Drzymala T, Jackiewicz-Rek W, Tomaszewski M, Kuś A, Galaj J and Šukys R 2017 *Procedia Engineering* **172** 256–263
[7] Ahmad A H H and Abdurrahman R B 2009 *Journal of Civil Engineering and Management* **17** 28–41
[8] Thomas T L 1979 *The effects of air content, water cement ratio, and aggregate type on the flexural fatigue strength of plain concrete* Master's thesis Iowa State University
[9] Goremikins V, Blesák L, Novák J and Wald F 2017 *Journal of Structural Fire Engineering* **8** 287–299
[10] Goremikins V, Blesák L, Novák J and Wald F 2016 *Acta Polytechnica* **56** 258–264
[11] Holan J 2019 Compressive strength of air-entrained concrete at high temperatures *PhD Workshop 2019* ed Horáková A and Petřík M (Prague, Czech Republic: Department of Concrete and Masonry Structures, Faculty of Civil Engineering, CTU in Prague)
[12] Novák J and Kohoutková A 2017 *IOP Conference Series: Materials Science and Engineering* **246** 1–7
[13] Tkalenko I, Tretyakov A, Wald F, Novák J, Štefan R and Kohoutková A 2017 *ce/papers* **1** 2678–2687
[14] Goremikins V, Blesák L, Novák J and Wald F 2015 To testing of steel fibre reinforced concrete at elevated temperature *ASFE 2015 Proceedings* pp 349–354
[15] Novák J and Kohoutková A 2017 *Procedia Engineering* **172** 784–790
[16] RILEM TC 129-MHT 1995 *Materials and Structures* **28** 410–414
[17] ČSN EN 12390-3 2009 *Testing Hardened Concrete Part 3: Compressive Strength of Test Specimens* (ÚNMZ)
[18] Saad M, Abo-El-Enein S A, Hanna G B and Kotkata M F 1996 *Cement and Concrete Research* **26** 669–675
[19] Ma Q, Guo R, Zhao Z, Lin Z and He K 2015 *Construction and Building Materials* **93** 371–383
[20] EN 1992-1-2 2004 *Eurocode 2: Design of concrete structures – Part 1-2: General rules – Structural fire design* (CEN)

Utilization of recycled plastic for plastic-based concrete

A Horáková and J Novák

Czech Technical university in Prague, Faculty of Civil Engineering,
Thákurova 2077/7, 166 29 Praha 6 - Dejvice, Czech Republic

E-mail: josef.novak.1@fsv.cvut.cz

Abstract. The environmental aspects of sustainable development in the construction industry consist in the utilization of secondary raw materials in a design and construction of new structures. The fact that China significantly reduced the import of plastic waste in 2017 raises the question of dealing with this waste in other states. A complicated recycling process of waste plastics puts pressure on scientific community to find new, more efficient ways of using this waste. The paper deals with the behaviour and mechanical properties of a completely innovative material called waste plastic-based concrete composed of natural aggregate and plastic waste which replaces cement as a binder. The completely unique composition of this concrete required to firstly test production technology and subsequently to conduct standard quasi-static experimental tests to obtain basic knowledge about the behaviour and mechanical properties of this composite. The obtained knowledge show that special attention must be paid to the production of the material which takes place under elevated temperature. The investigated composite has a relatively high tensile strength compared to conventional concrete and brittle fracture behaviour. In the next phases of the research the optimization of the production technologies and the composition of the composite will be provided.

1. Introduction

Currently, several studies deal with the utilization of recycled plastic as a binder for concrete. Recycled polyethylene terephthalate is used for this purpose. This polymer is currently used for production of beverage bottles and food packaging. Millions of tons of PET bottles are produced every year and it is desirable to deal with the utilization of this polymer, because PET is not biodegradable. The utilization as a binder in concrete is an effective way of recycling, which also leads to cement saving in the construction industry. This cement saving helps to reduce the consumption of raw materials and negative effects on the environment. Some recent studies [9,10] compared a polymer concrete made from virgin and recycled polymer. These studies reported that properties of these polymer concretes differ only insignificantly. Waste PET is usually during the processing depolymerized and unsaturated polyester resin (UPR) is obtained. Accelerator and initiator are added to the resin immediately before mixing with filler. A disadvantage is the expensive process of recycling of waste PET and its treatment before using as a concrete binder. Production costs are significantly lower, when another manufacturing process is selected – cut pieces of raw PET waste are heated together with the filler and melted. However, only few studies deal with this technology.

Several studies ([4],[8]) reported that the compressive and flexural strength of polymer concrete containing as a binder recycled waste PET were higher compared to conventional concrete containing as a binder cement. The seven-day compressive strength of samples containing 9 % of resin reached approximately 60 MPa [2]. The study [6] reported, that the compressive strength reached even 80 MPa, when the samples contained 10 % of resin.

A very fast strength development is also significant [9]. After 24 hours, 75% of its final compressive strength was obtained. After 4 days, 83% of its final strength was obtained. The compressive strength of conventional concrete reached after 24 hours approximately 20% of its final compressive strength. Therefore, this concrete is suitable for precast elements production, because the fast strength development allows to speed-up a manufacture. Some studies (e.g. [7]) observed a different failure mode of concrete containing plastic aggregate. Samples did not exhibit the brittle failure, which is typical for conventional concrete and did not separate in two parts.

The ratio of tensile and compressive strength of polymer concrete is higher, in comparison to conventional concrete. The ratio of tensile and compressive strength of conventional concrete is about 10 to 15 %. In case of polymer concrete, this ratio is about 30 – 50 % [1]. Polymer concrete differ from conventional concrete also in creep behavior. The study [2] report that the creep strains grow rather fast at early ages in comparison with ordinary concrete. The polymer concrete shows more than 20% of its long-term creep within the first two days, and about 50% during the first 20 days.

An advantage of polymer concrete is a lower porosity and water absorption. The study [3] report that the water absorption is only 0,47 %. These properties caused that this concrete is more resistant to chemical attack, compared to conventional concrete. Therefore, polymer concrete is suitable for sewage pipes surface or floors in chemical plants.

A disadvantage of this material is a lower heat resistance due to low melting point of the polymer (approximately 260 °C). Hence, the mechanical properties change at higher temperatures. The study [6] report that increasing the temperature from 25°C to 60°C leads to the reduction in the compressive strength by 40 % and to the reduction in the elastic modulus by 35-40 %.

The ratio of tensile and compressive strength of polymer concrete is higher, in comparison to conventional concrete. The ratio of tensile and compressive strength of conventional concrete is about 10 to 15 %. In case of polymer concrete, this ratio is about 30 – 50 % [2].

The mechanical properties of the material depend primarily on the polymer concrete composition - the filler and resin content, the type of aggregate and the admixtures. Study [10] report that an increase in compressive strength with increasing resin content. Specifically, the compressive strength of samples containing 5 % of resin reached 5 MPa. For 10 % resin content, the compressive strength increased to 12 MPa, for 20 % resin content even to 20 MPa. The study [11] reported that the compressive strength is also influenced by the maximum size of aggregate particle. The compressive strength increased with increasing particle size up to 10 mm particle size, when larger particles was used, the compressive strength reduced.

The study [3] deals with the influence of admixtures on the strength of polymer concrete. In the frame of this study, there was the compressive strength of samples containing fly ash, bitumen and nano-calcium carbonate investigated. The best results were obtained for the samples contained fly ash. For 15 % content of fly ash, the compressive strength increased by 32 %, the flexural strength even by 79 %, compared with samples without any admixtures. This is probably caused by a surface of fly ash particles with large negative charge, which supports the development of stronger links with polymer mix. Further cause is probably the increase of crystallization degree of PET and the improvement of the concrete workability. The addition of bitumen also improved the compressive strength of polymer concrete, but only up to content of 5 %. For this bitumen content, the compressive strength increased by 10,7%. When the bitumen content was higher than 5 %, the compressive strength decreased. The influence of nano-calcium carbonate content on the compressive strength was not significant.

Several studies [5,9] analyze dependence of mechanical properties on the amount and kind of accelerator and initiator, which are used during the polymer concrete production. The study [9] reported that properties of polymer concrete are significantly influenced by the amount ratio of accelerator, initiator and resin. Therefore, it is necessary to find the optimal amount ratio of these compounds. This study reported that the optimal content of initiator is 2% and optimal content of accelerator is 0,5%. In this study, methyl ethyl ketone peroxide (MEKP) was used as an initiator and

IOP Conf. Series: Materials Science and Engineering **596** (2019) 012033 doi:10.1088/1757-899X/596/1/012033

cobalt naphthenate was used as an accelerator. The study [5] compared the compressive strength of polymer concrete made by using two kinds of initiator - methyl ethyl ketone peroxide (MEKP) and benzoyl peroxide (BPO). According to this study, using methyl ethyl ketone peroxide (MEKP) as an initiator leads to higher compressive strength.

2. Experimental programme

2.1. Preparation of the samples

The samples were made of fine aggregate (aggregate size 0 - 4 mm) and waste PET which was cut into small pieces. The production consists in homogenizing the mixture of plastic waste and fine aggregate in the weight ratio of 23:77 at high temperature. At first, the aggregate was heated up to a temperature of about 430 ° C. After that, the PET waste was added to the aggregate and melted. After mixing the two components, the temperature decreased, and the mixture was homogenized. Optimal workability of the mixture was reached at about 310 ° C, when the consistency of the mixture was viscous. Then, the mixture was placed into the prepared forms. The forms were pre-heated to eliminate the thermal shock which could cause a surface disruption of the samples. After that, the moulds with the mixture were left at room temperature and after cooling, the samples were removed from the moulds. Overall, eight samples were made, four samples were cube shaped and four samples were beam shaped. The dimensions of cube shaped samples were 100 x 100 x 100 mm and the dimensions of beam shaped samples were 50 x 50 x 200 mm (figure 1).

Figure 1. Waste plastic-based concrete samples.

2.2. Experiments

The detection of basic mechanical properties of the waste plastic-based concrete was the aim of the experiments. The beam shaped samples were used for the three-point bending test, the distance of the supports was 180 mm. The deformation-controlled test was conducted with rate 0.02 mm/s. The loading force and the mid-span deflection were monitored. The obtained experimental data are demonstrated in flexural strength-deformation diagram. The diagram describes a relationship between flexural strength of waste plastic-based concrete and deformation. The cube shaped samples were used for the compression test. As in the flexural test, the loading force and the vertical deformation were monitored during compression test and subsequently used for the elaboration of a compressive strength-deformation diagram (figure 2). This diagram shows the relation between the compressive strength of waste plastic-based concrete and deformation.

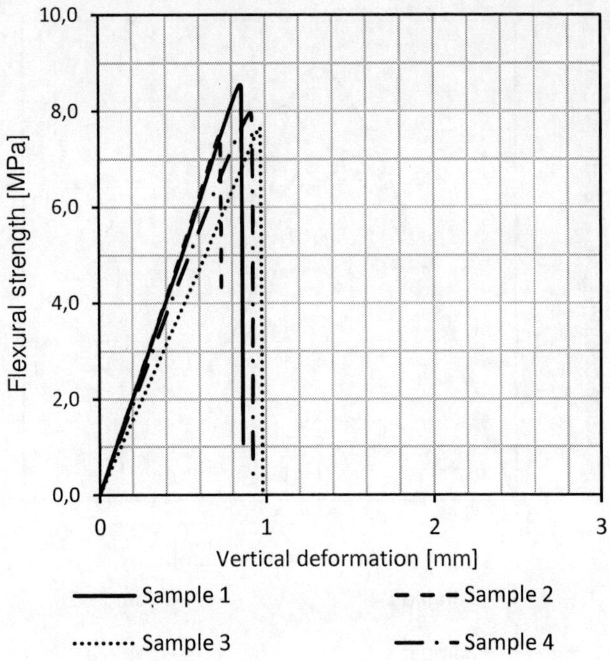

Figure 2. Flexural strength - deformation diagram of waste plastic-based concrete.

3. Results

The findings show that the mechanical behaviour of the waste plastic-based concrete is very similar to the mechanical behaviour of conventional concrete. The mechanical behaviour of the waste plastic-based concrete under the tensile stress is linear until a brittle failure occurs. The influence of ductile behaviour of plastic waste on the mechanical properties of the investigated composite is negligible. The reason may be either the low content of plastic waste in the composite mixture or the change in plastic properties caused by elevated temperature while being heated up during the production. As a consequence the waste plastic-based concrete is not possible to consider as a ductile material with post cracking tensile strength. This fact is limiting for the utilization of this material. On the other hand, the obtained values of the peak tensile strength ranged from 7.54 MPa to 8.55 MPa and the average value of tensile strength was 7.91 MPa. This value of tensile strength corresponds to the tensile strength of high performance cement-based concrete.

The behaviour of the waste plastic-based concrete under the compressive stress is also similar to conventional cement-based concrete. At first, the mechanical behaviour is linear. Later, under higher compressive stress, the material starts yielding until the peak compressive strength is reached. The measured values of the compressive strength ranged from 20.69 MPa to 27.03 MPa and the average value of compressive strength was 22.7 MPa. Based on the stress-strain diagram obtained from experiments, it seems that the elastic modulus of this waste plastic-based concrete is lower that elastic modulus of conventional concrete (figure 3).

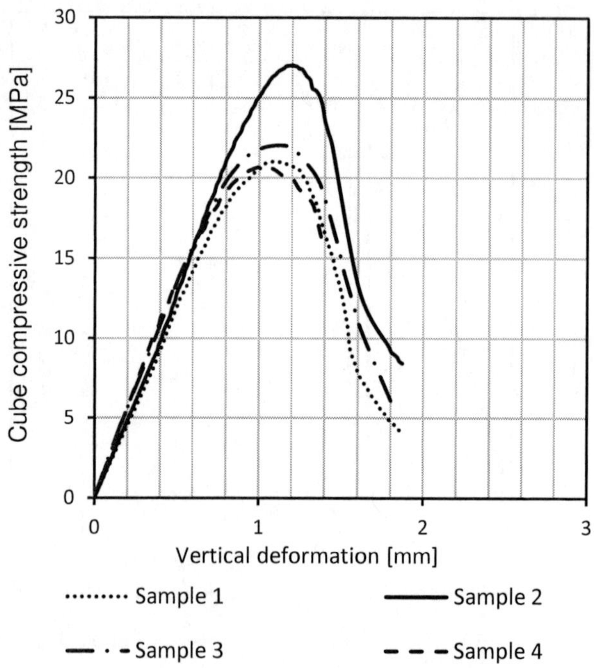

Figure 3. Compressive strength - deformation diagram of waste plastic-based.

4. Conclusion

The waste plastic-based concrete made of waste PET is a composite material which enables to utilize the waste material. The properties of this material differ not significantly from the properties of conventional cement-based concrete. The low ductility and the brittle tensile failure limit the utilization field of this material. However, the advantage of this material consists in its high tensile strength. Currently, there are not many findings about this material and only few experiments have been conducted so far. It is desirable to provide more experiments which verify the properties of the material.

Acknowledgments

This work was carried out within the framework of both the grant "Development of technological line for waste thermoplastic and recycled construction waste recovery for production of polymer concrete building components", No. TH02030412, sponsored by the Technology Agency of the Czech Republic and the grant "Durability of concrete structure and assessment of its life cycle", No. SGS19/149/OHK1/3T/11, sponsored by the internal CTU grant agency.

References

[1] Byung-Wan J, Seung-Kook P and Jong-Chil P 2008 Mechanical properties of polymer concrete made with recycled PET and recycled concrete aggregates *Construction and Building Materials* 22:2281–2291

[2] Byung-Wan J, Ghi-Ho T and Chang-Hyun K 2017 Uniaxial creep behavior and prediction of recycled-PET polymer concrete *Construction and Building Materials* 21:1552–1559

[3] Zhi G, Huang D, Sun R and Gao Z 2014 Properties of plastic mortar made with recycled polyethylene terephthalate *Construction and Building Materials* 73:682–687

[4] Ge Z, Wang H, Zhang K and Li P C 2012 Investigation on the Properties of Plastic Mortar *Journal of Shandong University* **1**:1–3

[5] Mahdi F, Abbas H and Khan A A 2010 Strength characteristics of polymer mortar and concrete using different compositions of resins derived from post-consumer PET bottles *Construction and Building Materials* **24**:25–36

[6] Rebeiz K S 1995 Time-Temperature Properties of Polymer Concrete Using Recycled PET *Cement & Concrete Composites* **17**:119- 124

[7] Reis J M L 2011 Effect of aging on the fracture mechanics of unsaturated polyester based on recycled PET polymer concrete *Materials Science and Engineering* **528**:3007–3009

[8] Siddique R, Khatib J and Kaur I 2008 Use of Recycled Plastic in Concrete: a Review *Journal of Waste Management* **28**:1835–1852

[9] Tawfik M E and Eskander S 2006 Polymer Concrete from Marble Wastes and Recycled Poly(ethylene terephthalate) *Journal of Elastomers and Plastics* **38**:65-79

[10] Miranda V J M, Narváez H L, Tapia L J I, Martínez F E E and Salvador H L 2014 Polymer mortars prepared using a polymeric resin and particles obtained from waste pet bottle *Construction and Building Materials* **65**:376–383

[11] Yao Z, Zhang Y, Ge Z, Pan X, Jin Z and Han J 2015 Mix Proportion Design and Mechanical Properties of Recycled PET Concrete *Journal of Testing and Evaluation* **43**:20140059

IOP Conf. Series: Materials Science and Engineering **596** (2019) 012034 doi:10.1088/1757-899X/596/1/012034

Bio-active concrete tile

M Husarčíková, P Štemberk, M Petřík and M Frantová

Department of Concrete and Masonry Structures, Faculty of Civil Engineering, Czech Technical University in Prague, Thákurova 7, 166 29 Prague 6, Czech Republic

E-mail: michaela.frantova@fsv.cvut.cz

Abstract. The aim of this work is to design and fabricate bio-active concrete tiles which encourage rapid plant coverage of building walls and urban spaces with vegetation. A design comprises two different types of tile, which one of them intends to be used as a planter for a variety of climbing vegetation. Through the process of designing and manufacturing suitable mould for tiles, a complex macro pattern was developed to ensure water retention on the structural surface. The results of this work provide an alternative solution to the existing green wall systems by implementation of a bioreceptive cementitious material.

1. Introduction

Today, more than half of the world's population live in the cities. With the rapid growth of the population, this number is still increasing and the situation in the urban areas are getting worse. Cities have become a huge built impervious environment, and they are facing enormous problems connected with the loss of natural green areas. A dense city infrastructure leaves no space for implementing the greenery on a ground level. This has resulted in the high demand for the green walls and roofs since they required little or no ground space. In particular, the living walls effectively uses vertical spaces as an opportunity for greenery. Their environmental and health benefits, as well as appealing aesthetics positively influence life in the urban area and lessen the environmental impact of the city. Unfortunately, the problem regarding these living walls rests on their usually very costly and demanding maintained system. The plants need a supporting structure with the complex watering system and sometimes the costs might outweigh the benefits. Therefore, an innovative new solution is needed. [1], [2], [3]

As opposed to the typical green walls a new approach of integrating the vegetation directly on the building surfaces has been developed. By altering the physical and chemical properties of traditional concrete a new type of biologically receptive concrete has been produced. This concrete encourages and sustain the growth of the microorganisms such as mosses lichens and algae directly on its surface and thus increases the cryptogamic cover of the material. The bioreceptive concrete supports a plant life to thrive on buildings in a way that is both more sustainable and more efficient than existing green walls. [4], [5] It brings an interdisciplinary approach of architecture, engineering and biology.

The aim of this work is to describe the design and the production technology of the bio-active concrete tile. The tile consists of two layers – the inner structural layer is made from concrete with the ordinary Portland cement and the outer bioreceptive layer is made from magnesium phosphate cement concrete to create an environment for plant growth. The production quality is also discussed and alternative solution is suggested.

IOP Conf. Series: Materials Science and Engineering **596** (2019) 012034 doi:10.1088/1757-899X/596/1/012034

2. Design and Production of the Tiles

2.1. Shape of the tile and numerical analysis

In order to maximize the green coverage and biodiversity, it was decided to combine two green walls technologies - green facades and green walls [6]. A vision to create a wall that will support climbing plants and at the same time serve as a module for pre-vegetated rich flora led to the decision of designing two types of tiles, Standard tile and Planter (Pocket) tile (figures 1 and 2). The size and shape of the tile units could be modified according to specific requirements of the area.

The tile represents an elevated 3D design for microflora to be established and proliferates on its surface. The tile consists of two-layered concrete. The structural layer is 10 mm thick. while the outer bioreceptive layer with a thickness of 10 mm consists of multileveled steps ranging from 5–20 mm. (figure 1)

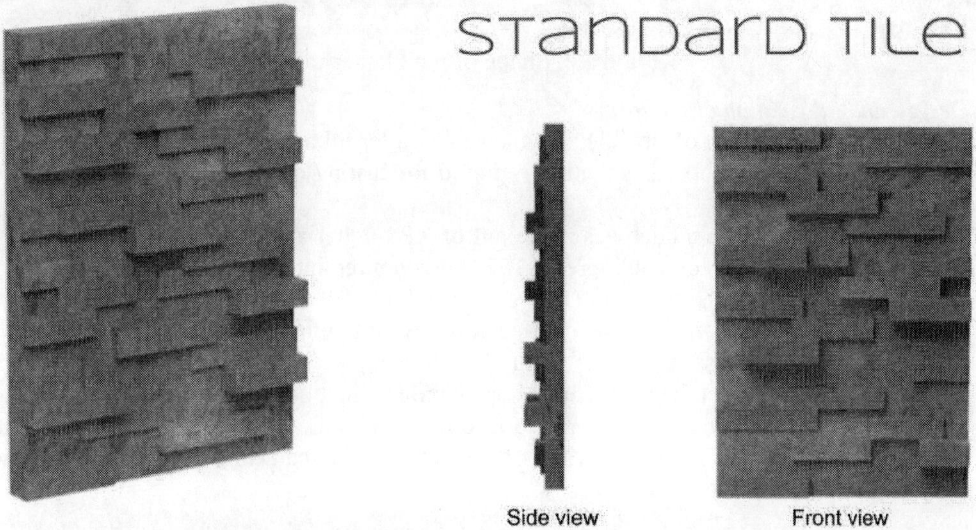

Figure 1. Shape of the Standard tile.

The planter tile or so-called pocket tile represents the same elevated 3D design, however, the pocket for planting is added in order to keep a wide variety of plants and to control the density of greenery covering the walls. A structural part creates a pocket for seeding with the thickness 10 mm. The front face layer is designed to be out of the bioreceptive concrete with the thickness varying from 10–30 mm. The volume of the pocket is 3 litres which are suitable for providing enough soil for a variety of climbing vegetation. At the bottom of the tile two drainage holes were designed as a precaution for over-watering, as well as for decreasing the risk of damage from freeze-thaw action (figure 2).

There are two ways of attaching tiles to the building structure, adhering and anchoring. Tiles could be attached directly to the wall structure by wall tiles adhesives. However, due to the weight of the concrete tiles, the installation would be more complicated and precaution of the slipping should be in concern. With mechanical anchoring, the tile would be attached to the wall by a suitable anchor.

Numerical modelling in program SCIA Engineering was used to analyse the behaviour of planter tile under the load action of soil and water inside the tile. The tile was considered as fully fix supported on the backside. The lateral soil and water pressure were applied on the three sides of the tile and the vertical load on the bottom of the tile. There were considered following load cases: permanent load (self-weight, soil pressure) and variable load (water pressure).

Numerical simulation results were obtained for the main stresses at both surfaces of the tile. Greatest tension strength was 0.3 MPa and 0.9 MPa at compression at the inner surface. For the outer surface, the highest tension was 0.9 MPa and 0.1 MPa at compression.

Figure 2. Shape of the Planter tile.

2.2. *Design and fabrication of tile forms*

The planter tile was made out of flexible silicone rubber after mixing with the catalyst at normal room temperature. This silicone rubber is suited as a mould for casting of various materials, in my case for concrete.

Firstly, the envisioned tile model was made out of XPS material (figure 3). The model was placed in enclosed space and the silicon (Lukopren N1522) was poured into it. The hardened silicon created a front face and side walls of the tile form (figure 4). The inner fill for the designed hollow space of the pocket tile was made out of XPS and its surface was covered with a fluid sealant of Lukopren S3782 as a separation layer. This assembly was secured with clamps for rigidity and tightness of the form. The mould made out of silicon benefits from its easy unforming due to its flexible structure. However, the replica model has to be made very precise in order to obtain an exact shape that we wished for. Special care should be during the mixing of Lukopren with catalyst, where exact amounts are necessary for suitable form structure.

Figure 3. Model of Planter tile.

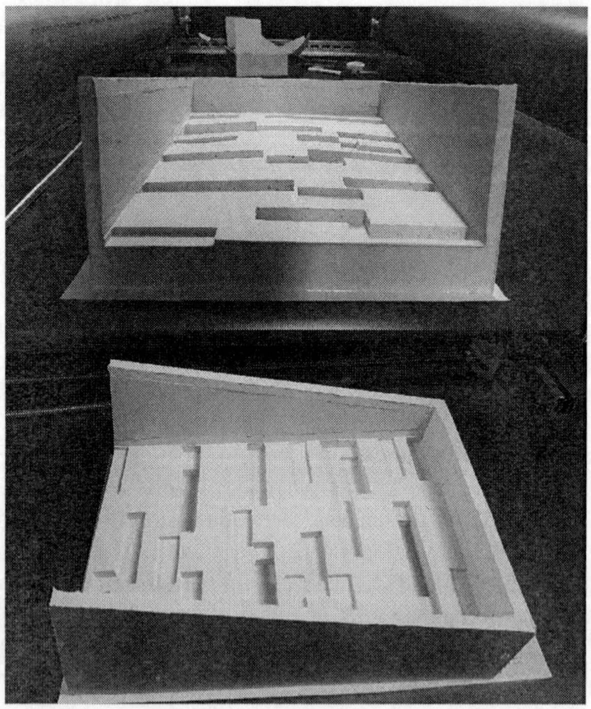

Figure 4. Silicon tile form.

IOP Conf. Series: Materials Science and Engineering **596** (2019) 012034 doi:10.1088/1757-899X/596/1/012034

Regarding the form for Standard tile, it was decided to use digital fabrication for this model. In a case of digital fabrication, digital data drives manufacturing equipment such as 3D printers, laser cutters and CNC machines, to form various geometrical shapes.

The first step was creating a virtual model of a tile form using Fusion 360 software (figure 5). The model served as an input for the fabrication tool, in my case CNC milling machine. (CNC -Konečný s.r.o) In the machine, the tool-paths were generated to guide the cutting tool (figure 6). A solid block of XPS polystyrene of thickness 80 mm was used for the fabricated model. The XPS block was shaped by removing material through cutting, drilling, boring and grinding (figure 7). The form was cut in the middle into two pieces for later easier unforming. The finishing layer of silicon sealant, Lukopren S3782, was used as a separation layer between XPS form and concrete.

Figure 5. Modelling of 3D form in Fusion 360.

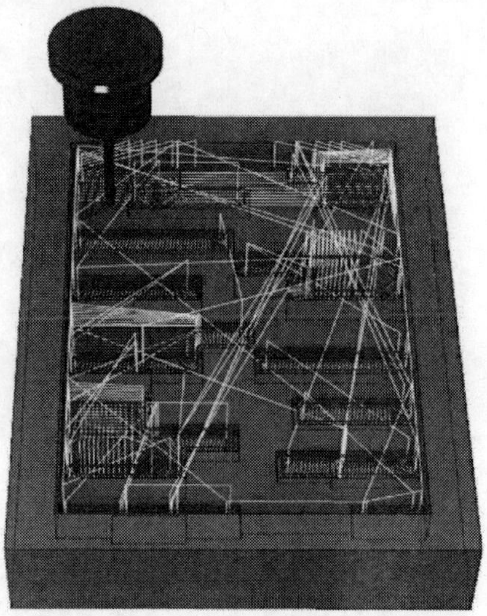

Figure 6. Simulating of tool-path for drilling machine.

Figure 7. Fabrication of tile model by using CNC milling machine.

One of the advantages of using digital fabrication is that it makes more precise shapes and details for a model which could be easily altered in the future if necessary. Work is more efficient and previewing and simulating the tool-paths prevents future problems during manufacturing. On the other hand, the design is limited to the size of the manufacturing machines and the thickness of the material.

By utilizing novel design and digital fabrication methods, multileveled surface and elevations were created to improve the facade performance. All of the designed formwork is repeatable and can be used for further series production.

2.3. Concreting of tiles
The concreting of the Standard tile was carried out in the two following phases. The first layer of magnesium phosphate cement concrete was poured. The substance had dry consistency and was spread across the tile form by hand, not to consolidate it much in order to maximize the porous structure. For the better bonding connection of two layers, the surface of magnesium phosphate cement was roughened. After that, the second layer of ordinary Portland cement concrete was poured on top of it. The anchor with bolt was embedded into the fresh concrete (figure 8).

Figure 8. Concreting procedure of Standard tile.

The Planter tile was concreted in a horizontal position. The first layer of magnesium phosphate cement concrete was placed into the silicon form. Right after, the XPS pocket form was placed and attached in the form, leaving a free space of 1 mm underneath. The mixture of ordinary Portland cement concrete needed to be very fluid in order to fill the space beneath the XPS form. The problem became when the concrete was filling the space under the pocket form and at the same time started to uplift form from its position. As a result, an alteration of the form was necessary and the four points, for pressing and holding the form in the right position, were installed. This has created the four holes on the tile backside. Another problem becomes while unmolding the tile. The inner part of the tile form was not able to be taken off. Therefore, it was necessary to dig it out (figure 9).

Figure 9. Concreting and removing of Planter tile form.

3. Results and Discussion of Production Quality

In the end, two types of bio-active concrete tile were fabricated - Standard tile (figure 10) and Planter tile (figure 11). Both tiles are made out of two-layered concrete. Layers can be distinguished due to their different colouring. The first, structure layer made out of ordinary Portland cement with the reinforcement of dispersed PVA fibers has grey colour adjusted by the presence of microsilica. The second, bioreceptive layer of magnesium phosphate cement mixture was successfully designed in order to support microorganism growth on its surface. The brown colour of the layer is due to the natural colour of magnesium phosphate cement. The colour of tiles could be altered by concrete pigments, or by addition of microsilica.

Figure 10. Standard tile.

Figure 11. Planter tile.

As a result of complications during the concreting of Planter tile, few imperfections were detected (figure 12). The ordinary Portland cement mixture did not entirely pass under the inner part of the tile form, creating a small area with only one layer of magnesium phosphate concrete.

Another unexpected result of concreting is the presence of four holes caused by fixing of inner form against the uplift. However, these holes could be used for anchoring of the tile as well for the drainage of water at the bottom of the tile.

To overcome the problems in the future, an alternative solution for tile manufacturing is suggested (figure 13.). The new inner part of tile form is hollow and made out of wood. The concreting of the tile will be proceeding in a vertical position for better form fixing and space controlling around the inner form.

Figure 12. Defects of Planter tile.

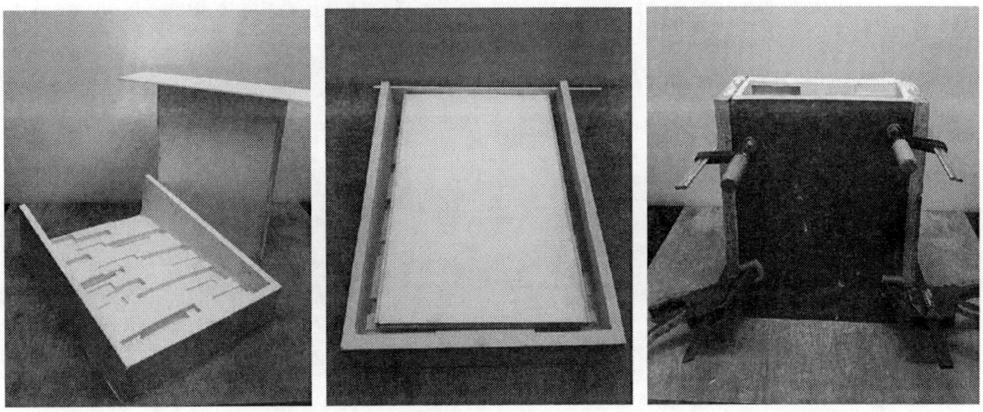

Figure 13. New form for Planter tile.

4. Conclusion

This work presents an alternative to the traditional green wall systems by designing and producing a bio-active concrete tile. The tile is made of two concrete layers, a structural (load-bearing) layer and a bioreceptive layer. Two types of a tile were designed. One is of the regular flat shape and the other serves as a planter box enabling growth of variety of plants.

An environmentally driven design with a multileveled surface structure creates suitable niches for plant embedding and at the same time increases the ability of the climbing plants to adhere to the surface of the wall. Moreover, the combination of various elevated steps helps to retain and accumulate water necessary for bio-colonisation.

Regarding manufacturing of the tile moulds, the digital fabrication, which utilized CNC milling for shaping the XPS-based moulds, proved to be very effective and precise compared to the hand-made silicon forming which required tedious modification due the its imperfection. A disadvantage of the digital fabrication was its manufacturing dimensional limitations. The advantage was the precision of

CNC milling which ensured desired angling the sides of the moulds so that the moulds could be reused for further concreting.

The tiles will be installed in the outdoor environment for observation of plant coverage development as well as to analysed the long-term performance of the tile. Further investigation would be necessary for evaluating the benefits of the entire wall composed out of the bio-active concrete tiles.

Acknowledgements
This work was financially supported by the Czech Technical University in Prague research project no. SGS19/038/OHK1/1T/11, which is gratefully acknowledged. The authors thank the companies Magnesia GmbH, Lehmann&Voss&Co.KG and Slovenské magnezitové závody akciová spoločnosť Jelšava for their support.

References
[1] Ottelé M 2011 *The Green Building Envelope - Vertical Greening* dissertation
[2] Lundholm J T 2006 Green Roofs and Facades: A Habitat Template Approach *Urban Habitats* **4** 87–101
[3] Velazquez A 2019 Green Walls Becoming More Popular In Singapore Buildings *Greenroofs.com*
[4] Guillitte O 1995 Bioreceptivity: a new concept for building ecology studies *Science of The Total Environment* **167** 215–20
[5] Manso S, Muynck W D, Segura I, Aguado A, Steppe K, Boon N and Belie N D 2014 Bioreceptivity evaluation of cementitious materials designed to stimulate biological growth *Science of The Total Environment* **481** 232–41
[6] Green Roofs for Healthy Cities 2008 Introduction to Green Walls *Greenscreen.com*

IOP Conf. Series: Materials Science and Engineering **596** (2019) 012035 doi:10.1088/1757-899X/596/1/012035

Mechanisms behind radiation-induced deterioration of concrete

Y Khmurovska and P Štemberk

Czech Technical University in Prague, Faculty of Civil Engineering, Department of Concrete and Masonry Structures, Thákurova 7, 166 29 Prague 6, Czech Republic

E-mail: yuliia.khmurovska@fsv.cvut.cz

Abstract. This paper describes the mechanism of radiation-induced deterioration of concrete and its components based on the literature review. The deterioration mechanism in different levels starting from the interaction between neutron and nucleus through the mineral metamictization and cement paste shrinkage up to the reduction of mechanical properties of concrete is explained in detail. All basic dependencies and patterns of volumetric change of minerals, aggregates, cement paste and finally concrete are also described in order to create a base for the future development of numerical models. Finally, the reduction of concrete mechanical properties is in correlation with its volumetric expansion. The radiation-induced volumetric expansion of concrete is affected by irradiation conditions (neutron fluence, neutron spectrum and temperature) and concrete composition (the amount, the proportion, the type and the structure of the minerals in the aggregate composition and the amount, the composition, the age and the structure of the cement paste).

1. Introduction

According to The World Nuclear Industry Status Report [1], the unit-weighted average age of the world operating nuclear reactors is rising continuously, and by mid-2017 stood at 29.3 years, up from 29.0 a year ago and 28.8 two years ago [2]. Despite the fact that the initial design operating period of nuclear power plants is 30 to 40 years, over a half of the total 234 units have operated for 31 years and more, including 64 that have run for 41 years and more. This means that the nuclear power plants which are nearing the end of their licensed operating period need to be either shut down or their operation time licenses have to be prolonged. The license renewal is preferable before the reactor shutdown because of the long nuclear power plant decommissioning process which is also associated with significant costs. For example, 84 of the 99 operating U.S. units had received a license extension with a further nine applications under review as of 1 July 2017.

The long-term exposure to radiation can cause the damage in the construction materials. Since the biological shield of the nuclear power plants most often consists of concrete, the possible damage of concrete due to radiation is of particular interest. In order to predict the damage of concrete and create a comprehensive numerical model, it is necessary to understand the mechanism of concrete deterioration and identify the main relations and patterns.

The research presented in this study is focused on review of the available literature in order to identify the main dependencies in the field of the effect of radiation on concrete for the future numerical modelling.

2. Deterioration mechanism

2.1. Neutron irradiation

Interaction of neutrons with matters occurs through inelastic scattering, elastic scattering and absorption depending on the neutron energy. The fast neutrons are scattered inelastically. When a high-energy fast neutron collides with a nucleus, it is absorbed and the compound nucleus is created. The compound nucleus decays into a lower-energy neutron and an excited nucleus. The nucleus gives up excitation energy by emitting one or more gamma rays (so-called secondary gamma-ray). The traveling direction of both, the neutron and the nucleus, are changed after collision. This process may repeat until the fast neutron becomes an intermediate neutron. It should be noted that inelastic scattering is typical for neutron collision with heavier atomic nuclei. The intermediate neutrons are scattered elastically. After intermediate neutron collision with the nucleus, its energy is transferred to the recoiling nucleus and both, the neutron and the nucleus, change their traveling direction, but the intermediate neutron does not have enough energy to cause the nucleus excitation and the subsequent gamma-ray release. The elastic collisions may repeat until the intermediate neutron becomes the thermal neutron. When the thermal neutron collides with the nucleus, its energy is too low to cause further scattering and its energy is absorbed in the collision.

Concrete is a composite material which consists of cement paste, aggregates (fine and coarse) and water, while in turn the aggregates are composed of minerals. The high energy neutrons cause defects in the lattice structure of the minerals by the elastic scattering and the inelastic scattering. According to [3], the threshold energy of the neutrons to cause the defects in the lattice structure of the minerals is equal to 10 keV. A vacancy at the original site and an interstitial defect at the new location (Frenkel pair) are created due to the neutron ballistic effect. These point defects can be accumulated and result in mineral metamictization (amorphization in some literature) which leads to the significant change in the properties of minerals (volume, density etc.) and consequently changes the properties of concrete (volume, strength, elastic modulus etc.). It is believed that the effect of the neutron irradiation on the cement paste is much smaller in comparison with the minerals and the aggregates and rather results in the cement paste shrinkage, because the calcium silicate hydrate (C-S-H) gel, the main component of cement paste, is amorphous and has very porous structure. The pore volume represents about 20 % of C-S-H volume. The threshold energy of the neutrons to cause the damage to the cement paste is much higher and equal to 0.8 MeV. However, [3] discusses the Portlandite metamictization due to the neutron irradiation and its subsequent reaction with the carbon dioxide presented in air. This leads to $CaCO_3$ formation (carbonation reaction), which can affect the properties of the cement paste. It should be noted that the high-energy neutrons may also excite water and thus provoke the radiation-induced water decomposition (radiolysis) with the consequent drying due to the gas release, which may mainly vary the properties of the cement paste since the aggregates have a much lower amount of water in their composition.

According to [3], the contraction of the lattice structure of the minerals due to the creation of the vacancy is lower than the lattice structure expansion due to the interstitial defects, see Figure 1, therefore the mineral metamictization due to the accumulation of the defects causes the radiation-induced volumetric expansion (RIVE). An uneven expansion of the different minerals in the aggregates causes concentration of stresses. When the stresses exceeds the strength of the material the cracks start to develop. Moreover, the cement paste may shrink under the exposure to neutron irradiation due to the radiation-induced water evaporation which in combination with the aggregate expansion causes development of cracks in the cement paste. The crack propagation in the aggregates and the cement paste leads to reduction of the elastic modulus and the compressive and tensile strength of concrete. Thus, the volume change of the aggregates and the cement paste can be considered as the primary factor of the radiation-induced damage of concrete.

The effect of neutron irradiation on creep of concrete is not yet understood. In the study presented in [4], the influence of neutron irradiation on concrete creep was studied and it appeared to be insignificant, nevertheless, the final conclusion is difficult to draw from this experiment. According to [3], the creep of the irradiated samples is approximately seven times higher in comparison with the control samples, however an increase in the creep is explained by the high temperature (240 °C) which accompanied irradiation in the test reactor. The research presented in [5] also concludes that the increase in the creep is equivalent to its increase under the elevated temperature.

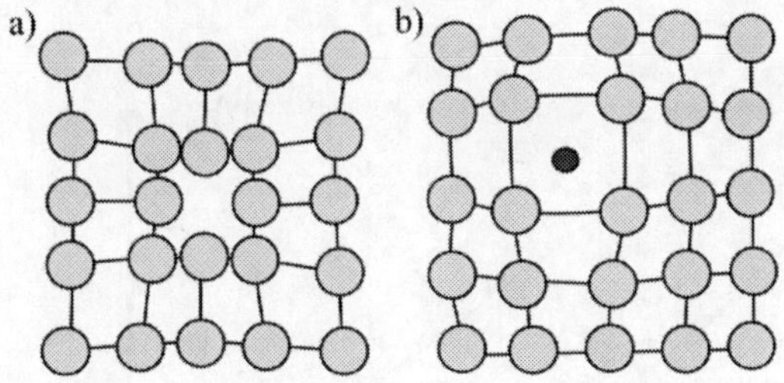

Figure 1. Lattice defects (a) vacancy, (b) interstitial.

2.2. *Gamma-ray irradiation*

The gamma-rays emitted during the inelastic neutron scattering cause the electron excitation and subsequently remove electrons from the atoms by the electron-positron pair production, the Compton scattering and the photoelectric effect that may also cause the water radiolysis, which leads to drying due to gas release [6]. The radiolysis is accompanied by the free radicals and hydrogen peroxide formation [7, 8], which may react with cement hydration products and may alter the cement paste properties and even provoke the carbonation reaction [9-11]. Finally, the gamma-ray is absorbed by the material as the thermal atom vibration, which causes the so-called gamma heating which can also affect the cement paste properties. The neutron radiation is always accompanied by the gamma-rays and thus also with an increase in temperature.

The effect of the neutron irradiation and the gamma irradiation on water can differ due to the difference in the transmitted to water energy during the interaction, [12].

It is believed that the gamma-rays affect only slightly the aggregate properties and that majority of the changes appear in the cement paste. According to [13], a significant reduction of the cement paste properties appears beyond the threshold of 200 MGy of the absorbed gamma-ray dose. Also, in [14] the hydrates decomposition within the absorbed gamma-ray dose range of 130 to 836 MGy can be found. On the other hand, the reduction in compressive strength at much lower absorbed gamma-ray dose (15 % reduction with the absorbed dose of 22 MGy and 10 % reduction with the absorbed dose of 0.3-0.5 MGy, respectively) was observed in [15-16]. However, the mechanism of this strength reduction is still unclear. According to [10] the strength of cement paste reduces due to its drying and heating under the gamma rays while the radiation-induced carbonation increases the strength of the cement paste, [11].

According to the only one available research of the effect of gamma-ray irradiation on the creep of the cement paste presented in [17], the gamma-ray reduces the creep and increases the autogenous shrinkage. However, the mechanism of this process is also not fully understood and demands additional experimental investigation.

3. Main patterns in radiation-induced volumetric changes

3.1. *Minerals*

The RIVE of the minerals, first of all, depends on the neutron fluence which is the accumulated number of the neutrons travelling through the unit area and also can be defined as the neutron flux integrated over the time period. The threshold fluence to cause the RIVE of minerals is in the order of 10^{19} n/cm^2. However, the neutron fluence by itself can not characterize the RIVE, since the same amount of the neutrons can cause different damage, because the high-energy neutrons can create more than one Frenkel pairs and cause the cascade of the lattice defects. Therefore, the RIVE depends not only on the neutron fluence but on the energy spectrum of the neutrons. With the same fluence the hard neutron spectrum (the spectrum with a higher amount of high energy neutrons) causes more lattice defects and consequently higher RIVE then the soft neutron spectrum (the spectrum with a lower amount of high energy neutrons), see Figure 2. The number of point lattice defects can be predicted by, for example, Kinchin-Pease model, [18].

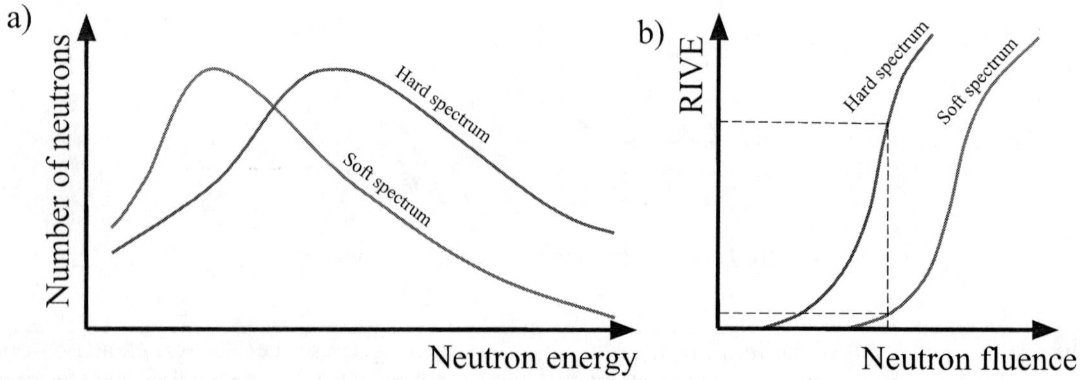

Figure 2. Comparison of RIVE due to hard neutron spectrum and soft neutron spectrum: a) neutron irradiation spectrum, b) RIVE, [5].

The number of lattice defects also depends on the temperature during irradiation. With higher temperature, the probability of the lattice defects recombination is higher, so some lattice defects can be annihilated. Therefore, the higher temperature results in lower RIVE, see Figure 3. However, the normal operation temperature of nuclear power plant is about 60 °C, which corresponds to relatively high RIVE. The silicate minerals are the most sensitive to the elevated temperature, see Figure 3(b) while the effect of elevated temperature on RIVE of carbonates is insignificant in the temperature range of 30-270 °C, [3].

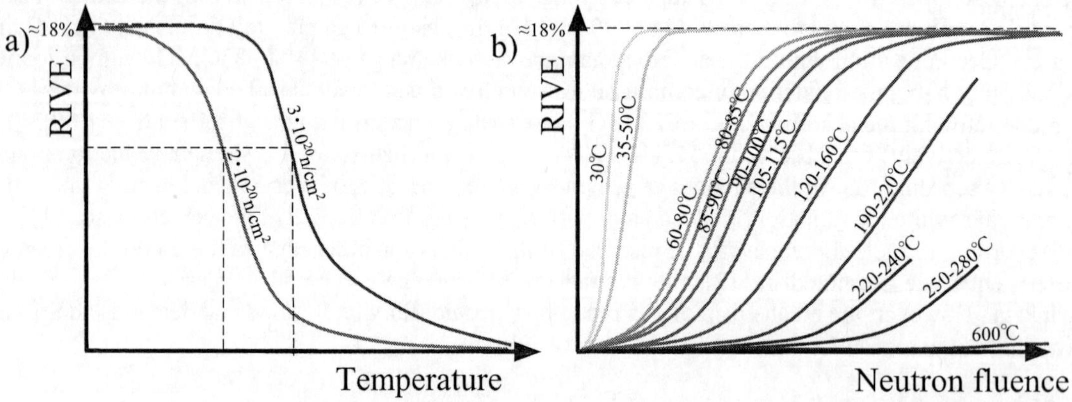

Figure 3. Influence of temperature on RIVE of α-quartz: a) RIVE versus temperature at different fluence [3], b) RIVE versus neutron fluence at different temperature [5].

The RIVE also depends on the type and the structure of the minerals. The highest RIVE is typical for the framework silicates (Tectosilicates) with the highest amount of silica and is equal to 17.9 % for quartz, [3]. The RIVE reduces with the decrease in the silica content up to 8 % for alkali-feldspar, [3]. Also the RIVE reduces with the decrease in the silicate degree of polymerization from the framework silicates to the sheet silicates (Phylosilicates), the RIVE is up to 5 % for mica, [3], and further to the chain silicates (Inosilicates), the RIVE of double-chain silicates is about 2.8 % and the RIVE of single-chain silicates is about 1.5 %, [3], and finally for the island silicates (Nesosilicates), the RIVE of olivine is about 0.9, % [3], see Figure 4.

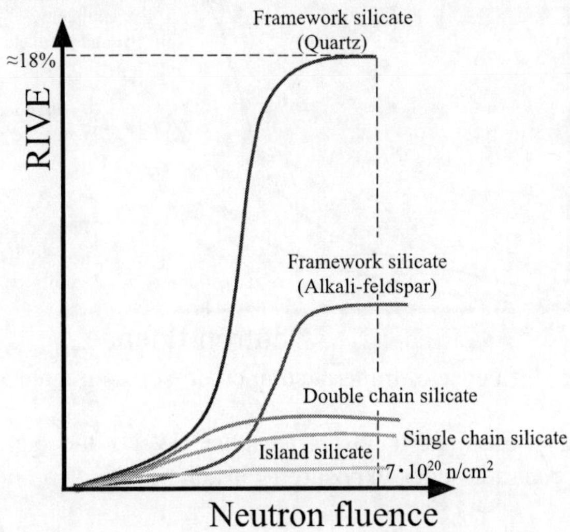

Figure 4. Influence of mineral structure on RIVE of silicates, [5].

The RIVE of the carbonates is typically much lower in comparison with the silicates and varies in the range of 0.5-4 %. The higher RIVE occurs in carbonates with complex composition and the lower structure symmetry, [3].

Moreover, according to [3], the RIVE depends on the heat of fusion of the minerals, which characterizes the energy required by the mineral for phase transformation, while according to [19], the RIVE depends on the bond-dissociation energy of the minerals, which characterizes the energy needed for chemical bond dissociation. Basically, these two parameters describe how strong and stable a lattice structure is, however, they affect the RIVE in the opposite way. With lower heat of fusion and/or with higher bond-dissociation energy, the RIVE becomes higher.

From the above, the following patterns can be highlighted:
The RIVE of mineral is higher with

- Higher neutron fluence
- Harder neutron spectrum
- Lower temperature
- Higher silicate degree of polymerization (for silicates)
- Higher silica content (for silicates)
- More complex composition (for carbonates)
- Lower structure symmetry (for carbonates)
- Lower heat of fusion and/or higher bond-dissociation energy

3.2. Aggregates

The RIVE of aggregates is defined by the minerals present in their composition and depends on the amount of the minerals with the highest RIVE, see Figure 5. The complex stress state due to the different level of the RIVE of the minerals results in development of cracks, therefore, the RIVE of aggregates is higher than the RIVE (up to 20-23 %) of the minerals due to additional volume of cracks. Also, the RIVE of coarse-grain aggregates is usually higher that the RIVE of fine-grain aggregates, [3, 5].

IOP Conf. Series: Materials Science and Engineering **596** (2019) 012035 doi:10.1088/1757-899X/596/1/012035

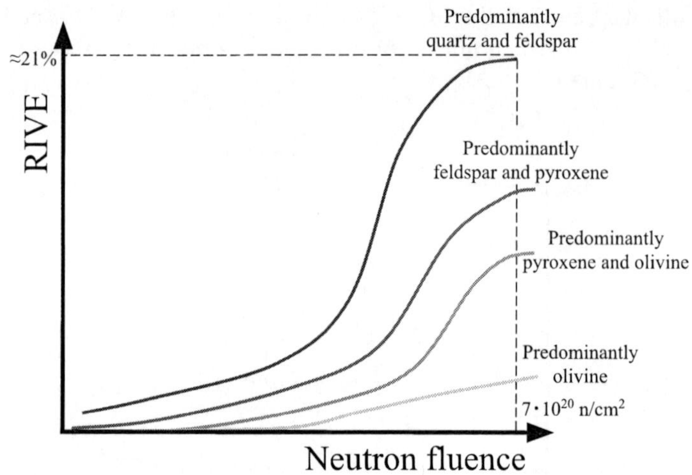

Figure 5. Influence of mineral composition on aggregate RIVE, [5].

The amount of the amorphous material also affects the RIVE of the aggregates since the amorphous materials, such as glass, contract under exposure to irradiation, [3, 5], and can reduce the aggregate RIVE.

Therefore, the RIVE of aggregates depends on

- Mineral composition of the aggregates
- RIVE of the minerals
- Grain size of the aggregates
- Amount of amorphous material

The reduction of the aggregates mechanical properties (strength, elastic modulus, etc.) correlates with their RIVE and the volume of cracks.

3.3. Cement paste

The Portland cement is the most commonly used type of cement and so the available experimental data are predominantly focused on the Portland cement.

The amorphous C-S-H gel is the main component of cement paste, therefore, instead of the typical magnitude of RIVE for crystalline materials, contraction of the cement paste (up to 9.5 % by volume at 400 °C, [3]) under irradiation is observed. The reason of the cement paste contraction is in the hydrate decomposition and the water radiolysis which leads to the loss of water and is accompanied by release of a significant amount of gas. The correlation between the cement paste shrinkage and the loss of water can be observed. According to [3], the cement paste can lose up to 64 % of water (included chemically combined and adsorbed water).

Similarly to the minerals, the effect of radiation on the cement paste increases with the increase in the neutron fluence and with the harder neutron spectrum. However, temperature has an opposite effect than in the case of minerals. The radiation-induced shrinkage of the cement paste is higher with higher temperature due to the water evaporation, see Figure 6 a).

The radiation-induced shrinkage of the cement paste reduces with an increase in the water-to-cement ratio due to the increase in carbonation. The carbonates fill in the pores and restrict the cement paste contraction. However, the early-age cement paste shrinks more than the hardened cement paste under exposure to neutron irradiation and its shrinkage increases with the increase in the water-to-cement ratio, [3].

It is believed that admixture and additives can vary the cement paste response to the neutron irradiation.

From the above we can conclude that the radiation-induced shrinkage of cement paste depends on:

- Neutron fluence

IOP Conf. Series: Materials Science and Engineering **596** (2019) 012035 doi:10.1088/1757-899X/596/1/012035

- Neutron spectrum
- Temperature
- Water-to-cement ratio
- Age of cement paste
- Presence of admixtures or additives

According to [5], the strength of cement paste reduces up to 40 % and according to [3] the strength reduction can reach 72 %, see Figure 6 b). The strength of the early-age cement paste may increase up to 45 % at the beginning of irradiation and then reduces up to 37 %.

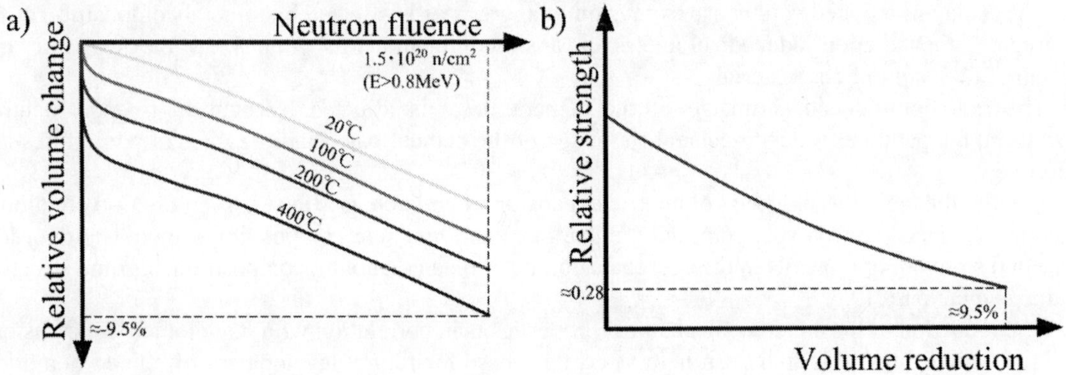

Figure 6. Radiation-induced changes in hardened cement paste: a) volume change, b) strength change, [3].

3.4. Concrete

The volumetric change of concrete depends on the volumetric change of the cement paste and the aggregates and its proportion in the concrete. The parameters on which the volumetric change of concrete depends can be divided into two groups:

- Irradiation conditions (neutron fluence, neutron spectrum and temperature)
- Concrete composition (amount, proportion, type and structure of minerals in the aggregates and amount, composition, age and structure of the cement paste)

The volumetric expansion of the concrete due to the neutron irradiation leads to reduction of strength and modulus of elastic due to the crack formation and the aggregate and the cement paste strength reduction, see Figure 7. The volumetric expansion of concrete is in correlation with reduction of its mechanical properties.

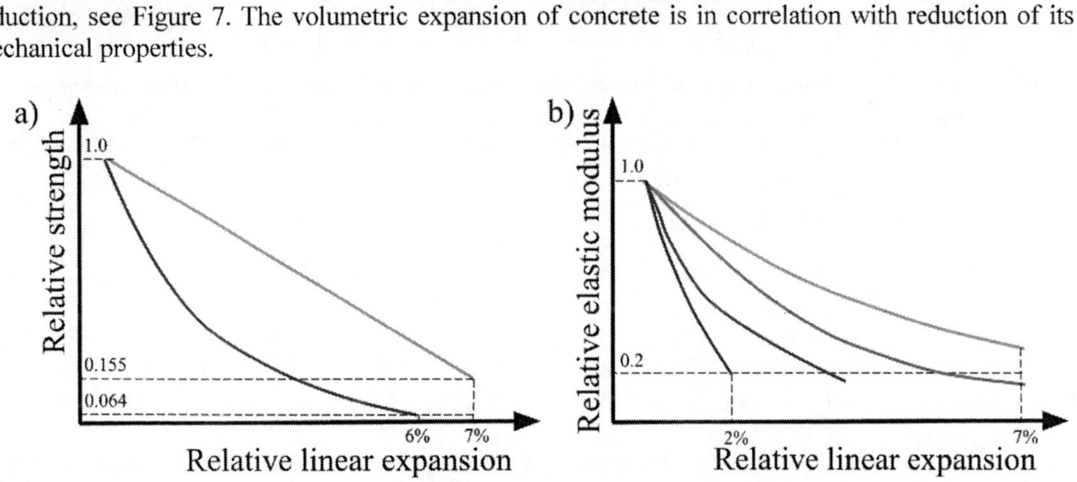

Figure 7. Reduction of mechanical properties of concrete with different composition: a) strength, b) elastic modulus, [3].

4. Conclusions

The mechanism of the radiation-induced deterioration of concrete and its components is described in detail based on the literature review.

All the basic dependencies and patterns of the volumetric change of minerals, aggregates, cement paste and concrete are also described.

The radiation-induced volumetric expansion of minerals is affected by neutron fluence, neutron spectrum, temperature, silicate degree of polymerization (for silicates), silica content (for silicates), complexity of composition (for carbonates), structure symmetry (for carbonates) and heat of fusion and/or bond-dissociation energy.

The radiation-induced volumetric expansion of aggregates is affected by mineral composition of the aggregates, the radiation-induced volumetric expansion of the minerals, grain size of the aggregates and amount of the amorphous material.

The radiation-induced shrinkage of the cement paste is affected by neutron fluence, neutron spectrum, temperature, water-to-cement ratio, age of the cement paste and presence of admixtures and additives.

Finally, the radiation-induced volumetric expansion of concrete is affected by irradiation conditions (neutron fluence, neutron spectrum and temperature) and concrete composition (amount, proportion, type and structure of minerals in the aggregate composition and amount, composition, age and structure of the cement paste).

The reduction of the concrete mechanical properties is in correlation with its volumetric expansion.

It is believed that this study can help to create a base for future development of related numerical models.

Acknowledgments

This work was supported by Ministry of Education, Youth and Sports of Czech Republic, project 8F17002, and the Czech Technical University, project SGS19/038/OHK1/1T/11, which are gratefully acknowledged.

References

[1] Schneider M, Froggatt A, Hazemann J, Katsuta T, Ramana M V, Rodriguez J C, Rüdinger A and Stienne A 2017 *The world nuclear industry status report 2017* (Paris: A Mycle Schneider Consulting Project)

[2] Schneider M, Froggatt A, Hazemann J, Katsuta T, Ramana M V, Fairlie I and Maltini F 2016 *The world nuclear industry status report 2016* (Paris, London, Tokyo: A Mycle Schneider Consulting Project)

[3] Denisov A, Dubrovskii V and Solovyov V 2012 *Radiation resistance of mineral and polymer construction materials* (Moscow: ZAO Publishing House "MEI") [In Russian]

[4] Gray B 1971 *The effect of reactor radiation on cement and concrete* (No. EUR--4751. UKAEA, Risley Engineering and Materials Lab., Culcheth, Eng)

[5] Dubrovskii V, Lavdanskii P and Engovatov I 2010 *Construction of nuclear power plants* (Moscow: Publishing Association of Civil Ingeeniring Universities)

[6] Bouniol P and Aspart A 1998 Disappearance of oxygen in concrete under irradiation: the role of peroxides in radiolysis *Cem. Concr. Res.* **28** 1669-1681

[7] Le Caër S 2011 Water radiolysis: influence of oxide surfaces on H2 production under ionizing radiation *Water* **3** 235-253

[8] Bouniol P and Bjergbakke E 2008 A comprehensive model to describe radiolytic processes in cement medium *J. Nucl. Mater.* **372** 1–15

[9] Vodák F, Vydra V, Trtík K and Kapičková O 2011 Effect of gamma irradiation on properties of hardened cement paste *Mater. Struct.* **44** 101-107

[10] Maruyama I, Kontani O, Takizawa M, Sawada S, Ishikawa S, Yasukouchi J, Sato O, Etoh J and Igari T 2017 Development of soundness assessment procedure for concrete members affected

by neutron and gamma-ray irradiation *J. Adv. Concr. Tech.* **15** 440-523
[11] Maruyama I, Ishikawa S, Yasukouchi J, Sawada, S, Kurihara R, Takizawa M and Kontani O 2018 Impact of gamma-ray irradiation on hardened white Portland cement pastes exposed to atmosphere *Cem. Concr. Res.* **108** 59-71
[12] Kontani O, Sawada S, Maruyama I, Takizawa M, and Sato O 2013 Evaluation of irradiation effects on concrete structure: Gamma-ray irradiation tests on cement paste *ASME 2013 Power Conference* pp V002T07A002-V002T07A002
[13] Hilsdorf H, Kropp J and Koch H 1978 The effects of nuclear radiation on the mechanical properties of concrete *Special Publication, ACI* **55** 223–254
[14] Lowinska-Kluge A and Piszora P 2008 Effect of gamma irradiation on cement composites observed with XRD and SEM methods in the range of radiation dose 0-1409 MGy *Acta Phys. Pol. A* **114** 399
[15] Soo P and Milian L M 2001 The effect of gamma radiation on the strength of Portland cement mortars *J. Mater. Sci. Letters* **20** 1345–1348
[16] Vodák F, Trtík K, Sopko V, Kapičková O, and Demo P 2005 Effect of γ-irradiation on strength of concrete for nuclear-safety structures *Cem. Concr. Res.* **35** 1447–145)
[17] McDowall D C 1971 *The effect of gamma irradiation on the creep properties of concrete* (No. EUR--4751. UKAEA, Risley Engineering and Materials Lab., Culchcth, Eng)
[18] Kinchin G H, Pease R S 1955 The Displacement of Atoms in Solids by Radiation *Reports on Progress in Physics* 18(1):1-51
[19] Le Pape Y, Alsaid M H F, Giorla A B 2018 Rock-forming minerals radiation-induced volumetric expansion – revisiting literature data *J. Adv. Concr. Tech.* **16** 191-209

IOP Conf. Series: Materials Science and Engineering **596** (2019) 012036 doi:10.1088/1757-899X/596/1/012036

Mechanical properties of concrete with small fibre for numerical modelling

Z Marcalikova[1], L Prochazka[2], M Pesata[2], V Bilek[2] and R Cajka[1]

[1] Department of Structures, Faculty of Civil Engineering, VSB-Technical University of Ostrava, Ludvíka Podéště 1875/17, 708 33 Ostrava-Poruba, Czech Republic
[2] Department of Building Materials and Diagnostics of Structures, VSB-Technical University of Ostrava, Ludvíka Podéště 1875/17, 708 33 Ostrava-Poruba, Czech Republic

E-mail: zuzana.marcalikova@vsb.cz

Abstract. The study specializes in the area of mechanical properties of select fibre reinforcement concrete. For the research fibre concrete reinforced with short steel fibres was used without bending. Fibre concrete is produced in the dosing 40 and 75 kg/m^3. Laboratory program includes a complete group of tests, i.e. concrete compressive strength in cubes, three-point bending test, and split tension strength perpendicular and parallel to filling direction. Results are processed in summary, and they also include fracture-mechanical parameters needed for structural modelling. The research carried out found the positive influence of fibre on tensile strength and fracture energy. However, with more fibre, only fracture energy increases. It has also been found that the effect on compressive strength is small. For numerical modelling, the suitability of using a 3D computational model in combination with the fracture-plastic material model for fibre reinforcement concrete was verified.

1. Introduction

Fibre reinforcement concrete ranks among the category of fibre-concrete composites [1]. Fibre reinforcement concretes may vary by concrete and especially by fibres used, i.e. by shape, length, diameter, and finishing [2, 3, 4, 5]. For building structures such as floors, foundation structures, tunnel linings or underground structures [6, 7, 8], the use of fibre concrete is often more beneficial that of the ordinary reinforced concrete or plain concrete. Concrete reinforcement by fibres increases the tensile strength [9, 10] and ductility of concrete, which is very small as compared with plain concrete. Other benefits of fibre concrete may include limitation of shrinkage cracks [11] and deformations of concrete, increased ductility and fatigue strength, improved consistency (no falling-off at marginal parts of concrete). For the design of structures from fibre concrete, it is important to know in details its properties, which are most frequently defined in laboratories on testing samples [12, 13, 14]. Fibre concrete itself and its testing are dealt with in the Czech Republic and globally on a long time basis. Research results can be found in many suggestions and national standards, norms, and technical conditions [15], [16]. There are also many associations dealing with fibre concrete, e.g. RILEM [17], DafStB [18], BS [19] and Model Code 2010 [20]. The topics in the area of fibre concrete being currently deal with in the CR include for example durability or definition and verification of mechanical properties.

IOP Conf. Series: Materials Science and Engineering **596** (2019) 012036 doi:10.1088/1757-899X/596/1/012036

2. Fibre concrete

For the experimental program, fibres Dramix OL13/20 [16] have been selected. General characteristics of these fibres are shown in table 1 and the fibres shape is shown in figure 1. The initial concrete matrix has a formula stated in table 2. Samples were prepared in the laboratory.

Table 1. General characteristics of fibres Dramix® OL13/20 [20].

General characteristic	Dramix® OL 13/20 (figure 4)
Length [mm]	13
Diameter [mm]	0.21
Tensile strength [N/mm^2]	2750
Impact on concrete strength [kg/m^3]	60
Modulus of elasticity [GPa]	200

The concrete mixture used for tested samples may be classified in the category of ordinary concrete of class C25/30.

Figure 1. Dramix® OL13/20 [20].

Table 2. General characteristics of concrete mixture.

Specification	Formula [kg/m^3]
Cement	CEM II/A-S 42.5
Min. cement content	320 kg
Water-cement ration: w/c	0.625
Aggregate 0/2 DTK Mankovice	525 kg
Aggregate 0/4 DTK Mankovice	420 kg
Aggregate 4/8 Tovačov	150 kg
Aggregate 8/16 HDK	820 kg
Water	200 l
Plasticizer STACHEPLAST	3.2 L

IOP Conf. Series: Materials Science and Engineering **596** (2019) 012036 doi:10.1088/1757-899X/596/1/012036

3. Experimental program – methodology

Initial testing of fibre concrete includes compressive strength in cubes. These are tests of cubes with the typical size of 150 x 150 x 150 mm. The testing diagram is shown in figure 2. The demonstration sample from the splitting tension test is in figure 3 and sample from the bending tensile strength test is in figure 4.

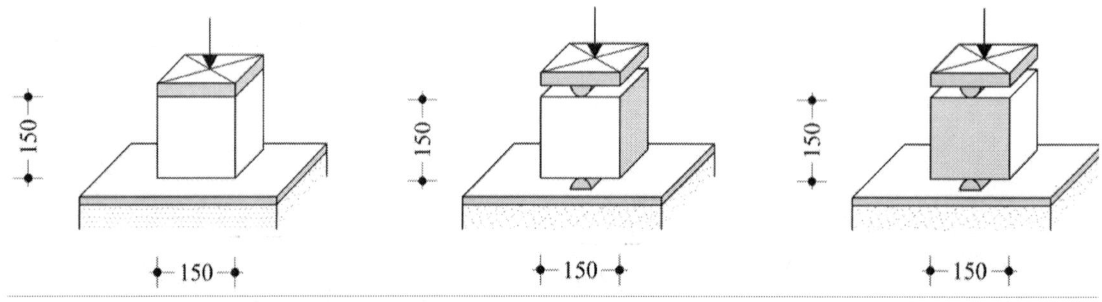

Figure 2. Compression and split tension tests.

Figure 3. Testing samples in a laboratory – split tension test.

Figure 4. Testing samples in a laboratory – bending tensile strength test.

Compressive strength testing in cubes is performed always perpendicularly to the filling direction. Compressive strength in cubes $f_{c,cube}$ is defined as:

$$f_{c,cube} = \frac{P_{\max}}{a^2} \tag{1}$$

where $P_{\max}$ is the maximum strength and a is the size of the cube side.

Next tests in the experimental program include split tension testing. Specifically, two variants have been chosen: perpendicular to the filling direction and parallel to the filling direction. The testing diagram is shown in figure 5. Split tension strength is defined as follows:

$$f_{ct,sp} = \frac{2 \cdot P_{\max}}{\pi \cdot l \cdot d} \tag{2}$$

IOP Conf. Series: Materials Science and Engineering **596** (2019) 012036 doi:10.1088/1757-899X/596/1/012036

where P_{max} is the maximum load, l is the length of contact line of the body, and d is the lateral dimension of the body. To verify cogency of the split tension test and define fractural-mechanical parameters, bending tension test is performed in its three-point version. The testing diagram is shown in figure 5.

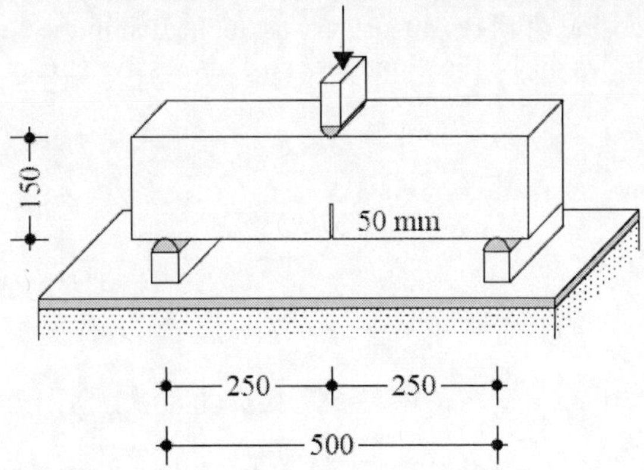

Figure 5. Typical three-point tests with notch 50 mm and span 500 mm.

Tensile strength of concrete in case of three-point configuration may be defined as:

$$f_{ct,ft,3B} = \frac{3 \cdot P_{max} \cdot L}{2 \cdot b \cdot (h - a_0)^2} \tag{3}$$

$$f_{ct,sp} = \frac{3 \cdot P_{max} \cdot L}{1.65} \tag{4}$$

where P_{max} is the maximum load; L, b and h are dimensions: span, cross-section width and height, and a_0 the shear height. Two testing variants are selected. It is important for wider classification of fragility to determine also fracture parameters of concrete that are necessary for examination of cracks occurrence and spread. The most crucial fracture parameter is the energy to fracture G_f. This energy is absorbed by the body until its total destruction occurs. To determine this parameter, it is most beneficial to carry out a three-point bending test on testing samples – joists, of the typical square section. Testing samples are provided with notch in the centre of their span, to be able to anticipate cracks occurrence in the middle of the concrete joist, not at the point of supports as experienced in the joist without notch. Notches are made most frequently up to 1/3 of the sample height. To be able to perform the three-point bending test correctly, it is necessary to use fibres of at least 60 mm to reinforce concrete samples, and use suitable formula of concrete with an aggregate fraction of max. 32 mm. Aggregate fraction size also greatly influences the energy to fracture G_f, value; the greater aggregate fraction is used, the greater energy to fracture we can anticipate. The energy to fracture G_f can be found from the load diagram. It is an integral that equals to the area under the load diagram, or the so-called work W needed for crack occurrence divided by unit surface of the testing sample without considering the notch depth. The fracture energy is defined as follows:

$$G_f = \frac{W}{h \cdot (b - a_0)} \tag{5}$$

4. Experimental program – Results

Compressive strength testing included three samples. Total of 6 tests were carried out. In case of the variant with fibre dosing of 75 kg/m^3, the compressive strength was higher. The difference was however relatively small. The spread of measured strengths is also well apparent from the diagram. The compressive strength for dosing of fibres 40 kg/m^3 was 36.4 MPa. In the variant with dosing 75 kg/m^3, strengths were very similar, i.e. 40.1 MPa. Summarized compressive strengths and volume densities are shown in figure 6.

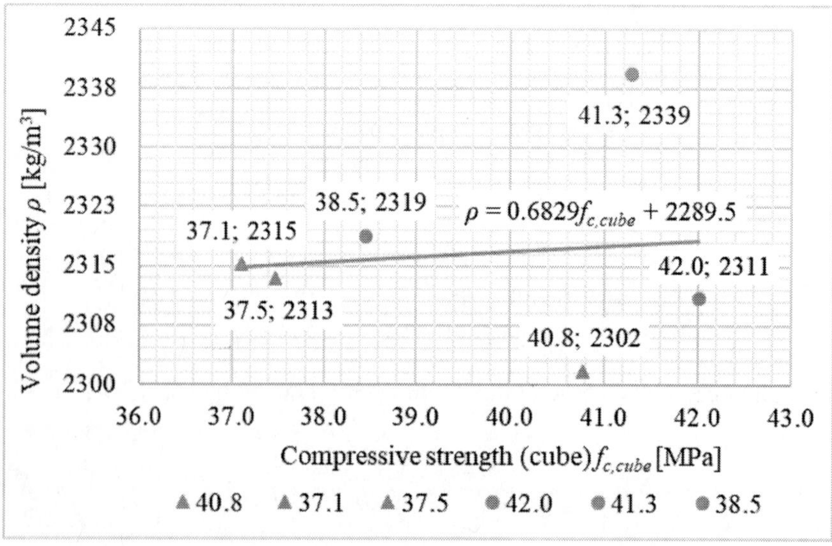

Figure 6. Linear dependence of compressive strength and density.

Split tension strength testing included 10 samples, i.e. 6 samples, the direction perpendicular to filling direction and 4 parallel to filing direction. The weight of these samples ranged about 8 kg. Based on the testing, split tension strengths were calculated. Tensile strength difference for dosing of 40 and 75 kg/m^3 was almost minimal and moved in hundredths. Tensile strengths were 3.07 and 3.08 MPa. In the case of a tensile test parallel to the filling direction, the strength was less, i.e. 2.7 MPa and 2.5 MPa.

Other selected testing options included the three-point bending test. Specifically, total of 6 joists of the size 150 x 150 x 600 mm and the notch in the middle were concreted. The notch depth was 50 mm and the span in the load test was 500 mm. The weight of samples ranged about 31 kg. Three joists were reinforced with the fibre content 40 kg/m^3, and the remaining 3 with 75 kg/m^3. After 28-day maturing period, a three-point bending test was carried out on joists, determining subsequently values of the bending tension strength of all test samples. The load test was applied up to the deflection of 8 mm. The bending tension strength was 4.73 and 4.98 MPa. Results are also shown in a diagram in figure 7 and table 3. The influence of the direction of filling is evident with the comparison of the results in table 4.

Table 3. Material properties of fibre reinforcement concrete.

Value	Fibre content [kg/m^3]	Volume density [kg/m^3]	Compressive strength (cube) [MPa]	Split tension strength (⊥) [MPa]	Split tension strength (‖) [MPa]	Bending tension strength [MPa]
Average	40	2310	38.5	3.1	2.7	4.7
	75	2323	40.2	3.1	2.5	5.0

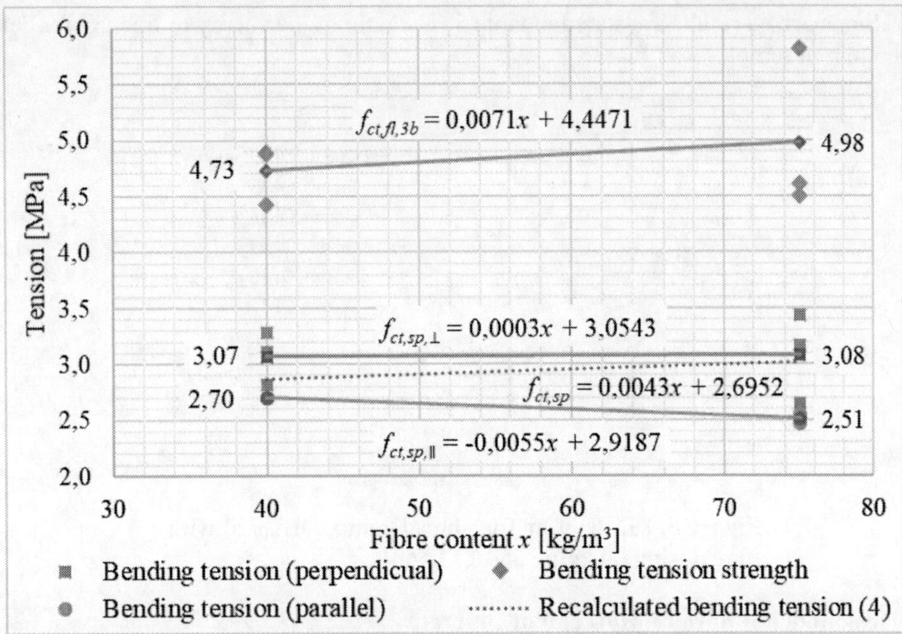

Figure 7. Split and Bending tension strength.

Table 4. Splitting strength ratio.

Fibre content [kg/m³]	Fibre type	Filling direction	Splitting tension test [MPa]	Ratio ⊥ / ‖ [-]
40	Dramix OL13/20	⊥	3.1	1.1
		‖	2.7	
75		⊥	3.1	1.2
		‖	2.5	

Also the assessment of all load diagrams shown in figure 8 was made, and the work W_f, or the energy to fracture G_f for deformations 5 mm and 8 mm were determined. By adding fibres to concrete, the load displacement diagram changes dramatically as compared with plain concrete.

Table 5. Comparing ratios of fracture energy adequate to deformations 5 and 8 mm.

Fibre content [kg/m³]	Fibre type	Deformation [mm]	Energy to fracture [N/m]	Fracture energy ratio (8 mm / 5 mm) [-]
40	Dramix OL13/20	5	1121	1.206
		8	1352	
75		5	1377	1.209
		8	1665	

In the load displacement diagram is see increase tensile strengths are in case of higher dosing. For detail description, it is however necessary to define energies to fracture. Average values of energies to fracture are shown in table 5. With respect to the fact that residual tensile strengths still remained at the end of testing, ratios of energies to fracture were therefore defined as a supplementary descriptive behaviour. Summarized results are stated in table 5. The ratio of the energy to fracture for deformations 8 mm and 5 mm in samples was 1.21.

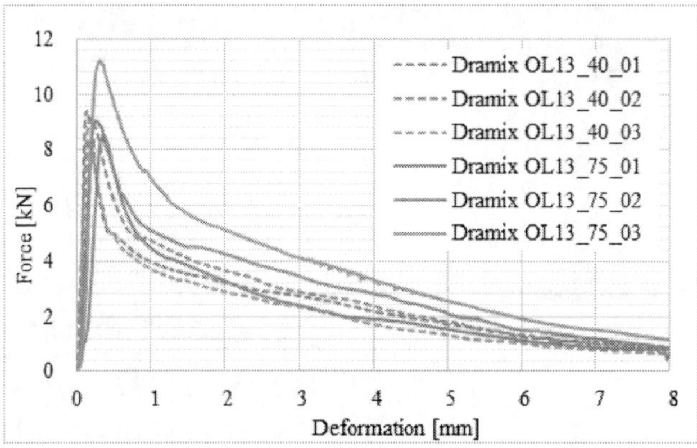

Figure 8. LD diagram for fibres Dramix OL13/20 with fibre dosing 40 kg/m^3 and 75 kg/m^3.

5. Numerical model of fibre reinforcement concrete

For analysis of concrete structures is currently typical used the finite element method. However, a non-linear solution is needed to take account of actual behaviour. The calculation must account the typical concrete failure. In pressure, the concrete is crushed and cracks are formed in the tension. These assumptions are met by the use of fracture theory and the Newton-Rapston nonlinear solver. For the numerical models and analysis using was used system Atena [21] with fracture plastic material model [22]. The specific problems of using nonlinear analysis of concrete structures are important to determine the input parameters of fibre reinforcement concrete [23]. The tensile softening is specific to each of fibre concretes recipe [24], [25].

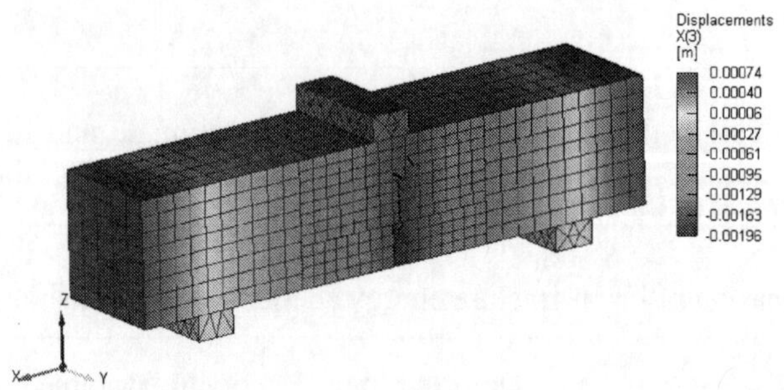

Figure 9. Numerical model in the Atena system for fibre dosing 75 kg/m^3 - displacements [m].

The determination of the specific fracture energy defined in the material model cannot be determined from bending tensile tests. For these reasons, it is necessary to use a parametric study [25] or inverse analysis to select the appropriate tensile softening approximation. It is also important to take into account numerical model uncertainties and respect recommendations [26] and [27]. A three-point bending tensile test was used to numerical modelling and determine the specific fracture energy. The typical 3D computational model is shown in figure 9. The modulus of elasticity was determined on the basic of class concrete and the used value for numerical analysis was 31 GPa. Compressive strength was used in the tests.

IOP Conf. Series: Materials Science and Engineering **596** (2019) 012036 doi:10.1088/1757-899X/596/1/012036

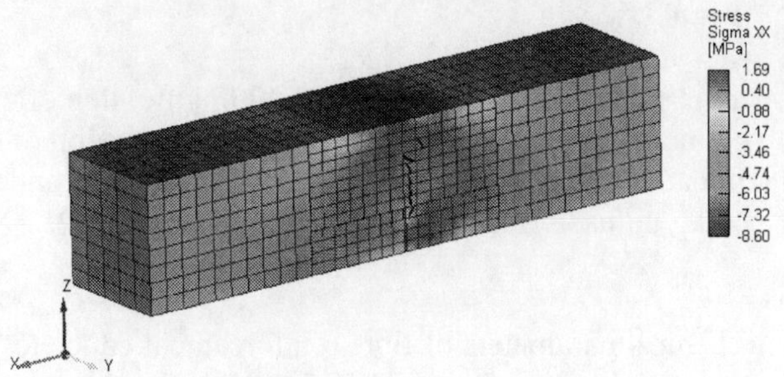

Figure 10. Numerical model in the Atena system for fibre dosing 75 kg/m³ – stress σ_x [MPa].

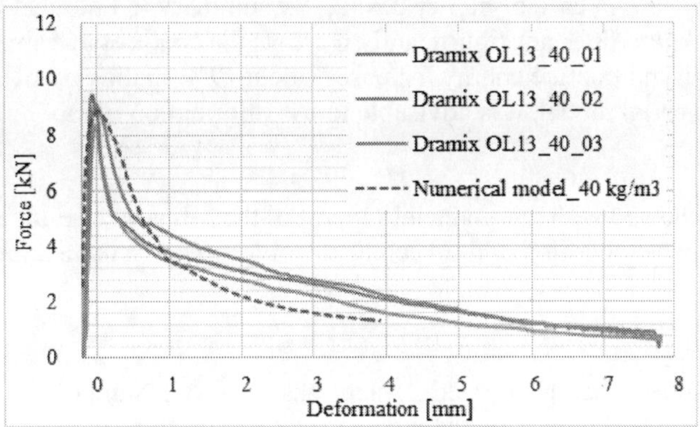

Figure 11. Load – displacement diagrams – fibre dosing 40 kg/m³.

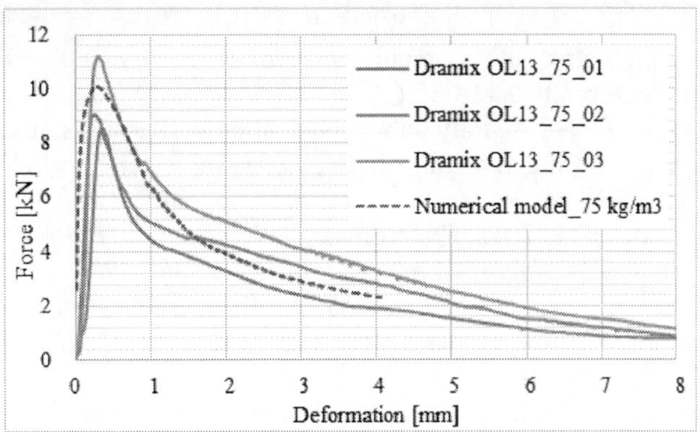

Figure 12. Load – displacement diagrams – fibre dosing 75 kg/m³.

Exponential crack law was used to approximate tensile softening in the interval specific fracture energy 300 - 1000 N/m. The initial tensile strength was also used from split tension testing and gradually decreased to a level of effective tensile strength. Figures 9 and 10 are used to illustrate numerical calculations. Figure 9 shows the deformations and figure 10 shows the normal stress σ_x at a load of 4.5 kN and a deformation of 1.5 mm.

The resulting set of specific fracture energy values are 400 N/m witch effective tensile tension of 0.87 MPa for a fibre of 40 kg/m^3 and 900 N/m witch effective tensile tension of 0.95 MPa for 75 kg/m^3. The resulting comparison of load displacement diagrams from experiments and numerical modelling is shown in figure 11 for the 40 kg/m^3 case, and for case 75 kg/m^3 in figure 12.

6. Conclusion

The research dealt with mechanical parameters of fibre reinforcement concrete, which differ by dosing 40 and 75 kg/m^3. Specifically, straight fibres Dramix OL13/20. Results showed positive impact on both, the compressive strength, fracture energy and tensile strength of concrete. The impact of fibres on the compressive strength is however small. It has also been demonstrated that the split tension strength parallel to filling direction is lower than in the direction perpendicular to filing direction. The difference in tensile strength both dosing types was small, about 15%. In the load displacement diagram is see increase fracture energy are in case of higher dosing. By summary to note that the increase of the fibre content from 40 to 75 kg/m^3 does not significantly increase the compressive strength nor tensile, but the residual tensile strength and fracture energy increases, about 19%. In the case of 3D numerical modelling and fracture-plastic material model, it is advisable to use information of laboratory tests.

Acknowledgement

This paper has been achieved with the financial support of the Ministry of Education, specifically by the Student Research Grant Competition of the Technical University of Ostrava under identification number SP2019/144.

References

[1] Brandt A M 2008 Fiber reinforced cement-based (FRC) composites after over 40 years of development in building and civil engineering *Composite Structures* vol **86** iss 1 pp 3–9 DOI: 10.1016/j.compstruct.2008.03.006

[2] Katzer J and Domski J 2012 Quality and Mechanical Properties of Engineered Steel Fibres Used as Reinforcement for Concrete *Construction and Building Materials* vol **34** pp 243–48

[3] Holschemachera K, Muellera T and Ribakov Y 2010 Effect of steel fibres on mechanical properties of high-strength concrete *Materials& Design* vol **31** iss 5 pp 2604-2615 DOI:10.1016/j.matdes.2009.11.025

[4] Koniki S and Ravi P D 2018 A study on mechanical properties and stress-strain response of high strength concrete reinforced with polypropylene–polyester hybrid fibres *Cement, Wapno, Beton* iss 1 pp 67-77

[5] Kasagani H and Rao C B K 2016 The influence of hybrid glass fibres addition on stress – Strain behaviour of concrete *Cement, Wapno, Beton* iss 5 pp 361-372

[6] Vaskova J and Cajka R 2018 Interaction of nonlinear numerical model of SFRC slab and nonlinear numerical subsoil model *International Journal of GEOMATE* vol **15** iss 47 pp 103-110 DOI: 10.21660/2018.47.3576

[7] Labudkova J and Cajka R 2017 3D numerical model in nonlinear analysis of the interaction between subsoil and sfcr slab *International Journal of GEOMATE* vol **13** iss 35 pp 120-127. DOI: 10.21660/2017.35.6689

[8] Sucharda O, Bílek V, Smirakova M, Kubosek J and Cajka R 2017 Comparative Evaluation Of Mechanical Properties Of Fibre-ReinforcedConcrete And Approach To Modelling Of Bearing Capacity Ground Slab *Periodica Polytechnica: Civil Engineering* vol **61** iss 4 pp 972-986 DOI:10.3311/PPci.10688

[9] Kurihara N, Kunieda M, Kamada T, Uchida Y and Rokugo K 2000 Tension softening diagrams and evaluation of properties of steel fibre reinforced concrete *Eng Fract Mech* vol **65** pp 235-245 https://doi.org/10.1016/S0013-7944(99)00116-2

[10] Tipka M and Vasková J 2017 Tensile properties of structural fibre reinforced concrete of Fiber Concrete 2017 *IOP Conf. Ser.: Mater. Sci. Eng.* vol **246** DOI: 10.1088/1757-899X/246/1/012052

[11] Sucharda O and Bilek V 2018 Aspects of Testing and Material Properties of Fiber Concrete *25th Concrete Days 2018* pp 1-6

[12] Zaborski A 2016 Constitutive model for restricted compression of fiber concrete *Cement, Wapno, Beton* iss 1 pp 46-52

[13] Kormanikova E and Kotrasova K 2011 Elastic mechanical properties of fiber reinforced composite materials *Chemicke Listy* vol **105** iss 17

[14] RILEM TC 162-TDF 2002 Test and Design Methods for Steel Fiber Reinforced Concrete – Design of Steel Fiber Reinforced Concrete Using the σ-w Method: Principles and Application. Materials and Structures/Matériaux et Constructions vol 35 iss 249 pp 262–267

[15] di Prisco M Colombo M and Dozio D 2013 Fibre-reinforced concrete in fib Model Code 2010: principles, models and test validation *Structural Concrete* vol **14** iss 4 pp 342-361 DOI:10.1002/suco.201300021

[16] RILEM 2011 About Rilem [Online] [Accessed on 4 May 2011]

[17] DAfStb guidelines 2011 DAfStb-Richtlinie Stahlfaserbeton. Deutscher Ausschuss für Stahlbeton – DAfStb, Berlin, German. (In German)

[18] BS EN 14721: Test method for metallic fibre concrete – Measuring the fibre content in fresh and hardened concrete. BSI, 2005, 2007.

[19] Model Code 2010 – Final Draft, fib, Bulletin No **65** and 66. 1-2. 2012

[20] https://www.bekaert.com/en/products/construction/concrete-reinforcement.

[21] Cervenka V, Jendele L, Cervenka J 2007 ATENA Program Documentation-Part 1: Theory. Praha, Czech Republic

[22] Cervenka J and Papanikolaou V K 2008 Three dimensional combined fracture-plastic material model for concrete *International Journal of Plasticity*, vol **24** iss 12 pp 2192-2220 DOI: 10.1016/j.ijplas.2008.01.004

[23] Sucharda O, Pajak M, Ponikiewski T and Konecny P 2017 Identification of mechanical and fracture properties of self-compacting concrete beams with different types of steel fibres using inverse analysis *Construction and Building Materials* vol **138**, pp 263-275 DOI: 10.1016/j.conbuildmat.2017.01.077

[24] Sucharda O, Konecny P, Kubosek. Done P J 2015 Finite element modelling and identification of the material properties of fibre concrete In *Conference: 23rd Conference of Italian Group of Fracture* Location: Sicily, Procedia Engineering, vol. **109**, pp. 234-239, 2015.

[25] Sucharda O, Lehner P, Konecny P and Ponikiewski T 2018 Investigation of Fracture Properties by Inverse Analysis on Selected SCC Concrete Beams with Different Amount of Fibres Ecf22 - Loading And Environmental Effects On Structural Integrity *Procedia Structural Integrity* vol **13** pp 1533-1538 DOI: 10.1016/j.prostr.2018.12.313

[26] Sucharda O and Konecny P 2018 Recommendation for the modelling of 3D non-linear analysis of RC beam tests *Computers and Concrete*, vol **21** iss 1 pp 11-20. DOI: 10.12989/cac.2018.21.1.011

[27] Sucharda O, Smirakova M, Vaskova, J, Mateckova P, Kubosek J and Cajka, R Punching Shear Failure of Concrete Ground Supported Slab *International Journal of Concrete Structures and Materials* vol. **12** iss. 1 Article number 36 DOI: 10.1186/s40069-018-0263-6

Underground ventilated wall based on TRC blocks

J Pazderka, M Nývlt and H Žáková

Department of Building Structures, Faculty of Civil Engineering, Czech Technical University in Prague, Thákurova 7, 166 29 Praha 6, Czechia

E-mail: jiri.pazderka@fsv.cvut.cz

Abstract. Ventilated wall is an air-insulating method, which principle is to create a ventilated cavity along the surface of remediated underground wall (usually masonry). The paper describes an innovative solution of this remediation method based on the precast blocks made form textile-reinforced concrete - TRC. The main advantages of the solution (compared to traditional constructions - masonry, hollow bricks filled by cement mortar, monolithic concrete) are low labour difficulty, fast montage and high durability. Since the blocks are not structurally connected or anchored to remediated structure, it is a non-invasive method, which is also suitable for installation in historically valuable buildings. The technical solution is protected by patent CZ307501(B6).

1. Introduction

By combining high-strength concrete and textile-reinforcement (TRC), there is possibility of extensive use of precast elements in the field of refurbishment of underground parts of buildings (in terms of protection against moisture). The biggest potential for the use of precast elements is in the field of air-insulating methods. These methods are often used remediation technology designed for reduction of moisture content in masonry - this applies to older buildings that lack functional waterproofing envelope of the substructure [1]. This group of remedial measures is designed to maximize the evaporation of water (in the form of diffusing water vapour) from the building structure or, alternatively, from the surface of the adjacent terrain. The result is a reduction of moisture content in building structures and preventing transport of liquid water into a higher part of masonry.

For effective evaporation of water from the structure, it is necessary to ensure an intense exchange of air in immediate vicinity of surface so as not to achieve 100% relative humidity of the adjacent air layer. Therefore, all air-insulating remediation systems are designed in a way that allows efficient airflow in the cavity created by the measure. At the same time, it is important that the supply air has the low relative humidity φ [%]. Therefore, air must be brought into the ventilated cavity from the outside. However, this may be a problem if the ventilated cavity is located at the interior surface of the structure. In winter, the temperature inside the ventilated cavity can causes undesired condensation on the surface of structure. Each design of the ventilated cavity must therefore be evaluated in terms of risk of condensation. From this point of view, it is a much simpler solution to place the ventilated cavity in the exterior so that the cavity is separated from the interior by the thermal insulating envelope of the building. For the right work of air insulating remediation measure, the surface of the remediated structure should be modified as diffusely open - allowing the movement of water vapor. At the same time, it should have as large a surface as possible. For this reason, it is advantageous to remove the original plaster from the masonry and to scrape the joints between the bricks to maximize the surface for evaporation.

Current air-insulating systems include: ventilated walls (can be made from the exterior or interior and above or below ground level), ventilated floors, ventilation ducts in the masonry (which are usually less efficient), ventilated plinths and ventilated horizontal tunnels. Some of these air-insulating systems have already been modified by the author using specially shaped reinforced concrete blocks [2,3].

The advantage of air-insulating systems is that they can completely replace the costly additional waterproofing envelope of substructure [4] (in the cases where the moisture level in masonry is not too high). An important advantage is also that it is a non-invasive method which is suitable for use in historically valuable buildings

2. State of the art, motivation

Ventilated wall is an air-insulating method, which principle is to create areal ventilated cavity along the surface of remediated underground wall (usually masonry). The ventilated wall differs from the ventilated horizontal tunnel by being installed all over the basement wall, while the tunnel is only a linear element with a small high of remediated area of structure. The ventilated tunnel is typically installed along the foundations at the non-basement buildings, not at the basement walls [3]. The ventilated wall can be located either on the interior or exterior side of the remediated perimeter wall. The interior ventilated wall can be realized from standard types of vertical non-load-bearing structures - such as gypsum boards on grids or masonry. The exterior ventilated walls placed in the contact with ground environment must be sufficiently rigid due to the pressure of the adjacent soil. At the same time, the exterior wall must be made from durable materials as they are exposed to the water contained in the surrounding soil environment. An important advantage of the external ventilated wall over the inner wall is the interruption of horizontal capillary transport of water from the soil environment to the basement wall. As a result, the effect of the exterior construction is significantly higher because there is a significant reduction of the moisture content in building perimeter wall that has the associated positive impact on its building-physical properties, such as increasing the compressive strength, reducing the thermal conductivity, reducing the biocorrosion and slowing the deterioration of the binder components. Another advantage of using the exterior wall over the inner wall is that there is no reduction of floor space in the basement.

External ventilated walls are mostly realised from masonry - usually from hollow bricks filled by cement mortar or form sand-lime bricks, exceptionally from the monolithic reinforced concrete (which is difficult and costly). In the case of a ventilated wall from bricks, the durability of this structure is limited due to direct contact with water and moisture in the soil environment. Another disadvantage of the masonry structure is the high risk of deformation due to the long-term effect of the ground pressure. Deformation occurs because the wall is not burdened with any other vertical load that would have a positive impact on the direction of the internal forces resultant (necessary for ensure the sufficient masonry stability even over a long period of time). In terms of mechanical parameters and stability, it is a much better solution a ventilated wall made from monolithic reinforced concrete but it is the most expensive variant - the necessity of extensive excavation work, complicated formwork in a confined space, technological pauses, solution of working gaps. A suitable alternative to the above-described contemporary design of exterior ventilated walls is a structure described in this paper.

3. Underground ventilated wall based on TRC blocks

The disadvantages of existing external underground ventilated walls based on bricks (low durability, risk of deformation, high labour difficulty) even on the basis of monolithic reinforced concrete (high price, difficult to implement) are removed by using of precast reinforced concrete wall from high-strength textile-reinforced concrete - TRC. This solution consists of a TRC blocks (Figure 1-2) which are placed side by side along remediated basement wall and thus create an air cavity allowing the evaporation of water from the masonry. Thanks to the use of TRC which includes a high-strength concrete technology, a small thickness of the block is possible. Using high-strength concrete also ensures

IOP Conf. Series: Materials Science and Engineering **596** (2019) 012037 doi:10.1088/1757-899X/596/1/012037

high durability of the element in permanent contact with the wet soil environment, since high-strength concrete has a closed microstructure that prevents the transport of liquid water inside the structure. Thanks to this, it is not necessary to design any additional protection against ground water - no additional waterproofing envelope is necessary, as in the case of brick walls.

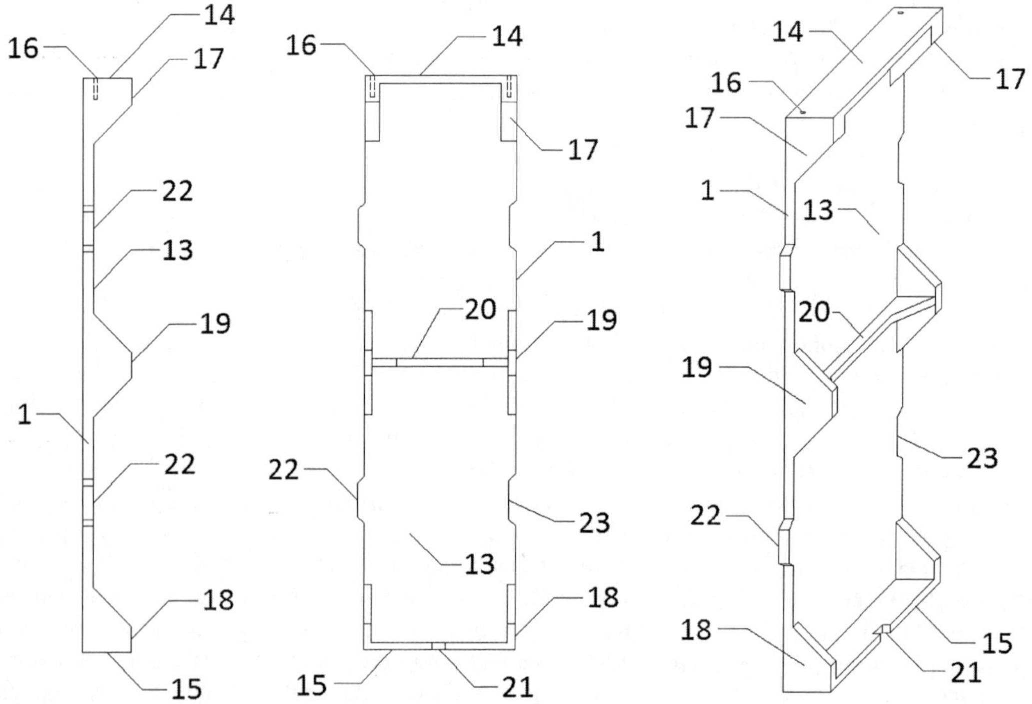

Figure 1. Special shaped underground block from TRC: 1 – whole block, 13 – basic plate, 14 – upper support element, 15 – lower support element, 16 - handle for manipulation, 17 - extended upper rib, 18 – lower rib, 19 – mid rib, 20 - transverse reinforcing rib, 21 - drain hole, 22,23 - tongue-and-groove system.

The block from TRC has a cross-sectional shape of the approximate letter "E" in its final position, the predominant dimension is height - the blocks are laid vertically. The block is made up of a basic plate, from which the perpendicular U-shaped support elements are drawn in the upper and lower parts. I the mid part of the block, there is another support element which stability is ensured by transverse reinforcing rib. The elements placed side by side creates continuous air cavity along the remediated wall.

In the middle of the lower U-shaped support element a small hole is placed. This hole serves as a drain for water that could eventually flow into the air cavity. The blocks are equipped with a tongue-and-groove system, which serves to ensure the uniform subsidence of the individual blocks. Both the tongue and the groove have a trapezoidal shape, the groove being slightly larger than the tongue so that the adjacent blocks can snugly snap into each other. The dimensions of the block do not allow manual handling due to too much weight - therefore the element is equipped by handle for manipulation in its upper part. The large dimensions of the block are due to the requirement for maximum mounting speed and the simplicity of solution.

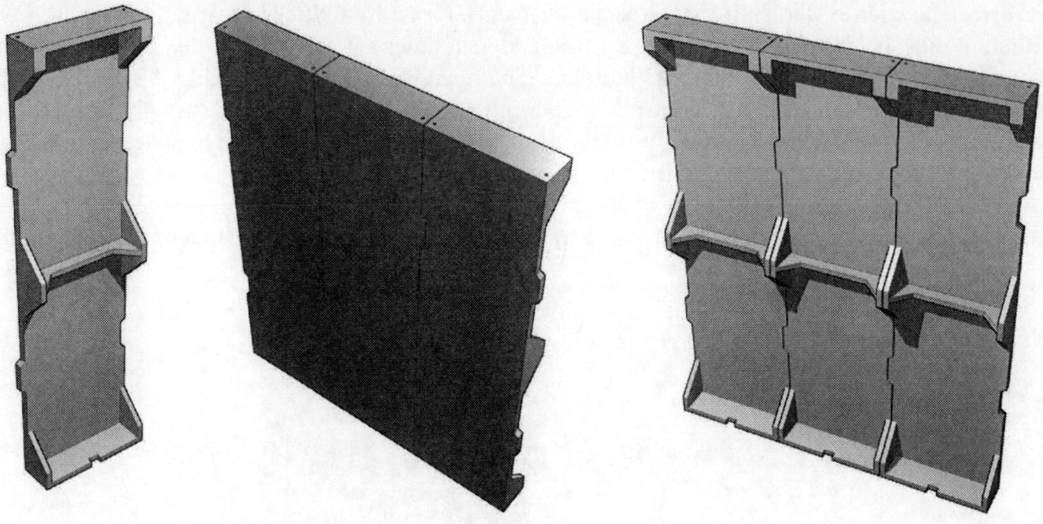

Figure 2. 3D model of concrete block: left - single block, middle - block assembly (external view), right - block assembly (internal view).

The basic system from the blocks is complemented by a specially shaped cover sheet that overlaps the upper part of the fitting and serves as a protection against the penetration of the leaking water into the cavity (Figure 3). The cover metal sheet is located close to the surface of the terrain and is shaped in an inclined gradient so that the leaking water flows into the adjacent landfill. In its upper part, the metal sheet is formed into a vertical position and is attached to the peripheral wall of the remediated object. In the case of specific aesthetic requirements, the required surface of the plinth can be glued to the sheet (ceramic tile, stone tile etc.). In the case of historically valuable buildings (protected monuments), there is the possibility to terminate the cover metal sheet under the surface (it is necessary to ensure its waterproof connection to the surface of the wall).

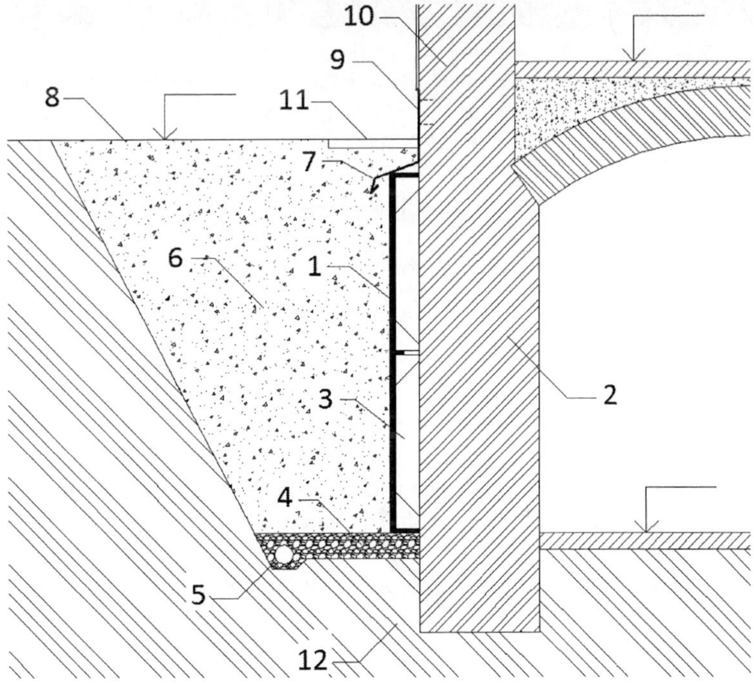

Figure 3. Example of block installation at the building: 1 – concrete block, 2 – remediated masonry, 3 – ventilated cavity, 4 – permeable gravel, 5 – drainage, 6 – landfill, 7 - cover sheet, 8 - terrain surface, 9 – plinth (part of the cover sheet), 11 – cover pavement, 12 – ground.

IOP Conf. Series: Materials Science and Engineering **596** (2019) 012037 doi:10.1088/1757-899X/596/1/012037

For correct function of the entire system, the air cavity formed by blocks must be ventilated. Cavity ventilation must be ensured by installing suction and vent holes - it can be provided by a ducts located in a vertical grooves in the masonry (Figure 4). The opening of the holes inside the cavity must be situated at different heights. To provide adequate ventilation efficiency, there is strongly recommended to equip the openings with axial fan of low power. In winter, it is recommended to cover the ventilation holes (to prevent access of the cold air from the outside) to avoid condensation of water vapour inside the cavity. To increase the efficiency of the remedial measure it is also recommended to scratch the joints of remediated masonry to a depth of 20 to 30 mm (this will increase the area of the evaporation).

Figure 4. 3D model of block installation at the building with marked vent holes equipped with a fan (cover sheet is not displayed, for clarity).

Blocks are assembled into the excavation which bottom is performed from compacted gravel. The gravel layer must be made as a drainage layer to prevent leakage of subsurface water into the space of the cavity created by blocks. From this point of view, it is highly recommended to install drainage pipes into the compacted gravel layer. After laying the block to the desired position, the block must be fixed on its place by using of temporary strut before the excavation is filled by soil (the strut is removed when the excavation is filling). Once finished, the block is fixed in its position by ground pressure - therefore no anchoring is needed. The absence of anchoring is advantageous in terms of acceleration of assembly and at the same time in the case of historically valuable buildings there is no breach of the original historical structure.

4. Discussion

The main advantages of the technical solution described above (compared to traditional constructions: masonry, hollow bricks filled by cement mortar, monolithic concrete) are low labor difficulty (fast montage) and high durability. Low labour difficulty is due to a simple installation of the system to the building, without the need for any wet processes. High durability is due to the waterproofing of used material (HPC) and the low risk of deformation from the adjacent soil (compared to the masonry variant). Since the blocks are not structurally connected or anchored to a rehabilitated structure, it is a non-invasive method, which is also suitable for installation in historically valuable buildings (this is also due to the resulting "invisibility" of the measure, which is common to all construction variants). The only one disadvantage is the necessity of using the lifting device for installation of the blocks into the excavation.

The technical solution is protected by patent CZ307501(B6), granted in 2018 [5].

5. Conclusion

The special shaped block from TRC designed for underground ventilated wall demonstrated that using of TRC precast elements opens up new ways of solution for traditional construction details in buildings. The advantages of this approach are evident: easier and faster assembly, higher durability and new additional functions. In the future, a further specialized precast systems based on TRC elements which will replace some traditional structures in buildings can be expected.

Currently, preparation for the production of the first prototype of the TRC underground wall block is performed. The abovementioned patent CZ307501(B6) which contain the technical solution presented in the paper, is valid only for Czech Republic, in other countries is free to use.

Acknowledgement
This work was supported by the research project SGS19/145/OHK1/3T/11.

References

[1] Franzoni E 2018 *Journal of Cultural Heritage.* **31** S3

[2] Pazderka J and Hájek P 2017 *9th International Conference on Fibre Reinforced Concretes (FRC), Textile Reinforced Concretes (TRC) and Ultra-High Performance Concretes (UHPC)* (Bristol: IOP Publishing) p 246

[3] Pazderka J, Hájková E and Jiránek M 2017 *Acta Polytechnica.* **57(5)** 331

[4] Pazderka J and Zigler R 2013 *CESB 2013 - Central Europe Towards Sustainable Building 2013* (Prague: Grada Publishing) p 163

[5] Pazderka J and Reiterman P 2018 *Patent CZ307501(B6)* (Prague: Industrial Property Office)

IOP Conf. Series: Materials Science and Engineering **596** (2019) 012038 doi:10.1088/1757-899X/596/1/012038

FRC origami bridge model

M Salák and M Frantová

Department of Concrete and Masonry Structures, Faculty of Civil Engineering, Czech Technical University in Prague, Thákurova 7, 166 29 Prague 6, Czech Republic

E-mail: michaela.frantova@fsv.cvut.cz

Abstract. This paper aims to describe the topic of origami folded plate structures and to design and built a small-scale model of such structure. To do so, firstly the properties of origami, in general, are discussed, then contemporary research of origami structures is described, and at last the concrete model is designed and built. During the work, the model was analysed with software and the process of building was experimentally tested on parts of the structure. The main outcome is a fibre reinforced concrete model of a bridge with a span of approximately 1.5 m, which competed in an international competition of FRC structures in Budapest in June 2019.

1. Introduction

Origami is world-wide spread ancient Japanese art of folding paper. The word comes from the end of the 19th century, but the folding of paper or other materials is much older and was existent all over the world.

There are many different types of modern origami, one of which is called origami tessellations. A tessellation is a collection of figures filling a plane with no gaps or overlaps. The most famous origami tessellation is probably Miura-Ori (see figure 1), this tessellation was invented to fold a solar panel into an aircraft and then to be easily spread once the satellite is deployed in the space. Miura-Ori has one degree of freedom; this means it is developable in one single motion executed by pulling any two parts of the tessellation apart. [1]

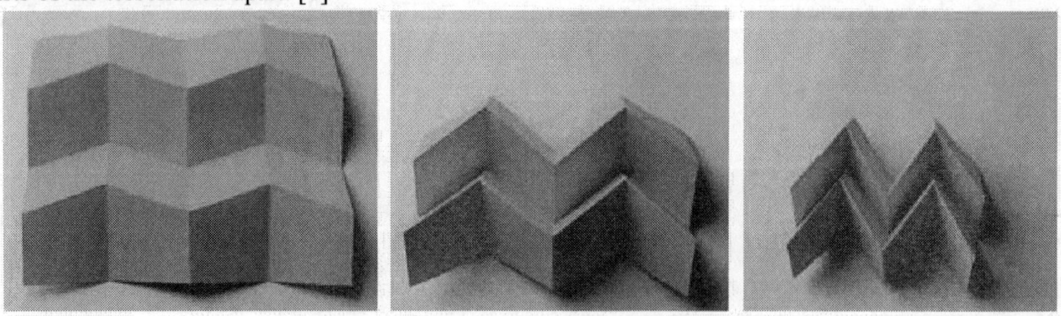

Figure 1. Folding of Miura-Ori tessellation.

Modern technology offers great opportunities for origami structures. Contemporary design techniques supported by CNC milling machines can very easily create formwork from timber panels using non-repetitive, creative shapes. Even though there are not many large concrete origami structures these days, it is clear that the research goes on and origami inspires architecture. In figure 2 you can see two structures inspired by origami; on the left there is Yokohama International Passenger Terminal by Foreign Office Architects built in 2002, using folded steel plates and concrete girders. On the right hand side, you can see wooden pavilion from Osaka built in 2008 by Ryuichi Ashizawi Architects. [2], [3]

Figure 2. Modern origami structures: Yokohama International Passenger Terminal (left) [2] and wooden pavilion in Osaka (right) [3].

Many universities research new ways of fabrication and usage of concrete origami. Among others, Martin Bechtold from Harvard built in 2006 experimental structure by subdividing the structure into pre-fabricated units that are poured flat over a simple formwork. The joint lines are left free when adding the cement matrix, but the mesh reinforcement continues through these joints. This allows the flat system to be later folded into its final three-dimensional shape. The flexible joints are then made rigid by filling them with expansion grout. [4], [5]

At RWTH Aachen University Rostislav Chudoba, Jan Dirk van der Woerd and Josef Hegger developed their concrete origami system. They tested different folding processes, one of them was using a crane and suction cups attached to the origami structure. Another tested possibility was to build the structure flat as before, then hang it upside down to form an arch to let the grout dry in the final position. [6], [7]

The aim of this work is to describe the design and the production technology of the innovative bridge model built at the Department of Concrete Structures and Masonry Structures, Faculty of Civil Engineering, CTU in Prague. The origami model was built for a 4th concrete bridge competition that took place in Budapest, Hungary on 21st June 2019, affiliated with periodical concrete canoe races. The bridge design is governed primarily by the rules of the competition. The critical parameter is the use of only non-rigid reinforcement. [8]

2. Bridge Design and Structure Analysis

The design process started from scratch by exploring origami tessellations that form an arch when folded (see figure 3). For further investigation, the Yoshimura pattern was chosen (figure 3, bottom right) for its simplicity and repetitiveness. This tessellation is flat foldable and developable, made by repeating an isosceles triangle.

The stiffness of the origami arch was analyzed using software Autodesk Fusion 360. Linear analysis of two arches with the same volume was done to compare them; one origami arch with a thickness of 20 mm and the other regular arch with a thickness of 26.5 mm, both with the same span of 4.5m and width of 0.9 m. They were both loaded with its weight only and the deflection in the middle of the span was measured as shown in figure 4. The origami arch deflection is an order of magnitude smaller then deflection of regular arch (0,3 vs 4,2 mm), thus the analysis demonstrates one of the key benefits of origami structures - higher stiffness. (figure 4)

Figure 3. Exploring and searching the shape of the arch.

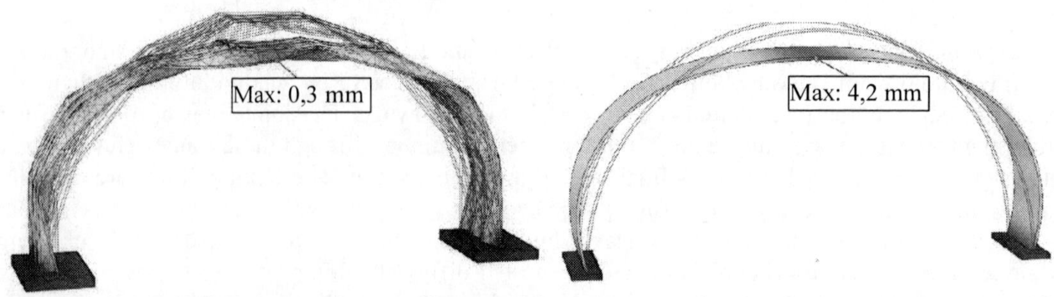

Figure 4. Comparing the stiffness of origami and regular arch using software.

The origami model was built according to the rules of the 4th concrete bridge competition that took place in Budapest [8]. The span is 1.5 m and the thickness of all components is 20 mm. The final design consists of three separated parts highlighted in figure 5 by different colors.

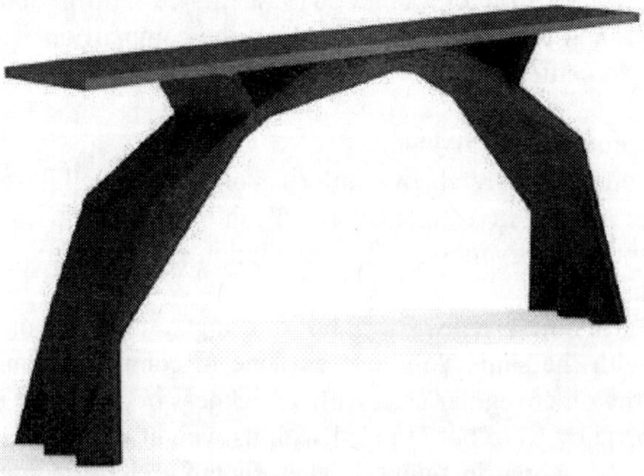

Figure 5. Model of FRC origami bridge.

The design was done in Fusion 360 by Autodesk, because it offers great ways of 3D modelling and also the simulation of static stress is a part of the software, as shown in figures 6 and 7. The model was then loaded in Fusion 360 by gravity and force located in the middle of the span to simulate loading during the competition. It is expected that the load of a 150 kg will cause the failure of the structure.

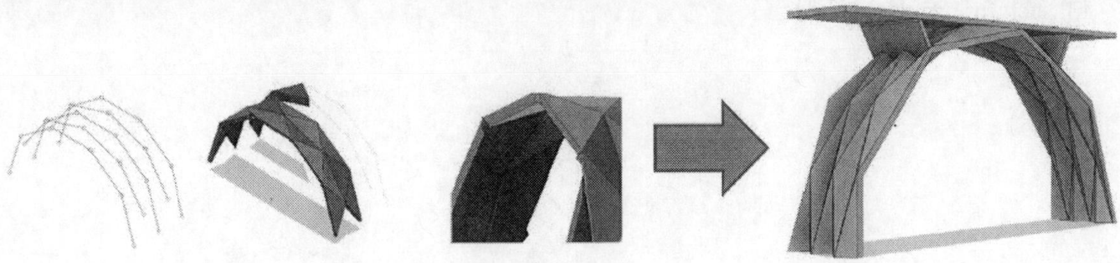

Figure 6. The design process.

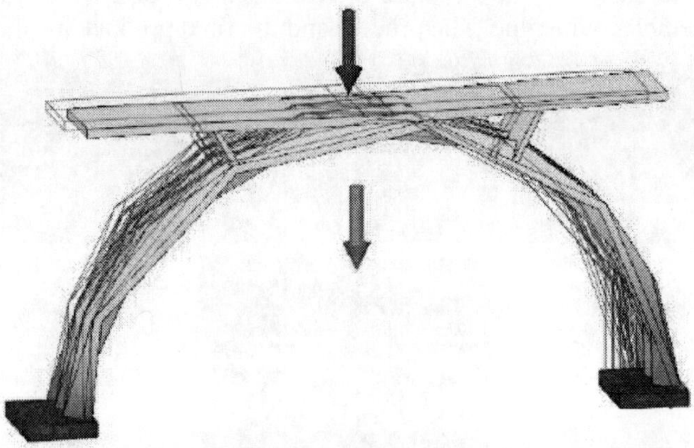

Figure 7. Structure analysis.

3. Realisation Process

3.1. Material properties

The material used to build the model was chosen to comply with the rules of the competition and to enable pouring the thin and curved concrete structure. Fibres of 12 mm length were used - MasterFiber 401 by BASF. The mixture is shown in table 1. The combination of mixture and fibres was successfully used in 2018 during concrete canoe fabrication [9]. The experimentally tested tensile strength of this fibre reinforced concrete is 8.1 MPa.

Table 1. Mixture design

CEM I 42,5 R		600 g
Microsilica		120 g
Lime		150 g
Water		215 g
	0 – 0,25 mm	150 g
Sand	0,25 – 0, mm	550 g
	0,5 – 1 mm	250 g
Plasticiser (Stachema 2000)		30 g
PVA fibres		10 g

3.2. Formwork and Process of Concreting

The model consists of three parts, each concreted separately. These three parts; the deck of the bridge (30x160 cm), two struts and the origami arch, will be then put together. The arch will be concreted laying on its side, as shown in figure 8. This process was tested to examine whether the mixture can fill such formwork properly.

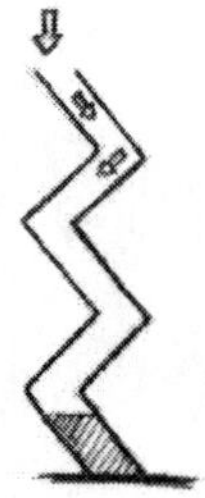

Figure 8. Process of concreting.

The testing was done on 1/9 of the structure of real size. The formwork was created from 4 mm thick plywood and assembled with glue. The process and the final product are shown in figure 9.

Figure 9. Testing the mixture and the formwork.

The testing was deemed successful, thus next work followed the same path. The outer and the inner parts of the formwork were built using the plywood, both consisting of 50 triangles. The wood was then treated to be suitable for concreting. (see figure 10)

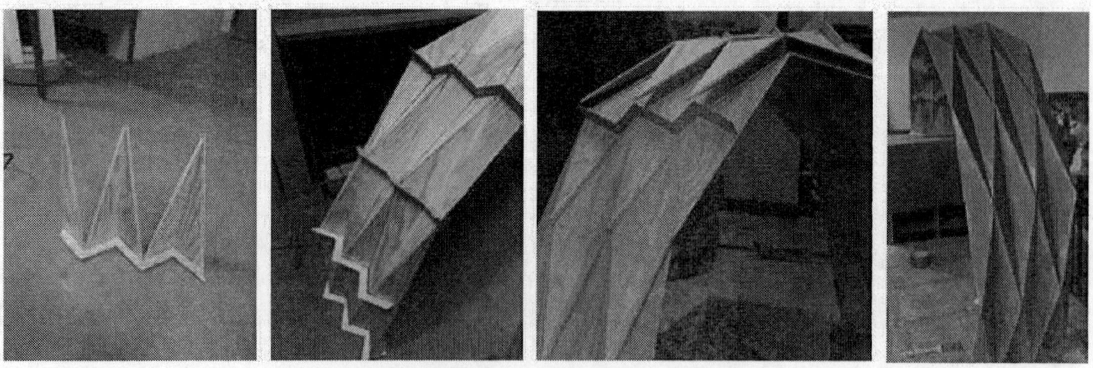

Figure 10. Building the formwork.

The formwork was then fixed on its side and it was concreted by pouring the mixture inside as shown in figure 11.

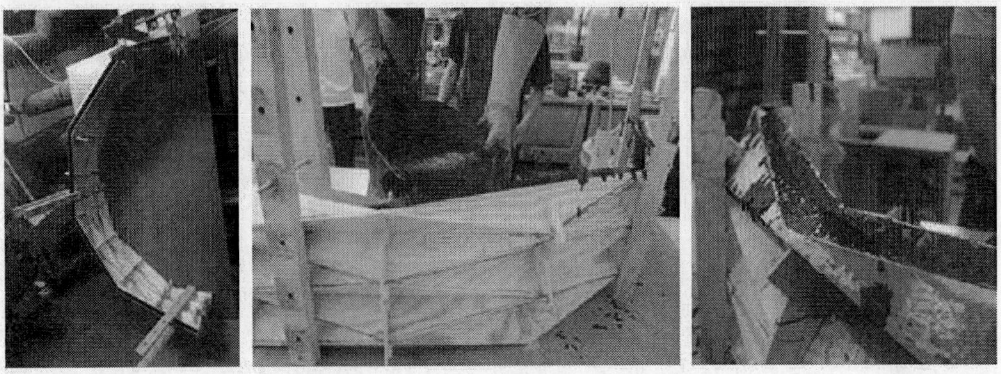

Figure 11. Concreting the origami arch.

The finished product is visible in figure 12. The arch was successfully built, slightly thicker then designed. The arch weighs 56.5 kg, the whole bridge weighs 79.5 kg, the inner span of the arch is 140 cm and the height is 80 cm.

Figure 12. FRC origami bridge.

4. Destructive test

The finished model took part in international competition in Hungary to represent CTU. As part of the competition a destructive loading test was performed which confirmed the calculated 150 kg maximum load. The bridge model collapsed at 149 kg load.

5. Conclusions and Future Work

This work aimed to describe contemporary research of origami structures and also to build such structure. The process of concreting and the formwork was chosen to fit the needs of the competitive model and it was successfully tested on a smaller part of the structure. 3D modeling and loading simulation of the model were performed. Results of the model were confirmed by a destructive test during an international competition. The FRC origami bridge model was awarded second place overall.

Acknowledgements
This work was financially supported by the Czech Technical University in Prague research project no. SGS19/038/OHK1/1T/11, which is gratefully acknowledged.

IOP Conf. Series: Materials Science and Engineering **596** (2019) 012038 doi:10.1088/1757-899X/596/1/012038

References

[1] Yutaka N 2012 *Int. J. of Pure and Applied Mathematics* vol **79** No- 2 269-279

[2] Langdon D 2018 *ArchDaily* (https://www.archdaily.com/554132/ad-classics-yokohamainternational-passenger-terminal-foreign-office-architects-foa)

[3] Ashizawa R 2009 *R A Architects* (http://www.r-a-architects.com/prj/2009/bambooforest.html)

[4] Bechthold M 2008 *Innovative surface structures: technology and applications* (New York: Taylor and Francis)

[5] Bechtold M 2019 *Concrete Origami* (https://www.gsd.harvard.edu)

[6] van der Woerd J D, Chudoba R and Hegger J 2013 *Proce. of the Int. Assoc. for Shell and Spatial Structures (IASS) Symposium 2013* (Wroclaw Univ. of Technology: Poland) p 1

[7] Chudoba R and van der Woerd J D 2018 Thin-walled folded plate structures made of cementitious composites. *Institute of Structural Concrete* (https://www.imb.rwthaachen.de)

[8] 2019 *Versenykiírás* (http://www.betonkenu.hu/versenykiirasrules/)

[9] Šonka Š, Frantová M, Štemberk P and Braun J 2018 *Design and Structural Analysis of CTU in Prague 2018 Concrete Canoe* Mechanika 2018 Proceedings of the 23rd International Scientific Conference (Kaunas: Kaunas University of Technology) pp 154-8

IOP Conf. Series: Materials Science and Engineering **596** (2019) 012039 doi:10.1088/1757-899X/596/1/012039

An investigation of the compatibility of different approaches to self-healing concrete: The superabsorbent polymers and microbially induced calcite precipitation

H Schreiberová, J Fládr, T Trtík, R Chylík and P Bílý

Faculty of Civil Engineering, CTU in Prague, Thakurova 7/2077, 166 29 Prague 6, Czech Republic

E-mail: hana.schreiberova@fsv.cvut.cz

Abstract. The concept of the self-healing concrete, thus an improvement of the durability of concrete structures, has become a widely investigated topic in recent decades. The aim of this study is to examine a possible combination of two existing approaches to the spontaneous crack-sealing – the biologically driven calcite precipitation and the addition of superabsorbent polymers (SAP) into the concrete matrix. Firstly, the paper compares the absorption rate of SAP in various solutions, including a nutrient-based solution which is used in the bio-concrete. The results show that the ionic composition of the liquid influence the swelling capacity greatly. In this study, the absorption rate of SAP in the nutrient-based solution is as low as 6% of the rate in distilled water. Thus, this finding could indicate a possible drawback of the combination of the proposed self-healing approaches. Further, this paper determines the impact of the SAP addition to cement paste with different water/cement ratios. In this experiment, the SAP is applied in various dosages (1% and 0.5% by c. w.) and states (a dry state or a fully swollen state). The conducted flowability and mechanical properties tests indicate a profound, generally negative, impact of the SAP addition on the cement paste characteristics. The paper then discusses possible adjustments of the SAP-cement paste composition and compares the present results with previous studies.

1. Introduction

The susceptibility of concrete to cracking and the resulting need for regular repairs are fundamental issues which significantly influence the durability of concrete structures. The early microcracks formation does not endanger the structure load-bearing capacity directly; however, if the crack widths exceed a certain level, the cracking might result in a failure of the reinforcement cover layer, thus possibly to a total failure of the structure.

This durability endangerment has led researchers to the idea of self-healing concrete. For the past decades, numerous studies have focused on the amplification of the natural autogenous crack-sealing capacity of concrete – a phenomenon when delayed cement grains hydration leads to crack closure [1]. This paper deals with a combination of two previously proposed approaches to self-healing concrete: the biologically driven calcification and incorporation of superabsorbent polymers.

The firstly mentioned method, the bio-based self-healing concrete, is based on the ability of certain microorganisms to produce calcium carbonate ($CaCO_3$) when supplied with appropriate compounds. In this novelty material, the crack sealing is a result of metabolic activity of bacteria which is incorporated in the concrete matrix. For the past two decades, numerous studies identified suitable bacterial genotypes and described their limitations [2–5]; they proposed different types of bacteria protection [2,6–10]; and they examined the self-healing efficiency through in-vitro [2,8–14] and in-situ experiments [12,15].

The second mentioned approach to self-healing concrete relies on the ability of superabsorbent polymers (ionic monomers with a lower cross-linking density than hydrogels) to absorb fluids up to a hundred times their own weight [16].

As summarized in a RILEM technical letter by V. Mechtcherine [17], superabsorbent polymers (SAP) have been extensively investigated in the last decade as a new and promising chemical admixture for concrete. In the overview [17], several proposed applications of SAP in concrete are presented: an application of SAP for mitigation of autogenous shrinkage [18–20] (i.e., internal curing by water absorbed in SAP), an application of SAP to improve frost resistance [21–23] (i.e. a replacement of air-entering agents to create air voids), rheological modification of shotcrete [24], waterproofing (i.e. blocking of water-flow by swollen SAP), and an application of SAP for self-healing.

The possible self-healing mechanism of concrete with SAP is multilayered [16,25,26]. Firstly, the swollen SAP present in cracks is blocking the outside water with harmful aggressive substances from penetration. Secondly, the barrier from SAP maintains the products of the autogenous healing inside the crack, thus promoting the crack-closing potential. Thirdly, the rate of autogenous healing is closely related to water/humidity presence. The water absorbed by SAP then provides the needed supply of moisture for the healing action. Furthermore, the SAP can be used as a carrier for healing substances. This SAP application combined with the biological concrete principles could be developed into a profitable technology as SAP possess a remarkable variety of features: great absorption and retention potential, sensitivity to different pH, temperature, light, and pressure which affect the release of the absorbed substance [16].

The possible effect of the application of SAP in the bio-based self-healing concrete is multiple. The SAP particles could be used as a carrier for the bacterial healing agent or to create air voids in the cement matrix in which bacteria would be protected from the destructive crystalline pressures. Further, the swelled SAP could serve as a needed water reservoir for bacterial metabolic activity.

A similar combination of the microbially induced calcite precipitation (MICP) and standard hydrogels, a significantly less absorbent material than SAP, has been investigated in several studies. Some of them performed experiments with bacterial spores encapsulated in hydrogel sheets [27,28]. These pilot studies proved the compatibility of the selected bacteria with hydrogels and reported successful encapsulation process and significantly promoted crack-sealing. However, the effect of the applied hydrogels to the compressive strength turned out to be unsatisfactory as the hydrogel addition led to a drastic drop of the strength [28]. Wang et al. [29] limited this drawback by an application of pH-responsive hydrogels which increase their swelling at lowering pH, thus the creation of voids after the swollen hydrogel was limited.

The combination of SAP and MICP as investigated in this paper has been also already proposed [30]; however, this novelty approach must be examined in further detail as numerous questions remain unanswered. This paper primarily focuses on the absorption rate of the selected SAP in different solutions (distilled water, tap water, and solutions with a nutritional part of the biological healing agent), the impact of SAP to cement flowability and mechanical properties and discusses the aspects which might influence the SAP and bio-concrete compatibility.

2. Materials and methods

2.1. The absorption rate of SAP in various substances

In the previous research, it has been demonstrated that the absorption rate of SAP is closely related to the ions present in the soaking solution [31]. In our study, the swelling capacity of SAP was measured in several different solutions to outline the possible applications of the material in the bio-concrete production.

The known quantities of SAP (a cross-linked acrylamide/acrylic acid copolymer, potassium salt obtained from Evonik Industries AG, Germany) were submerged in containers with tap water (TW), distilled water (DW), and two solutions with different concentrations of nutrients (CLYA and CLYB), which have been used as a part of the biological healing agent, description can be seen in **Table 1.** The SAP have been left to soak in the solutions for 24 h in sealed containers in 25 °C. After the end of the absorption period, the swollen SAP was filtered from the solutions using a fine sieve. Subsequently, the unabsorbed solutions and the swollen SAP were carefully measured, and the absorption capacity was evaluated.

Table 1. Solutions for the SAP immersion – an investigation of absorbency in different conditions.

TW	tap water
DW	distilled water
CLYA	distilled water, 80 g/l calcium lactate, 1 g/l yeast extract
CLYB	distilled water, 60 g/l calcium lactate, 17 g/l yeast extract

2.2. Mix design of cement paste with SAP and nutrients

As outlined in the Introduction, SAP have a striking absorption capacity. Although it has been shown that the liquid uptake in cement paste is decreased significantly when compared to water immersion [19,32], the effect of the SAP addition to concrete/cement paste workability is still pronounced. For this reason, researchers have been determining and testing various "extra" water-cement (w/c_e) ratios [19,20,33–36] which would make up for the water entrained by SAP particles and provide the needed moisture for the targeted internal curing. In our study, a variety of cement pastes with different composition was prepared and examined in greater detail.

In the conducted experiment, Portland cement CEM I 42.5 R and tap water were used (**Table 2**) for the cement paste preparation. The SAP was applied in the dosage of 1% by c. w. in the case of dry particles (S1 and S2) and approximately 0.5% by c. w. of particles which were immersed in tap water for 24 h prior to the mixing process (i.e. 456 g swollen SAP, S3).

Table 2. Mix designs for flowability test – the groups with CEM I and tap water.

	REF [g/l]	**S1** [g/l]	**S2** [g/l]	**S3** [g/l]
CEM I 42.5 R	586	586	586	586
tap water	293	293	293	0
w/c	0.5	0.5	0.5	0
extra tap water	-	-	87.9	-
w/c_e	-	-	0.15	-
sand (PR 0.1-2)	1758	1758	1758	1758
SAP dry	-	5.86	5.86	-
SAP wet (tap water)	-	-	-	456

In all cases, the SAP has been firstly added and mixed with dry cement prior to any other additions. As mentioned, the SAP has been applied in two forms – in a dry state and in a swollen state. In the case of the swollen SAP, no mixing water was further added into the cement paste.

2.3. The flowability test

A cement flow test was performed in an attempt to determine the impact of the investigated additions (SAP and nutrients) to the rheology of fresh cement paste and further optimize their content based on the workability of the pastes. Immediately after the mixing process, the cement pastes of all the series were submitted to the cement flow test using a cement flow table according to relevant standards.

IOP Conf. Series: Materials Science and Engineering **596** (2019) 012039 doi:10.1088/1757-899X/596/1/012039

2.4. The tensile and compressive strength tests
After the flowability tests, the prepared cement mortars were cast in molds to produce 40x40x160 mm mortar specimens. After 24 hours, the specimens were demolded and kept in room temperature and humidity for further examinations. After 7 days from casting, all of the series were submitted to mechanical testing. Firstly, a three-point bending test was performed. The bending tests were run in a deflection-controlled mode, and the maximal load values were recorded and analyzed using a controlling software.

Further, the compressive strength was determined on the specimen halves that were created after the three-point bending test. The specimens were evenly loaded with pressure, and the mode of failure was observed to exclude incorrectly damaged specimens. The maximum applied load was recorded and analyzed using a controlling software.

2.5. Dynamic modulus of elasticity
To further determine the effect of SAP (with/without the extra water addition) on mechanical properties of cement paste, non-destructive measurements of the dynamic modulus of elasticity (E_d) were performed on the prepared cement paste specimens at 7 days from casting.

3. Results

3.1. The absorption rate of SAP in various substances
As discussed previously, the ionic composition of the soaking solution may have a dramatic impact on the SAP absorbance capacity. It is evident from our results (**Table**) that they have a strong concordance with this prediction. The SAP immersed in distilled water (DW) could retain up to 246 g of the liquid per gram of the material, whereas in other solutions the absorption potential decreased dramatically. In tap water, the absorbance reached 157 g/g SAP, and in the nutritional solutions (CLY A and CLY B) the values were as low as 15 g/g SAP, which is only about 6% of the capacity of SAP immersed in distilled water.

Table 3. The average values of the absorption capacity of SAP immersed in various solutions.

Solution type	Absorption [g solution /g SAP]
TW	157
DW	246
CLY A	15
CLY B	16

3.2. The flowability test
In order to determine the impact of the SAP addition on the flowability of cement pastes with different compositions, the flowability tests were performed and the results compared with reference pastes. In general, the addition of dry SAP to cement paste dramatically affected the workability in all cases.

As it can be observed from **Figure 1**, the addition of dry SAP (1% by c. w.) without any increase of the mixing water (S1) resulted in a much drier mixture when compared to the reference sample (REF). Furthermore, even a considerable increase of the mixing water (w/c_e = 0.15, S2) did not make up for the water uptake by SAP in our experiment. On the other hand, the addition of a swollen SAP (0.5% by c. w.) with approximately 453 g of absorbed tap water (S3) resulted in a slightly more flowable mixture when compared with the reference paste (REF).

REF **S1** **S2** **S3**

Figure 1. Results of cement flow table tests – mix designs with CEM I, tap water and with/without the SAP addition.

3.3. The tensile and compressive strength test

It was suggested in the Introduction that the SAP addition to concrete may have a pronounced impact on the material's mechanical properties. To evaluate the effect, series of compressive and bending tests were performed.

Considering the results of flowability presented above, the impact of SAP addition on both the compressive and tensile strength is not surprising as the mechanical test results overall correspond with the paste consistency (**Figure 2** and **Figure 3**). The paste S1 with dry SAP 1% by c.w. and no water addition showed drastic drops of both strengths. The compressive strength fell down to 27% of the reference paste strength and the tensile strength to 45%. The extra water in the case of the S2 paste partially limited the negative impact of the SAP addition. The paste S2 reached 40% of the reference paste compressive strength and 60% of the reference paste tensile strength. The paste S3 (wet SAP 0.5% by c. w.), which had the closest consistency to reference paste, reached lower but not as severely decreased strengths values as the other pastes (73% of the reference compressive strength and 69% of the reference tensile strength).

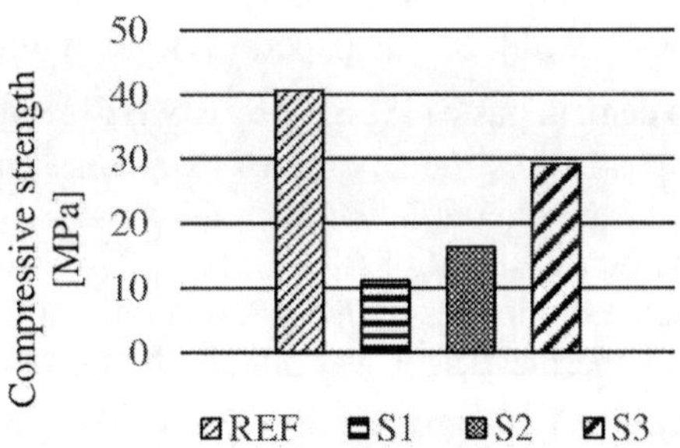

Figure 2. The compressive strength at 7 days from casting.

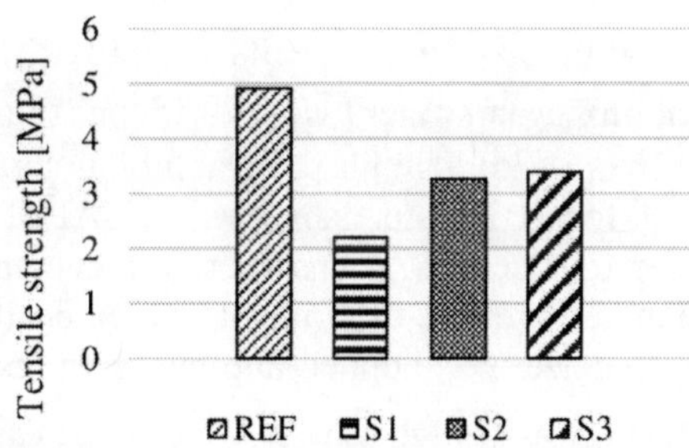

Figure 3. The tensile strength at 7 days from casting.

3.4. Dynamic modulus of elasticity

As it can be seen in **Figure 4**, the obtained dynamic modulus of elasticity of the cement pastes with various SAP and water additions are in line with the compressive strength values presented above. The findings suggest that the addition of wet SAP in the dosage 0.5% by c. w. (S3) had the lowest negative impact on the examined property (a decrease of 26% compared to the reference paste). The addition of dry SAP in the dosage 1% by c. w. with no water addition (S1) resulted in a 71% drop. However, the added extra water in the group S2 lowered the negative effect on E_d significantly (49% of the reference paste value).

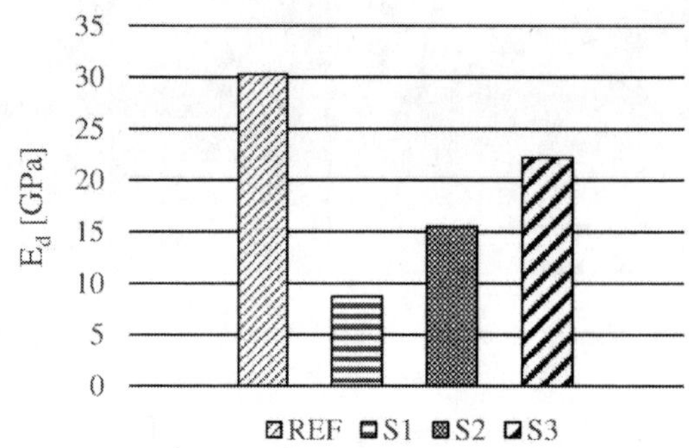

Figure 4. The values of dynamic modulus of elasticity measured on the cement paste specimens at 7 days from casting.

4. Discussion

The results describing the limits of the absorbance capacity of SAP in solutions with different ionic composition presented in this paper are in line with the existing studies [25,31]. The results of the significantly decreased absorption rates of SAP in nutritional solutions, used as a part of the healing agent in self-healing bio-based concrete, provide a valuable framework for future research of the SAP and bio-concrete combination. Possibly, the nutrients incorporated in the matrix could limit the effect of the SAP addition. Not only its internal-curing potential but also the role of SAP as a provider of the needed moisture for a bacterial activity could be restrained. However, more detailed knowledge of the SAP absorption kinetics in bio-based self-healing concrete should be explored and described in future studies.

The cement flow table tests verified the prediction that the dry SAP addition would lead to a significantly drier mixture as reported by previous studies [19,36]. In this paper, even the addition of 30% of mixing tap water ($w/c_e = 0.15$, i.e. 15 g/g SAP) to cement paste (1% SAP by c. w.) did not result in a mixture with a similar flowability as the reference paste, thus the amount of extra water did not make up for the moisture absorbed by SAP. This finding strongly contradicts with the most frequently considered value of SAP absorbency in cement paste between 12-13 g/g SAP [19,22,23,37]. However, in the majority of the experiments which dealt with cement composites with SAP, a certain amount of superplasticizer was applied into the mixtures. When no superplasticizer was added, the needed extra water ratio was as high as $w/c_e = 0.2$ in the case of 1% SAP by c. w. [34].

In our experiment, the addition of fully swollen SAP (0.5% by c. w. in a dry state) lead to similar flowable paste as the reference. It seems that the majority of the absorbed water by SAP was gradually released from the swollen particles during the mixing process, thus no further mixing water was needed.

It should be, however, noted that there are numerous possible sources of error when searching for a sufficient cement paste mix design with SAP. As presented earlier, the ions present in solution affect the absorbency greatly. Numerous studies also proved that the absorption kinetics significantly depends on the size of SAP particles [25,38]. Furthermore, we must keep in mind that the flowability of the cement paste with SAP would in all probability shift considerably in time, thus, in future studies, time-dependent flowability measurements are needed for deeper understanding of the mechanisms.

The results obtained from mechanical measurements show a distinctive tendency of SAP to decrease both the compressive and tensile strength of the material. With increasing SAP content, the negative effect seems to be more pronounced. This behavior was also reported by previous studies [19,33,34,36]; however, in our study, the mechanical tests were conducted at 7 days from curing only. This strength drop may decrease with time as the water captured inside SAP particles would be released at later ages, thus the internal curing process would be activated [19]. Furthermore, a number of researchers reported an effect of a curing mode (i.e. wet/air curing) on the strength development [35,36]. In our experiment no other method of curing was used except air curing; however, the phenomenon should be further explored as it seems to influence the material properties significantly.

5. Conclusion

In this section, the main observations gained from the conducted experiments and future proposes will be briefly concluded:

- The absorption capacity of SAP is dramatically reduced in solutions with a nutritional part of the healing agent used in bio-based concrete. This finding suggests that a sufficient encapsulation of the nutrients in SAP by simple soaking is not realistic.
- The absorption rate of SAP in different mortars (different cement types, nutrients additions, distilled/tap mixing water, etc.) should be further determined.
- The addition of soaked SAP seems to be also an applicable approach. Future research should explore the possibility of SAP with an absorbed bacterial solution (bacterial spores scattered in water). This impregnated SAP incorporated into the mortar could then provide both water for cement hydration and the needed protection of the spores against harmful crystalline pressures by creating of air voids.

Acknowledgments

This paper was prepared thanks to the support of the science foundation of the Czech Republic (GAČR), project "Self-healing of cementitious composites by bacteria-induced calcification" (no. 18-15697s) and the project "Durability of concrete structure and assessment of its life cycle" (SGS19/149/OHK1/3T/11).

References

[1] Edvardsen C 1999 *Aci Mater J.* 4 **96**
[2] Wang JY, Soens H, Verstraete W, De Belie N 2014 *Cem Concr Res.* **56**
[3] Wang J, Mignon A, Snoeck D, Wiktor V, Van Vliergerghe S, Boon N, et al. 2015 *Front Microbiol.* **6**
[4] De Muynck W, Debrouwer D, De Belie N, Verstraete W 2008 *Cem Concr Res.* 7 **38**
[5] Sierra-Beltran MG, Jonkers HM, Schlangen E 2014 *Constr Build Mater.* **67**
[6] Ersan YC, Boon N, De Belie N 2018 *Proc. of Final Conference of RILEM TC 253-MCI: Microorganisms-Cementitious Materials Interactions* vol.2 (Toulouse) p 475–84
[7] Wang JY, Snoeck D, Van Vlierberghe S, Verstraete W, De Belie N 2014 *Constr Build Mater.* **68**
[8] Wang JY, De Belie N, Verstraete W 2012 *J Ind Microbiol Biotechnol.* 4 **39**
[9] Chen H, Qian C, Huang H 2016 *Constr Build Mater.* **126**
[10] Tziviloglou E, Wiktor V, Jonkers HM, Schlangen E 2016 *Constr Build Mater.* **122**
[11] Wang J, Van Tittelboom K, De Belie N, Verstraete W 2012 *Constr Build Mater.* 1 **26**
[12] Paine K, Sharma T, Alazhari M, Heath A, Cooper R 2018 *Proc. of Final Conference of RILEM TC 253-MCI: Microorganisms-Cementitious Materials Interactions* vol.2 (Toulouse) p 387–94
[13] Tziviloglou E, Wiktor V, Jonkers H, Schlangen E 2015 *Proc 9th Int Conf Fract Mech Concr Concr Struct.* (Berkley) 1–6.
[14] Luo M, Qian CX, Li RY 2015 *Constr Build Mater.* **87**
[15] Sierra-Beltran MG, Jonkers HM, Ortiz M 2015 *Fifth Int Conf Self-Healing Mater.* (Durham, NC)
[16] Mignon A, Graulus GJ, Snoeck D, Martins J, De Belie N, Dubruel P, et al. 2014 *J Mater Sci.* 2 **50**
[17] Mechtcherine V 2016 *RILEM Tech Lett.* **1**
[18] Jensen OM, Hansen PF 2001 *Cem Concr Res.* 4 **31**
[19] Jensen OM, Hansen PF 2003 *Cem Concr Res.* 6 **32**
[20] Mechtcherine V, Gorges M, Schroefl C, Assmann A, Brameshuber W, Ribeiro AB, et al. 2014 *Mater Struct.* 3 **47**

[21] Reinhardt HW, Assmann A, Mönnig S 2008 *Proc. of 1st Int Conf Microstruct Relat Durab Cem Compos*. (Nanjing) p 313–22
[22] Laustsen S, Hasholt MT, Jensen OM 2008 *Microstruct Relat Durab Cem Compos*.
[23] Brüdern A-E, Mechtcherine V 2010 *Proc of Int RILEM Conf Use Superabsorbent Polym Other New Addit Concr*. (Lyngby)
[24] Hansen PF, Jensen OM 2000 *Water-entrained cement-based materials num. WO2001002317A1* World Intellectual Property Organization
[25] Snoeck D, Van Vliergerghe S, Steuperaert S, Dubruel P, De Belie N 2012 *J Intell Mater Syst Struct*. 1 **25**
[26] Snoeck D 2018 *RILEM technical letters* **3**
[27] Wang JY, Snoeck D, Van Vlierberghe S, Verstraete W, De Belie N 2014 *Constr Build Mater*. **68**
[28] Van Vliergerghe S, De Belie N, Wiktor V, Wang J, Boon N, Mignon A, et al. 2015 *Front Microbiol*. 6
[29] Wang J, Mignon A, Trenson G, Van Vlierberghe S, Boon N, De Belie N 2018 *Proceedings of final conference of RILEM TC 253-MCI*. (Toulouse) p 405–12
[30] Giriselvam MG, Poornima V, Venkatasubramani R, Sreevidya V 2018 *IOP Conf Ser Mater Sci Eng*. 1 **310**
[31] Kang S, Hong S, Moon J 2018 *Cem Concr Compos*. **88**
[32] Kang S, Hong S, Moon J 2017 *Cem Concr Res*. **97**
[33] Lura P 2016 *Proc of International RILEM Conference on Volume Changes of Hardening Concrete: Testing and Mitigation* (Lyngby) p 117–25
[34] Pelto J, Leivo M, Gruyaert E, Debbaut B 2017 *Smart Mater Struct*. **26**
[35] Jang S, Park W, Kim S 2018 *Magazine of Concrete Research* 8 **71**
[36] Sun B, Wu H, Song W, Li Z, Yu J 2019 *Constr Build Mater*. **204**
[37] Aziz FNAA, Elhibir MMO, Abu Bakar N, Safiee NA 2018 *Proc of 14th International Conference on Concrete Engineering and Technology* (Malaysia)
[38] Mechtcherine V, Hans-Wolf R 2012 *Application of Superabsorbent Polymers (SAP) in Concrete Construction* unedited version (Springer)

IOP Conf. Series: Materials Science and Engineering **596** (2019) 012040 doi:10.1088/1757-899X/596/1/012040

Visualization and analysis of concrete specimens damage after fire and blast experiments

R Štefan and M Foglar

Czech Technical University in Prague, Faculty of Civil Engineering, Department of Concrete and Masonry Structures, Thákurova 7, 166 29 Prague, Czech Republic

E-mail: radek.stefan@fsv.cvut.cz

Abstract. The paper is focused on visualization and analysis of concrete specimen damage after fire and blast experiments. A data visualization tool, based on an in-house MATLAB code, is described and its applicability and versatility are illustrated. In the first illustrative example, concrete spalling of fire exposed floor slab panels made of various types of concrete (without fibres, with polypropylene fibres, with steel fibres) is analysed. In the second example, the results of ultrasonic pulse velocity measurements of concrete bridge decks before and after blast experiments are shown in order to investigate the blast resistance of the decks made of different types of materials and to illustrate the damage of the decks in the form of the material degradation, surface spalling and puncture. The developed data visualization tool has been included in a freely available scientific computer program with a graphical user interface.

1. Introduction

Within an experimental investigation of concrete specimens subjected to fire or blast, data visualization is one of the methods that can be used for illustration of the measured results, their interpretation and subsequent analysis, both qualitative and quantitative – in connection with usual data analysis and descriptive statistical techniques. The data visualization approach has been employed by many researchers in order to display and analyse the amount of concrete spalling of fire exposed concrete specimens, e.g. [1–12]. The data (spalling depths, surface profile) can be obtained by manual measurements (e.g. by a depth slide gauge or by geodetic instruments) in a regular grid, e.g. [1–7], or by advanced photogrammetry, image processing, and 3D scanning methods, e.g. [11, 12]. The data can be displayed in the form of contour plots – so called "spalling maps", or other types of graphs, e.g. a surface plot. These figures provide information both about the spalling area and the spalling depths. Moreover, they can be used for descriptive statistical analysis in order to quantify the extent of spalling – e.g. by analysing the spalling area, volume, and depth, see e.g. [1–12].

The data visualization is also used in the diagnostics of structures by ultrasonic pulse velocity (UPV) measurements. Based on the pulse velocity, the material properties of concrete as well as its heterogeneity and damage (after fire, blast, etc.) can be assessed [13, 14].

Although many data visualization software tools, both commercial and scientific, are currently available, see the above references, we decided to develop our own tool in MATLAB [15] environment, based on the experience that we have gained during our research in the field of fire and blast test of concrete structures. This tool is described in this paper.

2. Data visualization procedure

The data visualization tool has been developed in MATLAB [15] environment. The entire procedure implemented in the code is described in algorithm 1.

Algorithm 1 Data visualization procedure implemented in the presented software tool

1: input a `.xlsx` file containing the measured data and the corresponding coordinates
2: set the geometry and dimensions of the analysed area
3: set the interpolation and extrapolation method
4: perform the interpolation and extrapolation
5: perform the statistical analysis
6: display the results in the form of contour plots or other types of graphs
7: export the figures

The individual steps of algorithm 1 are described below.

In **step 1**, an input `.xlsx` file is imported. The file contains both the measured data (e.g. the spalling depths, the UPV results, etc.) and the corresponding coordinates of the measurement points (their positions in the analysed area). In our approach, the data are obtained by manual measurements in a regular grid, see the examples described in sections 3 and 4.

In **step 2**, the analysed area of the specimen surface is defined by its geometry and dimensions.

In **step 3**, the method of interpolation and extrapolation of the data is specified. We use the built-in MATLAB function `scatteredInterpolant`, with the methods of interpolation: `linear`, `nearest`, or `natural`, and extrapolation: `linear`, `nearest`, or `none`, see the MATLAB [15] Documentation.

The interpolated and extrapolated data are obtained in **step 4** by MATLAB [15] `scatteredInterpolant` function.

A descriptive statistical approach can be employed in **step 5** in order to analyse the measured data (spalling depth, area, volume, UPV results, etc.) by usual statistical techniques (minimum, maximum, mean value, standard deviation, etc.).

In **step 6**, the results can be displayed in the form of contour plots ("maps" with isolines or isoareas) or other types of graphs (e.g. 3D surface plots, line graphs for selected sections of the specimen, etc.). The results of the statistical analysis can be displayed in a figure or exported separately in a text file or a `.xlsx` file.

In **step 7**, the created figure can be saved to a required format, such as `.eps`, `.emf`, `.jpg`, `.pdf`, `.png`, `.tif`, etc., see the MATLAB [15] Documentation.

3. Visualization of concrete spalling after fire test

3.1. Experiment description

Here, we focus on a blast and fire experiment described in our previous work [16, 17]. Within the experiment, six reinforced concrete floor slab panels of the dimensions of 3600 × 1000 × 150 mm were tested. They were made of normal concrete C30/37 (according to [18]) in three variants – without fibres (referred to as material 1), with polypropylene fibres (material 2), and with steel fibres (material 3). For each material, two panels were manufactured and one of them was subjected to blast loads before the fire test. Hereafter, the reference panels (without blast exposure) are denoted as "panels a", and the panels subjected to blast loads before the fire test are denoted as "panels b", see also [16, 17].

All panels were placed in a fire furnace, loaded by constant mechanical loads, and exposed to the standard fire exposure (ISO 834 fire curve, see e.g. [19]) for 135 minutes. During the fire experiment, the temperature evolutions within the panels as well as their deflections were recorded.

After a cooling period of one day, the panels were removed from the furnace and prepared for further investigation, including the measurement of concrete spalling depths, see figure 1.

A detailed description of the experiment can be found in [16, 17].

Figure 1. Reinforced concrete floor slab panels after the fire test. Concrete spalling on the heated (bottom) surfaces of the panels. Notation: 1 – concrete without fibres, 2 – concrete with polypropylene fibres, 3 – concrete with steel fibres; a – reference panels (without blast exposure), b – panels subjected to blast loads before the fire test. See also [16, 17].

As can be seen in figure 1, no spalling occurred for panels 2a and 2b made of concrete with polypropylene fibres. Hence, the following analysis and visualization of concrete spalling is focused on the other panels – 1a, 1b, 3a, 3b.

3.2. Measurement of the spalling depths

The spalling depths were measured manually by a depth slide gauge in a regular grid of 100 mm × 100 mm, see figure 2. The obtained data (spalling depths, coordinates of the measurement point) were stored in a `.xlsx` file for a subsequent analysis.

Figure 2. Measurement of spalling depths by a depth slide gauge in a regular grid of 100 mm × 100 mm, see also [16, 17].

3.3. Data visualization and analysis

The data visualization is conducted as described in algorithm 1. The analysed area (heated surface of the panel) as well as the measurement points are displayed in figure 3a. In this case, the measurement points lie on the boundary of the analysed area, and hence, no extrapolation of the data is needed. The interpolation methods mentioned above (`linear`, `nearest`, and `natural`, see the MATLAB [15] Documentation) are illustrated in figures 3b–c, where the data are visualized by filled contour plots (so called "spalling maps") using MATLAB [15] `contourf` function. Other types of graphs that can be used for the data visualization as well as their comparison with a photograph of the analysed panel are presented in figure 4.

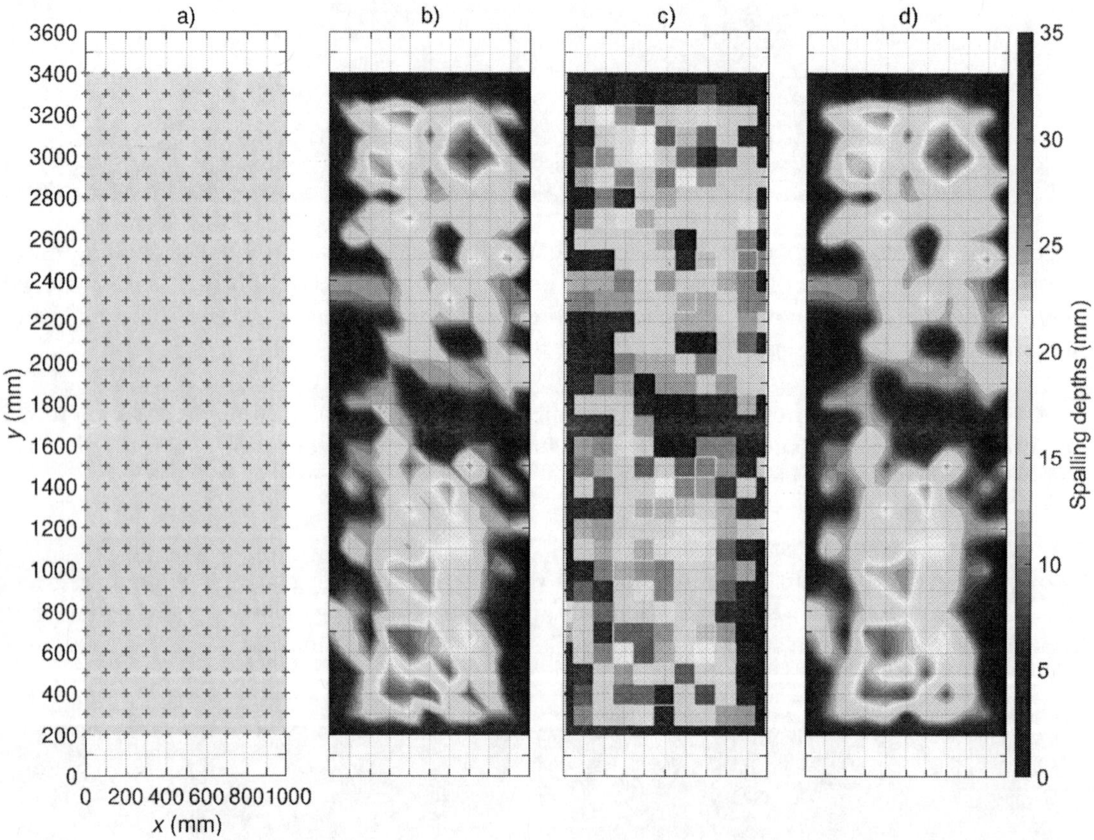

Figure 3. (a) Geometry of an analysed panel. The analysed area (heated surface) is highlighted in gray color; the measurement points are plotted by plus marks. (b–d) Example of the spalling maps on the 3a panel obtained by different interpolation methods: `linear` (b), `nearest` (c), and `natural`(d), see the MATLAB [15] Documentation.

As mentioned above, a descriptive statistical approach can be employed in order to analyse the measured data. Here, we can illustrate this approach by determining the maximum and mean values of the spalling depths for the anlysed panels, see table 1 (cf. e.g. [1–12]).

A more detailed study and discussion of the spalling behaviour of the analysed panels are out of the scope of this paper and will be done in our future work.

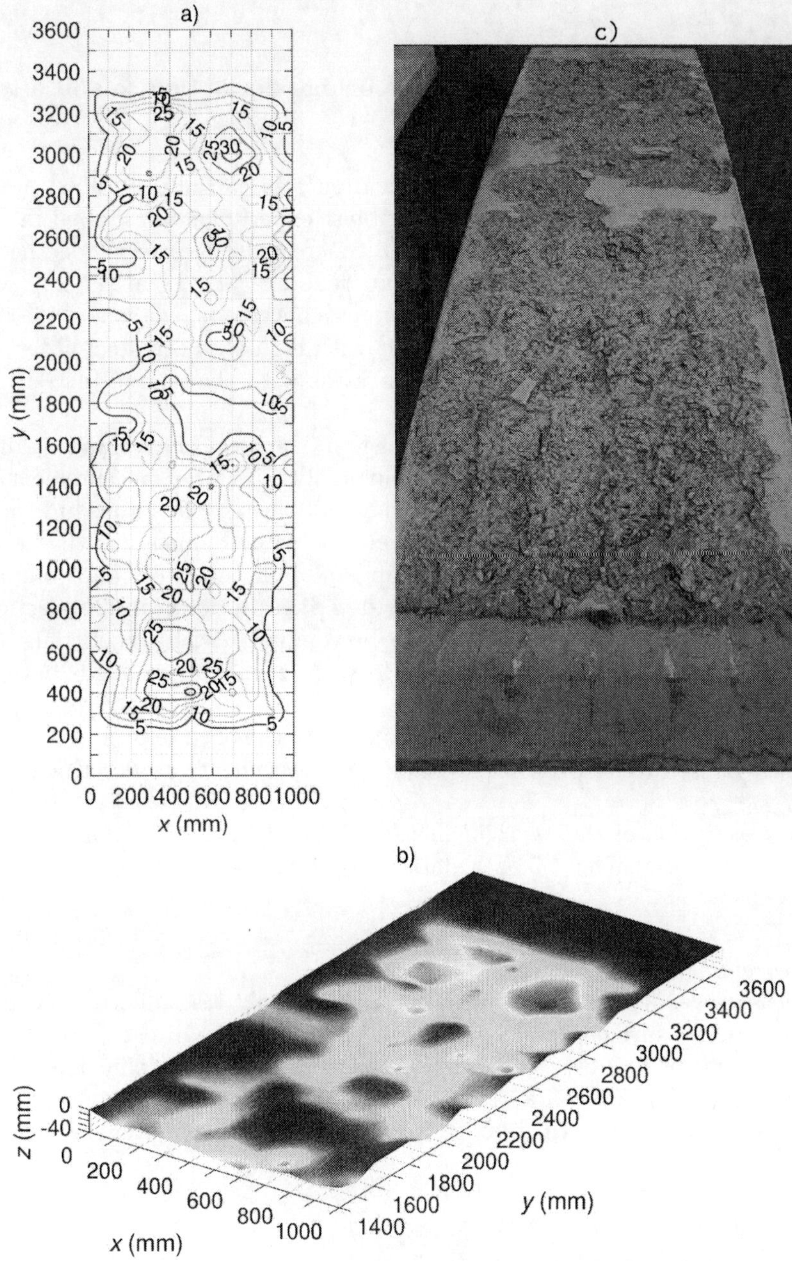

Figure 4. Spalling visualization on the 3a panel. (a) Spalling isoline map obtained by MATLAB [15] `contour` function. (b) Surface plot of a part of the panel obtained by MATLAB [15] `surf` function. (c) Photograph of the panel. See also [16, 17].

Table 1. Spalling depths of the analysed panels.

Spalling depth (mm)	Panel 1a	Panel 1b	Panel 3a	Panel 3b
Maximum	26	27	32	34
Mean	4.1	4.0	7.0	10.8

IOP Conf. Series: Materials Science and Engineering **596** (2019) 012040 doi:10.1088/1757-899X/596/1/012040

4. Visualization of UPV data of concrete bridge decks before and after blast exposure

4.1. Experiment description

The applicability of the presented data visualization tool can also be illustrated by analysing the UPV measurement results obtained during blast experiments described in [17, 20]. Within the scope of the experiments, six reinforced concrete bridge decks of the dimensions of $6000 \times 1500 \times 300$ mm were tested. They were denoted #12–17 in [17, 20], which we adopt here. The specimens were made of high performance concrete (specimens #12–15) or ultra-high performance concrete (specimens #16 and #17) with high-strength steel fibres.

Each deck was exposed to 25 kg TNT blast loads with the charge placed in the specimen centre near the upper surface.

After the blast exposure, the specimen damage in the form of cracking, spalling, deflection, delamination, and puncture was measured and analysed. The measurement of the surface damage (spalling, puncture) was conducted manually using the procedure described in the previous example in section 3 (manual measurement of the damage depths in a regular grid). The analysis of the damage and its scale – spalling and puncture areas and volume, see [20, Tab. 2], was performed by a semi-manual procedure described in detail in [21, Section 2.3, Fig. 4]. The results presented in [20, Tab. 2] can easily be compared with the outputs obtained by the automatic procedure presented in this paper. However, such comparison is beyond the scope of this paper and will be done in our future work.

In order to determine the material degradation due to the blast exposure, UPV measurements were performed on all decks before and after the experiment, as described in the following sections.

A detailed description of the experiment can be found in [17, 20]. Some preliminary results of the visualization and analysis of UPV data using the presented procedure have already been published in [22].

4.2. UPV measurements

The UPV measurements were performed by an ultrasonic test device with two transducers placed on the opposite surfaces of the specimen – i.e. using the direct method, see e.g. [13, 14]. The device is shown in figure 5. The measurement points were defined by a regular grid of 200 mm × 200 mm (specimen #13) or 300 mm × 300 mm (the other specimens), see [20, Figs. 7–11]. The obtained data (transit time, pulse velocity) were stored in a `.xlsx` file for a subsequent analysis.

Figure 5. Ultrasonic test device used for the UPV measurements, see also [16, 17].

4.3. Data visualization and analysis

The data visualization is conducted as described in algorithm 1. In this example, we use the `nearest` interpolation method and the `linear` extrapolation method, see the MATLAB [15] Documentation. The UPV data measured on specimen #12 before the blast exposure are shown in figure 6.

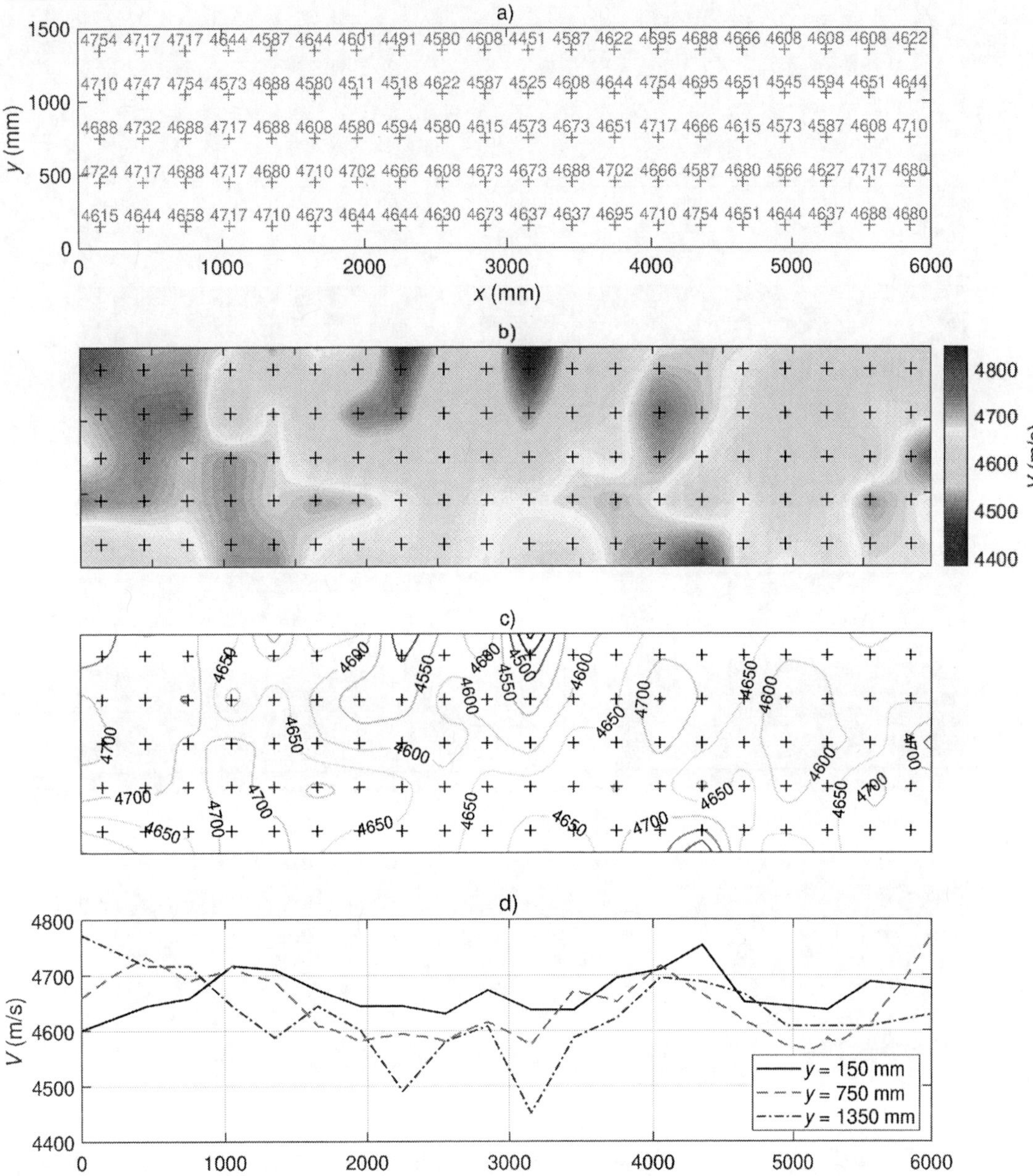

Figure 6. The UPV data measured on specimen #12 before the blast exposure: (a) Geometry of the analysed specimen, positions of the measurement points, measured values of the pulse velocity V (m/s). (b) Isoareas obtained by MATLAB [15] `contourf` function. (c) Isolines obtained by MATLAB [15] `contour` function. (d) Line graphs for selected sections. See also [22].

The UPV data measured on specimen #12 after the blast exposure are shown in figure 7a. Due to the damage (puncture) of the central par of the specimen, the UPV data for the measurement points in this area could not be obtained, see figure 7a.

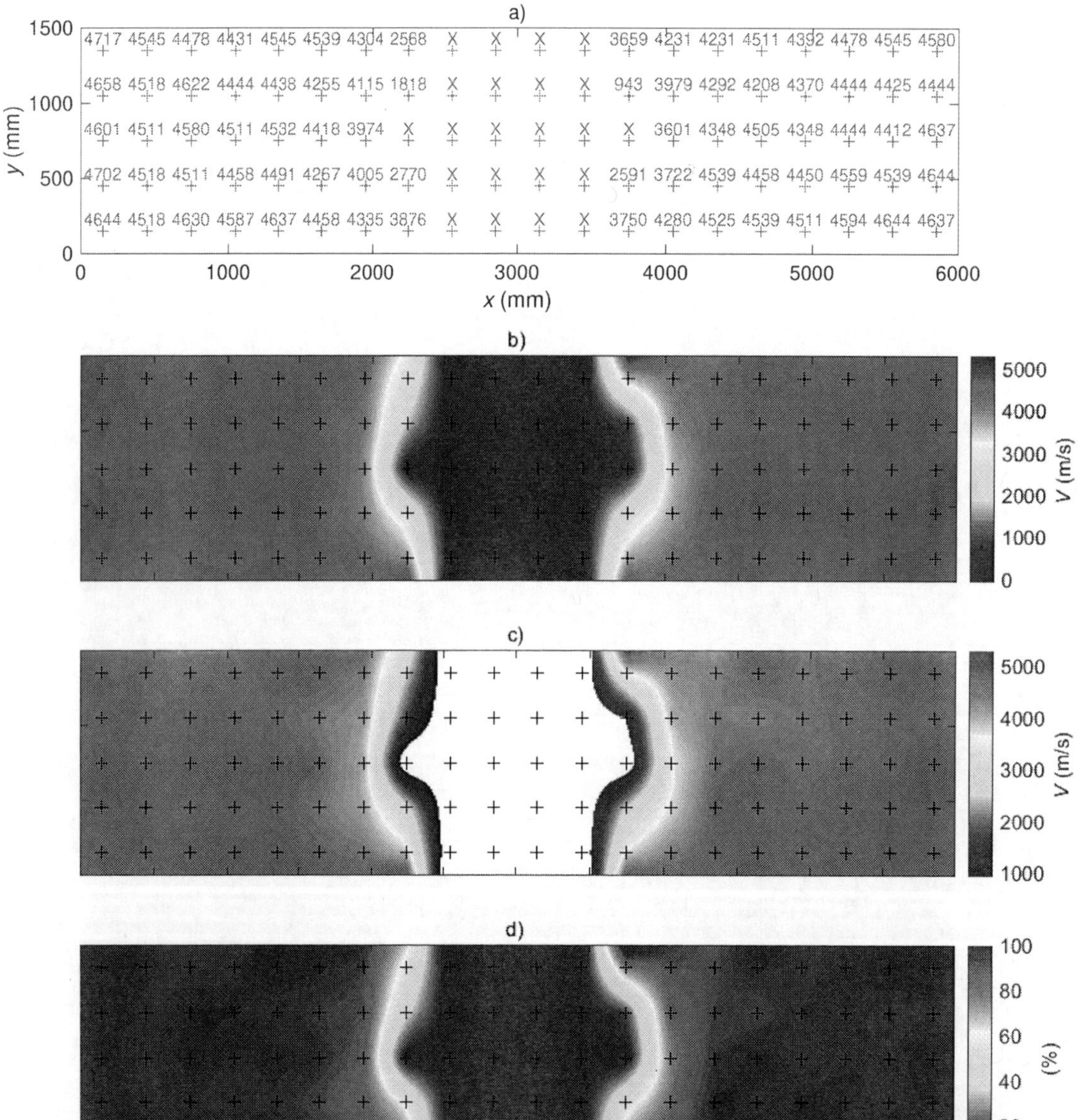

Figure 7. The UPV data measured on specimen #12 after the blast exposure: (a) Measured values of the pulse velocity V (m/s); for the points with the X-values, the data were not obtained. (b) Isoareas obtained by MATLAB [15] `contourf` function; the X-values are replaced by pulse velocity $V = 343$ m/s. (c) Isolines obtained by MATLAB [15] `contour` function; the X-values are replaced by pulse velocity $V = 343$ m/s; a limit for the visualization is set to the lowest measured value, $V = 943$ m/s. (d) Relative values of the pulse velocity. See also [22].

The lack of data for the damaged part of the specimen has to be solved during the visualization process. During the development of the presented data visualization tool, the following approaches have been employed:

- define the boundary of the damaged area, set it as the boundary of the analysed geometry in step 2 of algorithm 1, and perform the interpolation/extrapolation process only for the undamaged parts of the specimen; the boundary of the damaged area can be obtained by manual measurements or by an image processing method;
- use MATLAB [15] `NaN` (not a number) values for the measurement points in the damaged area and conduct the interpolation/extrapolation process for the whole specimen face area;
- set the pulse velocity in the damaged area to a specific value, e.g. $V = 0$ m/s, as used in [22], or $V = 343$ m/s (pulse velocity in normal air, see e.g. [23, Tab. 5.1]), and conduct the interpolation/extrapolation process for the whole specimen face area.

Here, we use the last-mentioned approach with $V = 343$ m/s for the measurement points in the damage area, see figure 7b.

For a better illustration of the damage (puncture), a limit can be set, e.g. to the lowest measured value. Below this limit, the values obtained by an interpolation/extrapolation process are excluded from the visualization, as shown in figure 7c.

The material degradation and damage of the specimen due to the blast exposure can be analysed by comparing the UPV data measured before and after the blast exposure. This can be expressed by the relative values of the pulse velocity defined as the pulse velocities measured after the blast exposure divided by the corresponding initial values (before the blast exposure), see figure 7d.

Comparison of the obtained isolines of the pulse velocity after the blast exposure with a photograph of the analysed specimen is presented in figure 8.

Figure 8. Comparison of the obtained isolines of the pulse velocity after the blast exposure with a photograph of the analysed specimen #12; author: J. Stöhr, see also [22].

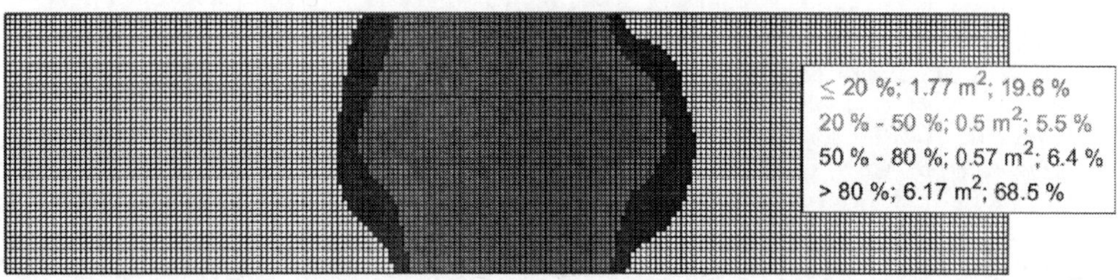

Figure 9. Areas with specific ranges of the relative pulse velocity for specimen #12 obtained by an automatic procedure, see also [22].

IOP Conf. Series: Materials Science and Engineering **596** (2019) 012040 doi:10.1088/1757-899X/596/1/012040

A quantitative investigation of the specimen properties can be conducted by descriptive statistical analysis (mean values, standard deviation) of the measured UPV data. The material degradation and the damage of the specimen due to the blast exposure can also be quantified by determining surface areas with specific ranges of the relative pulse velocity, see figure 9. Some preliminary results of this analysis have been presented in [22]. A more detailed study and discussion of the blast resistance of the analysed specimens will be done in our future work.

5. SVDult software

The developed data visualization tool has been included in a freely available scientific software SVDult [24]. The software, including its graphical user interface, has been developed in MATLAB [15] environment. The software will be described in detail in our future work.

6. Conclusions

In the paper, a data visualization tool, based on an in-house MATLAB code, was described and its applicability and versatility were shown on illustrative examples. The examples were focused on visualization and analysis of the data obtained from fire and blast experiments. The developed data visualization tool has been included in a freely available scientific computer program with a graphical user interface.

Acknowledgments

This work has been supported by the Czech Science Foundation, project No. GA17-23067S, which is gratefully acknowledged.

References

[1] Mindeguia J C, Pimienta P, Carré H and La Borderie C 2009 *1st International Workshop on Concrete Spalling due to Fire Exposure* pp 150–167

[2] Mindeguia J C, Pimienta P, Carré H and La Borderie C 2013 *European Journal of Environmental and Civil Engineering* **17** 453–466

[3] Guerrieri M and Fragomeni S 2013 *MATEC Web of Conferences* **6** 01002

[4] Guerrieri M and Fragomeni S 2016 *Journal of Materials in Civil Engineering* **28** 04016164

[5] Mohd Ali A Z, Sanjayan J and Guerrieri M 2017 *Journal of Materials in Civil Engineering* **29** 04017237

[6] Mohaghegh A M, Silfwerbrand J, Arskog V and Jansson McNamee R 2017 *Nordic Concrete Res.* **57** 89–102

[7] Hager I, Mróz K and Tracz T 2018 *MATEC Web of Conferences* **163** 02004

[8] Krzemień K and Hager I 2015 *Procedia Engineering* **108** 285–292

[9] Yan Z G, Shen Y, Zhu H H, Li X J and Lu Y 2015 *Fire Safety Journal* **71** 86–99

[10] Ozawa M, Uchida S, Kamada T and Morimoto H 2012 *Construction and Building Materials* **37** 621–628

[11] Chlepková M 2011 *INGEO 2011 – 5th International Conference on Engineering Surveying* pp 295–300

[12] Yi N H, Choi S J, Lee S W and Kim J H J 2015 *Fire Safety Journal* **71** 123–133

[13] Bungey J H, Millard S G and Grantham M G 2006 *Testing of Concrete in Structures* (Taylor & Francis)

[14] Ravi Kumar N S M, Barkavi T and Natarajan C 2018 *Journal of Building Pathology and Rehabilitation* **3** 6

[15] MATLAB 2015 Version 8.6.0 (R2015b) The MathWorks, Inc., Natick, Massachusetts, United States

[16] Štefan R, Foglar M and Fládr J 2016 *Beton TKS* **16** 42–48 [in Czech]

[17] Foglar M, Hájek R, Štefan R and Stöhr J 2016 *Performance Investigation of Cementitious Composites Subjected to Extreme Loading. Experiments, Modelling, Technology* (CTU in Prague) [in Czech]

[18] EN 206+A1 2016 *Concrete – Part 1: Specification, performance, production and conformity* (CEN)

[19] EN 1991-1-2 2002 *Eurocode 1: Actions on structures – Part 1-2: General actions – Actions on structures exposed to fire* (CEN)

[20] Foglar M, Hajek R, Fladr J, Pachman J and Stoller J 2017 *Construction and Building Materials* **145** 588–601

[21] Foglar M, Hajek R, Kovar M and Stoller J 2015 *Construction and Building Materials* **94** 536–546

[22] Stöhr J 2016 Non-destructive assessment of damage caused by blast overpressure to reinforced concrete panels *PhD Workshop 2016* (Dept. of Concrete and Masonry Structures, FCE CTU in Prague) [in Czech]

[23] Sinclair I R 2001 *Sensors and Transducers* 3rd ed (Newnes)

[24] Štefan R and Foglar M 2015–2019 *SVDult – Software for visualization of data obtained by ultrasonic analysis of concrete specimens* CTU in Prague. `http://people.fsv.cvut.cz/www/stefarad/software.html`

Methods of lighting of concrete structures for high-speed camera measurement

T Trtík[1], R Chylík[1], J Fládr[1], J Štoller[2] and I Broukalová[1]

[1] Faculty of Civil Engineering, Czech Technical University in Prague, Thákurova 7, 166 29 Praha 6, Czech Republic
[2] University of Defence, Kounicova 65, 662 10 Brno, Czech Republic

E-mail: tomas.trtik@fsv.cvut.cz

Abstract. In extensive research, damage to concrete panels by explosion is investigated. For detailed monitoring of particular phases of explosion, exploitation of a high-speed camera is intended. The paper presents testing of possibilities of lighting of the scene at detonation with explosive agents. Four alternatives of agents were tested – three varying amounts of aluminium powder and argon. Paper presents results of the tests – intensity of lighting for tested variants, delay between ignition of lighting agent and the main detonation, and evaluates convenience of tested explosive agents for illuminating of the scene captured by high-speed camera.

1. Introduction

Current high-speed cameras are used to record high-speed phenomena or to capture very short periods of time which cannot be seen with the naked eye or ordinary camera. High-speed camera was used in previous research, but without the primary light source [1,2]. Common frame rate is in the order of tens of thousands of frames per second, for special applications the frame rate exceeds one million frames per second. The subsequent slow-motion playback enables analysing of areas around exploded charge, as deformations of structures exposed to shock wave or propagation of the shock wave itself.

Before the actual start of the recording, the observed phenomena must be known to prepare the scene, lens, camera settings and select suitable lighting. Often a compromise between image resolution, frame rate and image luminance must be chosen. Varying conditions of the scene during recording must be considered, too. In explosion of the blasting agent, the scene is excessively lighted, what has to be taken into account in setting of lens and camera, to capture the desired action correctly and in the best possible quality.

1.1. Problem description

Blast resistance of concrete panels with dimensions 1000 x 1000 mm and thicknesses 100 mm, 150 mm and 200 mm was tested. The explosive was set on the upper surface of concrete panel and above the panel in various heights (figure 1). Cracking is an important feature followed in investigation of blast resistance. The principal cracks arise on the side opposite to the explosive. To prevent damage of the camera by flying concrete pieces impact and to enable observing the cracking, an inclined mirror serves to reflect the image. The record of cracking is used for comparison with the numerical model.

Figure 1. Picture of the scene.

The record was captured by a pair of high-speed cameras situated 25m from the monitored concrete panel. One camera recorded the overall scene and explosion (see figure 1), the other camera aimed at the mirror to (see figure 2) to monitor cracking. The development of blast cracks is a very fast process, which has to be monitored with higher frame rate than common image records. However, the higher frame rate requires better lighting conditions. A drawback is also a shadow on the mirror given by the concrete panel.

The explosion causes partial lighting which, however, proved to be insufficient (see figure 2). The shooting was performed under full sunlight. Lighting from primary source (see figure 3).

To improve the lighting conditions, an alternative was proposed where the set of mirrors should have been turned to illuminate the scene with sunlight. This way of lighting the scene proved to be inadequate and insufficient too. Due to the rotation of the Earth around its axis and around the Sun, the turning of the mirrors had to be constantly adjusted.

Figure 2. Figure of cracking after explosion with insufficient lighting.

Figure 3. Figure of cracking after explosion with primary source of lighting (aluminium powder).

Therefore, another primary light source was tried - aluminium powder (see figure 4). At a distance 1.6 meters from the concrete panel, 150g of aluminium powder and 150g of Semtex were placed to ignite the aluminium. All tests were lighted like this. The objective was to provide maximum possible illumination on the monitored side of the panel during development of cracks. Correct timing of the delay between ignition of the illuminating agent and the main detonation is very complicated and needs to be addressed in detail. With regard to this findings, an experiment was conducted, which explores the possibilities of illumination of the detonation, intensity and time course of the lighting for various types of lighting substances.

Figure 4. Shot of the explosion. On the left is the primary source of lighting – aluminium powder.

2. Experiment

2.1. Agents used for illumination

2.1.1. Aluminium powder. Burning of aluminium powder emits high light output. The burning rate is affected by the fineness of grinding of the used aluminium. The particle size analysis of the aluminium used in tests is shown in table 1 and the properties of the aluminium used are shown in table 2.

Table 1. Particle size distribution of the aluminium powder.

Retained on sieve	Percentage
0.1 μm	0.81 %
0.063 μm	25.14 %
0.045 μm	48.67 %
Passing 0.045 μm	51.33 %

IOP Conf. Series: Materials Science and Engineering **596** (2019) 012041 doi:10.1088/1757-899X/596/1/012041

Table 2. Properties of aluminium powder.

Specification	Value
Fe content	max. 0.14 %
Si content	max. 0.07 %
Cu content	max. 0.001 %
Humidity	max. 0.2 %
Melting point	661 °C
Ignition point of the powder	250 °C
Burning temperature	2500 °C

2.1.2. Argon. Argon under permanent pressure is a gas. A 1.5 litre bottle was filled with 1 bar overpressure.

2.1.3. Petrol. Bottle filled with 0.5 l of petrol was used in the test. The results of the test are not presented, as the full ignition of petrol came 60 ms after detonation of explosive. The increase of illumination intensity is much slower than start-up time of illumination of aluminium and argon. Inflammation of petrol is conditioned by dispersion of petrol in the air and reaching of a suitable compression ratio. Petrol was classified an improper illumination method and it has been eliminated from the experiment.

2.2. Description of the experiment

The camera was 10 m away from the light source, using 85 mm lens with a f/1.4 aperture diameter. The framerate was set to 10,000 fps, an image resolution of 1600 x 1200 px, and an electronic shutter of 33,339 ns. The electronic shutter speed was intentionally reduced (threefold) in order to observe the ignition of fumes, not the illumination as such. The same camera and lens settings were used for all records.

In the experiment, aluminium powder weighing 75 g, 150 g, 300 g and argon, which was filled into the bottle with 1 bar overpressure, were tested as a light source. Ignition of all light agents was provided by the same explosive (Trinitrotoluene) of 75 g weight. In contrary to original method of illumination described in the first chapter, the weight of the explosive was reduced by half, to reduce the shock wave magnitude.

2.3. Description of the testing

Representative images were selected to describe the monitored process. The first image captures the time just prior to the explosion (see figure 5, figure 9, figure 13 and figure 17). The second image captures a moment 1 ms (millisecond) after the explosion. At this moment, a fireball is formed what is accompanied by creation of explosion products. The third figure depicts moment 3 ms after explosion. At this moment the explosion products ignite and a shape of fireball is evaluated. The time of the last figure is different. The presented figures show moment, when a fireball is formed, all products have been ignited and a full lighting is provided. At this moment the fireball does not necessarily have the biggest volume; in all cases, the fireball has grown but the light intensity has not increased much.

3. Results

3.1. Lighting source – Aluminium powder 75 g

Figure 5. Just prior to explosion.

Figure 6. Time 1 ms after explosion.

Figure 7. Time 3 ms after explosion.

Figure 8. Time 9.5 ms after explosion - full illumination.

Initially, a big amount of explosive products is created (figure 6), which flash in the next phase (figure 7). Fireball is compact and emits light with high intensity for a long time (figure 8).

IOP Conf. Series: Materials Science and Engineering **596** (2019) 012041 doi:10.1088/1757-899X/596/1/012041

3.2. Lighting source – Aluminium powder 150 g

Figure 9. Just prior to explosion.

Figure 10. Time 1 ms after explosion.

Figure 11. Time 3 ms after explosion.

Figure 12. Time 8.5 ms after explosion - full illumination.

A time 1 ms after explosion, a horizontal strip of explosive products is formed (figure 10 and figure 11), which fully lights at time 8.5 ms (figure 12). The dose 75 g of activation explosive is sufficient to ignite all aluminium and the illumination intensity is higher than illumination given by 75 g of aluminium powder. The fireball is compact and during decrease of the light intensity, the source is divided in two parts.

3.3. Lighting source – Aluminium powder 300 g

Figure 13. Just prior to explosion.

Figure 14. Time 1 ms after explosion.

Figure 15. Time 3 ms after explosion.

Figure 16. Time 7 ms after explosion - full illumination.

The results show that the 4:1 ratio between the aluminium powder and the explosive is too high (figure 14). A great deal of the aluminium is dispersed around without ignition (figure 15). Intensity of lighting is higher than for 75 g of aluminium powder, but lower than for dose 150 g of aluminium powder (figure 16).

3.4. Lighting source – Argon

Figure 17. Just prior to explosion.

Figure 18. Time 1 ms after explosion.

Figure 19. Time 3 ms after explosion.

Figure 20. Time 10 ms after explosion - full illumination.

During detonation, horizontal expansion occurs first, and a cloud of explosion products is formed (figure 18 and figure 19). The explosion products ignite at time 10ms (figure 20). The resultant fireball shape is not a sphere and the intensity of lighting is lower than intensity of 75 g of aluminium powder.

The summary information and evaluation of test shows table 3. It describes the time course of lighting for each method, respectively each explosive agent.

Table 3. Time course of illumination.

Source of lighting	Creating of compact fireball	Full luminosity	Marked decrease of luminosity
Aluminium powder – 75 g	9.5 ms	20 ms	26 ms
Aluminium powder – 150 g	8 ms	13 ms	22 ms
Aluminium powder – 300 g	7 ms	11 ms	16 ms
Argon – 1 l	10 ms	12 ms	14 ms

Figures 21 – 24 show image characteristics using histograms, which depict number of pixels (vertical axes) related to luminance intensity (horizontal axes). The horizontal axis shows characteristics of pixels in grey scale from 0 (black) to 255 (white) for variant methods of lighting at the moment of maximum lighting of the fireball. Results show that the highest intensity of lighting was reached for aluminium powder of weight 150 gA, followed by aluminium powder of weight 300 g. Aluminium powder with weight 75 g embodies similar intensity of lighting as bottle with argon.

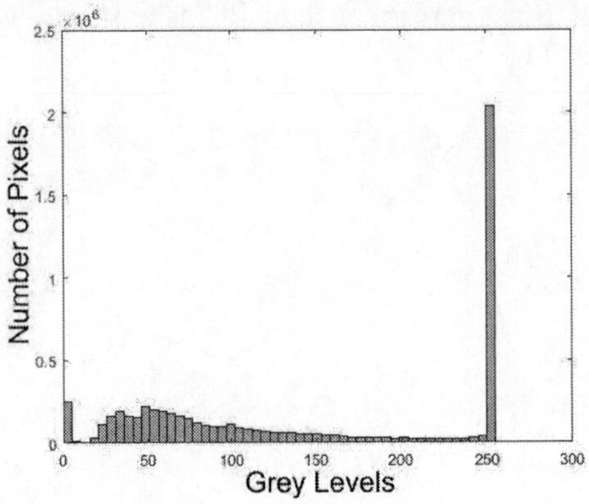

Figure 21. Pixel histogram of figure 8 which depicts explosion of aluminium powder of weight 75 g at time 9.5 ms after detonation.

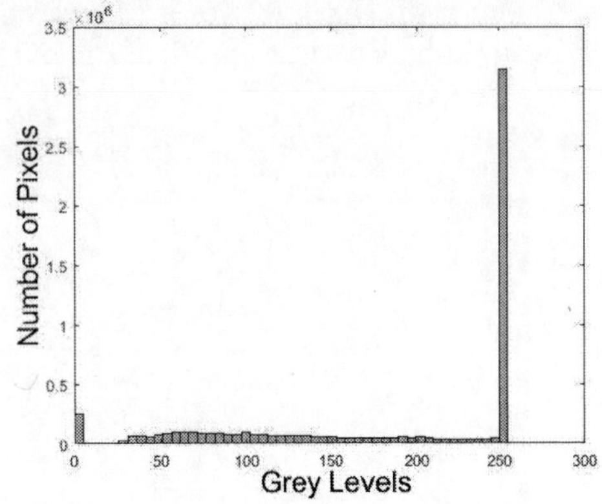

Figure 22. Pixel histogram of figure 12 which depicts explosion of aluminium powder of weight 150 g at time 8 ms after detonation.

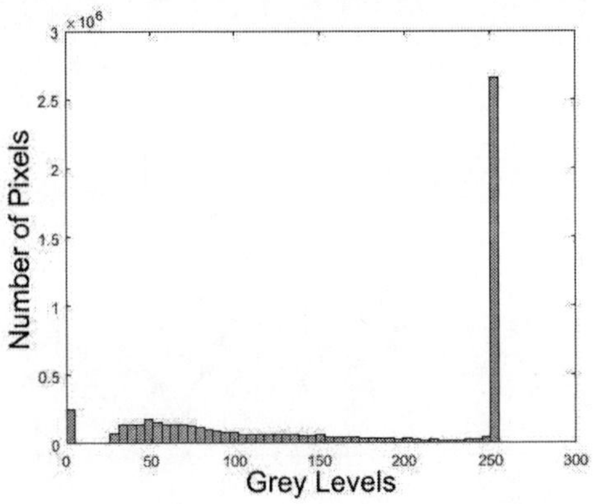

Figure 23. Pixel histogram of figure 16 which depicts explosion of aluminium powder of weight 300 g at time 7 ms after detonation.

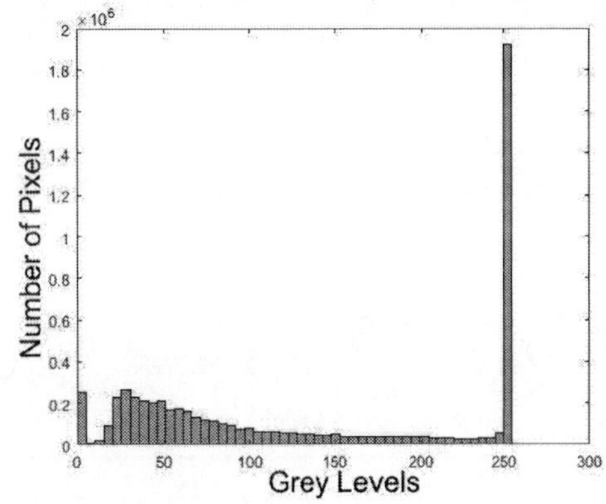

Figure 24. Pixel histogram of figure 20 which depicts explosion of bottle with argon at time 10 ms after detonation.

4. Conclusions

The experiment proved that aluminium powder is a stable light source which, in contrary to argon, can be accurately dosed. The optimal delay time between ignition of the illumination agent and the main detonation can be set to 8 ms. The time optimal for capturing the process by high-speed camera can be estimated at 6 ms. The most convenient illumination was achieved for combination of aluminium powder and 75 g of trinitrotoluene, when a compact fireball formed with intensity sufficient for lighting the scene throughout the duration of the monitored phenomenon. Luminosity of 1.5-liter argon bottle is comparable to 75 g of aluminium powder, but the lighting time is three times shorter. The results show that increasing of the aluminium powder dose does not enhance lighting of the scene.

Acknowledgments

This paper was prepared thanks to the support of the Technology Agency of the Czech Republic (TAČR), project (no TH02010888), project no. SGS19/149/OHK1/3T/11 and project no. SGS18/115/OHK1/2T/11.

5. References

[1] Foglar M, Hajek R, Fladr J, Pachman J and Stoller J 2017 Full-scale experimental testing of the blast resistance of HPFRC and UHPFRC bridge decks *Construction and Building Materials* **145** p 588-601

[2] Foglar M, Hajek R, Kovar M and Stoller J 2015 Blast performance of RC panels with waste steel fibers *Construction and Building Materials* **94** p 536-546

IOP Conf. Series: Materials Science and Engineering **596** (2019) 012042 doi:10.1088/1757-899X/596/1/012042

Influence of concrete additives on cement paste shrinkage

M Vích, J Fládr and I Broukalová

Department of Concrete and Masonry Structures, Faculty of Civil Engineering, Czech Technical University in Prague

E-mail: josef.fladr@fsv.cvut.cz

Abstract. The paper deals with the analysis of the effect of chosen latent hydraulic additives on concrete drying shrinkage. Variant dosage of latent hydraulic additives in cement paste was examined. Measuring of shrinkage in time was accompanied by testing of compressive strength and tensile strength at the age 28 days and 280 days. Results of investigations are summarized and compared.

1. Introduction

Shrinkage is a property of concrete, which fundamentally affects the long-term behaviour of concrete structures. Shrinkage and creep form a pair of rheological properties of concrete causing volume changes. Shrinkage affects the deflections of structures, the development of cracks and the durability of structures. In large prestressed structures, the shrinkage leads to significant losses in prestressing force. In certain types of structures – as retention tanks, waterproofing concrete basements, so called white tank systems, in which the maximum crack width limit governs the structural design, shrinkage crucially affects proportioning of reinforcement. The growing trend to use concrete additives with latent hydraulic properties has become incentive of the research focused on latent hydraulic additives.

2. Experimental work

The project investigated influence of chosen latent hydraulic additives – silica fume, metakaolin and fly ash on shrinkage. The shrinkage was examined on cement paste specimens 40 x 40 x 160 mm. Three reference specimens are from cement paste. Three sets differ in latent hydraulic additive content. The substitution of cement with the additive was 20 % and 50 %. Additionally to testing of shrinkage, the compressive tests and flexural tests were performed; first measuring of strengths was at the age of 28 days and the closing measuring of strength was performed after termination of shrinkage testing at the age 280 days.

2.1. Composition of mixes

The intention was to base the mixture design on the same water – binder ratio. Calculation of the water – binder ratio used the approach with a *k*-value correction according to relation (1) [1].

$$\omega_0 = \frac{w_a}{c_a + k.a} \tag{1}$$

where ω_0 is water – cement ratio
w_a is water content [kg/m^3]
c_a is cement content [kg/m^3]

a is content of additive [kg/m^3]

Originally, the k-value approach was derived for the design of concrete mixture and it assumes limitation of the additive content [2], [3]. In the research project, the replacement of cement was higher than the given additives limit.

The high silica fume content lead to pure workability of the mixture. In the silica fume samples with replacement of 20 % of cement, the lower workability did not impair the quality of specimens. But in samples with 50 % of silica fume, the low water content made impossible proper compaction and the elaborated specimens (figure 1) could not be used for testing and new specimens had to be prepared. The correction k-value was increased to $k = 2.4$.

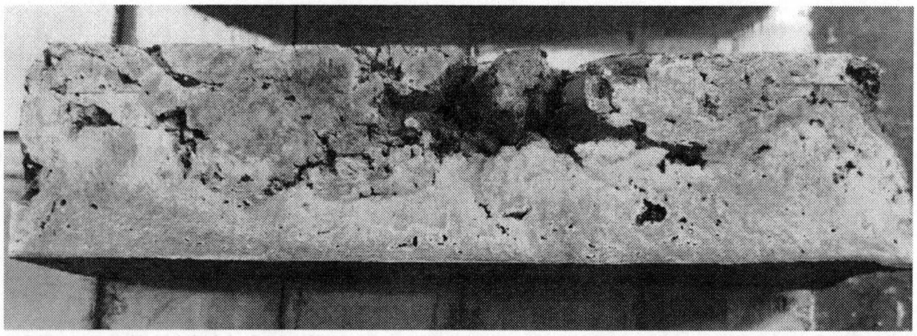

Figure 1. Defect sample with high replacement of cement by silica fume and low water – binder ratio.

Mixture compositions of cement pastes and labelling of mixes are summarised in a table 1.

Table 1. Cement pastes composition.

Component	REF	ME-20	ME-50	P-20	P-50	MI-20	MI-50
Cement CEM I 42.5	1375	1118	874	1164	946	1030	736
Metakaolin	-	223.6	437	-	-	-	-
Fly ash	-	-	-	232	472	-	-
Silica fume	-	-	-	-	-	206	369
Water	550	537	524	503	454	577	649
Aggregate type	-	-	-	-	-	-	-
Superplasticizer	-	-	-	-	-	-	-
Water/binder ratio	0.4	0.4	0.4	0.4	0.4	0.4	0.4
k-value	-	1	1	0.4	0.4	2	2.4

2.2. *Elaboration of specimens*

The manufacturing of cement pastes started with mixing of dry components – cement and additive, after 30 seconds, water was added. Total mixing time was two and half minutes. Compaction of cement pastes was performed on a vibrating table. After filling the moulds, the specimens were covered by stretch foil and stored in laboratory environment until the next day; after 24 hours, the specimens were demoulded.

Shrinkage measurement pins were inserted in the mould (figure 2 and 3). During demoulding of the first specimens the anchorage length of measuring pins appeared short and they were pulled out of the specimens. So that the anchor bolts had to be elongated, sufficient anchorage length was 20 mm.

Figure 2. Steel mould.

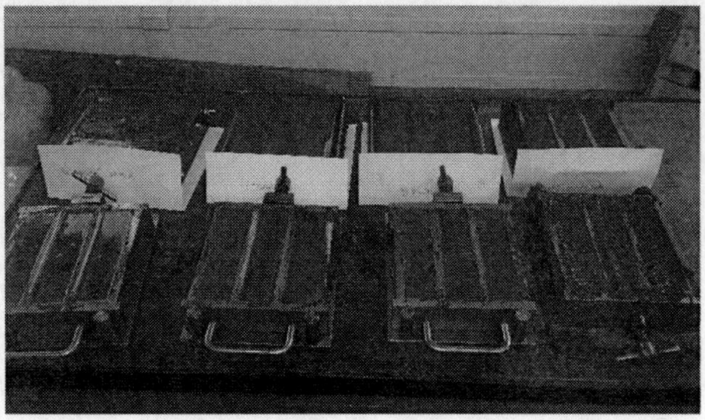

Figure 3. Specimens after concreting.

2.3. Measuring of shrinkage

The investigations focused in decrease of volume due to drying, so that during time when water which was not used for hydrating of cement is evaporated. Autogenous shrinkage, which causes decrease of volume due to chemical processes in first hours after mixing, is not covered by the research.

The first value of length was measured 24 hours after demoulding, the total time of recording the drying shrinkage was 280 days. The specimens were stored at the same place in the laboratory throughout the entire measurement. The samples were exposed to the environment of the laboratory where the temperature ranged between 22 °C and 28 °C. The relative humidity ranged from 33% to 49% and the absolute humidity was about 8-13 g/m^3. These values have been regularly monitored and their history is displayed in figures 4 and 5.

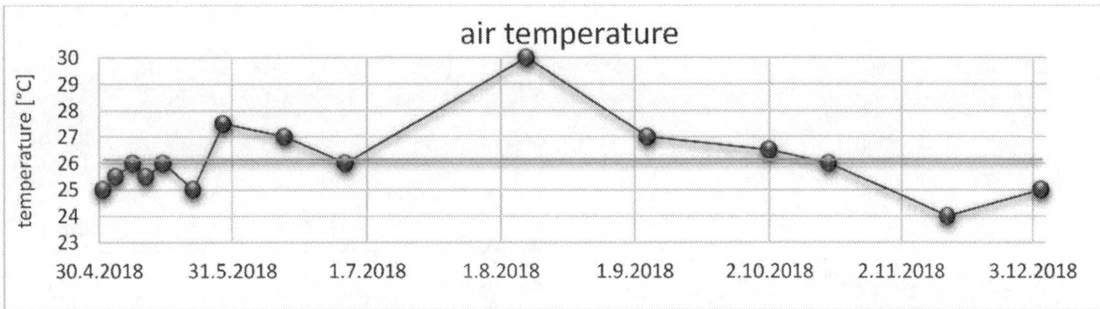

Figure 4. Air temperature of the environment.

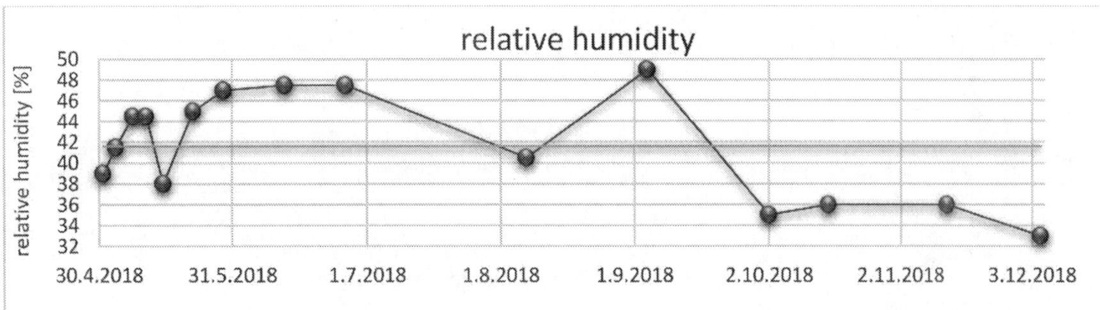

Figure 5. Relative humidity of the laboratory environment.

Table 2. Measurement frequency.

Age of specimens	frequency of measuring
1 to 16 days	every day
17 to 28 days	3 times a week
29 to 58 days	2 times a week
59 to 118 days	2 times in a month
119 to 280 days	every month

The measuring itself was carried out using a digital indicator. Data were measured to the nearest thousandth of a millimetre. Three prisms were elaborated for each mixture, and three values were recorded for each measurement. Hence, the resulting shrinkage value is an average of nine records. The measurement was divided into several periods, in which the frequency of measuring decreased in relation with deceleration of shrinkage – see table 2.

3. Results of shrinkage measurements

The results of measurements are arranged in graphs depicting time dependence of shrinkage in micrometres per meter run. The graphs (figure 6 to 10) compare:

- samples with same type and different amount of additive,
- samples with different additive type and same % of cement replacement.

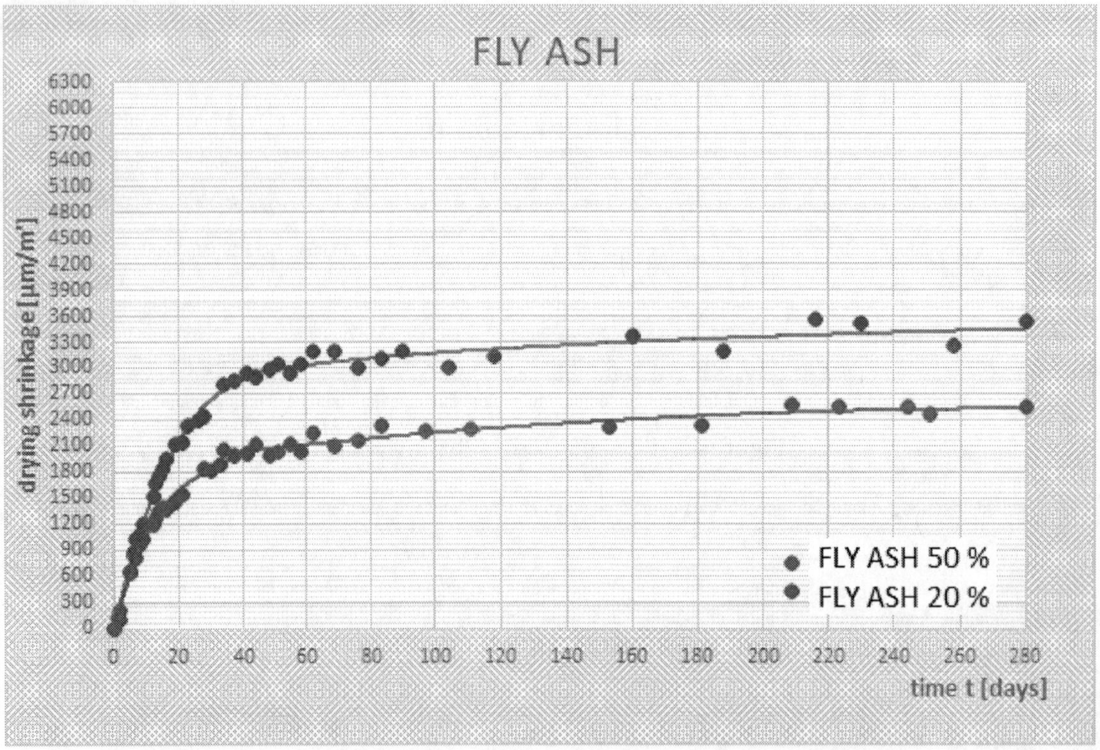

Figure 6. Drying shrinkage – fly ash.

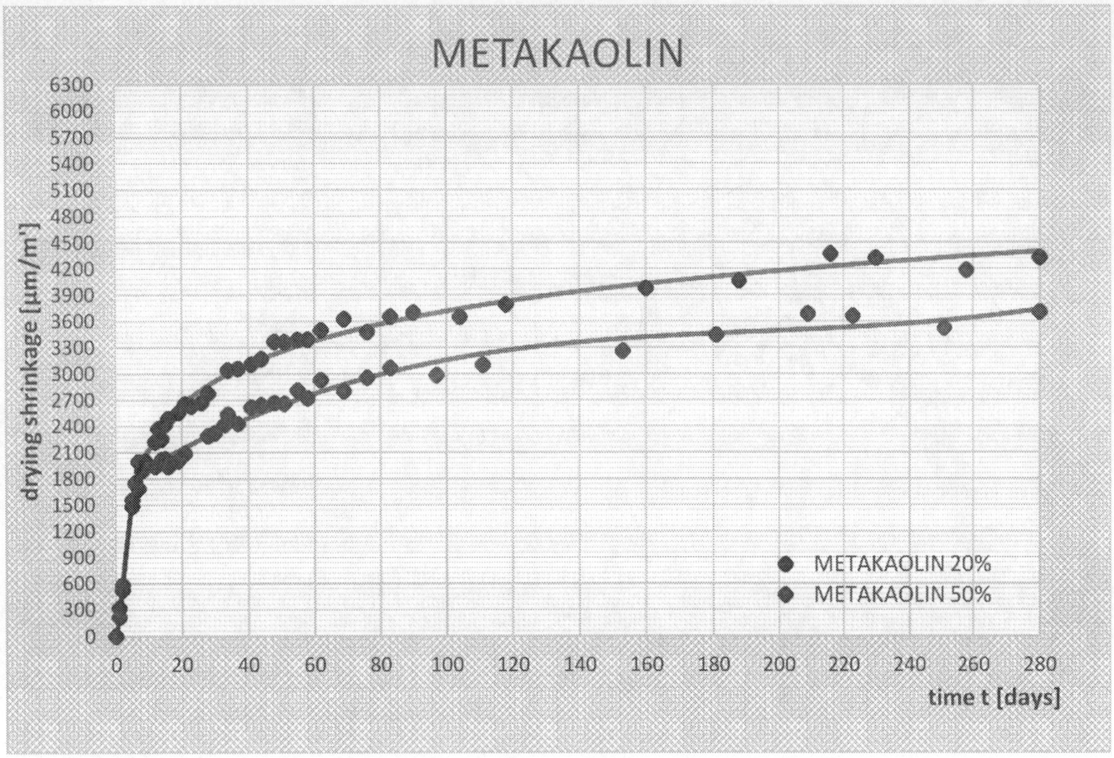

Figure 7. Drying shrinkage – metakaolin.

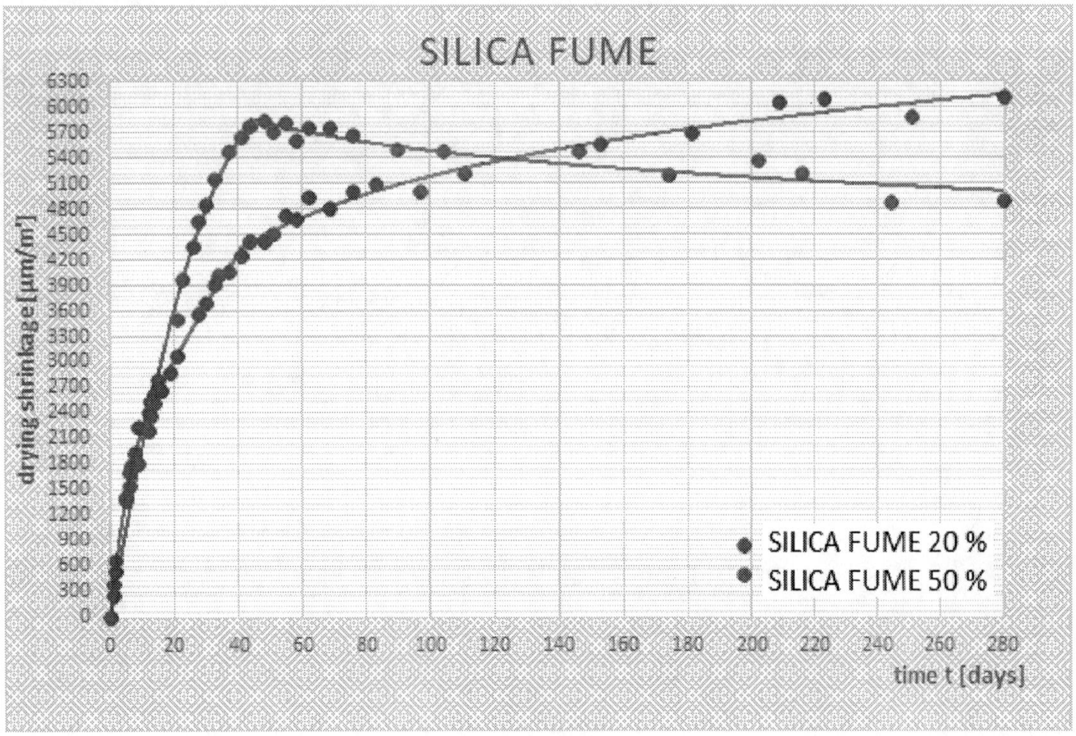

Figure 8. Drying shrinkage – silica fume.

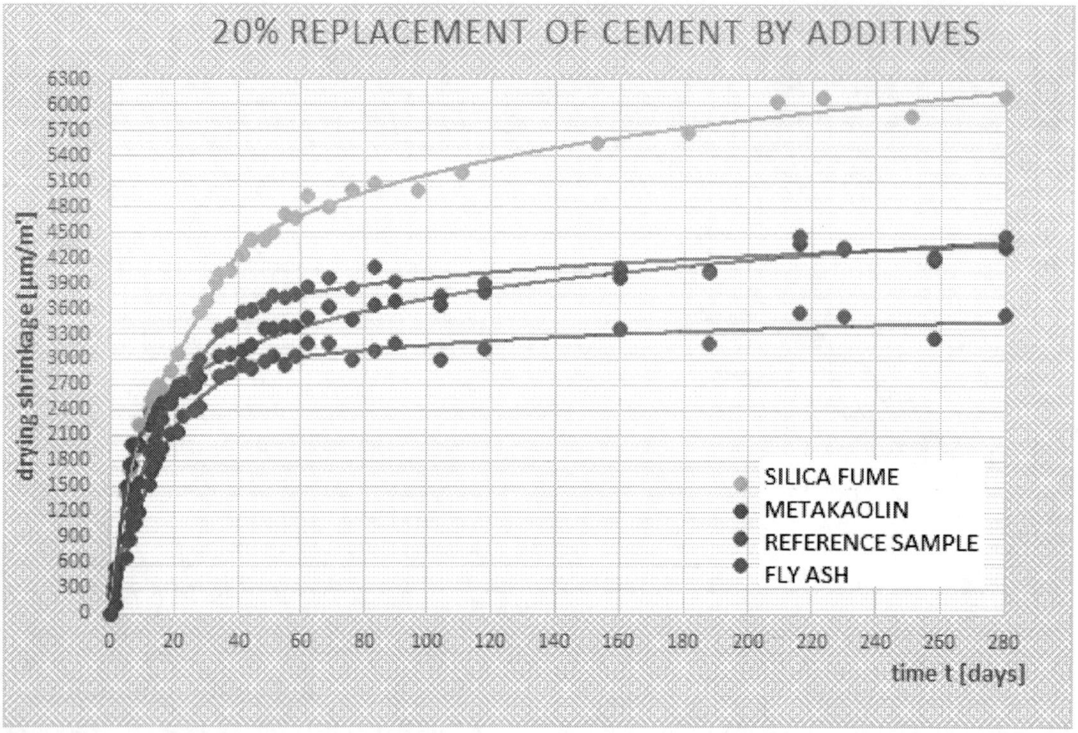

Figure 9. Drying shrinkage – 20 % of cement replaced by additives.

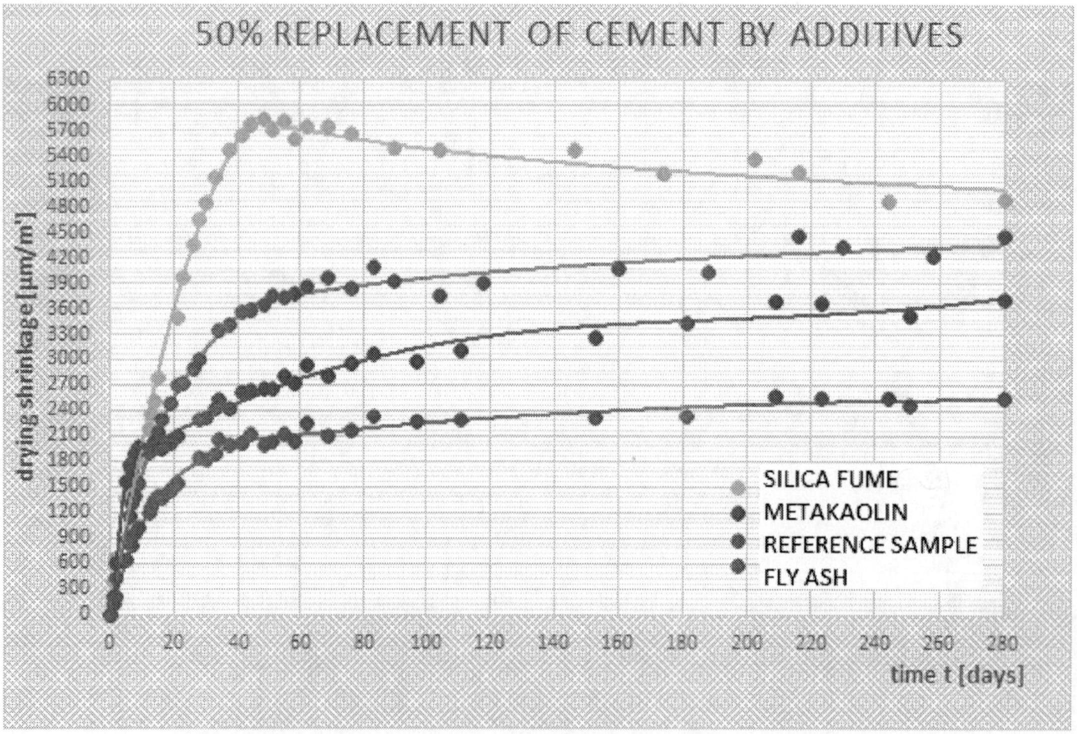

Figure 10. Drying shrinkage – 50 % of cement replaced by additives.

3.1. Results of strength testing
Strength testing was performed according to standard recommendations [4], [5]. Results are depicted in a form of column diagrams (figure 11).

3.1.1. Compressive strength

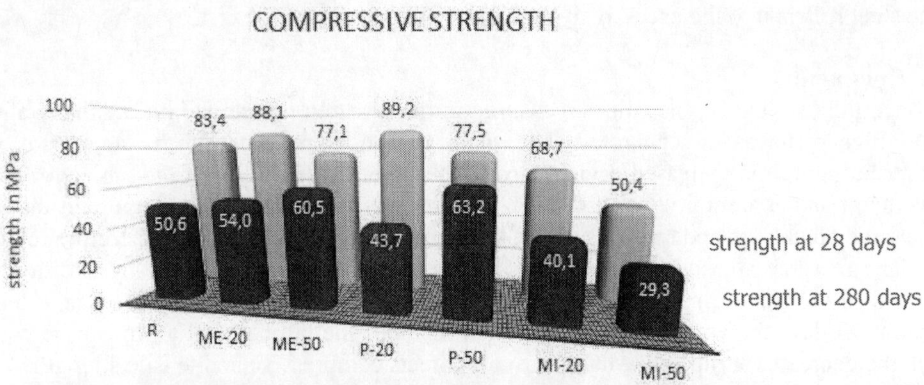

Figure 11. Comparison of compressive strength.

Comparison of compressive strength at 28 days and at 280 days show that shrinkage of cement pastes, i.e. concrete without aggregates, significantly reduces compressive strength.

This research demonstrated favourable effect of 50% replacement by metakaolin and fly ash on long term increase of compressive strength. For both additives the 28 days strength was lower than strength of reference sample. At the age of 280 days, the situation inversed and the samples with additives had strength by 25 % higher. Sample ME-20 (20% replacement by metakaolin) has almost the same strength as reference sample, thus the higher cost of additive does not provide any benefit.

In contrary, replacement of cement by lower dose of silica fume and fly ash had adversely affected terminal compressive strength. In case of pastes with silica fume, the reason of lower strength is obviously higher shrinkage. Extremely low strength of samples MI-50 (50% replacement by silica fume) is caused also by increasing of water/binder ratio, which was proposed to enable satisfactory workability without application of plasticiser.

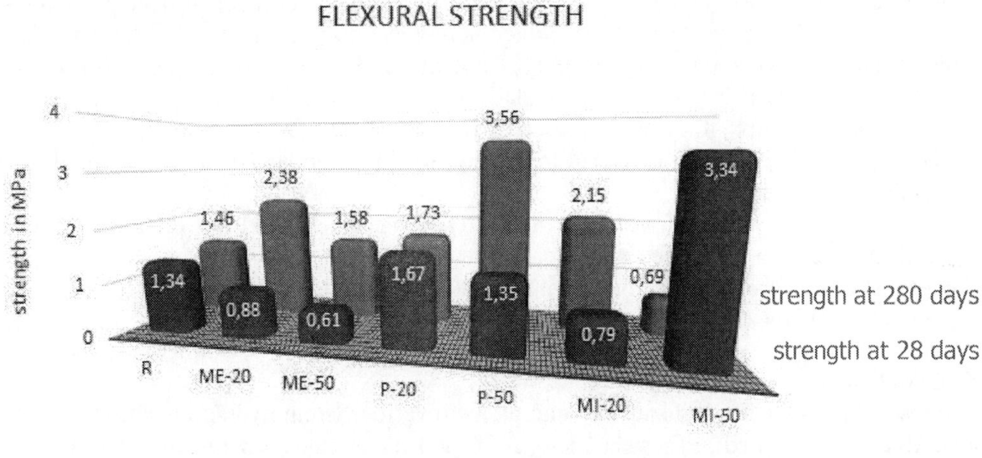

Figure 12. Comparison of flexural strength.

3.1.2. Flexural strength

The prism specimens with dimension 40/40/160 mm were tested in a three-point bending test with arrangement according to set-up given in [6].

Results display anticipated relation of shrinkage and flexural strength (figure 12). Non-standard results obtained for MI-50 sample (50% replacement by silica fume) can be explained by sinking of the steel rollers into the excessively cracked specimen during the test.

4. Comments

Concerning shrinkage, the values measured in the research project are higher than values for common and high-performance concretes. The main reason is absence of the aggregate. Another factor contributing to the increased drying shrinkage is the low humidity of the lab environment, where the average relative humidity was around 42%. However, in our case, it is important that all samples had the same conditions and relations of different cement pates shrinkage are therefore relevant.

The research approves previous studies about reduction of shrinkage by addition of fly ash. The record of shrinkage rises gently, and the peak is far below other additives. The graphs clearly show that at 90 days the volume changes reached 90 % of the ultimate value. In can be concluded that fly ash the decreases shrinkage, what positively affects compressive and tensile strength.

Metakaolin manifests the fastest initial increase in shrinkage compared to other additives. Nevertheless, the total shrinkage of samples ME-50 (50 % replacement by metakaolin) is still at an acceptable level.

From the results follow, that curing in the first 7–8 days would prevent the initial increase and thus effectively reduce the total shrinkage.

The graph comparing all additives indicates that replacing cement with 20% metakaolin to decrease shrinkage is not prosperous, as the shrinkage relation of ME-20 is almost identical with reference sample. However, for higher amount of metakaolin, the curve of shrinkage dependence moves towards more favourable values.

For metakaolin and fly ash, it can be concluded that increase of the additive leads to decrease of total drying shrinkage.

In contrast, influence of silica fume on volume changes is negative. The layout of the MI-20 shrinkage was in principle expected. The volume changes of the sample MI-50 was abnormal. The peak value of shrinkage was reached after 48 days, when a breakpoint came and according to measurements, the sample started swelling. It probably relates to high increase of internal stresses during initial 48 days, when progressive cracking led to disintegration of the specimen. Shrinkage measurement of one of the MI-50 was prematurely terminated because of damage of the specimen by the self-weight. The other specimen crumbled before testing of strength. Moreover, the tensile strength 0.69 MPa determined at one of the MI-50 specimens demonstrates high degree of drying.

Substitution of cement by silica fume shall be limited. Pastes with higher silica fume content have lower compressive strength, what is caused by high drying shrinkage.

The compressive strength is related to density; the lower is the density of the material, the lower is the compressive strength.

Low content of metakaolin and fly ash (20% replacement) did not retarded initial compressive strength compared to reference sample.

Flexural strengths are generally higher at age 280 days than tensile strength at the age of 28 days with the exception of silica fume.

5. Conclusions

The research deals with shrinkage of concrete with various latent hydraulic admixtures. Three types of admixtures were tested – fly ash, metakaolin and silica fume. In the initial phase of the project shrinkage of pastes was measured to reveal the tendencies and basic principles.

The research confirms findings of previous studies that fly ash reduces shrinkage. With increasing fly ash content, the shrinkage decreases, what at the same positively affected compressive and tensile strength.

Replacing cement with 20% metakaolin to decrease shrinkage is not prosperous. The effect of metakaolin on shrinkage manifested at higher dosage of this admixture.

For replacement of cement by silica fume, it is recommended to limit the substitution by 20 %. Higher dosage of silica fume leads to excessive drying shrinkage.

The paper presents result of the initial investigations of drying shrinkage of cement pastes with partial substitution of cement by admixtures with latent hydraulic properties. Future research will focus in shrinkage of concretes with aggregate and numerical analysis of drying shrinkage and verification of models for drying shrinkage (e.g. model B4).

Acknowledgments

This paper has been prepared with the financial support of the Technology Agency of the Czech Republic within the project TE01020168 Centre for Effective and Sustainable Transport Infrastructure (CESTI).

References

[1] Härdtl R 2010 The K-Value Concept Applied for GGBFS - Principles and Experiences, *International RILEM Conference on Material Science* 2010, ISBN: 978-2-35158-110-0, e-ISBN: 978-2-35158-111-7, pp 189 – 198.

[2] ČSN EN 206+A1 2017 *Concrete - Specification, performance, poduction and conformity.* ÚNMZ, Prague

[3] ČSN EN 197-1 2012 *Cement – Part 1: Composition, specifications and conformity criteria for common cements. ed. 2.* ÚNMZ, Praha,

[4] ČSN EN 12390-3 2009 *Testing hardened concrete – Part 3: Compressive strength of test specimens.* ÚNMZ, Praha,.

[5] ČSN EN 12390-5 2009 *Testing hardened concrete – Part 5: Flexural strength of test specimens.* ÚNMZ, Praha,.

[6] ČSN EN 1015-11+A1. *Methods of Test for Mortar for Masonry-Part 11-Determination of Flexural and Compressive Strength of Hardened Mortar.*

[7] Bažant Z P and Baweja S 2000 Creep and shrinkage prediction model for analysis ad design of concrete structures: Model B3, *ACI Special publication creep and shrinkage of concrete,* A. Al-Manaseer, Editor,

[8] Bažant Z P et al. 2015 RILEM draft recommendation: TC-242-MDC multi-decade creep and shrinkage of concrete: material model and structural analysis. Model B4 for creep, drying shrinkage and autogenous shrinkage of normal and high-strength concretes with multi-decade applicability. *Materials and structures 48* (2015).

Evaluation of the size effect of waste tyre rubber particles on properties of lightweight rubber concrete

M Záleská[1], M Pavlíková[1], D Čítek[2] and Z Pavlík[1]

[1] Department of Materials Engineering and Chemistry, Faculty of Civil Engineering, Czech Technical University in Prague, Thákurova 7, 166 29 Prague 6, Czech Republic
[2] Department of Experimental Methods, Klokner Institute, Czech Technical University in Prague, Šolínova 7, 166 08 Prague, Czech Republic

E-mail: martina.zaleska@fsv.cvut.cz

Abstract. Due to the increasing amount of waste tyres and also to their properties, it is necessary to look for suitable ways for their recycling. For production of rubber concrete, the crushed tyre rubber of fraction 0/4 and 4/8 was used for partial substitution of silica sand at the rate of 10 % by mass. For comparison, a reference concrete without rubber aggregate was also studied. Both natural and rubber aggregate were first assessed in terms of their physical parameters and thermal transport and storage properties. Evaluation of produced lightweight rubber concrete included measurement of basic physical properties, strength tests, and microstructural analysis. Thermal properties were studied from dry to fully water saturated state. The experimental results showed that concrete with a small amount of rubber aggregate had improved thermal insulation performance and can serve as an eco-friendly material for structural applications in civil engineering.

1. Introduction

Over 25 million tons of end-of-life tyres are generated per year all over the world and 31 % of this amount is landfilled, stockpiled or its treatment is unknown as is reported by World Business Council of Sustainable Development [1]. The valorisation of this waste tyre into cement concrete could be a partial solution for relieving their negative environmental impact [2]. Recently, many studies have been carried out on rubber concrete containing tire rubber-based aggregate as a replacement of fine or coarse natural aggregate [3-5]. Generally, researchers reported the reduction of strength parameters, weight unit and thermal conductivity of rubberized concrete in comparison with concrete prepared without rubber [6-8]. In order to evaluated the size effect of rubber-based aggregate particles on the rubber concrete properties, this paper is focused on the use of two size fractions of tyre rubber, which replaced the fine, coarse, and both fine and coarse natural aggregate in the weight ratio of 10 %.

2. Experimental

The materials used for preparation of test specimens were cement, water and both coarse and fine natural and rubber aggregate. Portland slag cement CEM II/A-S 52.5 N (Spenner Zementwerk Berlin GmbH & Co. KG, Germany) was used as a binder. Rubber particles, the size fraction of 0/4 mm and 4/8 mm (figure 1), were obtained from mechanical crushing of waste tyre rubber. Silica sand as natural aggregate of 0/4 mm and 4/8 mm fraction comes from the gravel-pit Dobříň, Czech Republic. Four concrete mixtures were prepared to study the rubber size effect. The rubber particles were used to replace the fine (mix CR4), coarse (mix CR8) and both fine and coarse (mix CR4+8) natural aggregate in the weight ratio of 10 wt%. The reference concrete without rubber (CRef) was prepared as well. Water to cement ratio (w/c) was 0.5 for all tested mixtures. Table 1 presents the composition of concrete specimens with incorporated rubber aggregate.

Table 1. Mix proportions of rubber concrete.

Notation	Mass (g)						w/c
	Cement	Sand 0/4	Sand 4/8	Rubber 0/4	Rubber 4/8	Water	
CRef	450	675	675	-	-	225	0.5
CR4	450	540	675	135	-	225	0.5
CR8	450	675	540	-	135	225	0.5
CR4+8	450	607.5	607.5	67.5	67.5	225	0.5

Figure 1. Rubber aggregate.

There were prepared prismatic samples with dimension of 40 × 40 × 160 mm and cubic samples with side dimension of 100 mm, which were after casting left at highly humidity environment (RH ≥ 98 %) and temperature (23 ± 2) °C. After 24 h samples were demolded and cured for next 27 days in water.

2.1. Characterization of rubber aggregate

The particle size analysis was made for both natural and rubber aggregate using the standard sieve method. There were used sieves with mesh dimensions of: 0.063; 0.125; 0.5; 1.0; 2.0; 4.0; 8.0 and 16.0 mm. The specific densities of aggregates were determined by automatic helium pycnometer Pycnomatic ATC (Thermo Scientific). In more detail were explored the thermal properties of both sand and rubber particles. Their thermal conductivity λ ($W \cdot m^{-1} \cdot K^{-1}$) and volumetric heat capacity C_v ($J\ m^{-3} \cdot K^{-1}$) as well as the powder density were measured in dependence on the compacting time because the compacted state of aggregate better reflects its performance in concrete structure. For compaction was used a high vibrating table VSB-15 (Brio). Thermal parameters were obtained using an ISOMET 2114 device (Applied Precision, Ltd.) working on a transient impulse method [9]. For the measurement, a needle probe was used.

2.2. Testing of rubber concrete

The hardened 28 days concrete samples were used for testing. The conducted tests included an optical microscopy analysis, assessment of structural, mechanical, thermal transport and storage properties.

An optical microscopy (digital light microscope VHX-200 D, Keyence) was used for study of hardened rubberized concrete microstructure.

The specific density was measured on the dried samples using a helium pycnometry (as described above). The bulk density was obtained on a gravimetric principle from measurement of the weight and dimensions of dried concrete specimens. The bulk density test was performed according to the standard EN 12390-7 [10]. The total open porosity was than calculated on the basis of specific and bulk density values. The relative expanded uncertainty of applied measuring method was approx. 5 %.

The flexural strength test was conducted following the standard EN 12390-5 [11]. The specimens were prism having dimensions of 40 × 40 × 160 mm. The compressive strength was determined according to the standard EN 12390-3 [12] on the fragments of samples from the flexural strength tests. The loading area was 40 × 40 mm. The relative expanded uncertainty of both compressive and flexural strength tests was 1.4 %. The dynamic Young´s modulus was measured using an ultrasonic device DIO 562 (Starmans Electronic) with working frequency of 50 kHz. Within this test, dried prism specimens of 40 × 40 × 160 mm were used. The expanded combined uncertainty of this test method was 1.5 %.

The thermal transport and storage properties of rubber concrete were investigated on the cubic specimens with the side dimension of 100 mm. Since the degree of saturation of concrete is one of the main factors influencing its thermal performance, the thermal conductivity and the volumetric heat capacity were determined in dependence on moisture content. An ISOMET 2114 device (see above) equipped with the surface probe was used for this measurement.

3. Results and discussion

The results of the sieve analysis performed for all used aggregates are shown in figure 2.

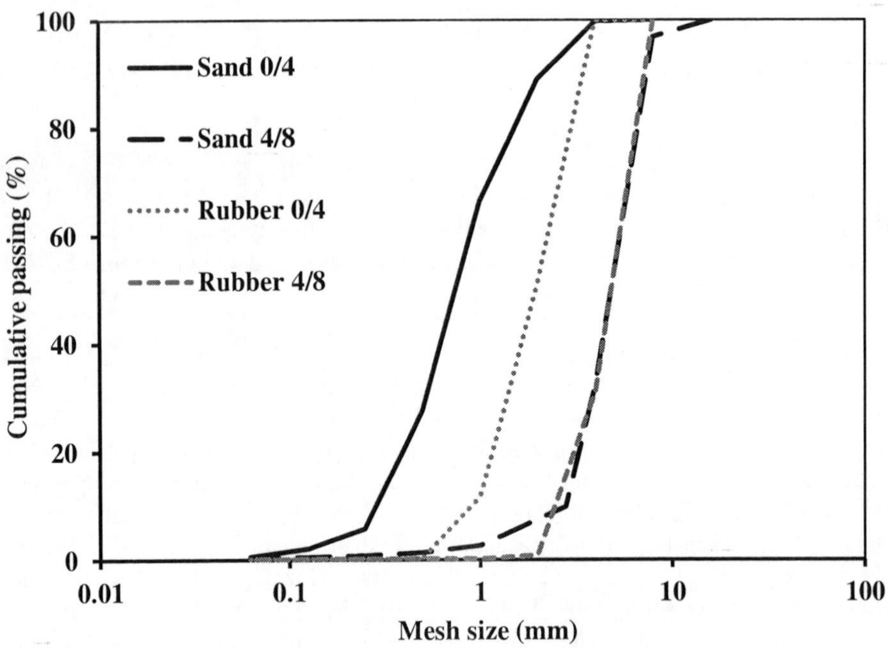

Figure 2. Sieve analysis of both natural and rubber aggregate.

In the table 2 there are introduced powder density and thermal properties of studied aggregates in dependence on the compacting time. Also aggregate specific density is presented.

Table 2. Physical and thermal properties of aggregates.

Aggregate	Specific density (kg·m^{-3})	Compaction time (s)	Powder density (kg·m^{-3})	λ (W·m^{-1}·K^{-1})	C_v (×10^6 J·m^{-3}·K^{-1})
Sand 0/4	2648	0	1604	0.306	1.522
		10	1775	0.390	1.561
		20	1775	0.399	1.570
		180	1799	0.403	1.576
Sand 4/8	2662	0	1408	0.315	1.404
		10	1539	0.348	1.420
		20	1581	0.369	1.426
		180	1603	0.389	1.443
Rubber 0/4	1154	0	468	0.093	0.326
		10	489	0.094	0.360
		20	506	0.095	0.380
		30	512	0.099	0.392
		180	524	0.100	0.412
Rubber 4/8	1174	0	445	0.095	0.253
		10	460	0.095	0.254
		20	465	0.096	0.256
		30	465	0.096	0.259
		180	465	0.096	0.262

We can see that both thermal conductivity and volumetric heat capacity of rubber were significantly lower compared to that of silica sand. This finding provides the assumptions for improved thermal insulation properties of rubber-containing concrete. Compaction reduced the air gaps between the grains of aggregate, which led to an increase in the thermal conductivity and volumetric heat capacity.

From the optical microscopy (figure 3), good cohesion between the rubber particles and cement matrix was clearly visible.

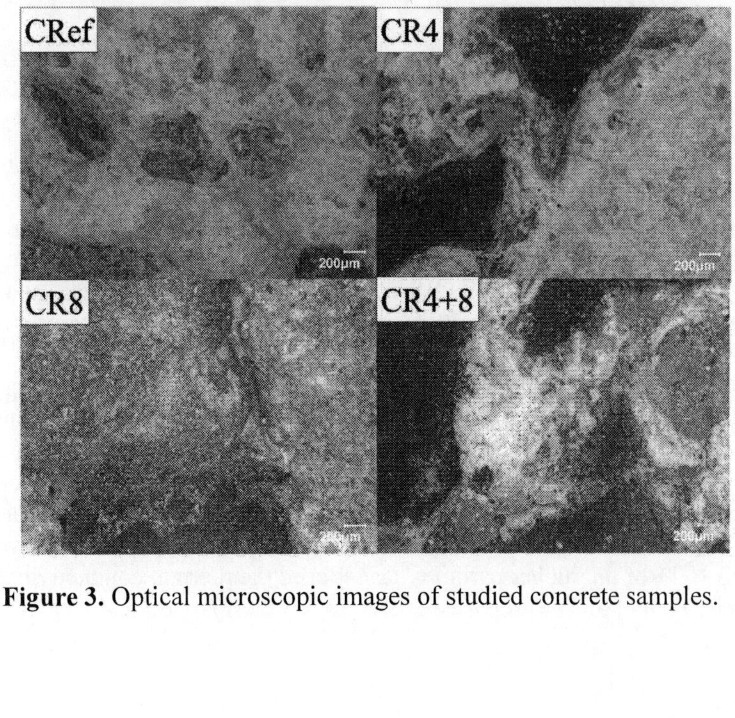

Figure 3. Optical microscopic images of studied concrete samples.

The bulk density, specific density, and the total open porosity of developed rubber concrete are given in table 3.

Table 3. Basic physical parameters of studied concrete samples.

Material	Specific density ($kg \cdot m^{-3}$)	Bulk density ($kg \cdot m^{-3}$)	Open porosity (%)
CRef	2453	2134	13.0
CR4	2281	1935	15.2
CR8	2249	1914	14.9
CR4+8	2281	1933	15.3

In comparison with the reference concrete, all rubberized concrete samples exhibited lower bulk density and specific density values, whereas the extent of differences between particular rubber concretes was not significant. The bulk density values of rubber concretes ranged from 1914 $kg \cdot m^{-3}$ to 1935 $kg \cdot m^{-3}$. They can therefore be classified into class LC 2.0 of lightweight concrete according to the standard EN 206-1 [13]. The open porosity of rubber concrete increased compared to reference sample. It can be explained by (i) the higher air entrapment during the preparation of the mixtures due to the presence of non-wetting rubber aggregate, (ii) the rubber particle shape, which differs from the spherical shape of the sand. Moreover, both size fractions of the rubber particles differ from each other, especially by the higher specific surface of the fraction 0/4, which can cause the higher porosity of concrete samples containing rubber fraction 0/4. Similar results in terms of density and porosity of rubber concrete were reported in literature [14].

The results of compressive strength, flexural strength, and dynamic Young´s modulus testing are summarized in table 4.

Table 4. Mechanical parameters of studied concrete samples.

Material	Compressive strength (MPa)	Flexural strength (MPa)	Young´s modulus (GPa)
CRef	61.5	6.8	39.0
CR4	28.0	5.7	18.1
CR8	26.0	5.1	17.0
CR4+8	31.0	4.9	18.6

The obtained mechanical resistance parameters were reduced with the addition of rubber aggregate. This finding was in line with previously published results [2], [5]. A decrease in compressive strength up to 54 %, 58 %, and 50 % for CR4, CR8 and CR4+8 respectively, compared to the reference concrete was observed. The CR4+8 concrete, where they were replaced both fine and coarse natural aggregate with rubber particles, exhibited the highest compressive strength and dynamic Young´s modulus among rubber concrete samples and, on the contrary, showed the lowest flexural strength. In [15] authors observed the worse performance in strength parameters for larger rubber particles from three sizes of rubber (3 mm, 0.5 mm, and 0.3 mm) used to replace of fine natural aggregate.

The thermal conductivity values measured for dry and fully water saturated samples are tabulated in table 5. It is clear from the results that 10% replacement of sand with rubber particles resulted in a decrease in the thermal conductivity value of 25-28 %. Part of the thermal conductivity reduction can be caused by the lower thermal conductivity of the rubber in comparison with the sand (see table 2) and part can be attributed to higher porosity of rubber concrete. It has been observed by Iqbal [3] that the concrete with 10 % of the rubber particles has reduced the thermal conductivity by 20 % relative to reference concrete. Hall et al. [16] reported a decrease in thermal conductivity of rubberized concrete compared to the reference concrete with an increase in a rubber substitution ratio and a mix design target slump value.

Table 5. Thermal conductivity of concrete samples in the dry and fully water saturated state.

Material	λ ($W \cdot m^{-1} \cdot K^{-1}$)	
	Dry	Saturated
CRef	1.78	2.38
CR4	1.31	1.82
CR8	1.28	1.90
CR4+8	1.33	1.97

Figure 4 shows the volumetric heat capacity in dependence on relative moisture content, which is defined as a ratio of the actual moisture content by weight to the saturated moisture content. The volumetric heat capacity decreased with the addition of rubber, the lowest value was obtained for the sample CR8. Since water is characterized by a higher volumetric heat capacity than concrete, the increase in water content in the samples also increased their volumetric heat capacity.

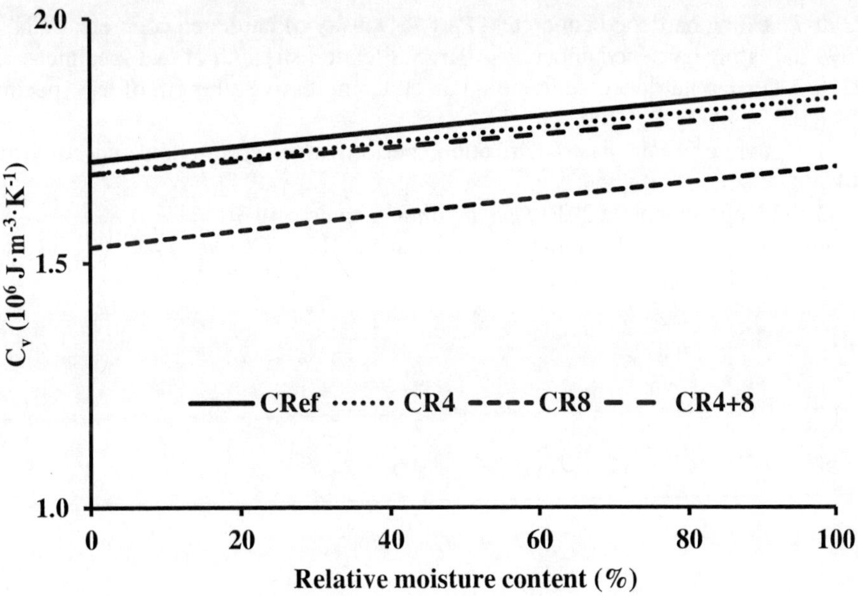

Figure 4. Moisture dependent volumetric heat capacity.

4. Conclusion

The experimental assessment of three types of rubberized concrete was presented. Modification of concrete with crushed tyre rubber decreased the mechanical resistance, unit weight and both thermal conductivity and volumetric heat capacity of investigated concretes. Moreover, the rubberized samples showed improved thermal-insulation properties even in the presence of moisture. Based on the obtained results it can be concluded that 10% substitution of fine, coarse, or both fractions of natural aggregate by crushed rubber particles of appropriate size, gave concrete with improved thermal-insulation performance that can find use in building industry in structural applications.

IOP Conf. Series: Materials Science and Engineering **596** (2019) 012043 doi:10.1088/1757-899X/596/1/012043

Acknowledgements
The authors gratefully acknowledge the financial support received from the Czech Science Foundation, under project No 17-04215S – Thermal insulation composites containing waste plastic based fillers. This research has been also partially supported by the Ministry of Education, Youth and Sport of the Czech Republic, under project No SGS17/166/OHK1/3T/11.

References
[1] WBCSD 2018 *Global ELT Management – A global state of knowledge on collection rates, recovery routes, and management methods* (www.wbcsd.org, accessed January 2019)
[2] Li G, Wang Z, Leung Ch K Y, Tang S, Pan J, Huang W and Chen E 2016 *J. Cleaner Prod.* **112** 797
[3] Iqbal M 2017 *Constr. Build. Mater.* **133** 516
[4] Aiello M A and Leuzzi F 2010 *Waste Manage.* **30** 1696
[5] Ganjian E, Khorami M and Maghsoudi A A 2009 *Constr. Build. Mater.* **23** 1828
[6] Thomas B S and Gupta R Ch 2016 *Renewable Sustainable Energy Rev.* **54** 1323
[7] Lv J, Zhou T, Du, Q and Wu H 2015 *Constr. Build. Mater.* **91** 145
[8] Pacheco-Torgal F, Ding Y and Jalali S 2012 *Constr. Build. Mater.* **30** 714
[9] Záleská M, Pavlíková M, Pokorný J, Jankovský O, Pavlík Z and Černý R 2018 *Constr. Build. Mater.* **180** 1
[10] EN 12390-7 Testing hardened concrete – Part 7: Density of hardened concrete, CEN, 2009
[11] EN 12390-5 Testing hardened concrete – Part 5: Flexural strength of test specimens, CEN, 2009
[12] EN 12390-3 Testing hardened concrete – Part 3: Compressive strength of test specimens, CEN, 2014
[13] EN 206-1 Concrete – Part 1: Specification, performance, production and conformity, CEN, 2014
[14] Girskas G and Nagrockienė D 2017 *Constr. Build. Mater.* **140** 36
[15] Su H, Yang J, Ling T Ch, Ghataora G S and Dirar S 2015 *J. Cleaner Prod.* **91** 288
[16] Hall M R, Najim K B and Hopfe K 2012 *Appl. Therm. Eng.* **33-34** 77

Fire temperature influence on the textile reinforced concrete with non-woven polypropylene fabric

J Žák and P Štemberk

Department of Concrete and Masonry Structures, Faculty of Civil Engineering, Czech Technical University in Prague, Thákurova 7, 166 29 Prague 6, Czech Republic

E-mail: jakub.zak@fsv.cvut.cz

Abstract. Textile reinforced concrete with non-woven polypropylene fabric has variable utilization for non-load bearing structures. One of the possible uses is production of facade panels or protective layers in the interior of buildings. For that reason it is important to determine fire resistance of this material. There was examined behaviour of specimens exposed to temperature in range 100 – 650°C for times from 5 to 60 minutes. Weight loss of the specimens caused by heating was also measured. Regardless to the relatively high moisture of material, the effect of spalling didn't occur.

1. Introduction

Textile reinforced concrete (TRC) with non-woven polypropylene fabric has potential for using in non-load bearing structures. One of many possibilities of its utilization is using it as protective layers in technical parts of buildings. Architectural concrete is currently very favored and TRC should be used for creation of facade panels. [1] Advantage of this solution is that in case of any damage, panel can be exchanged for a new one. Simultaneously there should be used external thermal insulation placed under the panels.

Fire resistance of building's facades is presently often discussed topic. Because of the technical requirements on the heat transfer during facade, using of external thermal insulation systems is mostly necessary. Separation distances between buildings are in the Czech Republic determined according to fire openness categorization of external walls. Because of this conservative approach separation distances are often bigger than it is necessary. This is primarily problem in the places with high density of urban structure. [2]

This is the main reason for the research of fire influence on TRC with non-woven polypropylene fabric.

2. Materials and methods

2.1. Test specimens

Test specimens were created by cutting up of one bigger slab to small specimens. This method of specimen's creation was chosen due to gain the most similar material properties as possible. Dimensions of specimens are 50 x 50 mm (figure 1). Material of the specimens is made as two layered with one interlayer reinforced by alkali-resistant glass fabric. Main layers are reinforced by isotactic polypropylene fabric.

Figure 1. Test specimens.

Specimens were divided into 6 sets contained 4 specimens. Sets were gradually exposed to the different temperatures.

Initial moisture was also detected by gravimetric method. Another four specimens were weighted and then they were dried at temperature 115°C. Then they were weighted again, and initial moisture of the specimens was determined. This detail is important because of the possible concrete spalling. For classic concretes this effect is considered if the initial moisture is higher than 3% wt. The other specimens were also weighted before the experiment starts but weight decrease does not have to be caused just by water loss. Weight loss should be caused also by polypropylene fabric evaporation.

2.2. Choice of material heating temperatures

It is well known that exposing of the structures made from commonly used materials to the high temperatures has negative influence on their mechanical properties [3-9]. Properties of classic concrete are stable to 200°C, after that they are decreasing. This effect is caused by changes in the material (table 1).

Table 1. Temperature effects on the composition of concrete [10].

Temperature [°C]	Temperature effect on concrete
100	Loss of evaporable water
100-850	Dehydration of hydrates
400-600	Dehydration of calcium hydroxide
574	Crystalline transformation of the quartz
600-900	Decomposition of calcium carbonate
1200-1500	Melting of Portland cement

Used reinforcing fabric was made of 100% isotactic polypropylene. Standard properties of this type polypropylene were used: softening the material occurs between 140 – 150°C and melting of the material occurs between 160 – 170°C.

Specimens were exposed to the surrounding air temperature sizes 100°C, 150°C, 200°C, 350°C, 500°C and 650°C. These temperatures approximately match to the temperature of nominal temperature-time curve for external fire. This curve represents exposure of the facade to the fire from window or to the external fire (figure 2). Because of that, these temperatures are important for using TRC with non-woven polypropylene fabric as a facade panels.

Result of the experiment can be also applied for nominal temperature-time curve for slow heating during its first 30 minutes (figure 2). This curve can be used for elements with properties dependent on a high intensity of heating – for example external thermal insulation.

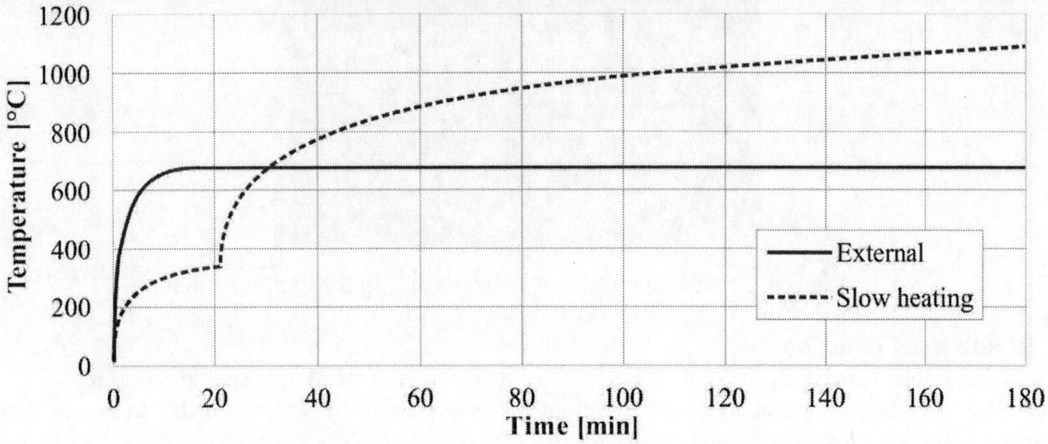

Figure 2. Nominal temperature-time curves.

Measured results should be also applied for internal fire in residential buildings before the flashover starts. This phase is ended by achieved the air temperature range 400 – 700°C. Specific temperature depends on type of objects placed in the room. Temperatures after flashover are in the range 600 – 1200°C (figure 3). For undeveloped fire the temperature doesn't increase so much as for fully developed fire.

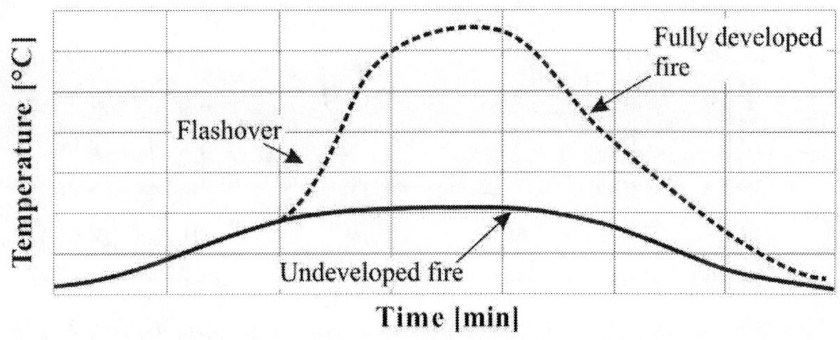

Figure 3. Temperature development in the fire zone for fully developed and undeveloped fire.

2.3. *Material heating process*

Four specimens from one set were heated by air flow of the specific loading temperature from the bottom side (figure 4). After the specific temperature was achieved the time was measured and gradually the specimens were displaced outside the heat source. The first specimen was exposed to the temperature for 5 minutes; the second one specimen was exposed to the temperature for 10 minutes; the third specimen was exposed to the heating for 30 minutes and heating of the last specimen was ended by 60 minutes after the experiment starts. After chilling of the heated specimens they were weighted again and weight loss was determined.

For finding out of the visual changes the digital microscope was used. Width of the cracks on the specimen's surface was measured also by using the digital microscope.

Figure 4. Placement of the specimens during the experiment.

3. Results and discussion

The weight loss caused by heating process was determined for all the specimens (table 2). It is impossible to reliably determine how big weight loss was caused by water loss due to polypropylene fabric evaporation.

Table 2. Weight development of the specimens before and after heating on the specific temperature.

No.	T [°C]	Weight [g] Before	Weight [g] After	Weight loss [%]	No.	T [°C]	Weight [g] Before	Weight [g] After	Weight loss [%]
1	100	31.14	30.56	1.9	13	350	28.16	27.92	0.9
2		32.36	30.12	7.4	14		30.18	29.98	0.7
3		29.88	28.12	6.3	15		35.98	34.02	5.8
4		30.10	27.98	7.6	16		29.92	28.14	6.3
5	150	28.12	27.94	0.6	17	500	36.14	32.00	12.9
6		30.14	28.08	7.3	18		28.14	25.86	8.8
7		27.86	26.12	6.7	19		30.28	28.06	7.9
8		27.14	25.68	5.7	20		39.82	33.92	17.4
9	200	34.16	34.02	0.4	21	650	31.96	29.96	6.7
10		34.44	31.88	8.0	22		32.12	29.88	7.5
11		31.78	29.62	7.3	23		31.88	28.02	13.8
12		30.02	28.06	7.0	24		31.28	28.02	11.6

In ordinary concretes loaded by fire the effect of spalling of surface layers can appear when the moisture is higher than 3% wt. Initial moisture was detected for 4 specimens of that reason. Average moisture of these 4 specimens was 9.1% wt. There is an assumption that all of the specimens had similar initial moisture. Nevertheless, the spalling effect wasn't occurred for any of the specimens.

Both surfaces of the specimens were photographically documented before and after heating (figure 5).

IOP Conf. Series: Materials Science and Engineering **596** (2019) 012044 doi:10.1088/1757-899X/596/1/012044

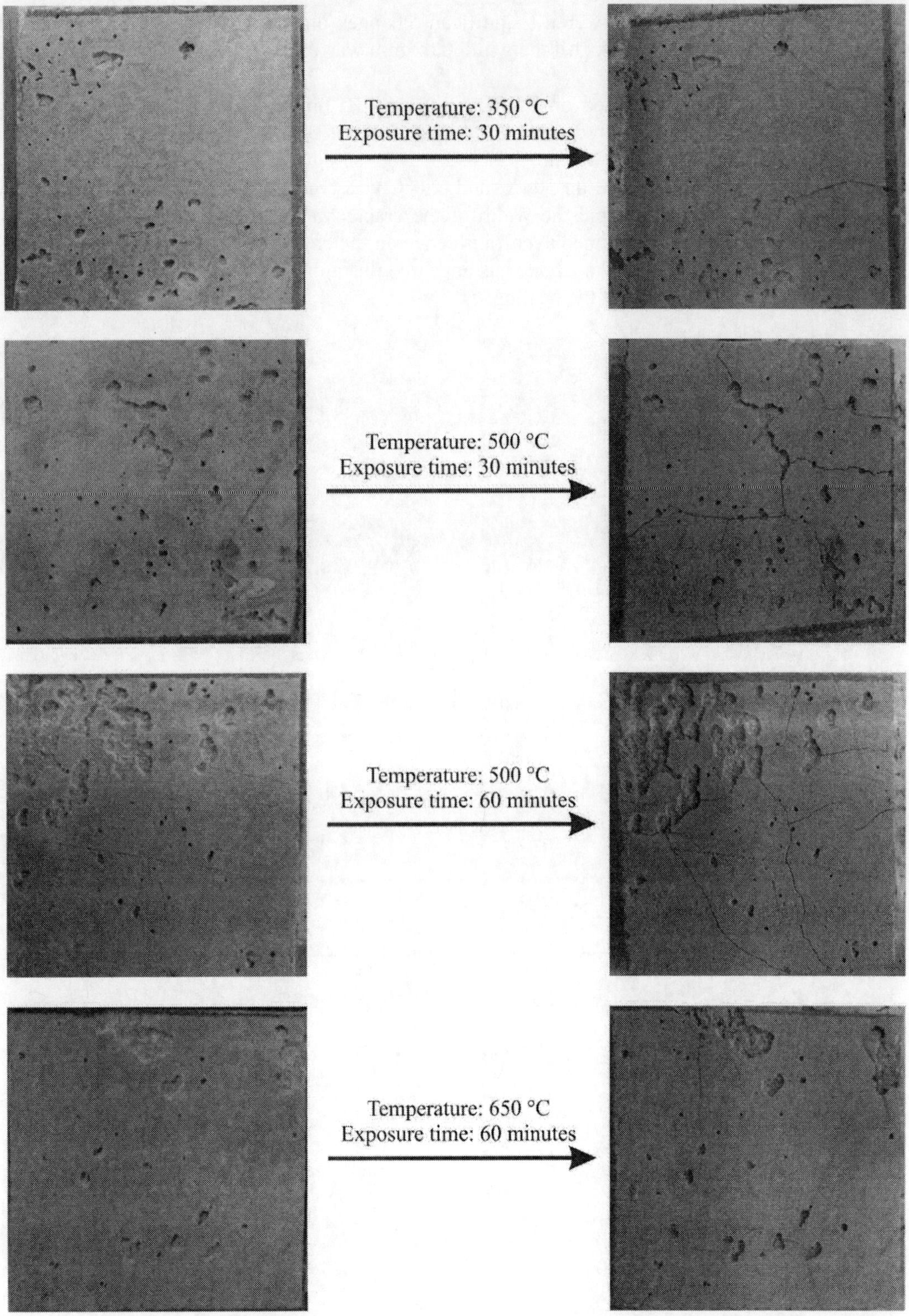

Figure 5. Surface of the specimens: left – before heating, right – after heating.

IOP Conf. Series: Materials Science and Engineering **596** (2019) 012044 doi:10.1088/1757-899X/596/1/012044

For temperatures up to 200°C weren't significant changes on the surface even after 60 minutes. Just development of the microcracks till the width 0.02 mm was occurred. Fibers of the polypropylene fabric are still intact in the pores.

Surface of the specimens heated on 350°C for 5 and 10 minutes was as like as for the lower temperature – just with small microcracks. For exposure times 30 and 60 minutes the cracks on the surfaces are wider – about 0.1 mm width.

Surface of the specimens heated on 500°C and 650°C was cracked with cracks 0.1 mm wide after 5 minutes. For longer exposure time the width of the cracks wasn't change. It can be said that for higher temperatures the exposure time haven't influence on the cracks width (figure 6).

For specimens heated on 650°C a smoke has appeared during first 15 minutes. After that there was no other smoke effect till ending of the heating.

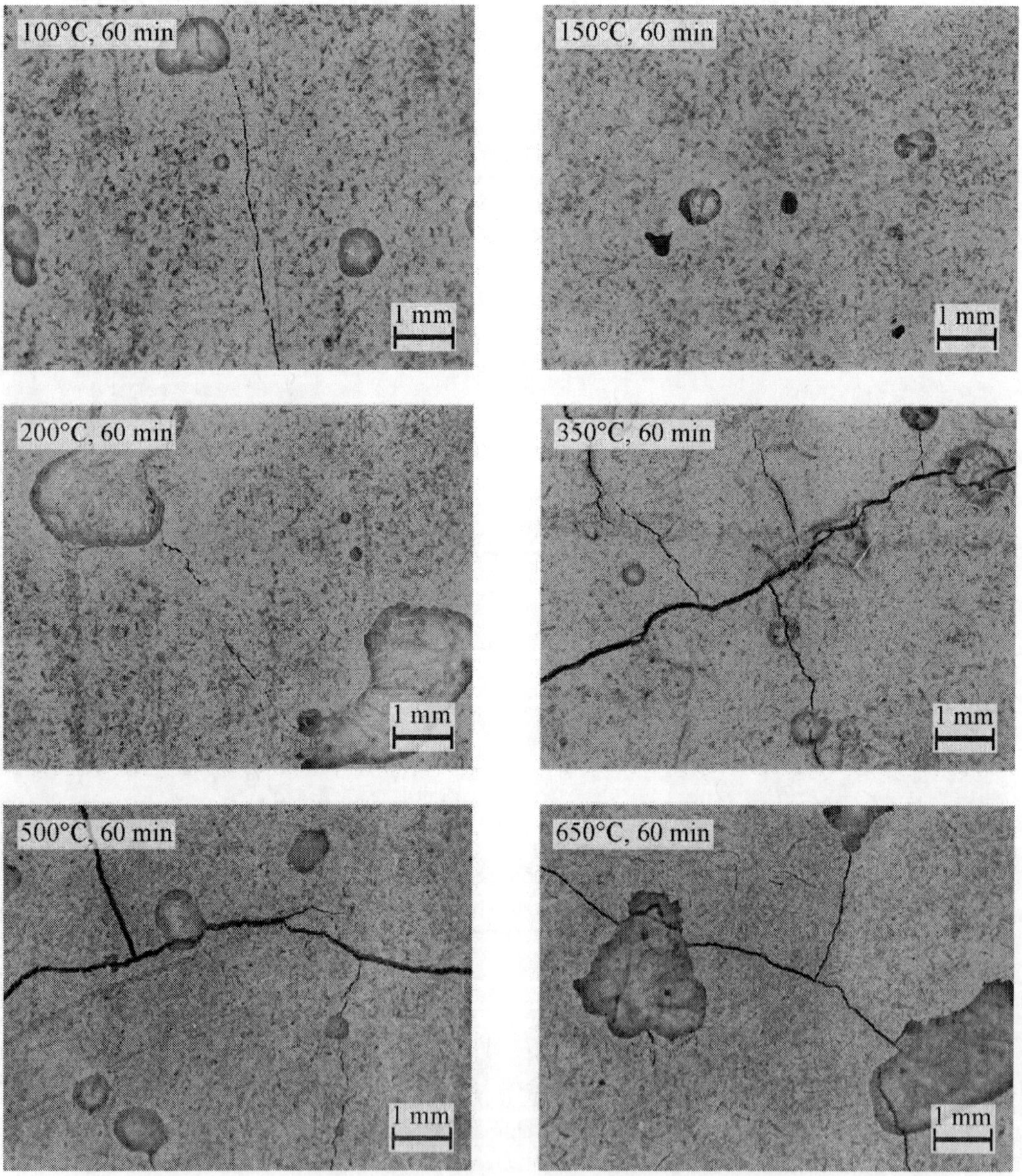

Figure 6. Surfaces of the specimens for different temperatures; *DigiMicro Profi II.*

4. Conclusion

Fire resistance of TRC with non-woven polypropylene fabric was examined. Chosen temperatures were between 100 – 650°C for exposure times 5 min, 10 min, 30 min and 60 min. There was significant crack development on the surfaces of the specimens exposed to temperatures higher than 200°C. Nevertheless, spalling of the surface layers didn't occur even though initial moisture of the specimens was about 9.1% wt.

Next step should be to research properties of bigger specimens for longer exposure time.

Acknowledgements

This work was financially supported by the Czech Technical University in Prague research project no. SGS19/038/OHK1/1T/11, which is gratefully acknowledged.

References

[1] Žák J, Štemberk P and Vodička J 2017 Production of a textile reinforced concrete protective layers with non-woven polypropylene fabric *IOP Conf. Ser.: Mater. Sci. Eng.* **246** 012054

[2] Hejtmánek P, Najmanová H and Váchal T 2018 Experimental assessment of separation distances of a load-bearing straw-bale construction *J. Phys.: Conf. Ser.* **1107** 042013

[3] Yermak N, Pliya P, Beaucour A-L, Simon A and Noumowé A 2017 Influence of steel and/or polypropylene fibres on the behaviour of concrete at high temperature: Spalling, transfer and mechanical properties *Construction and Building Materials* **132** 240–50

[4] Li Y, Tan K H and Yang E-H 2018 Influence of aggregate size and inclusion of polypropylene and steel fibers on the hot permeability of ultra-high performance concrete (UHPC) at elevated temperature *Construction and Building Materials* **169** 629–37

[5] Jackiewicz-Rek W, Drzymała T, Kuś A and Tomaszewski M 2016 Durability of High Performance Concrete (HPC) Subject to Fire Temperature Impact *Archives of Civil Engineering* **62** 73–94

[6] Pothisiri T, Chou S and Sektheera C 2016 Effect of Polypropylene Fibers and Wire Mesh on Fire Performance of Precast Concrete Walls *ACI Structural Journal* **113** 393–403

[7] Bagherzadeh R, Sadeghi A-H and Latifi M 2011 Utilizing polypropylene fibers to improve physical and mechanical properties of concrete *Textile Research Journal* **82** 88–96

[8] Waheed F, Khaliq W and Khushnood R A 2018 High-Temperature Residual Strength and Microstructure in Air-Entrained High-Strength Concrete *ACI Materials Journal* **115** 425–35

[9] Lam E S et al. 2012 Monotonic and Cyclic Behavior of High-Strength Concrete with Polypropylene Fibers at High Temperature *ACI Materials Journal* **109** 323–30

[10] Sevón T et al. 2005 *Molten core-concrete interactions in nuclear accidents. Theory and Design of an Experimental Facility* (VTT-TIED--2311) Finland

IOP Conf. Series: Materials Science and Engineering **596** (2019) 012045 doi:10.1088/1757-899X/596/1/012045

Using of TRC for research of crack evolution and the effect of autogenous healing

H Žáková and J Žák

Department of Building Structures, Faculty of Civil Engineering, Czech Technical University in Prague, Thákurova 7, 166 29 Praha 6, Czech Republic

E-mail: hana.zakova@fsv.cvut.cz

Abstract. Cracks in the concrete part of substructure are a problem in terms of penetration of water into the building. The best way is to prevent their development or support autogenous healing of concrete as much as possible. It is suitable to create many small specimens for research of crack evolution in various boundary conditions. TRC is applicable material for this research due to its compact dimensions and the ability to create a lot of cracks in relatively small area of the specimen. There are two basic methods how to create cracks in this material, tensile and bending loading. Each method provides different crack shapes.

1. Introduction

Concrete structures almost always have cracks (figure 1). In the substructure of the buildings this problem is often solved by using of some admixtures which causes secondary hydration in concrete. Primary purpose of these admixtures is to support autogenous healing of concrete. If it doesn't work, then some remediation method is applied. The best way should be defend the structure from the moisture penetration. For making research focused on admixtures which causes secondary hydration in concrete, the biggest problem is to determine boundary conditions. It is also appropriate to evaluate suitable boundary conditions for support of the autogenous healing and then apply it to the real structure. Research like that requires creation of a large number of specimens saved in many different conditions. It is advantageous to create the smallest specimens with the largest number of cracks as possible. The way should be using textile reinforced concrete (TRC) with non-woven polypropylene fabric [1-12].

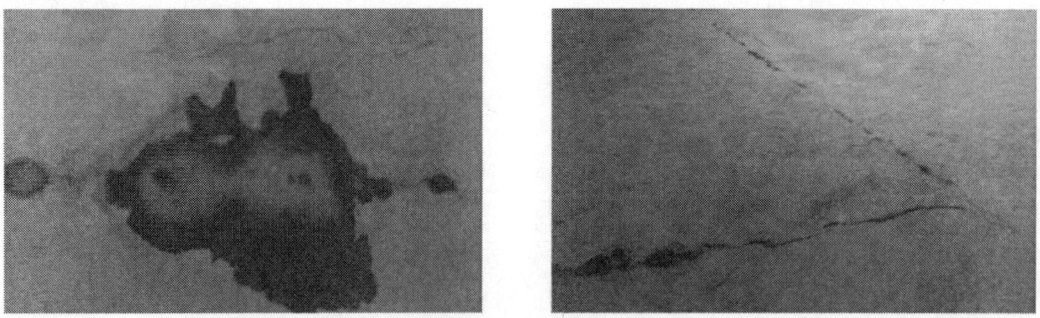

Figure 1. Moist cracks in substructure.

IOP Conf. Series: Materials Science and Engineering **596** (2019) 012045 doi:10.1088/1757-899X/596/1/012045

2. Materials and methods

For research of the crack evolution two possibilities were considered. The first one is using standard test cubes with edge length 150 mm or 100 mm. The second one possibility is based on using of TRC with non-woven polypropylene fabric (figure 2).

Using of the cube specimens should be more similar to the situation in the real buildings than using of TRC. The biggest advantage is the possibility to use the same concrete mixture as used for real structure.

The largest aggregates which can be used for TRC production is 0.25 mm. Also the water-cement ratio is often higher than in the real structure and it is also necessary to use plasticizer for create the TRC specimens [4]. It is suitable to examine some trends of autogenous healing in concrete and then apply it to specimens made of the same mixture as it is used in the specific structure. Exception is a situation, where the examined structure is made from fiber-reinforced concrete (FRC) or textile reinforced concrete (TRC) – there should be used similar mixture.

Figure 2. Test specimens: left – cube specimens; right – TRC specimens.

2.1. Crack-creation in cube specimens

Creation of the realistic cracks in the cube specimens is a problem, which can be solved for example by using of fiber reinforcement. It is suitable to use non-metallic fibers, polypropylene or PVA fibers are recommended. Thanks to the presence of fibers, the cracks should be created without complete destroying of the specimens.

This procedure allows to create just one crack in one test specimen (figure 3). Only external proportion of the crack can be measured, internal geometry of the crack is unknown. Cracks in the cube specimens are about 0.1 – 0.6 mm wide.

Figure 3. Cube specimens: left – storage of the specimens during the experiment; right – crack in the specimen, width of the crack is about 0.5 mm.

IOP Conf. Series: Materials Science and Engineering **596** (2019) 012045 doi:10.1088/1757-899X/596/1/012045

2.2. Crack-creation in TRC specimens

Creating of many cracks and micro-cracks is feature of textile reinforced concrete with non-woven polypropylene fabric. Thanks to this property, this material is not suitable for load-bearing structures [13,14]. But it is usable for protective layers, or for the research of the autogenous healing of cracks in concrete.

It means that there is a large number of cracks in relatively small test specimen compared with standard cube specimens.

There are two ways, how to create cracks. One way is to use tensile loading on the specimen. Problem of this process is that cracks are closed after unloading. The second way is to use bending moment. Cracks stay open after unloading, but the specimens are permanently deformed (figure 4).

Cracks created in the TRC specimens using the bending moment are about 0.03 – 0.15 mm wide. This method was chosen for testing of the crack evolution in different boundary conditions (figure 5).

It is suitable to create two-layer specimens due to better crack formation.

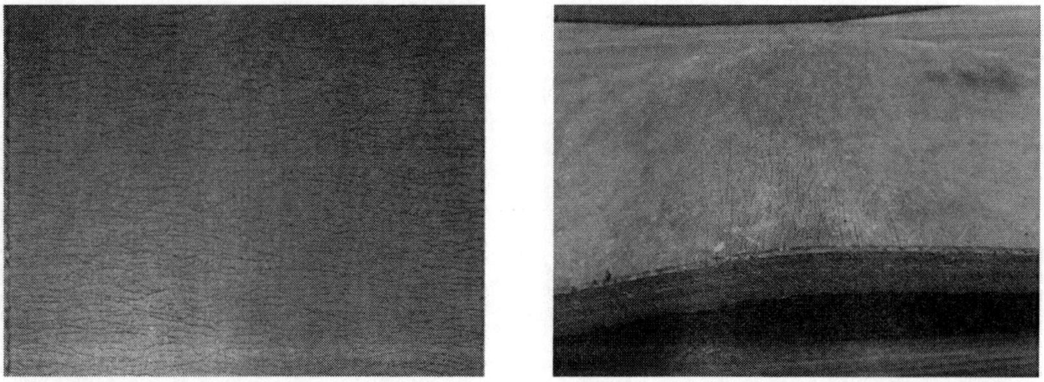

Figure 4. Cracks in the TRC: left – caused by tension; right – caused by bending moment.

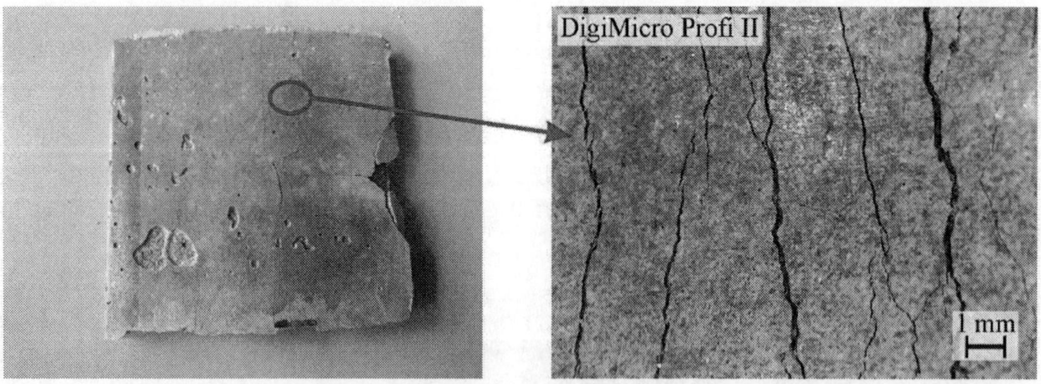

Figure 5. Cracks in the TRC test specimen, width of the cracks is about 0.04 – 0.15 mm.

2.3. Comparison of the use of test specimens

Using of each type of the test specimens has its own advantages and disadvantages (table 1). Cube specimens can be produced from the same concrete mixture as like as the real structure. Permeability test can be simply done alike it is applied on classic cube specimens. But only one crack can be made in one specimen. Also there is a greater storage requirement for the cubes than storage requirement for TRC specimens. Manipulation with the cube specimens is harder due to their greater weight.

For creation of the TRC specimens the concrete mixture is completely different than used in the real structure. It means that comparison with cube specimens is recommended. But there are a large number of cracks in one specimen and manipulation with the specimens is easy. These specimens are good choice for research of the trends for crack-healing in concrete.

Table 1. Comparison of the use of test specimens.

Cube specimens	TRC specimens
+ The same mixture as in the structure	– Different mixture than in the structure
– Greater storage space requirement	+ Less storage space requirement
– One crack in one specimen	+ Many cracks in one specimen
– Problem with crack-creation	+ Easy crack creation
– Bigger cracks (0.1 – 0.6 mm)	+ Smaller cracks (0.05 – 0.15 mm)
– Complicated manipulation with specimens	+ Easy manipulation with specimens
– Consumption of more material	+ Consumption of less material
+ Easy testing of the permeability	– Hard testing of the permeability

3. Production possibilities of the TRC specimens

For the specimens the edge length was chosen as 100 mm. There are two ways, how to create specimens with these dimensions.

The first one possibility is to create a larger slab and cut it up to smaller specimens (figure 6).

Figure 6. TRC slabs: left – bigger slab; right – cutting of the slab on the specimens.

The second one possibility is to create specimens with exact edge length (figure 7). Smaller space storage requirement during curing compared with bigger slabs is advantage of this method.

Figure 7. Curing of the test specimens in the formwork.

The main difference between these methods of creation is in the quality of the edges. Edges of specimens cut up from bigger slab are more equal than are on the specimens made with exact length (figure 8). For the crack research this detail is not important.

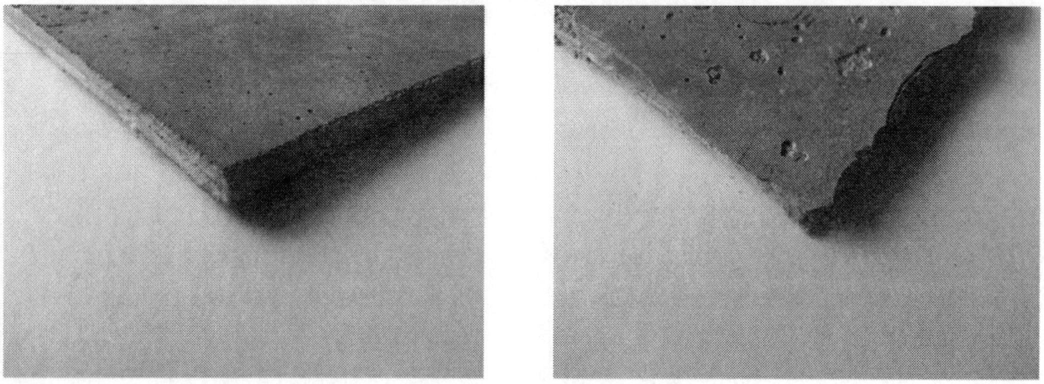

Figure 8. Difference in the quality of the specimen's edges: specimen cut up from bigger slab; right – specimens made with exact length.

4. Results and discussion

Specimens for research of the evolution of cracks in concrete and autogenous healing were made from TRC with non-woven polypropylene fabric. The specimens were squared with edge length 100 mm. These dimensions were chosen due to relatively small storage space requirements and small material consumption for the creation.

There are two ways how to create cracks in the specimens. The first one is using tensile loading which ensures that the specimen stays flat. But there should be a problem for crack research after unloading, because TRC with non-woven polypropylene fabric has a tendency to reclose cracks in the specimen. This property is more useful for elements in real structure than for research of the cracks. The second way is to load the specimen by bending moment. After unloading the specimen stays deformed but the cracks stay open, which is necessary property for the crack research.

Cracks in the specimens used for the autogenous healing research were created using bending moment to prevent their reclosing. Because of the small width of the cracks, digital microscope for the control and crack-width measuring was used (figure 9).

This method allows to investigate a large number of cracks in concrete specimens, which were saved in different boundary conditions.

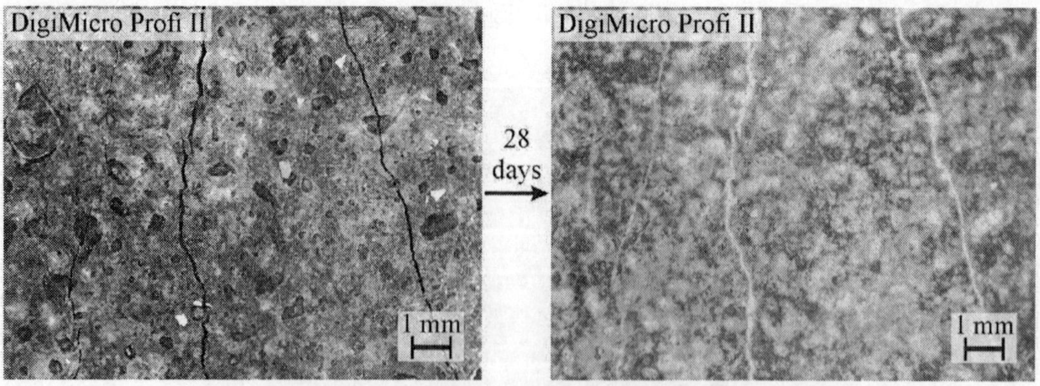

Figure 9. Research of the crack evolution using TRC specimen.

5. Conclusion

Using TRC with non-woven polypropylene fabric specimens is good choice for research and description of the trends of the crack evolution and autogenous healing process in concrete. Huge advantage of this material is that it is possible to examine a large number of cracks in relatively small specimens. It means, that there is less amount of consumed material for the specimens creation, small space storage requirements and last but not least small specimens are more ecological terms to waste disposal then classic cube specimens.

Acknowledgements

This work was financially supported by the Czech Technical University in Prague, research projects no. SGS19/145/OHK1/3T/11 and SGS19/038/OHK1/1T/11, which are gratefully acknowledged.

References

[1] Pazderka J and Hájková E 2016 Crystalline Admixtures And Their Effect On Selected Properties Of Concrete *Acta Polytechnica* **56** 306–11
[2] Pazderka J and Hájek P 2017 Two innovative solutions based on fibre concrete blocks designed for building substructure *IOP Conf. Ser.: Mater. Sci. Eng.* **246** 012046
[3] Pazderka J, Hájková E and Jiránek M 2017 Underground Air Duct To Control Rising Moisture In Historic Buildings: Improved Design And Its Drying Efficiency *Acta Polytechnica* **57** 331–9
[4] Pazderka J 2016 Concrete with Crystalline Admixture for Ventilated Tunnel against Moisture *Key Engineering Materials* **677** 108–13
[5] Rahhal V et al. 2009 Scheme of the Portland cement hydration with crystalline mineral admixtures and other aspects *Silicates Industriels* **74** 347-52
[6] Zhou M R et al. 2011 Study on Experiment of Concrete Compounding XYPEX and Steel Fiber *Applied Mechanics and Materials* **105-107** 1755–9
[7] Pazderka J and Hájková E 2017 The speed of the crystalline admixture's waterproofing effect in concrete *Key Engineering Materials* **722** 108-12
[8] Edvardsen C 1999 Water Permeability and Autogenous Healing of Cracks in Concrete *ACI Materials Journal* **96** 448–54
[9] Mechtcherine V and Lieboldt M 2011 Permeation of water and gases through cracked textile reinforced concrete *Cement and Concrete Composites* **33** 725–34
[10] Lieboldt M and Mechtcherine V 2013 Capillary transport of water through textile-reinforced concrete applied in repairing and/or strengthening cracked RC structures *Cement and Concrete Research* **52** 53–62
[11] Brückner A, Ortlepp R and Curbach M 2006 Textile reinforced concrete for strengthening in bending and shear *Materials and Structures* **39** 741–8
[12] Žák J, Štemberk P and Vodička J 2017 Production of a textile reinforced concrete protective layers with non-woven polypropylene fabric *IOP Conf. Ser.: Mater. Sci. Eng.* **246** 012054
[13] Hegger J, Will N, Bruckermann O and Voss S 2006 Load–bearing behaviour and simulation of textile reinforced concrete *Materials and Structures* **39** 765–76
[14] Žák J and Štemberk P 2017 Experimental Investigation of Tensile Behavior of Textile Reinforced Concrete with Nonwoven Polypropylene Fabric *Mechanika 2017 Proceedings of the 22st Int. Sci. Conf.* (Kaunas: Kauno technologijos universitetas), p. 399-401